AF771840

Mechanochemistry in Nanoscience and Minerals Engineering

Peter Baláž

Mechanochemistry in Nanoscience and Minerals Engineering

Springer

Peter Baláž
Institute of Geotechnics
Slovak Academy of Sciences
Watsonova 45
043 53 Košice
Slovakia
balaz@saske.sk

ISBN: 978-3-540-74854-0 e-ISBN: 978-3-540-74855-7

Library of Congress Control Number: 2008933295

© 2008 Springer-Verlag Berlin Heidelberg

This work is subject to copyright. All rights are reserved, whether the whole or part of the material is concerned, specifically the rights of translation, reprinting, reuse of illustrations, recitation, broadcasting, reproduction on microfilm or in any other way, and storage in data banks. Duplication of this publication or parts thereof is permitted only under the provisions of the German Copyright Law of September 9, 1965, in its current version, and permission for use must always be obtained from Springer. Violations are liable to prosecution under the German Copyright Law.

The use of general descriptive names, registered names, trademarks, etc. in this publication does not imply, even in the absence of a specific statement, that such names are exempt from the relevant protective laws and regulations and therefore free for general use.

Cover design: deblik, Berlin

Printed on acid-free paper

9 8 7 6 5 4 3 2 1

springer.com

Dedicated to the memory of my son Pali
who passed away on one Spring sunny day

Preface

There are three keywords in title of this book. *Mechanochemistry* is a branch of science concerned with chemical and physico-chemical changes of solids due to the influence of mechanical energy. Very frequently, the small particles are product of this mechanical treatment. *Nanoscience* is a branch of science dealing with particles less than the size of 100 nm, giving to nanostructures built from them extraordinary properties. It has been experienced that mechanochemistry and nanoscience have impact on several technologies. *Minerals engineering* as well as many others applications serves as an good example.

The first chapter *Mechanochemistry in Nanoscience* deals with both disciplines. History, theories and models and synthesis routes are described. Special attention is devoted to nanogeoscience and application of nanoparticles in medicine which is hot topic for scientists and technologists.

The second chapter *High-Energy Milling* is devoted to energetically intensive treatment with which special structures in mechanochemistry and nanoscience are created. Various mills are described, the process variables which govern their mechanical effect as well as important phenomena accompanying the milling process.

Special techniques needed for investigation and characterization of solids in mechanochemistry and nanoscience are described in the third chapter named *Selected Identification Methods*.

The fourth chapter *From Minerals to Nanoparticles* show many examples how it is possible to obtain nanoparticles from minerals.

Mechanochemistry in Minerals Engineering is exclusively described in the fifth chapter. Here, various combinations of mechanochemical processing is illustrated for extraction of elements from minerals as well as their behaviour in leaching and sorption operations.

The largest sixth chapter *Applied Mechanochemistry* is devoted to applications of mechanochemistry in various technological fields. The effect of high-energy milling on particles (very frequently in nanodimensions) in scaled-up processes is illustrated for technological applications in mineral processing, extractive metallurgy, chemical engineering, materials engineering, coal industry, building industry, agriculture, pharmacy and waste treatment.

Košice, Slovakia *Peter Baláž*
Spring 2008

Contents

1 Mechanochemistry and Nanoscience 1
1.1 Mechanochemistry of Solids 1
1.1.1 History of Mechanochemistry 1
1.1.2 Theories and Models in Mechanochemistry 8
1.1.3 Mechanical Activation 14
1.1.4 Thermodynamics in Mechanochemistry 16
1.1.5 Kinetics of Mechanochemical Reactions 19
1.2 Introduction to Nanoscience 29
1.2.1 Historical Outline 31
1.2.2 Classification of Nanostructures 32
1.2.3 Theoretical Fundamentals 34
1.2.4 Synthesis Routes (with the Exception of Mechanochemical Approach) 39
1.2.5 Mechanochemical Approach 44
1.2.6 Properties 69
1.3 Nanogeoscience 76
1.3.1 Mineral Nanoparticle Surfaces 77
1.3.2 Nano- Versus Mechano-Approach: Analogy 78
1.3.3 Clusters and Aggregates Formation 80
1.4 Nanoparticles in Medicine 82
1.4.1 Size of Biological Objects 82
1.4.2 Nanomaterials for Cancer Therapy and Diagnosis 83
1.5 Opportunities and Other Challenges 90
1.6 Conclusion 91
References 92

2 High-Energy Milling 103
2.1 Introduction 103
2.2 Basic Phenomena 103
2.3 Energy Requirement 105
2.4 Milling Equipments 107

2.4.1 Vibration Mills ... 107
2.4.2 Planetary Mills ... 111
2.4.3 Mixer Mills ... 113
2.4.4 Attritors (Stirred Ball Mills) ... 114
2.4.5 New Mill Design ... 116
2.4.6 Process Variables ... 117
2.5 Accompanying Phenomena ... 119
2.5.1 Temperature Effects ... 119
2.5.2 Contamination ... 121
2.5.3 Amorphization ... 121
2.5.4 Mechanochemical Equilibrium ... 121
2.5.5 Agglomeration and Aggregation ... 122
2.6 Surfactants ... 126
References ... 129

3 Selected Identification Methods ... 133
3.1 Introduction ... 133
3.2 Infrared Spectroscopy ... 133
3.3 X-Ray Photoelectron Spectroscopy ... 138
3.4 Scanning Electron Microscopy ... 143
3.5 Scanning Tunnelling Microscopy ... 146
3.6 Atomic Force Microscopy ... 148
3.7 Transmission Electron Microscopy ... 150
3.8 X-Ray Diffraction ... 152
3.9 Mössbauer Spectroscopy ... 162
3.10 Electron Paramagnetic Resonance ... 167
3.11 Conclusion ... 171
References ... 172

4 From Minerals to Nanoparticles ... 177
4.1 Introduction ... 177
4.2 Solid–Gas Reactions Stimulated by Mechanical Activation ... 177
4.2.1 Stibnite Sb_2S_3 ... 179
4.2.2 Cinnabar HgS ... 180
4.2.3 Galena PbS ... 181
4.2.4 Sphalerite ZnS ... 182
4.3 Mechanochemical Reduction via Solid–Solid Reactions ... 186
4.3.1 Redox Reactions ... 187
4.3.2 Nanocomposites ... 188
4.3.3 Oxides ... 190
4.3.4 Wolframates ... 209
4.3.5 Titanates ... 212
4.3.6 Sulphides ... 216
4.3.7 Other Minerals ... 245
References ... 249

5 Mechanochemistry in Minerals Engineering 257
5.1 Introduction 257
5.2 Mechanochemical Solid–Solid Processing 259
5.2.1 Mechanochemical Reduction 259
5.3 Mechanochemical Solid–Liquid Processing 263
5.3.1 Milling and Leaching as Separated Processes 263
5.3.2 Oxidative Leaching 264
5.3.3 Acidic Leaching 266
5.3.4 Alkaline Leaching 268
5.3.5 Selective Leaching of Bismuth 273
5.3.6 Non-Cyanide Leaching of Precious Metals 276
5.3.7 Simultaneous Milling and Leaching (Mechanochemical Leaching) 283
5.4 Sorption 288
5.4.1 Zinc Sorption on Calcite 288
5.4.2 Sulphides 290
References 292

6 Applied Mechanochemistry 297
6.1 Introduction 297
6.2 Mineral Processing 297
6.2.1 Sulphides 298
6.2.2 Other Minerals 300
6.3 Extractive Metallurgy 302
6.3.1 Mechanical Activation as Pretreatment Step for Leaching . . 303
6.3.2 Attritors in Hydrometallurgy 318
6.3.3 Mechanochemical Processes for Ore Treatment 321
6.4 Chemical Engineering 341
6.4.1 Crystal Engineering 341
6.4.2 Heterogeneous Catalysis 345
6.5 Materials Engineering 352
6.5.1 Thermoelectric Power Generator Materials 354
6.5.2 Room Temperature Solders 354
6.5.3 Bearings 355
6.6 Coal Industry 355
6.6.1 Humic Acids 355
6.7 Building Industry 358
6.7.1 Mechanical Activation of Quartz SiO_2 359
6.7.2 Mechanical Activation of Cement 360
6.7.3 Solidification of Silicate Concretes with Mechanically Activated Precursors 363
6.7.4 Modification of Concrete Properties by Application of Waste Materials After Milling 363
6.7.5 Manufacture of Exterior Wall Panels with Intervention of Mechanical Activation 364

6.8 Agriculture 365
6.9 Pharmacy 370
6.9.1 Solubility of Drugs 370
6.9.2 Polymorphism and Amorphization 373
6.9.3 Biological Activity of Drugs 375
6.9.4 Mechanochemical Synthesis of Drugs 377
6.10 Waste Treatment 380
6.10.1 Mineral Waste 380
6.10.2 Materials Waste 387
6.10.3 Organic Waste 389
References 392

Author Index 407

Subject Index 409

Acknowledgements

The completion of this book would not have been possible without influence of many people on my professional life. Wish to thank my first teacher in science Ivan Žežula, the learner of the only Czechoslovak Nobel laureate Jaroslav Heyrovský, who arouse my enthusiasm for science. Klára Tkáčová, a founder of the Slovak mechanochemical school, supervisor of my PhD. thesis first introduced me to the charm of mechanochemistry.

As early workers in the field of mechanochemistry it was pleasure to personally meet such pioneers as Peter Adolf Thiessen, Gerhard Heinicke, Hans-Peter Heegn and Eberhard Gock of Germany, Vladimir Vjačeslavovič Boldyrev, Pavel Yurievič Butyagin and Evgenij Grigorievič Avvakumov of Russia, Emmanuel Gutman of Israel, Zoltán Juhász of Hungary and Mamoru Senna and Fumio Saito of Japan. Of course, the list is not complete and several hundreds of contributions from laboratories of active mechanochemists throughout the world can be traced at the end of each chapter of the book. My best thanks is extended to all mechanochemists whose contributions created this work.

The completion of this book would have been impossible without the tremendeous help of Erika Dutková, my post-doc student, who gave the final shape to the manuscript.

The preparation of this book was partly supported by the project APVV-0347-06 from the Research and Developing Agency of Slovakia.

I would like to thank my wife Ela, for her encouragement, patience, and love.

Chapter 1
Mechanochemistry and Nanoscience

1.1 Mechanochemistry of Solids

Mechanochemistry is a branch of chemistry which is concerned with chemical and physico-chemical changes of substances of all states of aggregation due to the influence of mechanical energy.

1.1.1 History of Mechanochemistry

The above mentioned definition is based on theoretical considerations of *Ostwald* (Nobel Prize 1909) concerning the relationship between chemical and mechanical energy. The German scientist significantly contributed to the development of modern chemistry by systemization of chemical disciplines. Ostwald (Fig. 1.3) was engaged in the systematization of chemical sciences from the energetic point of view. The term mechanochemistry has been used for the first time in his early works [Ostwald 1887, 1909]. He understood mechanochemistry in a wider sense when compared with the present view, regarding it as a part of physical chemistry like thermochemistry, electrochemistry or photochemistry.

From twenties, the further activities in mechanochemistry can be observed. The colloidal mill was developed (Fig. 1.1) and practically forgotten book on mechanochemistry was published [Pierce 1928]. Here author defines mechanochemistry as "the new science of mechanical dispersion involving the use of principles in physical chemistry". He hesitated to name this science because of his remark "... we shall call it mechanochemistry for lack of a better name, as it involves dispersion or defflocculation by mechanical means, thereby bringing about so-called colloidal dispersions".

However, the fact that mechanical activation can lead to chemical consequencies was not a discovery. Since the first attempts by man to obtain fire by friction and to the more recent data on the possibility of ignition and detonation of certain solid explosives, it has come to light that chemical reactions may be initiated by mechanical means [Boldyrev 1986]. That is a reason to look for the beginnings of the written history of mechanochemistry in the very early times.

© Springer-Verlag Berlin Heidelberg 2008

Fig. 1.1 Colloid mill Plauson-Oderberg for wet milling [Ostwald 1927]

Several studies has been published to find the very early traces of mechanochemistry [Takacs 2000, 2003, 2004; Baláž 2001].

1.1.1.1 Mechanochemistry in Antic Greece and Old Europe

The beginning of mechanochemistry of sulphides might be probably dated to antic area. *Theophrastus of Eresus* (Fig. 1.3) was student of Aristotle and served as his successor as the head of the Lyceum in Athens from 322 B.C. until his dead. His book, "On Stones or De Lapidibus" is the earlier surviving scientific book on minerals. In this book the sentence ... "native cinnabar was rubbed with vinegar in a copper mortar with a copper pestle yielding the liquid metal ..." is given. This is a very clear description of a mechanochemical process. The first described mechanochemical reduction probably followed the reaction

$$HgS + Cu \rightarrow Hg + CuS \qquad (1.1)$$

and vinegar was smartly used to prevent the side effects which usually accompany dry milling on air. It remains a mystery why the mechanochemical preparation of mercury from its sulphide according to reaction (1.1) was forgotten during the Middle Ages.

However, as published recently, examples of other mechanochemical reactions between 300 B.C. and the end of the 18th century can be also traced in medieval literature [Takacs 2000]. *Agricola* documented several examples of chemical reactions under influence of mechanical action which can be connected with mining and metallurgical operations [Agricola 1546, 1556].

It is interesting to note that in the 17th century *Bacon* (Fig. 1.3) referred to four treatments that, in essence, are still among the most important procedures to prepare active solids: one of them is milling [Bacon 1658]. It was *Wenzel* who stressed out the fact that by heterogeneous reactions the degree of conversion depends mainly on the surface area of reacting solids and is not proportional to their amount [Wenzel 1777].

1.1.1.2 Faraday's Contribution to Mechanochemistry

Faraday, famous English physicist noticed in 19th century that certain hydrated salts dehydrated spontaneously when mechanically treated [Faraday 1834]. However, as stated recently his contribution to mechanochemistry started earlier [Takacs 2007]. In his book published in 1827, a twenty-page chapter was dedicated to mortars and comminution [Faraday 1827]. A very direct reference to a mechanochemical process was published in 1820 on the decomposition of silver chloride [Faraday 1820]. The reaction proceeds according to the equation

$$2AgCl + Zn \rightarrow 2Ag + ZnCl_2 \qquad (1.2)$$

and the experiments applying "mortar milling" have been repeated with Sn, Cu and Fe metals. According to Faraday's description, the reaction between silver chloride and zinc is fast and highly exothermic, raising the possibility a mechanochemically induced self-sustaining reaction [Takacs 2002].

The reaction (1.2) was studied recently with modern mechanochemical tools. The mechanically induced self-sustainity of the process has been verified [Takacs 2007]. However, Faraday's contribution to solid state chemistry exceeds the frame of mechanochemistry. He contributed also to nanoscience (see later in this chapter). One of the first bridges between mechanochemistry and nanoscience has been built.

1.1.1.3 Carey Lea, the First Mechanochemist

The interesting papers concerned with the effect of mechanical energy on properties of substances were published by American scientist *Lea* [Lea 1891a, b, 1892–1894]. His contribution to the mechanochemistry has been analyzed recently by contemporary mechanochemists [Baláž 2001; Takacs 2003].

Lea published many works in his brilliant carrier, starting with the groundbreaking papers on colloidal silver [Lea 1866, 1891a, b]. He discovered, what he called three allotropic forms of silver: "soluble silver", "insoluble silver", and "gold-yellow or copper-coloured silver". He recognized that these allotropic forms, which formed colloids containing particles too small to be seen in any microscope of the day, suggested numerous applications [Smith 1972; Whitcomb 2006].

After studying the transformations of silver metal, he turned attention to halides of Ag, Hg, Pt and Au. From halides, chlorides, bromides and iodides were studied. Investigation of these compounds were summarized and a paper was red before American National Academy of Science in April 1892 when Lea was elected to membership. In case of AgCl and AgBr two modes of mechanical effect were applied: simple pressure and shearing stress. The observation recorded proved the existence of perfect uniformity in the action of both kinds of mechanical energy on the halides. When heated, AgCl melts without decomposition, but under the effect of low stress it decomposes with the formation of elemental silver

$$2AgCl \rightarrow 2Ag + Cl_2 \tag{1.3}$$

The reaction probably preceeds through silver subchloride formation.

On the other hand, mercurous chloride Hg_2Cl_2 sublimes under effect of temperature. However, under pressure (a glass rod was used in these experiments) the decomposition reaction leading to elementary mercury and chlorine was observed

$$Hg_2Cl_2 \rightarrow 2Hg + Cl_2 \tag{1.4}$$

This is one of Lea's frequently cited results, the first example of a mechanochemical reaction that brings about an outcome different from the effect of heat [Takacs 2004].

These initial results were followed by systematic investigations published in a series of articles during 1893–1894. As an example can serve the reaction

$$2NaAuCl_4 \rightarrow 2Au + 2NaCl + 3Cl_2 \tag{1.5}$$

The reaction has been performed under influence of mechanical stress. However, the decomposition cannot be produced by heat. Many others compounds were studied by Lea under the effect of sharp stroke of the pestle like silver tartrate, carbonate, citrate, oxalate, arsenate, sulphite, salicylate, orthophosphate and ferricyanide. The main objective of these studies was the initiation of endothermic reactions, specifically the decomposition of compounds with negative heat of formation, by the application of mechanical force.

Lea has shown for the first time the parallelism between the action of electricity, heat, light, chemical action and mechanical force on the silver halides and allotropic silver. He experimentally proved the theoretical ideas of Ostwald's who in his textbooks on general chemistry understood the equivalency of electrochemistry, thermochemistry and photochemistry with mechanochemistry. There is an interesting note in Ostwald's book "Die Welt der vernachlässigten Dimensionen" about Lea investigations [Ostwald 1927]. Ostwald appreciated him as one of the most known American researchers in the field of colloid chemistry (colloid chemistry studies small particles including nanoparticles).

This is one of the first documented traces of evidence the mutual connection between mechanochemistry and nanoscience.

1.1.1.4 From Simple Decompositions to Solid State Reactions in a Mill

Another American chemist *Parker* also appreciated the Lea's results but on the other hand critized that almost entirely his research was devoted to single compounds [Parker 1914]. Parker studied the solid-solid reaction (1.6) by trituration

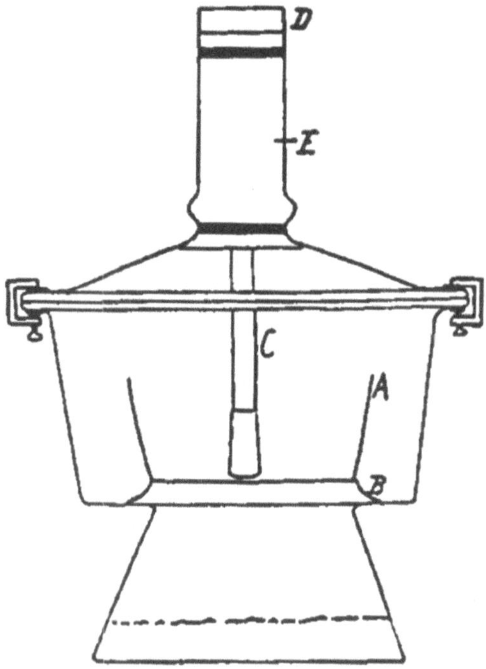

Fig. 1.2 The first mechanochemical reactor: A – mortar, B – iron collar, C – pestle, D – handle, E – rubber tubing [Parker 1914]

$$Na_2CO_3 + BaSO_4 \rightarrow Na_2SO_4 + BaCO_3 \quad (1.6)$$

However, during milling (5–30 min) the sodium carbonate was transformed only very slightly: yield of reaction (1.6) was only 1–3%. In further study he has shown that under certain conditions it is possible to bring about interactions between solid substances by means of a shearing stress, at ordinary temperatures, or at least very greatly to increase the velocity of these reactions [Parker 1918]. In accordance with Lea he concluded that shearing stress, such as could be applied by hand between a pestle and mortar, is widely different in its effects from simple pressure, and that one of the main reason why it is able to bring about reactions between apparently solid substances is that local or surface fusion of the reacting substances is occasioned.

Parker published as the first the picture of mechanochemical reactor (Fig. 1.2). Together with Lea they drew attention on the importance of controlling the atmosphere during milling process. In order to avoid the side effects in the reactor, the pestle and mortar as milling means were fixed in the desiccator.

1.1.1.5 Mechanochemistry in Europe: Traces in a Previous Century

In several review papers on mechanochemistry [Boldyrev and Tkáčová 2000; Boldyrev 2002, 2006] the Russian scientist *Flavickij* is quoted frequently with his contributions in solid state reactions performed by milling [Flavickij 1903, 1909].

As for the time scale his papers were published before Parker's contributions and their works can be appreciated as the pioneer works on solid state reactivity induced by milling.

Later on, a big gap in mechanochemistry of inorganic solids can be traced in literature. Particularly in western countries little seems to have been done except of one area of mechanically initiated explosion reactions in solids, an interest stimulated by the last World War [Fox 1975]. These investigations are linked with researches performed in England, France and Russia [Bowden and Tabor 1958; Bowden and Yoffe 1952, 1958]. The investigations resulted in the elaboration of hot spots theory (see later) which explained the initiation and development of explosion by local increase of temperature at the contact of two solids under mechanical action. Later, this theory was expanded for other theories, like oxidation of metals [Boldyrev and Tkáčová 2000].

In the 1920s the effect of mechanical energy on metals was investigated [Tammann 1929]. *Tamman* working in Göttingen found that not all energy is transformed into the heat but 5–15% of the expended energy remains in the metal as potential energy and increases the thermodynamic potential of the solid. As a consequence of the treatment a displacement of the thermodynamic potential and a significant increase of dissolution velocity occurs. Later Tamman and his co-workers studied the powder reactions in oxide and carbonate systems. They observed that by mechanical contact of quartz (SiO_2) with calcite ($CaCO_3$) the monomolecular layer of Na_2SiO_3 is formed

$$Na_2CO_3 + SiO_2 \rightarrow Na_2SiO_3 + CO_2 \tag{1.7}$$

The old alchymist dogma (unfortunately accepted also for a long time later) that the solid state reactions can not proceed without liquid or gas medium was definitively overcome. *Hedvall*, another outstanding German scientist though not working in mechanochemistry supported by his observations Tamman's results [Hedvall 1938].

The other field of mechanochemistry has its traces in Tamman's time. In twenties, mechanochemistry of organic molecules was first elucidated. The first published papers [Wanetig 1921, 1922, 1927] were inspired by the needs of the pulp and paper industry. The benefitial effect of milling on cellulose solubility by the mechanical destruction of the macromolecule and creation of new hydrophilic groups in the place of discrupted valence bonds was explained [Boldyrev and Tkáčová 2000]. The research into mechanochemistry of organic molecules practically started with works of *Wanetig*. The mechanical degradation of polymers illustrated on polystyrene case started at about the same time. It was found that increase of the reactivity of polymers, e.g. of solubility is parallel to the decrease of the molecular weight [Staudinger and Dreher 1936; Hess et al. 1942]. The investigations in this field were continued in Germany [Berlin 1958; Grohn et al. 1962; Grohn and Paudert 1963] as well as later [Baramboin 1970; Simionesku and Oprea 1971].

In subsequent 3 decades *Fink* and *Bowden* and *Tabor* in particular investigated oxidation reactions of metals, decomposition reactions and conditions of rolling and sliding friction [Fink and Hofmann 1932, 1933; Bowden and Tabor 1958]. Thus during the rolling friction of iron rolls on the stressed surface tarnishing layers come into existence within a few minutes, which without mechanical stress would normally require about 10^{17} years for their growth. Fink excluded the influence of temperature for the interpretation of this effect and discussed plastic deformation as a cause of the acceleration of the reaction [Fink and Hofmann 1932; Heinicke 1984]. The experiments were made with steel as well as with pure metals like iron, nickel and copper. Bowden and Tabor alledge that temperatures over 700°C can be observed at the contact of solid substances exposed to frictions. These high temperatures, however, last only 10^{-4}–10^{-3} s.

The transformations of solids caused by milling were studied as early as 1940 by *Clark* and *Rowan* (1941) followed by *Dachille* and *Roy* (1960). These studies showed that both PbO transformations (massicot→litharge, litharge→massicot) can take place as a result of milling in a ball mill, depending on the amount of hydrostatic pressure generated in the mill. They suggested that pure hydrostatic pressure causes the litharge-massicot transformation, whereas bond breakage at low pressures, due to the action of displacive shear, causes its reverse.

The small excursion into history of mechanochemistry is not exhausted by the above given paragraphs and photographs of important players (Fig. 1.3), of course.

Fig. 1.3 Photogalery of important players in history of mechanochemistry

Table 1.1 Review papers on history of mechanochemistry

Title	Author(s)
Mechanochemische Reaktionen	Peters 1962
Review of the Phase Transformation and Synthesis of Inorganic Solids obtained by Mechanical Treatment (Mechanochemical Reactions)	Lin and Nadiv 1970
Mechanically Initiated Chemical Reactions in Solids	Fox 1975
Mechanochemistry of Inorganic Solids	Boldyrev 1986
Accelerating the Kinetics of Low-Temperature Inorganic Synthesis	Roy 1994
Colloid-chemical Aspects of Mechanical Activation	Juhász 1998
Mechanochemistry of Solids: Past, Present and Prospects	Boldyrev and Tkáčová 2000
Mechanochemistry in Extractive Metallurgy: The Modern Science with Old Routes	Baláž 2001
M. Carey Lea, the First Mechanochemist	Takacs 2004
Mechanochemistry: The Mechanical Activation of Covalent Bonds	Beyer and Clausen-Schaumann 2005
Mechanochemistry and Mechanical Activation of Solids	Boldyrev 2006

After these fundamental works the investigations continued as is in the following decades summarized in a series of review papers (Table 1.1), starting with the first pioneer review on mechanochemistry published by *Peters* and presented on the 1st European Symposium on Size Reduction which was held in Germany [Peters 1962].

1.1.2 Theories and Models in Mechanochemistry

1.1.2.1 Hot-Spot Theory

The first trial in mechanochemistry to explain the reason of mechanical initiation of chemical reactions was developed by Bowden, Tabor and Yoffe. They found that with friction processes for 10^{-4}–10^{-3} s temperatures of over 1000 K on surfaces of about 1 μm^2 can occur, and that these represent and important cause of mechanically initiated reactions [Bowden and Yoffe 1952, 1958; Bowden and Tabor 1958]. These temperatures can also be found near the tip of a propagating crack [Weichert and Schönert 1974]. Later this theory was expanded for other processes, like oxidation of metals. However, the controversy concerning the possibility of an increase in temperature at or on the friction surface has still not ended. Judging from indirect features, centers with a high temperature exist only under extreme conditions. These include, for example, the instant when cavitation bubbles are flooded following the absorption of ultrasound in liquids [Suslick 1990; Butyagin 1994]. The phenomenon was experimentally proved and interpreted for three substances as given in Table 1.2.

Table 1.2 The maximum temperatures at the propagating crack [Weichert 1976]

Material	Glass	Quartz	Sugar
Temperature (K)	3200	4700	2500

To study the chemical processes occurring at the tip of a propagating crack a set-up has been used which allow a crystal to be cleaved directly in the vicinity of the ionic source of a mass spectrometer [Fox and Soria-Ruiz 1970]. Their experiments have shown that during the cleavage of inorganic crystals (calcite, magnesite, cerussite, lead and sodium azides) gas evolution can be substantial which suggests that high temperatures exists at the crack tip. However, since in brittle substances the rate of crack propagation can approach the sound velocity, i.e. 10^3 m/s, the life time excitation on the chemical bond must be 10^{-13} s. It is evident that under these conditions the term temperature can be used only symbolically. Later it was proved that the processes occurring at the tip of a crack during the cleavage of a crystal may proceed by different mechanisms depending on the speed of crack motion [Boldyrev et al. 1990; Boldyrev 1993].

1.1.2.2 Magma-Plasma Model

In the sixties, the first model in mechanochemistry – *the magma-plasma model* was proposed [Thiessen et al. 1967]. According to this model a great quantity of energy is set free at the contact spot of colliding particles. This energy is responsible for formation of a special plasmatic state which is characterized by emission of fairly excited fragments of solid substance, electrons and photons over a short time scale (Fig. 1.4).

The surface of contact particles is rather disordered and local temperatures can reach more than 10000 K. Thiessen distinguishes the reactions which occur in the plasma from the reactions taking place at the surface of particles during the significantly excited state, or immediately after its expiration. These considerations led to

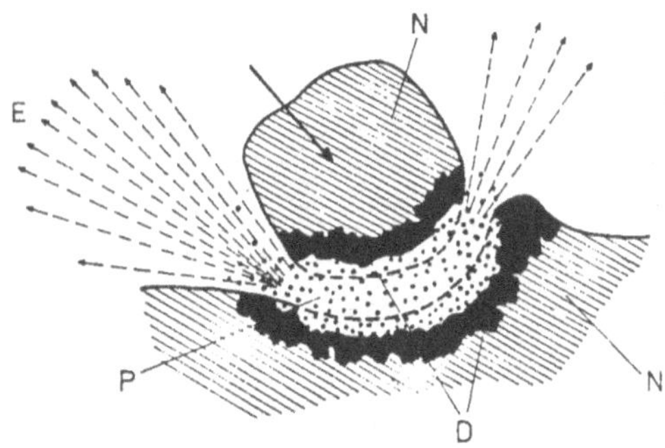

Fig. 1.4 Magma-plasma model: E – exo-electrons, N – undeformed solid, D – highly deformed surface layer, P – plasma [Thiessen et al. 1967]

an important conclusion which is valid for mechanically activated reactions – these reactions do not obey a single mechanism.

The German school elaborated the concept of "hierarchy" of energetic states which appeared to be very important for analysis of the processes induced by the effect of mechanical energy [Thiessen et al. 1967; Thiessen 1979; Thiessen and Sieber 1979a, b; Heinicke 1981]. In this concept, a large number of excitation processes occur due to mechanical activation which are characterized by different relaxation times (Table 1.3).

Heinicke shown the value of classifying tribochemical reactions according to the excitation processes initiating the individual reactions and to subdivide these excitation processes in their temporal courses. This way one will arrive at *a hierarchic model*, in which the most highly excited states having the shortest excitation times stand at the beginning and the numerous other states with smaller excitation incorporate themselves according to their temporal occurrence in the dissipation phase into the model [Heinicke 1981].

Long-lived states including defects and dislocations have been analyzed by equilibrium thermodynamics [Heegn 1989]. Mechanically disordered solids have an actual free energy content G which is higher than the equilibrium free energy G*:

$$G = G^* + G^E \tag{1.8a}$$

The excess free enthalpy is made up of the contributing defects and their disturbance enthalpy

$$G^E = \sum_i c_i H_i^* + Tk_b \, \mathrm{NA} \sum_i C_i \ln c_i \tag{1.8b}$$

Table 1.3 Relaxation times of excitation processes in mechanically activated solids [Heinicke 1981]

Excitation process	Relaxation time
Impact process	$> 10^{-6}$ s
Triboplasma	$< 10^{-7}$ s
Gaseous discharge	$\sim 10^{-7}$ s
"Hot spots"	10^{-3}–10^{-4} s
Electrostatic charging	10^{2}–10^{5} s
Emission of exoelectrons	10^{-6}–10^{5} s
Triboluminescence	10^{-7}–10^{3} s
Lattice defects	10^{-7}–10^{6} s
Dislocation motion	10^{5} cms^{-1}
Lattice vibrations	10^{-9}–10^{-10} s
Fracture formation	10–10^{3} cms^{-1}
Fresh surface	at 1.3.10^{-4} Pa: 1–10^{2} s
Lifetime of excited metastable states	at 10^{5} Pa: $< 10^{-6}$ s
	10^{-3} s
	10^{-2} s

with major contributions from step and screw dislocations, outer specific surface, particle size of primary crystallites and amorphous regions and different phases [Hoffmann et al. 2005].

1.1.2.3 Spherical Model

During one of the first conferences on mechanochemistry in Berlin 1983 Thiessen demonstrated the different stages of an impact stress by *a spherical model* (Fig. 1.5). By this simplifying model it could be shown that the impact stress is combined with the appearance of different species. This state is limited to very small spaces and very short times and qualified as triboplasma [Thiessen et al. 1967]. The concept of triboplasma was later developed from thermodynamic and kinetic point of view.

The short life of triboplasma causes no Maxwell-Boltzmann distribution so that an equilibrium temperature cannot be given and the chemical process taking place in this excitation phase cannot be described by the laws of thermodynamics. The conversions in triboplasmas are of a stochastic nature. The highest stage of energy excitation changes dynamically into the next stage, characterized by the relaxation of the plasma states and termed "edge-plasma" and "post-plasma". A step diagram of the energy dissipation was composed for the total process in the form of a hierarchy of the energy states (Fig. 1.6). A number of physical processes (see Table 1.2) take place in this step, such as the recombination of plasma products, the propagation of dislocations, fracture processes, the propagation of photons and the emission of electrons and photons which have important functions for the initiations of chemical reactions [Heinicke 1984].

1.1.2.4 Dislocation and Phonon Theory

The authors of *dislocation theory* allege that the mechanical action on solid substances gives rise to dislocations which come to the surface and subsequently become areas with increased chemical activity [Gutman 1974]. The motion of dislocations in solid substance is accompanied by the formation of phonons due to interactions between dislocations and other dislocations, defects, admixtures or interfaces. *The phonon theory* then emphasizes the distribution, mutual effect and origination of phonons in the course of disordering of solid substances by milling [Bertenev and Razumovskaya 1969].

1.1.2.5 Theory of Short-Live Active Centers

The essence of the theory consist in the idea that new surface arising during mechanical treatment is unable to stabilize in the 10^{-9}–10^{-11} s of thermal excitation. During the 10^{-4}–10^{-7} s required for stabilization, chemical bonds are liable to

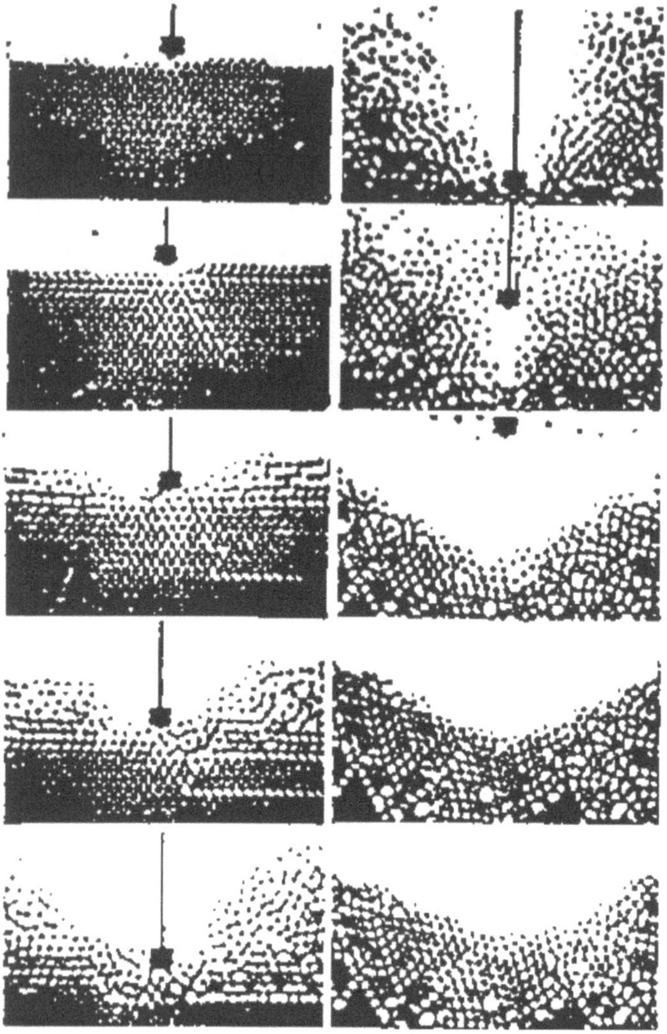

Fig. 1.5 Different stages of the impact stress schematically shown by *the spherical model* (↓ – penetration into the lattice of the solid, ↑ – decay phase up to the condition of frozen lattice distortion) [Heinicke 1984]

rearrangement, the electric surface relief is formed and further relaxation processes proceed. The decay of short-lived centres is related to the relaxation of excess energy. In vacuo, this relaxation is due to rearrangement of chemical bonds, whereas the interactions of short-lived centres with the molecules of surrounding medium are responsible for relaxation in chemically active medium. This is a case of exothermic process which can be accompanied e.g. by luminiscence or other phenomena involving radiation of energy [Butyagin 1973].

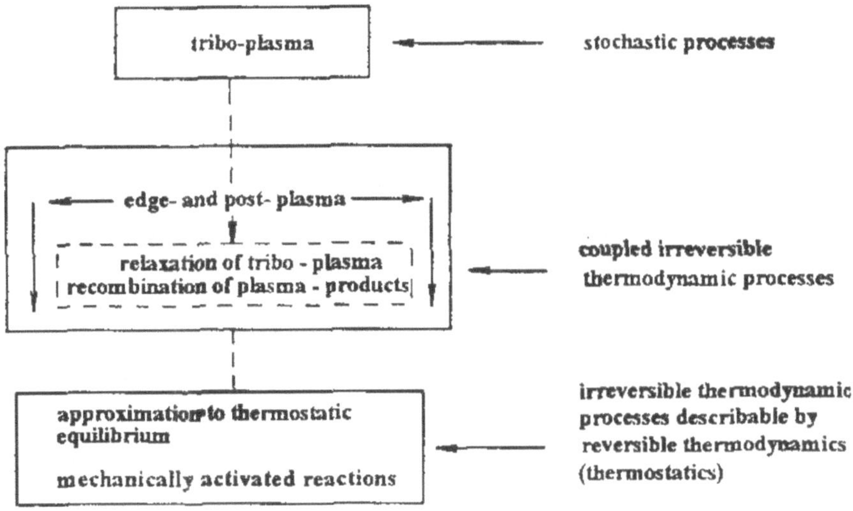

Fig. 1.6 Step diagram of the energy dissipation in solids stressed by impact [Thiessen and Sieber 1979a]

1.1.2.6 Kinetic Model

The specific features of mechanochemical effects have been analyzed from the viewpoint of limiting stages of the process [Boldyrev 1972]. The decomposition process of solid substance can be evaluated according to activation, deactivation and proper chemical reaction. We can discern two boundary cases: the decay is limited either by the processes of excitation and bond splitting (e.g. thermal decomposition) or by following stages (e.g. transformation of intermediates arising in the primary stage). Two characteristic features of mechanochemical effects have been determined: pulse action and local character of action. Both these features form the grounds of the kinetic model of mechanical activation. The kinetic model proposed by Boldyrev was experimentally verified.

1.1.2.7 Impulse Model

The model is based on the idea that the kinetics of the reaction is determined by the time in which substance is liable to be in contact with balls owing to the impulse effect of milling balls on solid substance [Lyachov 1984]. This time is different from the overall time of milling and is connected to the temperature rise during the impulse. The author succeeded in determining the equivalent temperatures corresponding to the contact of milling balls with solids and verified this idea [Boldyrev 1972, 1986].

In mills, the formation of stress field and its relaxation occur at regular intervals by subjecting a solid to a sequence of mechanical pulses following one after the

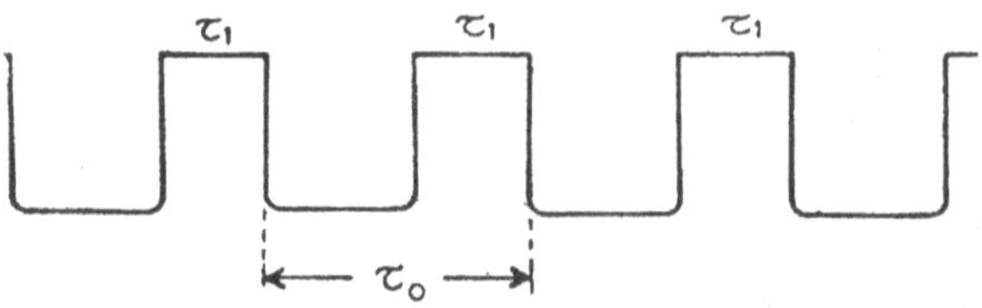

Fig. 1.7 Impulse character of mechanical action. The left side of the impulse – stress field formation, the right side – stress relaxation [Boldyrev 1986]

other. A schematic representation of these pulses is shown in Fig. 1.7. Every pulse characterizes the occurrence of stress field in the region being treated (the left-hand side of the pulse) and its relaxation leading to various physical and physico-chemical consequences (the right-hand side of the pulse) [Boldyrev 1986].

1.1.2.8 Theory of the Energy Balance

In this theory it was shown that the individual mill parameters as well as the different mill types lead to characteristic changes of the crystal lattice of solids [Heegn 1989, 1990].

1.1.2.9 Analogy Model

This model has been based on similarity of energy transfer in mill with the energy transfer in an electric circuit [Tkáčová et al. 1988]. The concept was experimentally verified with a great deal of minerals and has enabled a description of the mechanical action for variable specific energies of structural disordering of solids.

1.1.3 Mechanical Activation

The term mechanical activation was introduced by Smékal who regarded it as a process involving an increase in reaction ability of a substance which remains chemically unchanged [Smékal 1942]. Provided the activation brings about a change in composition or structure, it is a mechanochemical reaction. In this case, the mechanical activation procedes the reaction and has no effect during the course of this reaction. The definitions of mechanical activation published later were always dependent on the observed effect.

It was Butyagin who contributed to a certain unification. He considered the behaviour of the solids exposed to the effect of mechanical energy from the view-point of three main aspects: structural disordering, structure relaxation and structural mobility. Under real conditions, three factors simultaneously affect the reactivity of a

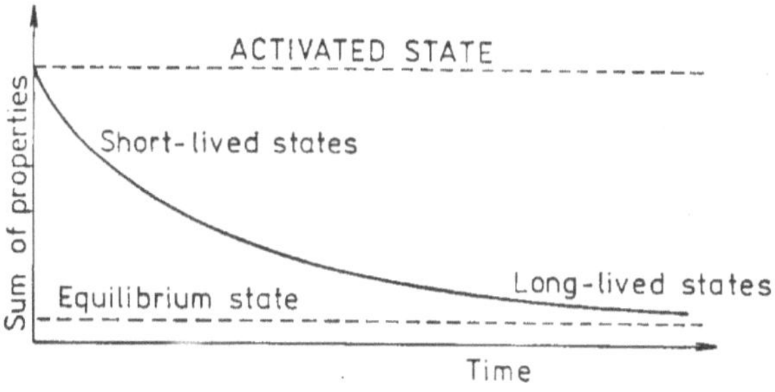

Fig. 1.8 A generalised relaxation curve of mechanically activated state [Lyachov 1993]

solid. The mechanical activation is defined as an increase in reaction ability due to stable changes in solid structure [Butyagin 1984].

However, structural relaxation plays the important role in mechanical activation. The concept of slowly changing states after interrupting the action of mechanical forces by activation has been described [Lyachov 1993]. Lyachov published a generalised relaxation curve for activated solids where individual parts of the curve correspond to processes with different characteristic times of relaxation (Fig. 1.8).

By this theory there is no possibility of influencing the reactivity of activated solids via states whose relaxation time is less than characteristic time of reaction itself. On the contrary, some long-lived states (e.g. surface area) may be regarded as constant during the course of a reaction and their influence has to be a subject of mechanical activation studies.

As for the kinds of relaxation processes various processes were described: heating, formation of a new surface, aggregation, recombination, adsorption, imperfections, chemical reaction between adjoing particles, etc. [Boldyrev 1986; Juhász and Kolláth 1993]. The rate of these relaxation processes may be vastly different and the processes can change from one way of relaxation to the other (Fig. 1.9).

Thus, mechanical activation can be regardered as a multi-step process with changes in the energetic parameters and the amount of accumulated energy of solids in each step. The four processes, namely, the accumulation of defects, amorphization, the formation of metastable polymorphous forms, and chemical reaction, are united by the term mechanical activation [Boldyrev and Tkáčová 2000].

Juhász proposed that processes under the influence of mechanical activation can be subdivided into primary and secondary ones [Juhász 1974, 1985, 1998; Juhász and Kolláth 1993]. The primary process (e.g. increase of internal and surface energy, increase of surface area, decrease of the coherence energy of solids) generally increase the reactivity of the substance. The secondary processes (e.g. aggregation, adsorption, recrystallization) take place spontaneously in activated systems and may appear even during milling or after milling has been completed.

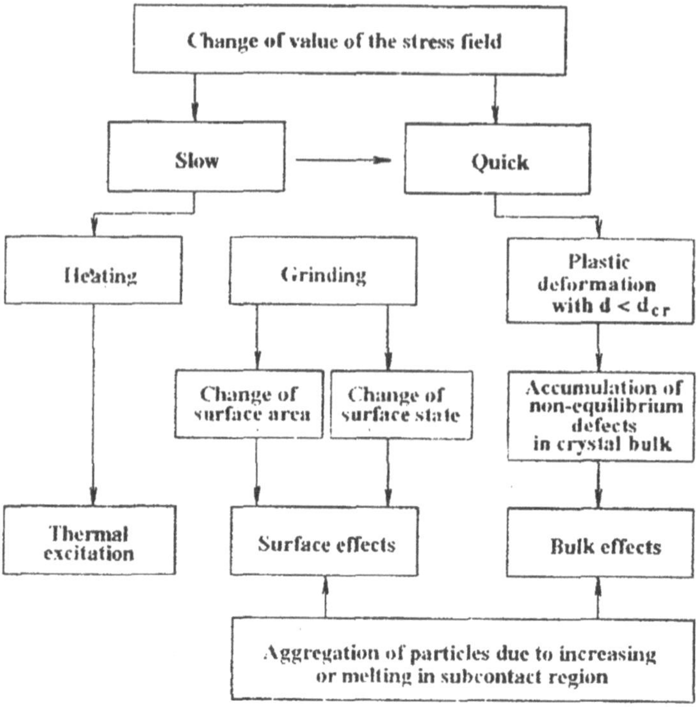

Fig. 1.9 Flowsheet of changes of relaxation processes [Boldyrev 1986]

1.1.4 Thermodynamics in Mechanochemistry

Chemical interactions between solids proceed through the consumptions of educts and the formation of products. These processes occur in agreement with physico-chemical laws and can be characterized using the fundamental thermodynamical equations. According to the Gibbs-Helmholtz equation it holds

$$\Delta G = \Delta H - T\Delta S \tag{1.9}$$

where ΔH is enthalpy and ΔS is entropy. If entropy ΔS is small (crystal structure is preserved and its disordering is slight), the term TΔS is small and the Gibbs free energy ΔG is predominantly determined by change of enthalpy ΔH. For highly disordered solids the entropy ΔS becomes significant and the term TΔS cannot be omitted.

A principal possibility for one or another chemical reaction to occur is determined by the sign of ΔG. According to the laws of thermodynamics, the process occurs spontaneously in the direction of decreasing the free Gibbs energy. A reaction between the substances proceeds with the formation of products, if

$$\Delta G_{\text{reac}} = \sum \Delta G_{\text{final products}} - \sum \Delta G_{\text{initial products,}} \Delta G < 0 \tag{1.10}$$

and takes the back direction if $\Delta G > 0$. In case when a system is in equilibrium $\Delta G = 0$. For example, in the reaction $A_{solid}+ B_{solid} = AB_{solid}$, Gibbs energy is estimated as

$$\Delta G_{reac} = \Delta G^0_{reac} + RT \ln a_{AB}/(a_A a_B) \tag{1.11}$$

where a_A, a_B, a_{AB} are the activities of the initial and final products. However, since the activity of solids is equal (or close) to unity at normal pressure and room temperature, then

$$\Delta G_{reac} = \Delta G^0_{reac} \tag{1.12}$$

Thus, in order to determine the possibility for a reaction to proceed, it is sufficient to know ΔG^0_{reac} in standard state ($p = 0.1$ Mpa, $T = 298$ K) which is calculated as the difference between Gibbs energies of the formation of final and initial products in standard state. These values are listed in reference books, see for example [Kubaschewski and Alcock 1979].

If a reaction in a mixture of solids is accompanied by the formation of gas or fluid phases (melts, solutions), solid solutions, or by the generation of defects, then, for a more strict thermodynamic forecast, it is necessary to take into account the changes of entropy and specific heat capacity during phase transitions of the components (melting, vaporization, dissolution), changes of volume and other parameters. If these factors are not taken into account, one can come across the contradictions between experimental data and thermodynamic calculations [Avvakumov et al. 2001].

The interpretation of the relation between the state of the solid before and after mechanical activation requires data for describing this state to be available. In contrast to gases and liquids, these solid particles are not able to be characterized with only a few thermodynamic factors related to these states. Even at the absolute zero point temperature, activated solids possess a finite zero point entropy caused be the disorder of lattice components. The inner thermodynamics equilibrium is first attained above the melting temperature. The state of any lower temperature depends considerably on the preliminary treatment, e.g. kind, intensity and duration of mechanical stress. The type and concentration of defects (Fig. 1.10) resulting from this determine the thermodynamic state [Heegn 1989].

The new definition of activated solid state has been postulated [Hüttig 1943]. Hüttig defined this state as a thermodynamically and structurally unstable arrangement at temperatures exceeding the melting point. He characterized the activated state of solid by "residual" Gibbs energy ΔG

$$\Delta G = G^*_T - G_T \tag{1.13}$$

where G_T*, G_T and T are the free enthalpy of activated solid, free enthalpy of this substance in non-activated state and temperature, respectively.

The Gibbs energy ΔG was analyzed by several authors and a simplified term was derived

$$\Delta G = \Delta G^*_1 + \Delta G^*_2 \tag{1.14}$$

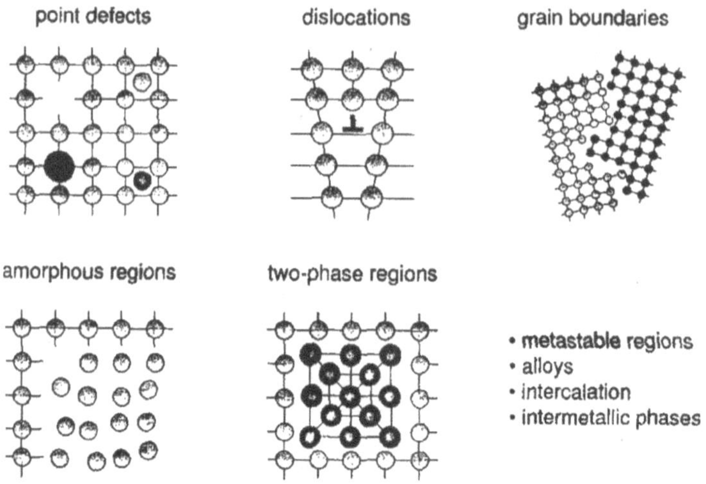

Fig. 1.10 Defects created by mechanical activation of solids [Hoffmann et al. 2005]

where $\Delta G_1{}^*$ is the residual surface energy and $\Delta G_2{}^*$ is the energy of lattice defects formation [Zelikman et al. 1975].

It holds for the surface energy in thermodynamics

$$\Delta G_1 = \sigma \Delta S \tag{1.15}$$

where σ is specific surface energy and ΔS in the change of overall surface of a solid. It was estimated for mechanical activation of ionic crystals that the surface energy ΔG_1 corresponds approximatelly to 10% of overall Gibbs energy ΔG [Schrader and Hoffmann 1973].

Schrader estimated the dependence of calcite milling efficiency η on milling time (Fig. 1.11). Based on presented plots one can obtain the maximum values $\eta = 0.19\%$ (1) and 0.006% (2), respectively (1-total milling efficiency, 2-milling efficiency consumed for increase of surface area). By comparison of these values, only 3% of total energy is consumed for increase of surface area of $CaCO_3$.

However, it has been found that there are reactions, for which the equilibrium thermodynamics does not afford favourable conditions and in spite of it they successfully proceed. The oxidation of gold is governed by reaction

$$4Au + 3CO_2 \rightarrow 2Au_2O_3 + 3C \tag{1.16}$$

From the point of view the equilibrium thermodynamics the reaction (1.16) has to be only hypothetic because of positive ΔG values (Table 1.4).

But the reaction proceeds by milling and from the mechanochemical point of view gold is not so much noble as we know from its chemistry [Thiessen et al. 1970]. In such cases the relationships of irreversible thermodynamics seemed to be more suitable [Heinicke and Sigrist 1971].

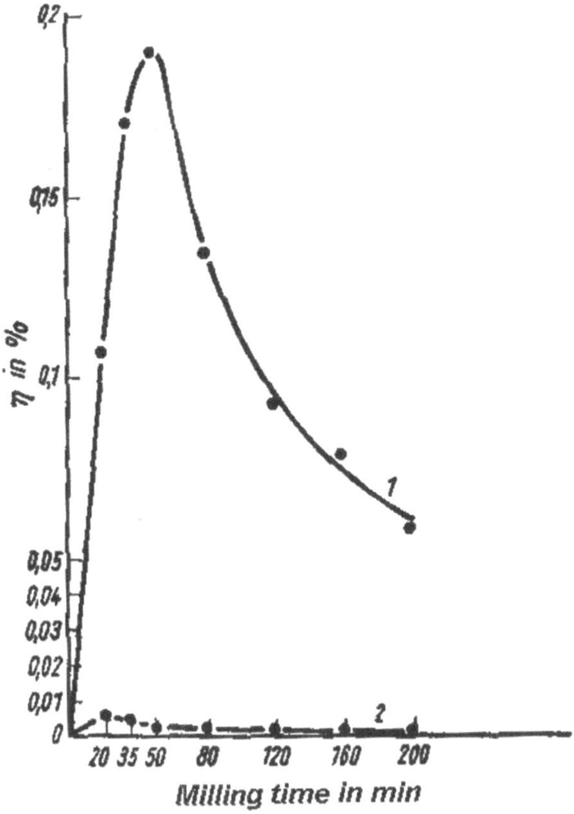

Fig. 1.11 Efficiency of calcite $CaCO_3$ milling, η in dependence on milling time [Schrader and Hoffmann 1973]

Table 1.4 The value of ΔG for reaction (1.16) [Heinicke 1984]

T (K)	ΔG (kJ mol^{-1})
298	377
1000	477
2000	615

1.1.5 Kinetics of Mechanochemical Reactions

1.1.5.1 Mechanochemical Pecularities

The general idea of a mechanochemical reaction course under effect of mechanical energy on reacting system can be illustrated by Fig. 1.12.

The reaction occurring before the beginning of the mechanical treatment (phase 1) is determined by the thermal excitation and is therefore a function of the reaction

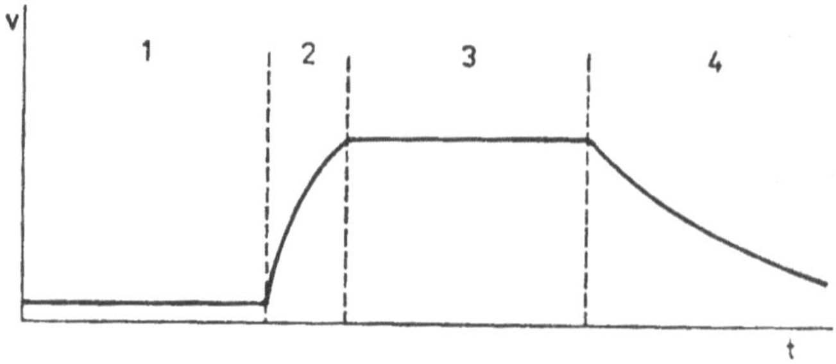

Fig. 1.12 General diagram of a mechanochemical reaction course, v – reaction velocity, t – reaction time, 1 – reaction on untreated solid, 2 – rising reaction, 3 – steady-state reaction, 4 – decaying reaction [Heinicke 1984]

temperature. However, at room temperature most solid state reactions proceed immeasurably slowly. The application of mechanical energy generally results in a significant increase of the reaction velocity (phase 2). After having passed through the rising phase a constant reaction velocity appears under external constant conditions (phase 3). After the interuption of the treatment, the decay phase comes into the existence (phase 4) [Heinicke 1984]. It was shown later that the mechanochemical reaction course described by Fig. 1.12 had a general character. However, the course of reaction is not only determined by the type of reaction but also by the kind and the intensity of applied mechanical energy, since these factors also determine the formation of the defects mainly responsible for the solid state reactions.

Probably, the first experimental observation of a change in reactivity caused by the introduction of defects in crystalline solids was that of Faraday who noticed that certain hydrated salts, dehydrated spontaneously when scratched with a pin [Faraday 1834]. Since then numerous reports have appeared on the correlation of dislocation generation and bulk and surface reactivity [Tompkins 1963]. As stated later, much of this work suffers from the drawback that insufficient attention has been paid to the research for a direct one to one correlation between change in reactivity and dislocation density. For example it is usually not sufficient to observe that a change in reactivity occurs when the solid is subjected to some treatment which is assumed to change the concentration of defects, unless it can be shown that no other accompanying change can explain the observed effect [Fox 1975].

However, even after more than 30 years the situation is sometimes the same. Very often, the results obtained are interpreted in the very standard way, something like "the mechanical activation gives rise to the accumulation of defects in reacting crystals thus increasing their reactivity", which is rather meaningless and worthless [Boldyrev 2006].

1.1.5.2 Kinetic Description

In literature, several attempts have been described to apply principles of solid state heterogeneous kinetics for the mechanochemical systems [see e.g. Avvakumov

1986]. However, only few common features with "classical" heterogeneous kinetics can be found (e.g. solid state diffusion) and the special approaches accepting the pecularities of mechanochemistry were needed.

Due to the impulse character, space non-uniformity and changes of the conditions for chemical interactions during a mechanochemical process its kinetic description is a complicated task. Therefore, there is no general approach possible and only some particular models are feasible.

The changes in conditions for a solid state reaction can be illustrated by surface area changes on Fig. 1.13 [Avvakumov 1986; Avvakumov and Kosova 1991]. The stage I of this interaction is connected with the progressive growth of surface area during activation. Real solids are polycrystals or they contain the nuclei of microcracks. Therefore, their destruction occurs via separation into crystallites. The reaction proceeds at the contact of particles. At stages II and III, the processes involving plastic deformation of disordered particles are developed. Dispersion process is overlapped by the formation of secondary particles, while the rate of the latter process is comparable with dispersing rate; thus the surface area remains constant (see also the effect of mechanochemical equilibrium described in Chap. 2). Chemical reactions take place inside secondary aggregates at the contact between particles. At the stage III, the crystallization of the products from the solid phase may occur, as well as repeated amorphization, till some stationary state between these two is achieved. The relations between the duration of stages depend on the amount of mechanical energy loading. If it is low, the process can stop at the stage II; at high energies, stages II and III occur [Avvakumov et al. 2001].

Since dispersing process and activation develop as statistically probable processes, this allow to consider chemical interaction from the view point of collision theory. In this case group collisions of surface atoms of two different particles in the zone of mechanical action are considered. The limiting stage of this process

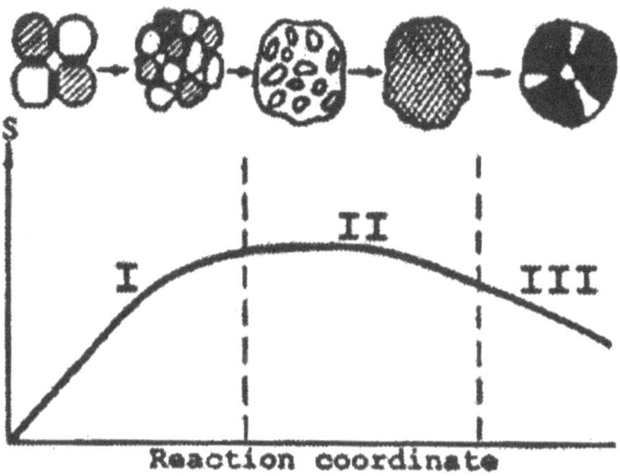

Fig. 1.13 The change of specific surface area, S in the course of a mechanochemical reaction [Avvakumov 1986]

is the probability of formation of contacting regions between solid particles of the components and a collision with a milling body [Avvakumov et al. 2001].

According to the mentioned statements, the rate of reaction in a mill can be estimated as

$$v = K_m x S_n \tag{1.17}$$

where K_m is a constant characterizing the probability of reaction to occur at a given mechanical action pre unit contact (energy constant determining thermodynamic reaction parameters T and p), x is a probability of contacting particles to come across a collision with milling body, S_n is an area of contacting regions of A and B components during mechanical action [Avvakumov 1986].

Dispersion brings to an increase of S_n, since the surface of interacting components increase. The process of new surface area formation can be described by equation

$$S = S_m(1 - e^{-kt}) \tag{1.18}$$

where S is specific surface after milling time t and S_m is maximum specific surface [Chodakov 1972]. Constant k implies the significance of rate constant of new surface formation. It is assumed that the initial specific surface value is small and can be neglected (see also Sect. 2.5.4 in this book). Equation (1.18) describes the processes in which the formation of new surface is limited by milling equilibrium after a certain time of milling.

Since, as it follows from simple geometric considerations, S_n is proportional to the total surface of particles, a similar equation can be also written for the contact surface

$$S_n = S_{nm}(1 - e^{-kt}) \tag{1.19}$$

On the other hand, as the reaction proceeds, the amount of initial reagents decreases; the probability for reagent particles to meet each other decreases. Similar to the law of mass action, we assume that the interaction rate can be written as

$$\frac{\mathrm{d}\alpha}{\mathrm{d}t} = K_m x S_n (1 - \alpha)(1 - \varepsilon\alpha) \tag{1.20}$$

where α and $\varepsilon\alpha$ are the transformation degrees for substances A and B, respectively, $(1 - \alpha)$ and $(1 - \varepsilon\alpha)$ are the amounts of non-reacted substances A and B, $\varepsilon = n/m$ is a coefficient of a mixture composition, where n is a stoichiometric coefficient representing the ratio of A and B moles in the product, and m is a real molar ratio of A and B in the initial mixture [Avvakumov and Strugova 1974; Avvakumov and Kosova 1991; Avvakumov et al. 2001].

Integration of the Eq. (1.20) taking into account the Eq. (1.19) brings to the equation

$$\frac{1}{\varepsilon - 1} = \ln \frac{1 - \alpha}{1 - \varepsilon\alpha} = K\left[t - \frac{1}{k}\left(1 - e^{-kt}\right)\right] \tag{1.21}$$

where

$$K = K_m x S_{nm} \tag{1.22}$$

At stoichiometric components ratio $n = m$, $\varepsilon = 1$, the integration gives the following equation

$$\frac{\alpha}{1-\alpha} = K\left[t - \frac{1}{k}\left(1 - e^{-kt}\right)\right] \tag{1.23}$$

In the case when specific surface area is linearly dependent on milling time (i.e., $S_t = kt$), the Eq. (1.20) transforms into

$$\frac{\alpha}{1-\alpha} = Kt^2 \tag{1.24}$$

In the case when specific surface area of the mixture is not changed during the milling, i.e., $S_t = \text{const}$ (it occurs when the initial substances are taken in highly disperse state or when a stationary state is achieved at which the substances are not milled but only are subjected to plastic flow), the integration of (1.20) gives an equation similar to that describing by bimolecular reaction

$$\frac{\alpha}{1-\alpha} = Kt \tag{1.25}$$

However, the physical meaning of the constant K is quite different from that in molecular kinetic theory of chemical reactions [Avvakumov et al. 2001].

A large number of papers on the kinetics of mechanochemical reactions report S-shaped experimental curves. The equations given in this paragraph provide a good description of this S-like shape (see e.g. Figs. 4.1 and 4.2 in this book).

Two types of mechanochemical reactions were described by Eq. (1.23)

$$FeS_2 \rightarrow FeS + S \tag{1.26}$$

and

$$FeS_2 + Fe \rightarrow 2FeS \tag{1.27}$$

In these investigations was shown that the course of reaction is not only determined by the type of reaction but also by the nature and intensity of the applied mechanical energy, since these factors also determine the formation of centers responsible for the reaction [Avvakumov et al. 1972].

Kinetics of the massicot→litharge (massicot-orthorombic PbO, litharge-tetragonal PbO) polymorphous transformation in a rotation ball mill has been studied [Lin and Niedzwiedz 1973; Lin and Somasundaran 1972]

$$\alpha - PbO \rightarrow \beta - PbO \tag{1.28}$$

In the case of PbO transformations the mechanochemical reaction proceeds in two successive stages: (i) increase of the litharge volume fraction in accordance with a modified logistic growth function, and (ii) establishment of "mechanochemical equilibrium" at a high transformed volume fraction. A modified logistic growth function of the form

$$C = \frac{1}{1+\Theta^{-n}} \tag{1.29}$$

where C is the volume fraction of the transformed new phase at time t, n is the time exponent and $\Theta = t/t_C = 0.5$, was found to fit the results of the first stage of transformation, up to a certain high value of transformed volume fraction: at this value the second stage result deviate from the linear plot of $\ln[C/(1-C)]$ vs. $\ln\Theta$. The rate dC/dt decreases and tends to dC/dt = 0, where the system stabilizes at a certain mixture value of C. This value is specific to the milling regime. Justification for the use of Eq. (1.29) for the PbO phase transformation is based on the claim that the process is activated by displacement shear [Lin et al. 1975]. This point was confirmed experimentally by high-pressure experiments [Dachille and Roy 1961]. The authors showed that although massicot is a high-temperature and high-pressure polymorph with higher density than litharge, which is a low-temperature and low-pressure polymorph, both phases may exist, in a metastable form, each in the field of the other equilibrium state. Both will transform into the stable polymorph once replacement shears are produced during milling which means that a small amount of mechanical action triggers the nucleation of other polymorph.

In the second stage of transformation, the rate of the litharge volume-fraction decreases rapidly to a "mechanochemical equilibrium" at which the phase transformation of the mixture remains constant under the given mechanical regime. It is believed that the deviation from the modified logistic growth function at the end of the first stage results in the initiation of the litharge→massicot back transformation [Lin et al. 1975].

The mechanochemical solid-state reduction of hematite with magnesium has been performed [Sherif El-Eskandarany et al. 2001]

$$2Fe_2O_3 + 3Mg \rightarrow 2Fe + 3MgO \tag{1.30}$$

The kinetics of this reaction and the growth mechanism of new phases with nano-dimension crystals have been studied assuming the nucleation and crystal growth according to the Avrami equation [Avrami 1941; Erofeev 1946]

$$-\ln(1-x) = kt^n \tag{1.31}$$

where x is conversion degree, n is the order of the reaction and k is the reaction rate constant.

The fraction x transformed at any time t can be calculated from the area under the Mössbauer spectra of the produced α-Fe phase. Figure 1.14 illustrated the variation of the fraction x as a function of milling time. The order of the reaction n was calculated from the slope of the ln[-ln(1-x)] as a function of ln(t) (Fig. 1.15), the value was found to be 0.902 while k was 0.0147 ks^{-1}. From the calculated value of n and the behaviour of the experimental data ($n = 1$) it is clear that this reaction is a first order reaction [Sherif El-Eskandarany et al. 2001]. Such behaviour indicates that the reaction proceeds with one-dimensional growth and is diffusion controlled with an agglomeration of the product on active centers [Kotkata and Mahmoud 1982].

The kinetics of nanophase iron carbide formation at composition $Fe_{75}C_{25}$ was investigated making use of the high sensibility of ^{57}Fe Mössbauer spectroscopy to the atomic environment of iron [Miani and Maurigh, 2004]. For kinetic data treatment the modified Avrami equation has been applied in the forms

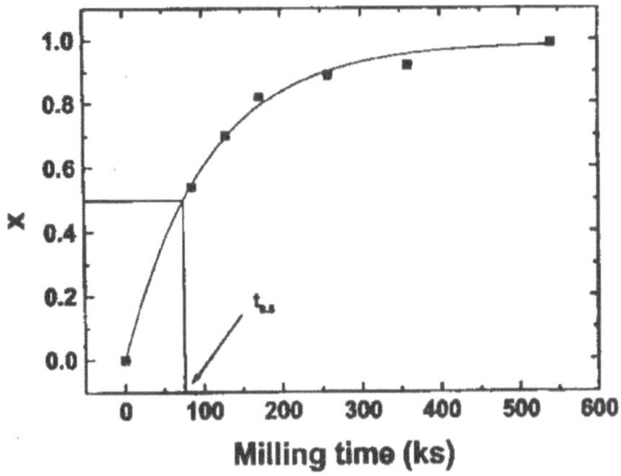

Fig. 1.14 The influence of ball-milling time on the fraction x for the reaction (1.30) [Sherif El-Eskandarany et al. 2001]

$$y = \%\mathrm{Fe} = K\exp\left(-bt^n\right) \tag{1.32}$$

$$y = \%\mathrm{Fe} = \left[1 - \exp\left(-b/t^n\right)\right] \tag{1.33}$$

where t stands for the milling time in hours, b, K and n are constants represent well the iron atomic presence in the milled powders. Some 90 samples were investigated and the kinetics followed a sigmoid-type curves. A correlation was also established between the time at which the maximum iron conversion rate is obtained and the ball-to-powder weight ratio.

Mechanochemical solid-state reduction (see also Chap. 4) of sulphides with Fe as reducing metal to prepare nanocrystalline products can be schematically expressed by general equation

$$Me_1\mathrm{S} + 2\mathrm{Fe} \rightarrow Me_1 + 2\mathrm{FeS} \tag{1.34}$$

where Me_1 is reduced metal (Cu, Pb and Sb).

Avrami-Erofeev equation

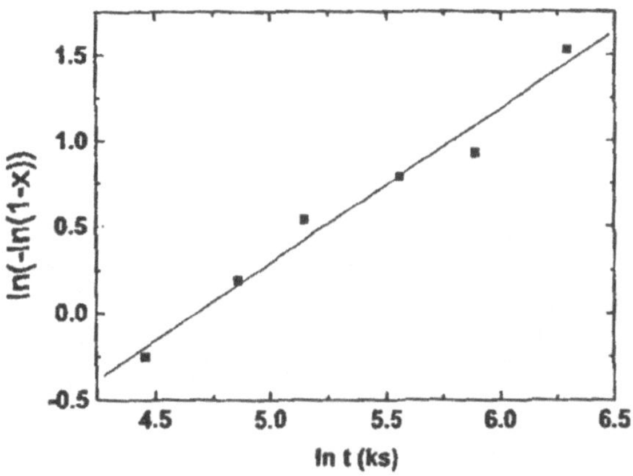

Fig. 1.15 ln[–ln(1–x)] versus ln t plot showing the application of Avrami equation for the reaction (1.30) [Sherif El-Eskandarany et al. 2001]

$$-\ln(1-x) = kt^n \tag{1.35}$$

has been applied for processing of kinetic data. In Eq. (1.35) n is the order of reaction and k is the reaction rate constant. The conversion parameter x can be taken from the magnetometry data. Mathematical transformation of Eq. (1.36) leads to the equation

$$\ln(-\ln(1-x)) = n\ln k + n\ln t \tag{1.36}$$

and allows to calculate the order of the reaction n from the slope of the $\ln(-\ln(1-x))$ as a function of $\ln t$.

SEM images of synthesized nanocomposites are seen in Fig. 1.16. According to the observed surface morphology individual nanoparticles have tendency to form nanoparticle agglomerates during milling process. The solid state recombination of nanoparticles into agglomerates is a general phenomenon which reflects a tendency of nanoparticulate systems to compensate unsaturated surface forces via surface reconstruction.

The kinetics of the solid-state reactions (1.34) has been studied assuming the nucleation and crystal growth take place at nanostructures formation. The Avrami-Erofeev equation (1.35) has been applied for the kinetics description. The same procedure has been applied in paper [Sherif El-Eskandarany et al. 2001] for single-step displacement reduction of hematite with magnesium.

The kinetic data for calculation of parameters k and n of Eq. (1.35) have been obtained from the magnetometry data presented for all three systems under study in Fig. 1.17. The calculated parameters are given in Table 1.5.

The rate of the nanometals preparation is in the order Cu > Sb > Pb. Kinetic equation (1.35) has been derived on assumption of three-dimensional growth of nuclei and is well suited for description of our processes. The parameter k with some approximation has meaning of the overall rate constant [Barret 1978]. The parameter n is function of nuclei number and their shape, composition of initial reactants and products as well as gives information about reaction mechanismus [Christian 1965]. The value of this parameter enables to resolve kinetic ($n \geq 1$) and diffusion ($n \sim 0.5$) regimes of solid-state reactions [Boldyrev 1983]. According to this statement, our experimental data in Table 1.5 show that the kinetic regime is the rate-determining step for all the systems under study.

The concept is based on the quantitative difference among factors influencing the course of mechanochemical solid state reactions (see also details in this chapter).

Fig. 1.16 SEM images Me/FeS systems: (**A**) Cu/FeS, (**B**) Pb/FeS, (**C**) Sb/FeS

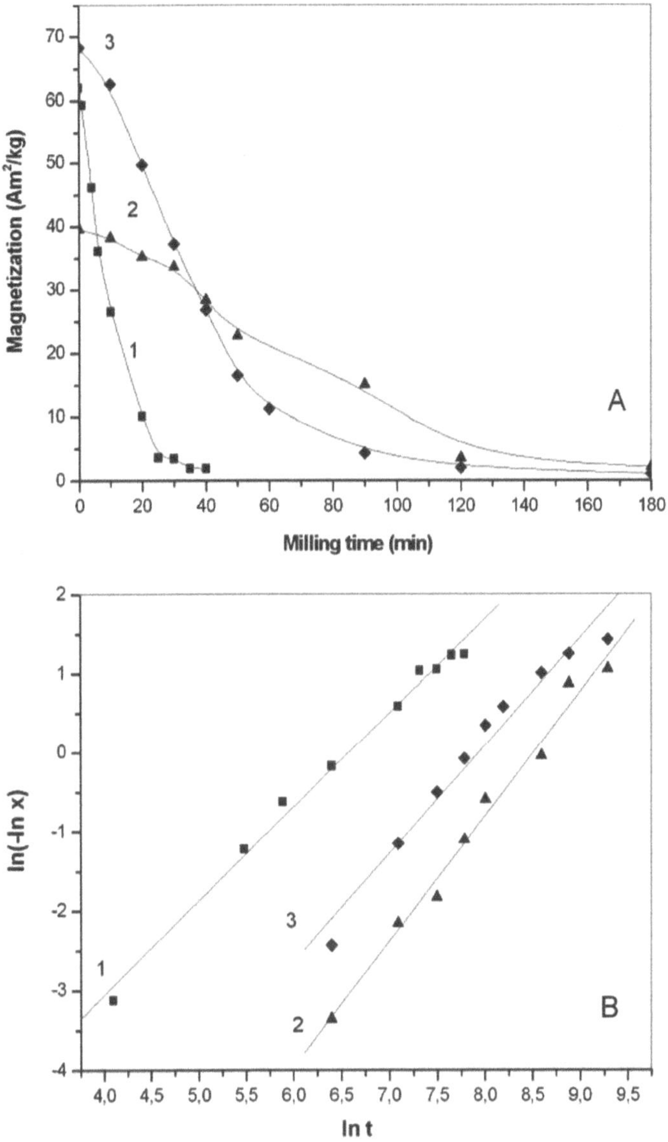

Fig. 1.17 (**A**) Kinetics of the mechanochemical synthesis of Me/FeS nanoparticles, Me=1–Cu, 2-Pb, 3-Sb; (**B**) ln(−ln$x$) vs. ln$t$ plots showing the application of Avrami-Erofeev equation

These factors may be subdivided into two main classes: the first one may be characterized as an "extensive" (specific surface area, particle size distribution, shape of particles, etc.) while the other one may be assigned to the class of "dynamic" factors (energy storage, stress relaxation, local temperature and pressure, etc.) The difference between these two groups is obvious: the life time of new states obtained as a result of mechanical activation is not limited for "extensive" parameters whereas

Table 1.5 Calculated parameters of Avrami-Erofeev equation (1.35) for mechanochemically induced solid-state reactions of metal sulphides with iron

Reaction	Parameter	
	K [s^{-1}]	n
$Cu_2S + Fe \rightarrow 2Cu + FeS$	0.0014	1.1885
$PbS + Fe \rightarrow Pb + FeS$	0.0002	1.5787
$Sb_2S_3 + 3Fe \rightarrow 2Sb + 3FeS$	0.0004	1.3650

for "dynamic" ones it is usually comparable with the reaction time. To illustrate this idea, in Fig. 1.18 a general relaxation cure for activated solid is shown where different parts are selected corresponding to processes with different characteristic times of relaxation. There is no chance to influence reactivity of activated solids for those states which have relaxation time less than characteristic time of reaction itself, expressed as $\tau_{ch}=1/k$ (short-lived states). On the contrary, some long-lived states with $\tau_i >> \tau_{ch}$ may be considered as constant in the course of reaction and their influence has to be a subject of mechanical activation studies [Lyakhov 1993].

The concept of "two kinetics" elaborated by Lyakhov is based on idea that the effect of mechanical activation on the rate of heterogeneous reactions depends on the relation between characteristic relaxation times of the processes under study. It is reasonable to suppose that in some cases chemical reaction may promote or inhibit physical relaxation but solids never react in all points of their bulk at the same time. For this reason the concept of "two kinetics" is more or less adequate to real situation in mechanically activated systems [Lyakhov 1993].

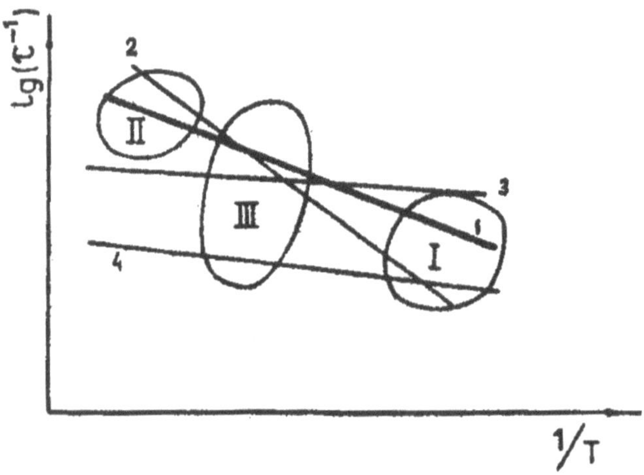

Fig. 1.18 Arrhenius plots for chemical reaction (1) and for relaxation processes: (2) – surface, (3) – internal stress, (4) – point defect concentration [Lyakhov 1993]

With the separation of relaxation times in two classes the kinetic equation in its general form may be written as

$$\frac{dn}{dt} = k\,(T_1, p_1 \ldots, p_n) * f\,(\alpha) \tag{1.37}$$

where $p_1 \ldots p_n$ are parameters of activation (*) depending on time and temperature

$$p_i = p_i(0) * \exp\,(-t/t_i)^* \tag{1.38}$$
$$t_i = t_{i0}^* \exp\,(+E_i/RT) \tag{1.39}$$

It is to be noted that for mechanochemical reactions all $p_i \ldots p_n$ are maintained on some average steady-state level during the reaction until it is finished whereas for mechanical activation some of p_i with $t_i << t_{ch}$ may be taken as equilibrium and some others with $t_i >> t_{ch}$ may be assumed to be constant. The Eq. (1.37) written in this case in the form

$$\frac{d\alpha}{dt} = k\,(T, p_i^e \ldots, p_k^e, p_{k+1}, \ldots, p_n) * f\,(\alpha) \tag{1.40}$$

where $p_i^e, \ldots, p_k^e$ are equilibrium values of corresponding parameters and all $p_{k+1}, \ldots, p_n$ are constant. It is of importance to underline that relaxation of any physical property to its "normal" state may always occur without any chemical reaction: only time and temperature are necessary for that [Lyakhov 1993].

Taking into account this separation of two relatively independent kinetics one can estimate all the complexity of reacting of mechanically activated compounds. As an example some of possible combinations are represented in Fig. 1.18, where the different temperature dependence of physical and chemical processes may result in different consequences of mechanical activation. Even this imaginary and very simple example shows the effect of mechanical activation on chemical reactions can not be observable if the temperature range is chosen with no relation to the relaxation process. Real processes are much more complex because mechanical activation modifies many properties at a time and any of them have its own kinetics of relaxation.

It is evident from the above mentioned analysis that much more effective should be activation of reaction instead of reaction precursors. Mechanochemistry allows us to use at some conditions also short-lived states whereas mechanical activation deals only with long-lived slowly relaxing states [Lyakhov 1993].

Recently, the new models for description of kinetics in mechanochemical systems have been published [Smolyakov et al. 2007, 2008].

1.2 Introduction to Nanoscience

The prefix *nano* means a billionth (1×10^{-9}) of a meter or 1/75,000th the size of a human hair. The world "nano" means dwarf in Greek language. It has been said that a nanometer is "a magical point on the length scale, for this is the point where the smallest man-made devices meet the atoms and molecules of the natural world" [Wong 1999].

Nanoscale is usually accepted in 1–100 nm range and this range is depicted in Fig. 1.19 in comparison with the dimensional scale of the earth sciences. Arrows depicted along this dimensions include: STM image of Pb and S atoms on a galena surface (10^{-10} m); crystallization nucleus of calcite (10^{-9}–10^{-8}m); bacterial cells (10^{-6}m); a single crystal of quartz (10^{-2} m); a typical open pit mine, Nevada, USA (10^{2}–10^{3} m); Mt.Fuji, Japan (10^{4} m); the Red Sea from space (10^{5} m wide and 10^{6} m long); Earth (10^{7} m); the Earth-Moon system as seen from Apollo 11 (4×10^{8} m) [Hochella, Jr. 2002].

NASA recently suggested the following definition for nanotechnology: "The creation of functional materials, devices and systems through control of matter on the nanometer length scale (1–100 nm) and exploitation of novel phenomena and properties (physical, chemical, biological) at that length scale [http://www.ipt.arc.nasa.gov/nanotechnology.html].

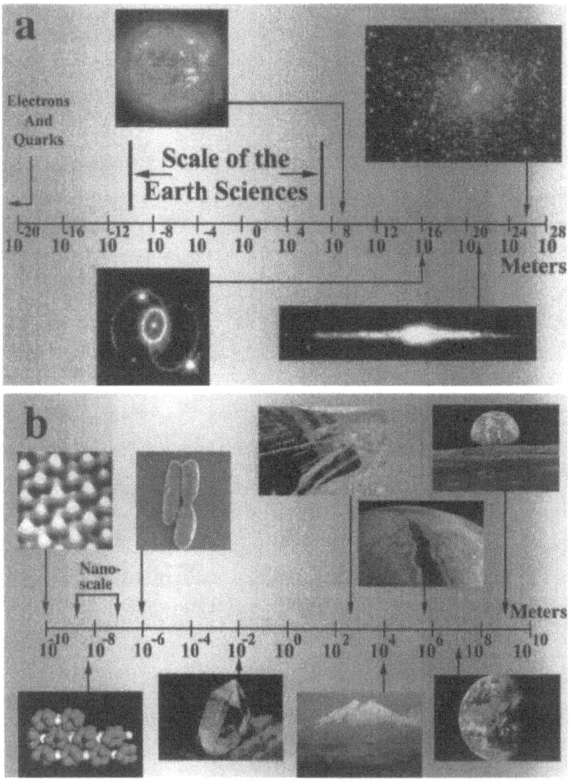

Fig. 1.19 Nanoscale among the dimensional scale of earth sciences [Hochella, Jr. 2002]

1.2.1 Historical Outline

For a long time the properties of small particles have been exploited in various fields. Especially, fabrication of colour glasses in antic times is connected with using of nanoparticles. It is known that in fourth-century A.D. Roman glassmakers were fabricating glasses containing nanosized metals. An artifact from this time called the Lycurgus cup resides in the British Museum in London. The cup is made from soda lime glass containing silver and gold nanoparticles. The colour of the cup changes from green to a deep red when a light source is placed inside it [Poole and Owens 2003].

The preparation of the individual metal particles has a very long history. It began with the popular experiments of *Faraday* on gold nanoparticle aqueous dispersions (hydrosols) in the 19th century [Faraday 1857]. Faraday discovered that nanometer-sized particles create the colour variations of metal-particle colloids through surface charge effects. He found that colloids of 6 nm gold particles are red and those of 12 nm are blue [Klabunde 2001]. One of his main interest at that time was to explain how metal particles affect the colour of church windows.

Mie was the first to provide an explanation of the dependance of the colour of glasses on metal size and kind [Mie 1908]. Mie's theory may be used to calculate the absorption coefficient versus the wavelength of the light. The Mie theory is still valid and the scattering behaviour of small spherical particles with light can be described.

In nanoworld, the legendary vision lecture of *Feynman* (Fig. 1.20) entiled "There is Plenty of Room at the Bottom" is frequently cited. The speech was given in 1959 at the Annual American Physical Society Meeting and Feynman speculated on the possibility and potential of nanosized materials. He proposed manipulating individual atoms to make new small structures having very different properties. Like many of present day nanotechnology researchers, he recognized the existence of nanostructures in biological systems. He also predicted such sci-fi accomplishments as writing 24 volumes of Encyclopedia Brittanica on the head of a pin, and even more amazingly, the complete reproduction of every book ever produced to fit within a small handheld pamphlet of less than 40 pages [Fahlman 2007].

The other American visionary, *Smalley* (Fig. 1.20) quoted: "Just wait-the next century is going to be incredible. We are about to be able to build things that work on the smallest possible length scales, atom by atom. These little nanothings will revolutionize our industries and our lives" [Smalley 1999].

There are further milestones in history of nanoscience and nanotechnology, e.g.

- The first observation of quantum confinement (*Ekimov* and *Onuschenko*, 1981)
- The discovery of the third form of carbon, known as buckminsterfullerene (*Smalley*, *Curl* and *Kroto*, Nobel Prize 1985)
- The design of scanning tunneling microscope (STM) (*Binnig* and *Rohrer*, Nobel Prize 1986)
- The synthesis of carbon nanotubes (*Iijima*, 1991)
- The developing of "electrochemical paint brush" circuit that uses an STM probe to manipulate Cu atoms on a Si surface (*Cohen*, 1998)

Fig. 1.20 Richard Feynman (1918–1988, Nobel Prize 1965) – *left*, Richard Smalley (1943–2005, Nobel Prize 1996) – *right*

1.2.2 Classification of Nanostructures

In spite of tremendeous effort in nanoscience and technology there are several approaches how to classify nanostructures.

Zero-dimensional nanostructures are usually denoted as OD structures. Here belong nanoparticles, nanoclusters and nanocrystals. These particles have usually diameter less than 100 nm and these three names are used synonymously. *Nanoparticles* are sometimes defined in a wider range, e.g. 1–1000 nm and can be single crystallites, aggregates of crystallites or amorphous. The range 1–1000 nm corresponds to *colloidal particles*. *Nanoclusters* are collection of units (atoms or molecules) of up to about 50 units. A special case of *nanocrystal* that is comprised of a semiconductors is known as *quantum dot*. Typically, the dimensions of quantum dots lie in the range 1–30 nm [Fahlman 2007].

One-dimensional nanostructures are referred as 1D structures and are characteristic for those materials that are equivalent in all but one direction. Here belong nanotubes, nanofibres, naowires and nanorods. The common feature among all these structure is that their diameters must be within the 1–100 nm range. Typically their lengths are within the micron (or layer). A *nanotube* is a 1D structure that contains a hollow core, whereas the other three forms are solid throughout (Fig. 1.21)

The term *nanofiber* is reserved for 1D nanostructures that are amorphous and usually nonconductive. By contrast, a *nanowire* designates a structure that is crystalline, with either metallic or semiconducting properties. A *nanorod* is typically a crystalline 1D nanostructure, with an overall length comparable to its width (i.e. both dimensions are < 100 nm). Nanorods are sometimes called "rod-like nanocrystals" [Fahlman 2007].

Nanocrystalline materials can be classified into four groups according to their dimensionality: zero-dimensional atom clusters, one-dimensional modulated multilayers, two-dimensional ultrafine-grained overlayers, and three-dimensional nanocrystalline structures (Fig. 1.22).

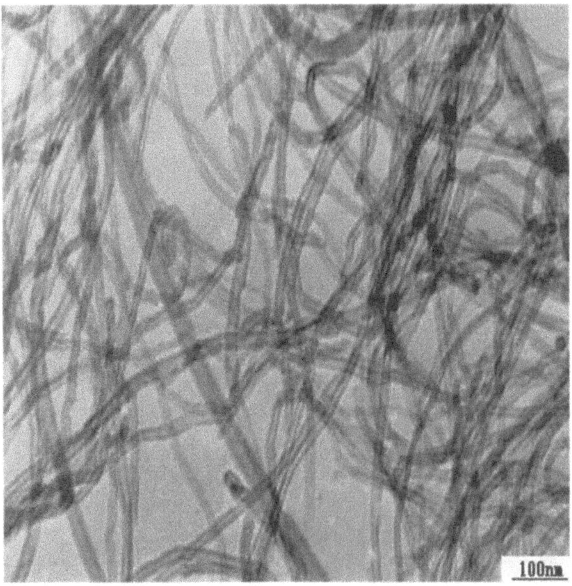

Fig. 1.21 Multi-walled carbon nanotubes (length 10–50 μm, outer diameter 8–15 nm) [www.cheaptubesinc.com]

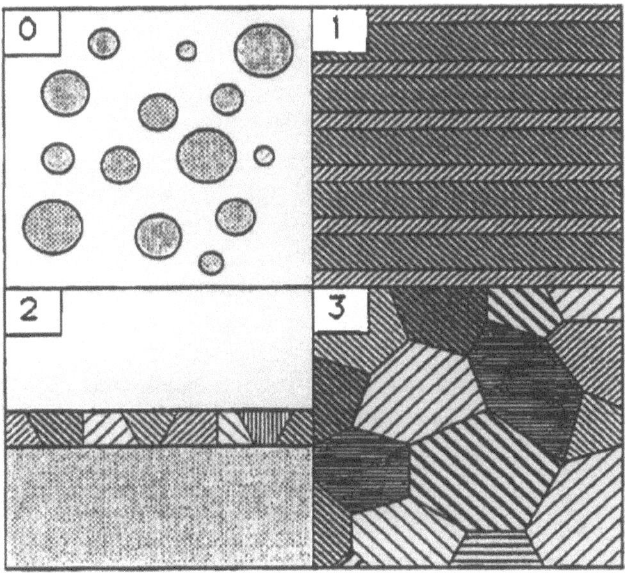

Fig. 1.22 Classification of nanostructures according to the dimensionality [Siegel and Fougere 1994]

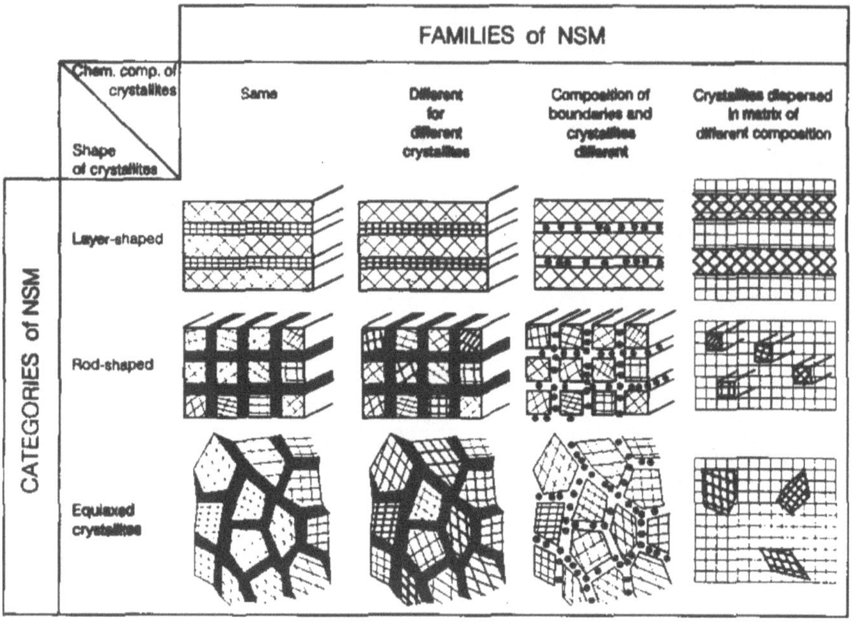

Fig. 1.23 Classification of nanoclustures according to their chemical composition and the dimensionality [Gleiter 1995]

Classification of nanostructures in twelve groups according to their chemical composition and the dimensionality is given in Fig. 1.23.

According to the *shape of the crystallites*, three categories of nanostructured materials (NSM) may be distinguished: Layer-shaped, rod-shaped and NSM composed of equiaxed nanometer sized crystallites. Depending on the *chemical composition* of the crystallites, the three categories of NSM may be grouped in for families. In the simplest case all crystallites and interfacial regions have the same chemical composition (e.g. semicrystalline polymers). NSM belonging to the second family consist of crystallites with different chemical compositions (indicated) in Fig. 1.23 by different hatchings (e.g. multiplayer structures). If the compositional variation occurs primarily between crystallites and the interfacial regions, the third family of NSM is obtained (e.g. nanometer Cu crystals with Bi atoms segregated to the grain boundaries). The fourth family is formed by nanometer sized crystallites (layers, rods, equiaxial crystallites) dispersed in a matrix of different chemical composition (e.g. precipitation hardened alloys) [Gleiter 1995].

1.2.3 Theoretical Fundamentals

One of the special feature of nanoparticles is the surface/bulk ratio of atoms. For bulk materials, the surface atoms form a negligible part of the total number of atoms.

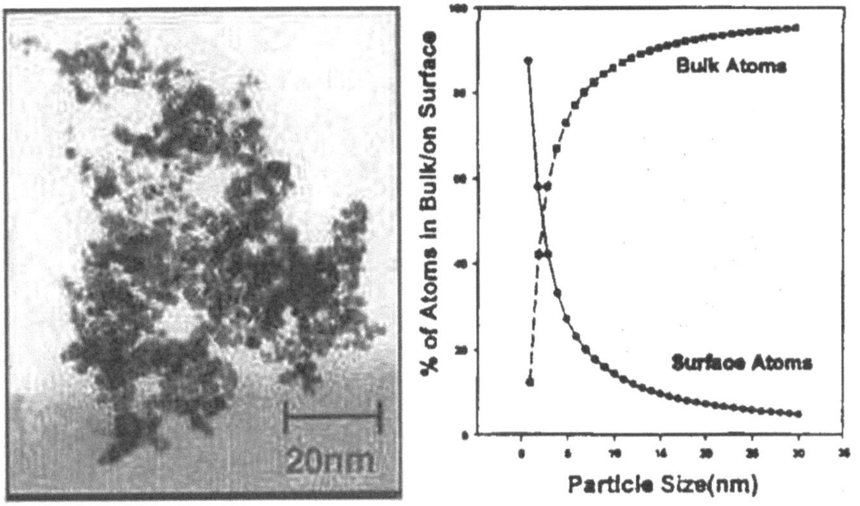

Fig. 1.24 TEM of iron particles (*left*) and calculated surface to bulk atoms ratios for these particles [Klabunde et al. 1996]

The smaller a particle becomes, the more proportion of surface atoms increases. A spherical iron particle of 30 nm in diameter has only 6% of surface atoms. However, the same particle of 4 nm has more than 40% of surface atoms (see Fig. 1.24).

The majority of metals form hexagonal or cubic close-packed structures (hcp or fcc) with coordination number of 12, except of surface atoms where it is 9 or smaller, depending on whether faces, and which kind of faces, edges, or corners are considered [Klabunde 2001].

Collection of atoms of up to about 50 units is called *cluster*. It has been shown that the growth of nanoclusters proceeds through formation of "magic number" (or closed shell) clusters that exhibit unusual electronic stability. For fcc or hcp transition metals, stable clusters contain 13, 55, 147, 309, 561 ... metal atoms, where the number of surface atoms in the nth shell is given by $(10n^2+2)$ where $n = 1,2,3,4\ldots$ [Teo and Sloane 1985]. The high yield of magic number nanoclusters is a consequence of kinetically controlled surface growth.

Metal nanoparticles consist of several tens or hundreds of metal atoms in each one. Thanks to this limitation of particle size and number of metal atoms, nanoparticles show their own properties, which can be classified by the terms *quantum size effect* or *nanosize effect*. This size limitation introduces the high population of atoms located on the surface area.

For example, 1.5 nm sized noble metal nanoparticles have 55 metal atoms in each particle (Fig. 1.25).

In this case, 42 atoms (76% of the total atoms) are located on the surface area but only 13 atoms (24% of the total atoms) are located in the inner core. When metal nanoparticle is formed from 309 atoms, then 52% of the total atoms are located on the surface area and the rest (48%) are located in the inner core. Atoms located on

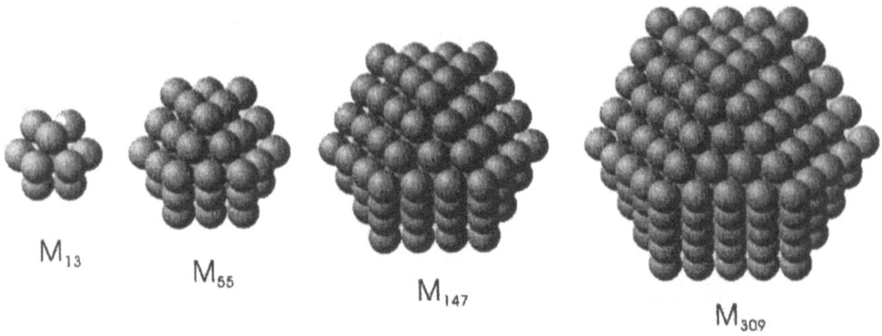

Fig. 1.25 Magic numbers of nanoparticles M_n obtained by surrounding a given atom by successive shells of atoms [Yonezawa 2004]

the surface area are chemically unsaturated and they are dominant in nanoparticle systems [Yonezawa 2004].

Generally, when the grain size of nanoparticles is below a critical value (10–20 nm), more than 50 vol. % of atoms is associated with grain boundaries or interfacial boundaries. Controlling the size, shape and structure of metal nanoparticles is technologically important because of strong correlation between these parameters and optical, electrical and catalytical properties [Tjong and Chen 2004].

A number of properties of materials composed of micrometer-sized grains, as well as nanometer-sized particles depend strongly on the surface area. For example, the electrical resistivity of a granular material is expected to scale with the total area of the grain boundaries. The chemical activity of a conventional heterogeneous catalyst is proportional to the overall specific area per unit volume, so the high areas of nanoparticles provide them with the possibility of functioning as efficient catalyst [Poole and Owens 2003].

The specific surface area is reported in the units of m^2g^{-1} and general expression for theoretical calculation is

$$S = \frac{(\text{area})}{\rho\,(\text{volume})} = \frac{A}{\rho V} \tag{1.41}$$

where ρ is the density (gcm^{-3}). Assuming that all particles are spherical and dense, a sphere of diameter d has the area $A = \pi d^2$ and the volume $V = \pi d^3/6$, then we obtain the relation for calculation of particle size

$$d = \frac{6}{\rho S} \tag{1.42}$$

The special case for particle size-dependent properties can be identified for quantum dots (with sizes below 30 nm). These particles have many potential applications in the area of demonstration devices, such as light-emitting diodes, photocatalysts, electrochemical cells and means for biomedical applications (e.g. biological in vivo imaging and drug delivery). When an electron is promoted from the valence to conduction band, an electron-hole pair known as an *exciton* is created in the bulk lattice. The physical separation between the electron and hole is referred to as the *exciton Bohr radius* (r_B) that varies depending on the semiconductor composition. In a bulk semiconductor crystal, r_B is significantly smaller than the overall size of the crystal; hence, the exciton is free to migrate throughout the lattice. However, in a quantum dot, r_B is of the same order of magnitude as the diameter (d) of the nanocrystal, giving rise to *quantum confinement* of the exciton. Empirically, this translated to the strongest exciton confinement when $d \leq 2r_B$ [Fahlman 2007].

For a semiconductor crystal, electron excitation involves a loosely bound electron-hole pair, usually delocalized over a length much longer than the lattice constant. As the diameter of the semiconductor crystallite approaches this exciton Bohr diameter, its electron properties start to change. This is so-called *quantum size effect* as named earlier [Klabunde 2001].

Table 1.6 gives the r_B values for selected semiconductors whose bandgap can be easily fine-tuned by simply changing the diameter of the quantum dot, as long as the dimensions are smaller than r_B.

The restriction of the quazi-freely mobile electrons in a piece of bulk can be reached not only by reduction of the volume to the minute size of a "zero-dimensional" (OD) quantum dot but can also be gained by reducing the dimensionality from 3 to 2 or 1. Figure 1.26 summarizes the relations between bulk, 2D, 1D and 0D situations with respect to the change of electronic characteristics [Klabunde 2001].

In the 1980s, the size-dependent electronic properties of quantum dots based on relations between the band gap E_n and $1/R^2$ (R is the radius of quantum dots) were described [Efros and Efros 1982]. Due to confinement of both electrons and holes, the lowest energy optical transition from the valence to conduction band will increase in energy, effectively increasing the bandgap

$$E_n = E_g + \frac{h^2\pi^2}{2\mu R^2} \qquad (1.43)$$

Table 1.6 Calculated exciton Bohr radii for various semiconductors [Fahlman 2007]

Semiconductor	r_B (nm)
Si	5.5
CdS	31.5
CdSe	6.1
CdTe	10
ZnO	1.8
ZnS	5.0
PbS	20.4
PbSe	46

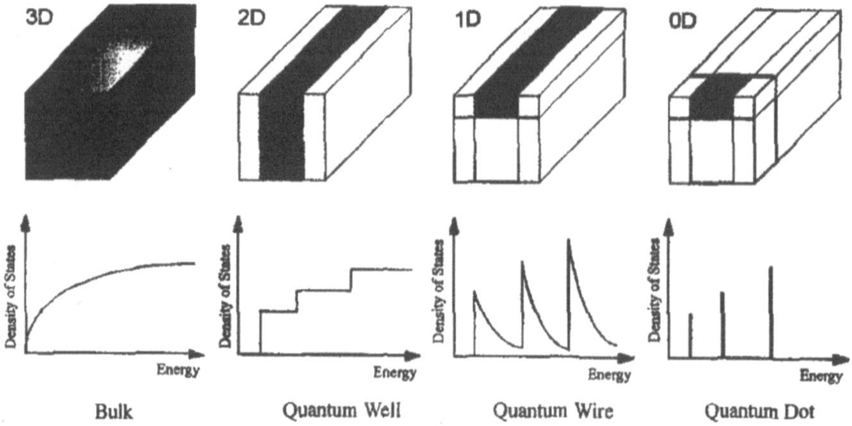

Fig. 1.26 Formation of zero-dimensional quantum dot from bulk structure [Klabunde 2001]

where E_g is the bandgap of the bulk semiconductor, h is Planck's constant, R is the radius of the quantum dot and μ is the reduced mass of exciton given by $m_e m_h/(m_e + m_h)$. Here m_e and m_h are effective masses of the electron and hole, respectively.

This model was expanded by Brus to include Coulombic interaction of excitons and the correlation energy E_R [Brus 1984]

$$E_n = E_g + \frac{h^2\pi^2}{2\mu R^2} - \frac{1.786e^2}{\varepsilon R} + 0.284E_R \tag{1.44}$$

where E_R is the Rydberg (spatial correlation) energy of bulk semiconductor

$$E_R = \frac{\mu e^4}{2\varepsilon^2\varepsilon_0^2 h^2} = 13.56\frac{\mu}{\varepsilon^2 m_e} \tag{1.45}$$

where ε_0 is the permitivity, ε is the dielectric constant of the bulk semiconductor and m_e is the mass of electron. The Brus model relates particle size to the bandgap energy of a semiconductor quantum dot.

Hence, the absorption energy of quantum dots will shift to higher frequency with decreasing diameter of dots, with a dependance of $1/R^2$. This is really observed from the reflexed colors of quantum dots with varying diameters, shifting from red to blue with decreasing size (blue shift) [Fahlman 2007].

Figure 1.27 displays the blue shift in the absorption spectra as a function of the crystal size for nanocrystalline CdS (dispersed in a transparent dielectric matrix). Being specific for this system, if a light quantum is absorbed in CdS, an electron is transferred from the valence band into the conduction band, leaving behind a positive hole. In small particles like nanoparticles the wave functions of the electron and the hole are confined to the particle volume. Hence, if the particle size

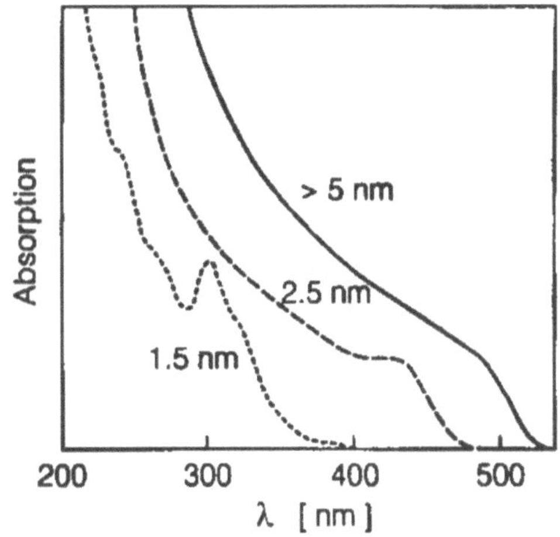

Fig. 1.27 Blue shift of CdS nanoparticles as a function of particle size [Gleiter 1995]

becomes enough small, the confinement increases the energy required for creating an electron-hole pair. The increase shift the absorption spectrum towards shorter wavelengths (blue shift) [Gleiter 1995]. For a semiconductor that has a bulk bandgap in the near-IR, its visible color can be tuned from black (bulk) to red, to yellow, to white, depending on nanoparticle size [Weller 1993].

1.2.4 Synthesis Routes (with the Exception of Mechanochemical Approach)

The synthesis of nanoparticles has received considerable attention in view of the potential for new materials and unique properties. The novel properties have encouraged many researchers to invent and explore the methods, both chemical and physical, by which such materials can be prepared [Klabunde 2001].

Various processing routes have been developed for the synthesis and commercial production of nanoparticles including vapour, liquid, solid state processing routes, and combined methods (Table 1.7).

Table 1.7 Classification of the techniques to synthesize nanoparticles [Tjong and Chen 2004]

Route	Processing
Vapor	Physical vapor deposition (PVD), chemical vapor deposition (CVD), aerosol processing
Liquid	Sol-gel process, wet chemical synthesis
Solid	Mechanical alloying, mechanochemical processing
Combined	Vapour-liquid-solid approach

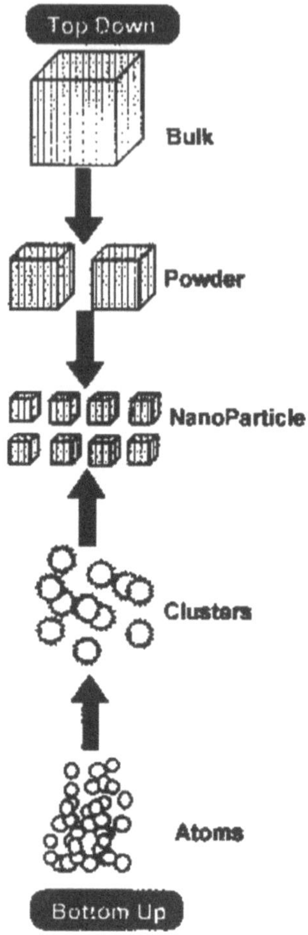

Fig. 1.28 The top down and the bottom up techniques for synthesis of nanoparticles [Dutta and Hoffmann 2003]

In literature, *the top down* and *the bottom up* approaches are used to synthesize nanoparticles i.e. either to break or dissociate solids into finer pieces or assemble atoms together (Fig. 1.28).

The synthesis is interdisciplinary work which needs chemists, physicists and materials scientists to work together in order to obtain sophisticated nanostructure. Several approaches are illustrated in further text [Gleiter 1989; Klabunde 2001].

1.2.4.1 Physical Vapor Deposition

Physical vapor deposition (PVD) is a versatile synthesis method and capable of preparing thin film materials with structural control at the atomic or nanometer scale by careful monitoring the processing conditions. PVD involves the generation of vapor phase species either via thermal evaporation, sputtering, laser ablation or spray pyrolysis [Tjong and Chen 2004].

1.2.4.2 Thermal Evaporation

Thermal evaporation, using different types of heat sources was the earliest method for achieving supersaturated vapor. Preparation of nanoparticles from supersaturated vapor produced by thermal evaporation was first established in 1930 to prepare nanoparticles of elements [Pfund 1930]. In principle also electron evaporation is possible where atoms are removed from the source by electron means.

1.2.4.3 Sputtering

Sputtering is a more convenient method of evaporation and has advantages over thermal evaporation techniques. When ions of a suitable substance (for example, those of Ar or Kr), accelerated to high energies, are directed towards a surface, atoms and clusters are ejected. As an example, nanocrystalline ZrO_2 powder was synthesized using a sputter source that has a zirconium target of 75 nm diameter positioned 100 nm from the cold surface [Hahn and Averback 2001].

1.2.4.4 Laser Ablation

This technique uses high-power pulsed lasers to ablate a sample target located in an inert gas chamber. The resulting plum of metal, believed to be mostly neutral atoms, is entrained and cooled by the carrier gas, resulting in the required supersaturation and cluster growth. The wavelength of the laser light has to be adjusted to the material. The temperature at the focusing spot can exceed 10,000 K and vaporize any substance. Virtually any material that can be fabricated into an appropriate target may be used. As an example, nanoparticles, of SiC and different titanium carbides were prepared by laser ablation of Si and Ti in a mixture of helium and isobutene [El-Shall et al. 1994; Baraton and El-Shall 1995].

1.2.4.5 Chemical Vapor Deposition

Chemical vapor deposition (CVD) is a process where one ore more gaseous adsorption species react or decompose on a hot surface to form stable solid products. CVD is a more complex method of forming thin films and coating than PVD. CVD is widely used for the deposition of metallic, ceramic and semiconducting thin films [Tjong and Chen 2004].

1.2.4.6 Spray Processing

The starting materials in this process are chemical precursors, usually appropriate salts, in solution, sol or suspensions. The process involves the generation of aerosol droplets by nebulization or "atomization" of the starting precursors. Aqueous solutions are usually used because of their low cost, safety and the availability of a wide range of water soluble salts. A wide variety of the nanoparticles have been prepared using this technique, including MgO, Al_2O_3, ZrO_2, SiO_2, TiO_2, Bi_2O_3, etc. The transformation of the aerosol droplets into particles involves various processes including solvent evaporation, precipitation of dissolved precursor and thermolysis of precipitated particles.

1.2.4.7 Sol-Gel Process

The sol-gel process (solution-gelation) process is a versatile solution-based process for making nanomaterials. In general, the sol-gel process involves the formation of a sol (colloidal suspension of ca. ≥ 200 nm solid particles) and subsequent crosslinking to form a viscous gel. The most common starting materials (precursors) used in the preparation of the sol are water-sensitive metal alkoxide complexes $M(OR)_x$ (where R-alkyl group). The process is typically carried out in the presence of polar solvents which facilitate the two primary reactions of hydrolysis and condensation [Fahlman 2007]

$$M-OR+H_2O \rightarrow M-OH+ROH \tag{1.46}$$

$$M-OR+M-OH \rightarrow [M-O-M]_n+ROH \tag{1.47}$$

Subsequent condensation eliminates either water or alcohol to produce metal oxide or hydroxide linkages. Under acidic conditions, three-dimensional solid phase networks consisting of extended linear M-O-M chain polymers are developed. When all hydroxide species are linked in one networklike structure, gelation is achieved and a dense porous gel is obtained. The gel is a polymer of a three-dimensional skeleton surrounding interconnected pores [Klabunde 2001].

Sol-gel process has been useful for synthesizing of nanometer-sized metal oxides and ceramic structures. A significant advantage in comparison to methods involving high temperatures (e.g. calcinations, evaporation) is the low temperature of the method. The disadvantage is the high cost of the alkoxide precursors. Depending on the post treatment for the sol, a wide variety of materials may be synthesized: ultra-fine powders, thin film coatings, ceramic fibres, microporous inorganic membranes, ceramics and glasses, or extremely porous materials (Fig. 1.29). Thin films are easily generated on a substrate through simple spin-coating or dip-coating of the gel, followed by slow evaporation to prevent extensive cracking. Alternatively, the gel may be retained in a mold and heat treated to convert the material into a dense ceramic or glass [Fahlman 2007].

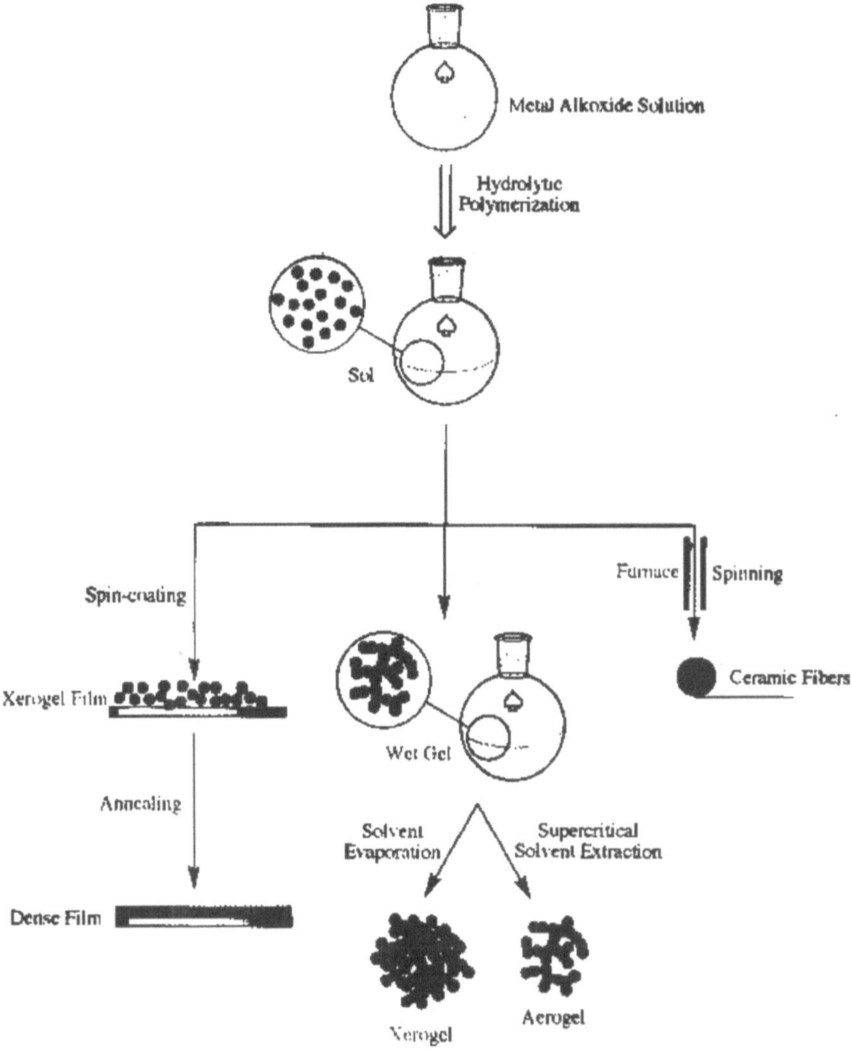

Fig. 1.29 Sol-gel method for preparation of small particles [Fahlman 2007]

1.2.4.8 Micelles Method

The micelle approach is one of the recent promising routes to nanocrystalline materials. Surfactants dissolved in organic solvents form spheroidal aggregates called *reverse micelles*. In the presence of water, the polar head groups of the surfactant molecules organize themselves around small "water pools". Water pool is characterized by the water/surfactant molar ratio, $w = [H_2O]/[S]$. For reverse micelles a low amount of water is characteristic, $w < 15$. Reverse micelles methods is used for synthesis of nanocrystalline semiconductors [Pileni 1993, 1997]. There are several advantages to using this method, including the preparation of very small particles and the ability to control the particle size. Disadvantages include low production yields and the need to use large amount of liquid [Khaleel and Richards 2001].

1.2.4.9 Precipitation

One of the conventional methods for preparing nanoparticles is the precipitation method [Gao et al. 1999]. This process involves dissolving a salt precursor, usually a chloride, oxychloride or nitrate; for example $AlCl_3$ to make Al_2O_3, $Y(NO_3)_3$ to make Y_2O_3, and $ZrCl_4$ to make ZrO_2. The corresponding metal hydroxides usually form an precipitate in water on addition of a base solution such as NaOH or NH_4OH. Sometimes hydrolysis approach is applied as in the case of nanocrystalline TiO_2 powders [Hoffmann et al. 2001]

$$TiCl_4 + 2H_2O \rightarrow TiO_2 + 4HCl \tag{1.48}$$

The resulting chloride salts – NaCl or NH_4Cl – are then washed away and the hydroxide is calcined after filtration and washing to obtain the final oxide powder. One disadvantage of this method is the difficulty in controlling the particle size and size distribution. Fast (uncontrolled) precipitation often takes place, resulting in large particles [Khaleel and Richards 2001].

1.2.5 Mechanochemical Approach

Various methods used to synthesize nanostructured materials have been described in previous paragraph. Unlike many of the above mentioned methods, mechanochemical approach produces its nanostructures not by cluster assembly but by the structural decomposition of coarser-grained structures as the result of severe plastic deformation. This has become a popular method to make nanocrystalline materials because of its simplicity, the relative inexpensive equipment (at least on the laboratory scale, see mills in Chap. 2) needed, and the applicability to essentially all classes of materials. The major advantage often quoted is the possibility for easily scaling up to tonnage quantities of material for various applications [Koch 1997].

The phenomenology of the development of nanocrystalline microstructure by high-energy milling has been summarized into three stages, namely

- stage 1 – deformation localization in shear bands containing a high dislocation density,
- stage 2 – dislocation annihilation/subgrain structure with nanoscale dimensions – further milling extends this structure throughout the sample,
- stage 3 – orientation of the grains becomes random, i.e. low angle grain boundaries → high angle grain boundaries – grain boundary sliding, rotation is likely [Fecht 1995].

1.2.5.1 Metals

Much of the systematic research on the evolution of nanocrystalline single phase materials by high-energy milling has been carried out at California Institute of Technology. The crystal size as a function of milling time is given for Ru and AlRu in Fig. 1.30.

The decrease in average crystal size with milling time as observed in this case is common to all the studies which follow the microstructural evolution of grain size with milling time. The lattice strain measured from the X-ray diffraction line broadening also increased with milling time but reached a maximum and then decreased at the longest milling times as illustrated in Fig. 1.31.

The authors suggested that the strain was mainly due to the diclocation density in the grains which decreased as the grains become very small. The development of the nanocrystalline structure by high-energy milling is due to the plastic deformation induced. The grain size decreases and the lattice strain increases (or goes through the maximum) with milling time. Once the grain size reaches a given constant size, further refinement ceases for a given set of milling conditions. For the fcc metal studied this minimum grain size scales inversely with the melting temperature of the metal [Eckert et al. 1992].

The lattice strain values available from the literature were plotted against reciprocal grain size, 1/d in Fig. 1.32.

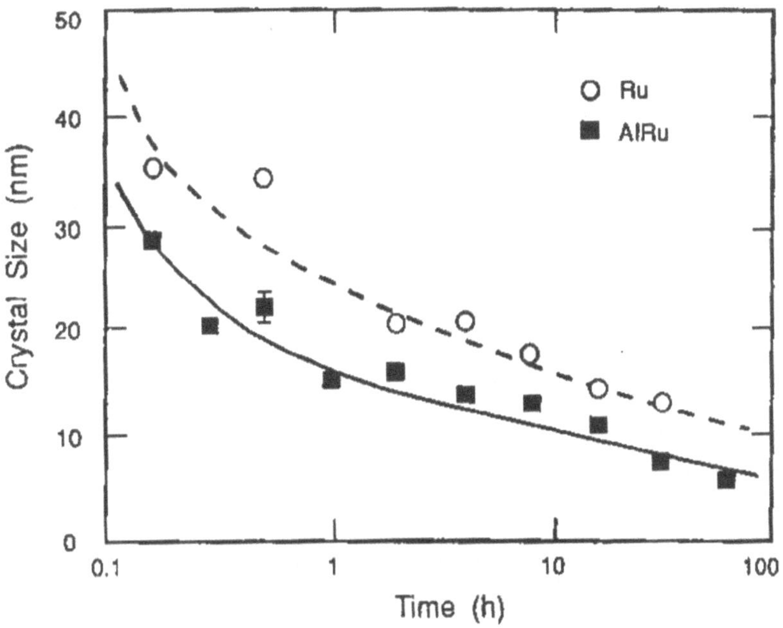

Fig. 1.30 Crystallite size vs. milling time [Hellstern et al. 1989]

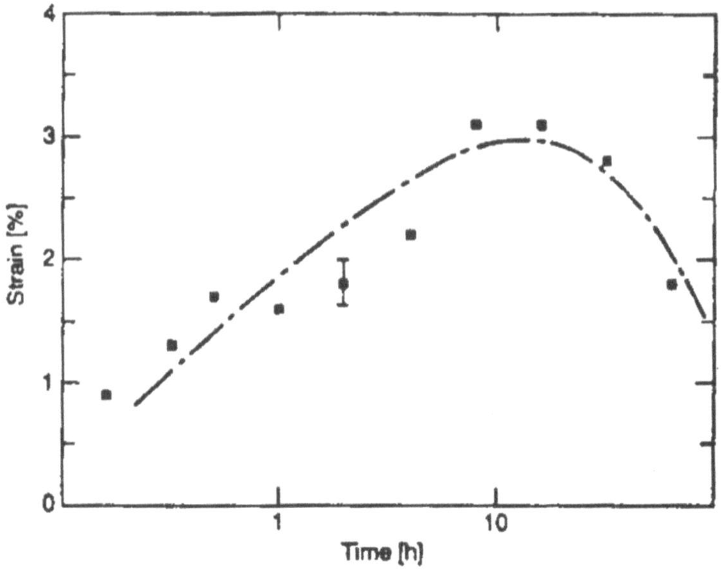

Fig. 1.31 Average strain in AlRu vs. milling time [Hellstern et al. 1989]

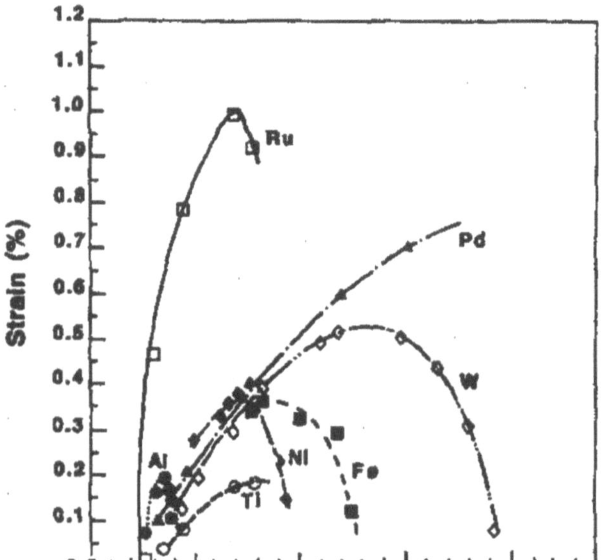

Fig. 1.32 Lattice strain vs. reciprocal grain size for various metals [Enzo et al. 1989; Hellstern et al. 1989; Eckert et al. 1992; Oleszak and Shingu 1996]

With the exception of Ru, the data for increasing lattice strain with 1/d appear to fall on a common relative narrow band before decreasing from the maximum strain values. However, these data are from several groups using mills with various energy levels and possible differences in milling temperatures [Koch 1997].

Additional information to help explain the mechanism of nanocrystalline formation comes from measurements of stored enthalpy. Maxima in stored enthalpy vs. 1/d plots are typically observed [Hellstern et al. 1989; Eckert et al. 1992; Oleszak and Shingu 1996]. However, the maximum in stored enthalpy is usually found at smaller grain size than the strain maximum, as illustrated for W in Fig. 1.33.

Here the maximum in strain occurs at $d = 8.3\,\mathrm{nm}$ while the maximum in stored enthalpy is at $d = 5.5\,\mathrm{nm}$. It is stated that the stored enthalpy comes mainly from X grain boundaries [Eckert et al. 1992; Oleszak and Shingu 1996] and grain boundary strains. Stress relaxation may be responsible for the maxima [Oleszak and Shingu 1996] but, as above, the strain and stored enthalpy maxima do not necessarily coincide [Koch 1997].

The silver powder was synthesized in a mechanochemical process by inducing a solid state reduction reaction between AgCl and Na metal. The reduction was

$$\mathrm{AgCl} + \mathrm{Na} \rightarrow \mathrm{Ag} + \mathrm{NaCl} \tag{1.49}$$

The enthalpy change in this reaction is $-284\,\mathrm{kJ\,mol^{-1}}$ at room temperature and this high exothermal reaction makes it possible to achieve combustion during milling. In X-ray diffraction pattern the peaks characteristic of Ag and NaCl can be seen directly. In addition, The pattern shows traces of an intermetallic Ag_2Na phase.

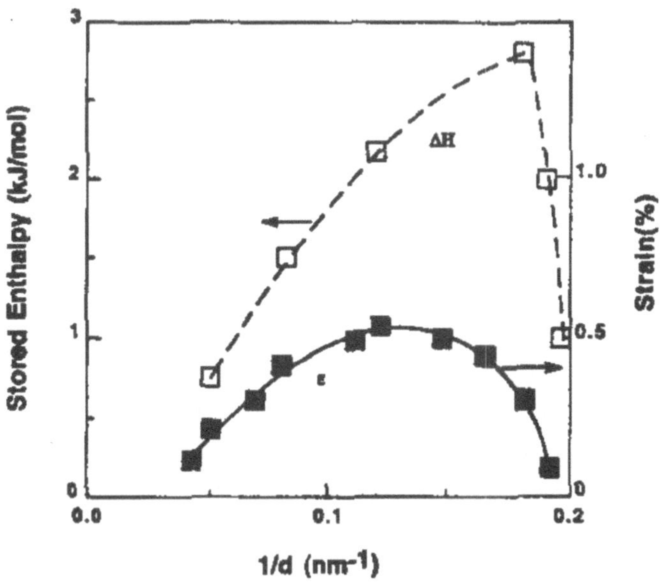

Fig. 1.33 Stored enthalpy vs. reciprocal grain size 1/d for W [Oleszak and Shingu 1996]

The smallest particles that can be found by TEM were below 50 nm. In the main, the size of the particles falls in the range 100–500 nm [Keskinen et al. 2001].

Mechanochemical approach has been successfully applied to the synthesis of other ultrafine metals [Gaffet and LeCaër 2004].

The formation of Fe is due to the reduction of $FeCl_3$ by Na via the reaction [Ding et al. 1995]

$$FeCl_3 + 3Na \rightarrow Fe + 3NaCl \quad (1.50)$$

In the case of ultrafine Cu particles prepared by the mechanochemical process [Ding et al. 1996], the milling induced solid state reaction is

$$CuCl_2 + 2Na \rightarrow Cu + 2NaCl \quad (1.51)$$

Depending on the reaction modes (continuous or combustion ones), two district morphologies of Cu particles have been observed; Cu particles with sizes uniformly distributed in the range 20–50 nm are formed by a steady state reaction during milling, whereas larger particles were observed in the combustion reaction case [Gaffet and LeCaër 2004].

1.2.5.2 Semiconductors

Crystalline inorganic solids can be divided electronically into three well-known classes: metals, semiconductors, and insulators. In these extended solids, atomic orbitals overlap to give nearly continuous electronic energy levels known as bands [Hannay 1959]. Metals are electronically characterized by having a partially filled band. Semiconductors have a filled band (the valence band) separated from the (mostly) empty conduction band by a bandgap E_g corresponding to the familiar HOMO-LUMO energy gap for small molecules. Insulators are conceptually the same as semiconductors in their electronic structure, except that the bandgap is larger in insulators (Fig. 1.34). Here the shaded boxes represent the filled valence

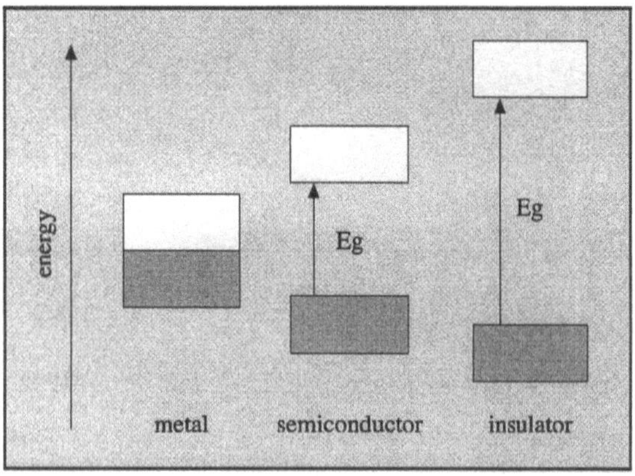

Fig. 1.34 A simplified energy level diagram for metals, semiconductors and insulators [Murphy and Coffer 2002]

bands. The empty boxes represent the empty (at temperature 0 K) conduction bands. The arrows represent the band energy E_g.

If the particle size of a bulk inorganic crystalline solid is on the order of nanometers, it is now well-known that interesting optical and electronic effects may result. Semiconductors with all three dimensions in the ~1–10 nm (or up to 30 nm) size range are referred to as quantum dots (see also Sect. 1.2.3). In this size range electrons exhibit quantum mechanical effects [Steigerwald and Brus 1990]. In the literature, semiconductor quantum dots are also known as semiconductor nanocrystals or nanoparticles. In Fig. 1.35 energy level diagram comparing a bulk semiconductor to its molecular analog and a quantum dot is given.

Here, the semiconductor's electron are in bands. The molecule's electrons are in molecular orbitals (bonds). The vertical arrow denotes the bandgap E_g for the bulk semiconductor, and the highest occupied molecular orbital-lowest unoccupied molecular orbital (HOMO-LUMO) energy gap in the molecule. On the nanometer scale, the electronic structure of a semiconductor quantum dot is in the intermediate regime between bands and bonds [Murphy and Coffer 2002].

The driving force for the increased interest in semiconductor nanoparticle research is both new physics as well as novel technological applications. Growth of nanoparticles of variety of semiconductor materials of well controlled dimensions and compositions is the important first-step towards the realization of these objectives [Balamurugan et al. 1999].

II-VI semiconductors (ZnO, ZnS, CdS, CdSe and CdTe and others) can be prepared by various methods such as solution growth techniques, electrodeposition, spray pyrolysis, chemisorption, sputtering, successive ionic layer deposition, inert gas condensation, hydrothermal route, chemical capping, vacuum evaporation, etc. Mechanochemical synthesis belongs among the synthesis routes which can effectively control and regulate the course of solid state reactions [Baláž 2000; Suryanarayana 2001].

ZnS, CdS and $Zn_xCd_{1-x}S$ are II-VI direct bandgap semiconductors widely investigated as bulk or thin film semiconductors due to their wide applications in

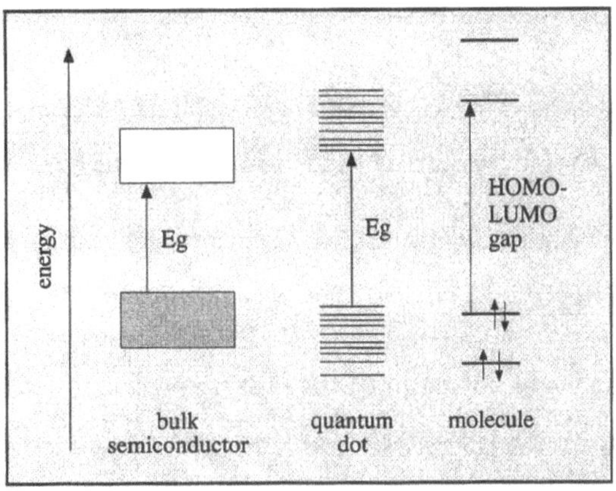

Fig. 1.35 A simplified energy level diagram for bulk semiconductor, quantum dot and molecule [Murphy and Coffer 2002]

optoelectronics. These systems have unique size-dependent properties which include optical properties such as absorption. The size dependence of the band levels of semiconductor particles results in a change of optical properties [Lin et al. 2005].

1.2.5.3 Zinc Sulphide ZnS

Zinc sulphide as a very important direct wide-bandgap semiconductor with the highest E_g value (3.66 eV) among all II-VI semiconductors, has been attracting extensive interest in material science due to its special electronic and optical properties for a variety of applications including optical coatings, solid-state solar cell windows, electrooptic modulators, photoconductors, field effect transistors, sensors, transductors, and light-emitting applications [Lan et al. 2003]. Therefore, much research on ZnS particles and their physicochemical properties has been carried out and many methods have been used for the preparation of these nanoparticles [Baláž et al. 1997; Chen and Liu 1999; Dhas et al. 1999; Wang and Hong 2000; Pawaskar et al. 2002; Chen et al. 2004] even as one-dimensional nanostructures [Velumani and Ascencio 2004].

ZnS nanoparticles were prepared by high–energy milling of zinc acetate and sodium sulphide according to the reaction

$$(CH_3COO)_2Zn.2H_2O + Na_2S.9H_2O \rightarrow ZnS + 2CH_3COONa + 11H_2O \quad (1.52)$$

The reaction is thermodynamically feasible at ambient temperature, as the enthalpy change is negative. The value $\Delta H^{\circ}_{298} = -171\,\mathrm{kJmol^{-1}}$ was calculated from thermodynamic data [Wagman et al. 1982]. After the completion of reaction (1.52) the synthesized ZnS nanoparticles have been washed, decantated and dried according to the procedure described in [Baláž et al. 2003]. The properties of mechanochemically synthesized ZnS(M) were compared with the chemically precipitated ZnS(C).

The XRD patterns of the chemically precipitated ZnS(C) and mechanochemically synthesized ZnS(M) are shown in Figs. 1.36 and 1.37, respectively.

The XRD analysis of the ZnS(C) confirmed the presence of hexagonal wurtzite, $\alpha - ZnS$ (JCPDS 36-1450) and cubic sphalerite, $\beta - ZnS$ (JCPDS 5-566). Clearly, the occurrence of a weak reflection peak (200) at 33.4° and a very small peak at 69.9° indicates the cubic phase (sphalerite). In addition, the investigation of the relative intensity and comparison with those in JCPDS database revealed the relatively higher intensities of reflections associated with both phases. However, it should be noted that the wurtzite is the main phase in the sample and only a small amount of sphalerite is present in the ZnS(C).

In contrast, the XRD pattern of the ZnS(M) show mainly the reflections of cubic phase, which is also supported by relative intensity. No reflection peaks were found at 39.6° and 69.9°, which shows only the hexagonal form of ZnS. The (200) reflection of the cubic phases at about 33.4° is probably masked due to a large broadening of the reflection (111). The peaks associated with hexagonal phase disappear with milling and only the cubic structure is evident in the diffraction pattern of the milled sample. The mechanochemical transformation of wurtzite to sphalerite can be

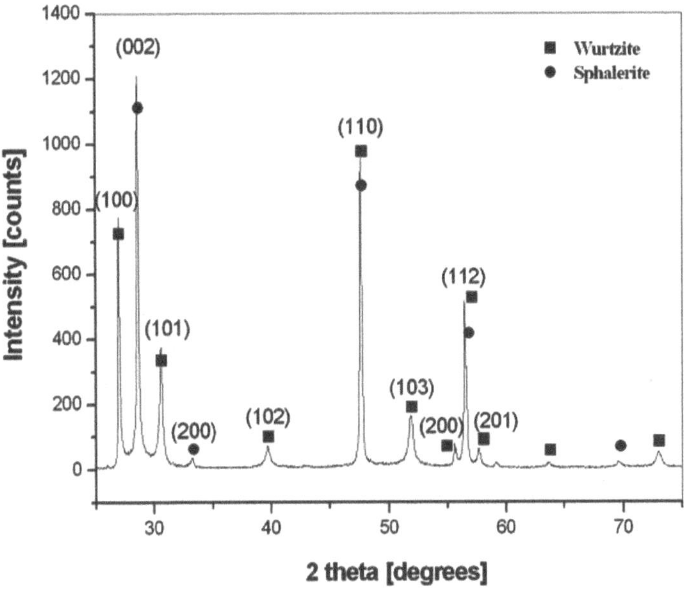

Fig. 1.36 XRD pattern of chemically precipitated ZnS(C)

attributed to the motion of dislocations in the activated solid state. The results agree with the previous observations [Imamura and Senna 1984]. The higher background on the XRD pattern of ZnS(M) implies the formation of some amorphous material. The amorphization is in fact a highly distorted periodicity of lattice elements, and it

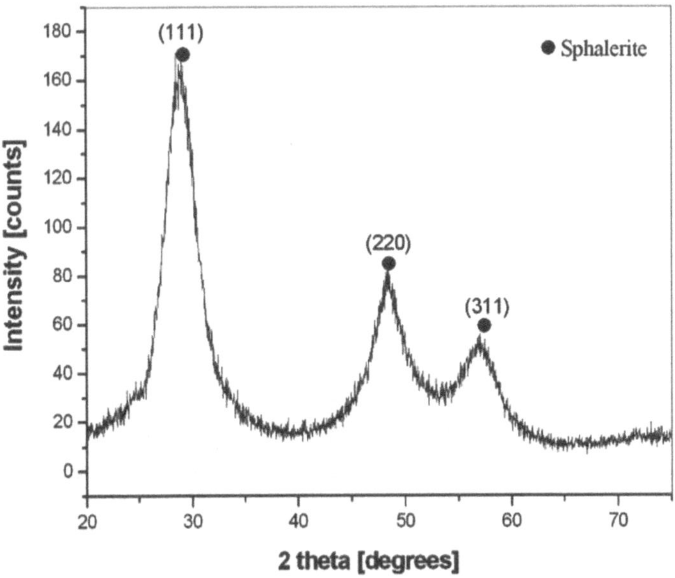

Fig. 1.37 XRD pattern of mechanochemically synthesized ZnS(M)

Table 1.8 The volume weighted crystallite sizes for chemically synthesized ZnS(C)

Reflections	Scherrer (nm)	Warren-Averbach (nm)
(100)	42.0	46
(002)	33.2	39.8
(101)	18.1	18.5
(110)	47.5	11.9
(103)	11.2	12.5
(200)	35.4	36.9
(112)	35.2	38.5
(201)	21.6	24.2
Average	30.5	28.5
(102)	3.7	11.5

is often characterized as a short range order in contrast to the long order of a fully crystalline structure.

The Warren-Averbach analysis provides detailed information regarding to crystallite size, lattice strain and their distributions [Warren and Averbach 1950]. The results for ZnS(C) using the Warren-Averbach and Scherrer methods are given in Table 1.8. Since the ZnS(C) is free of strain, the Scherrer equation can be applied successfully for calculation of the weighted crystallite size. From the Warren-Averbach method, the average surface weighted crystallite size was estimated as 29 nm, corresponding to the volume weighted crystallite size of 31 nm obtained using Scherrer equation. The results of the Scherrer equation shows directly good agreement with the results of the Williamson-Hall plots because both the methods measure the volume weighted crystallite size. Hence the obtained results from the Warren-Averbach method and Williamson-Hall plots are comparable.

Regarding the ZnS(M), the presence of non zero slope and intercept reveals that both size and strain components exist in the sample (Fig. 1.38). The plot shows high correlation coefficient ($r = 0.96$) and negligible scatter in the β_f^{*2}, suggesting uniform crystallites. The ZnS(M) yields larger broadening compared to the ZnS(C). The volume weighted crystallite size and maximum lattice strain calculated are 4 nm and 7.5×10^{-2}, respectively.

For the ZnS(M) (cubic phase), the Warren-Averbach method was applied to calculate both strain and crystallite size components. The three intensive reflections were used for calculation. The average surface weighted crystallite size obtained was 1.8 nm and the root mean square strain (RMSS) at $L = 1$ nm,$< \varepsilon^2_{L=1nm} >^{1/2}$, accounted for 2.6×10^{-2}. As discussed earlier the results of the Williamson–Hall plots and Warren-Averbach methods with regard to their relationships are in a good agreement [Pourghahramani and Forssberg 2006a, b]. The specific surface area of ZnS(C) ($7\,m^2g^{-1}$) is lower in comparison with ZnS(M) where this value attains 97–128 m^2g^{-1} with an increase with milling time, which is a further advantage of the mechanochemically synthesized ZnS nanoparticles.

Surface morphology of the ZnS(M) with estimated size from 10–20 nm is depicted in Fig. 1.39a forming irregular particles. From the surface analysis using

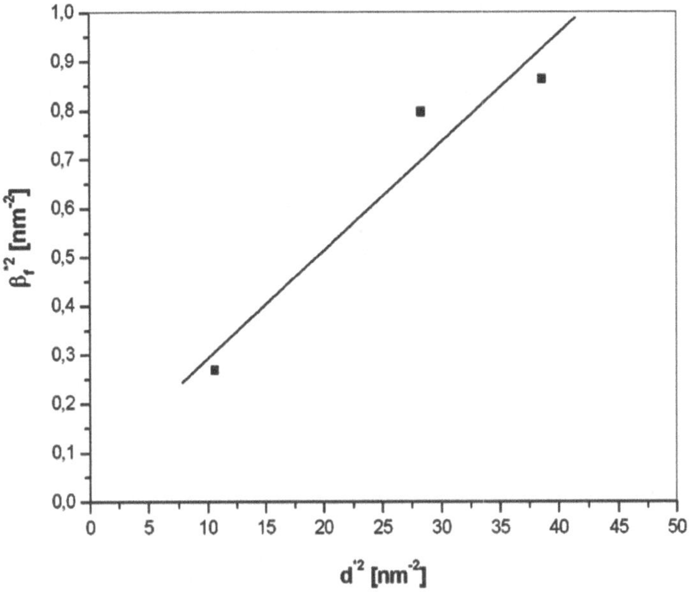

Fig. 1.38 Williamson-Hall plot of mechanochemically synthesized ZnS(M)

SEM it is revealed that there is a homogeneous distribution of the particles and the surface is smooth. Individual nanoparticles have tendency to form nanoparticle agglomerates during milling process and both entities can be clearly seen.

In order to analyze the surface characteristics in more detail, sampling methods of AFM in contact mode was used. Figure 1.39b shows a three-dimensional representation of the image obtained from the surface of the ZnS(M) nanoparticles. A homogeneous distribution of crystalline domains can be observed from the image. The deflection image of the ZnS(M) nanoparticles surface is shown in Fig. 1.39c, from which the crystalline distribution was observed to be uniform and regular. From the comparison of the deflection image (Fig. 1.39c) and the altitude topography (Fig. 1.39b) it is observed that the crystallites were of the same size.

HRTEM technique allows the determination of size of the nanoparticles [José-Yacamán et al. 2001], the type of structures produced [Ascencio et al. 1998] and also the possible induced morphologies [Ascencio et al. 2006]. In Fig. 1.40 four different micrographs of ZnS are shown. In Fig. 1.40a an area of 16 nm × 16 nm is observed, where several clusters are clearly identified, and particularly three of them are measured with sizes of 2.6, 3.7 and 3.4 nm. In fact the corresponding fast Fourier transformation (FFT) technique denotes a polycrystalline material, which must be composed by the nanocrystals. Higher magnification allows determining the lattice distance of the material as in case of Fig. 1.40b, where it is found a square contrast in the center of the micrograph with interplanar distances of 0.27 and 0.28 nm that implies a region with an axis zone near to the (001). Figure 1.40c shows a well defined cluster of ~4 nm with an hexagonal profile and rhombic

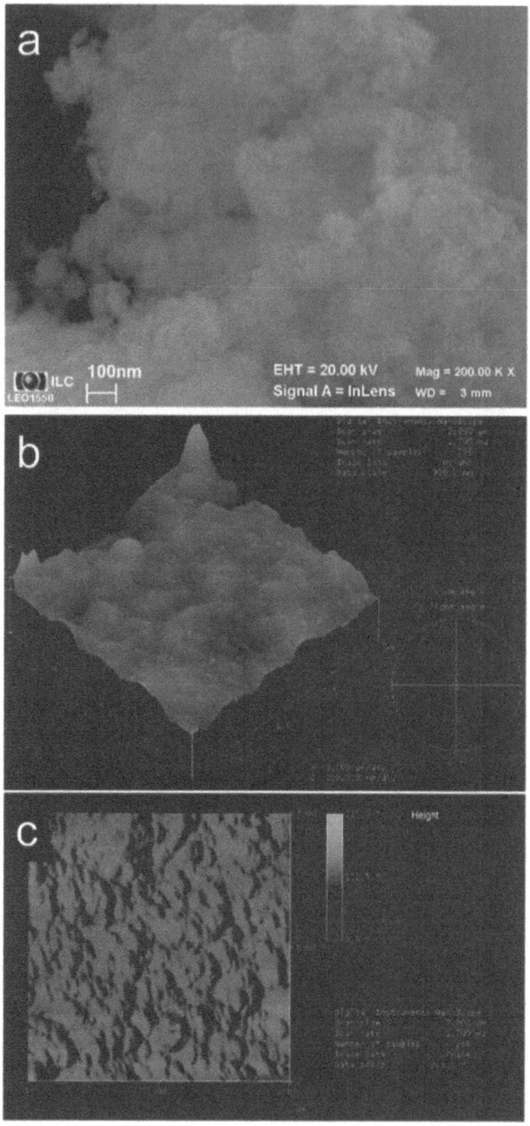

Fig. 1.39 SEM (**a**) and AFM analysis (**b, c**) of mechanochemically synthesized ZnS(M) [Dutková et al. 2008a]

internal contrast that are characteristic of a truncated octahedron particle observed in the (011) zone axis. The HRTEM images allow finding defects in the nanocrystalline material as marked with an arrow in Fig. 1.40d, which have been reported for the mechanosyntesized nanostructures, and these can be representatives of the processes that induce the reduction of the nanocrystal size and the corresponding properties.

A chloride route has been applied for mechanochemical synthesis of metal and sulphide nanoparticles [McCormick 1995].

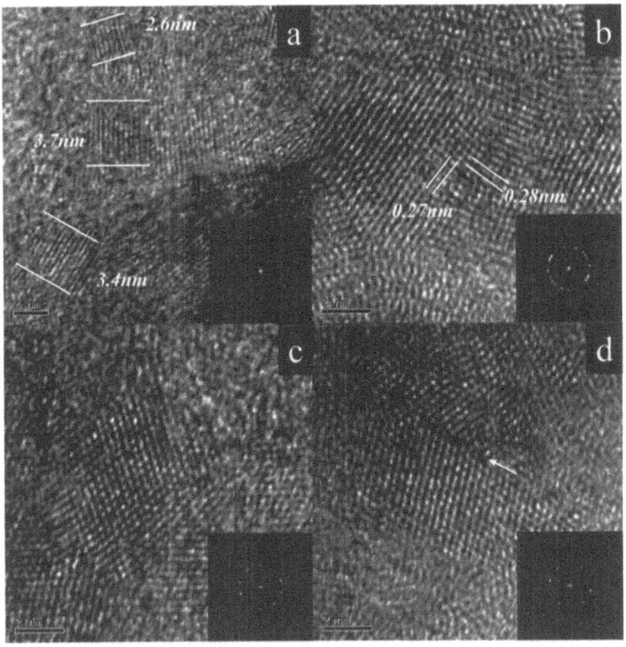

Fig. 1.40 HRTEM micrographs of mechanochemically synthesized ZnS(M): (**a**) identification of nanoparticle with size around 3 nm, (**b**) determination of structure with help of the interplanar distance measurement, (**c**) hexagonal profiles for fcc-like nanoparticles, (**d**) example of fracture induced in the nanoparticles [Dutková et al. 2008a]

The reaction

$$ZnCl_2 + CaS \rightarrow ZnS + CaCl_2 \tag{1.53}$$

was studied to prepare nano-sized ZnS particles. Milling was performed with a Spex 8000 mixer/mill till 36 h. The CaS peaks decreased in intensity and broadened somewhat with increasing milling time, and disappeared after milling for 36 h. Broad peaks corresponding to the cubic ZnS phase first appeared after milling for 6 h, and increased in intensity with increasing milling time. The average crystallite size of ZnS calculated from the width of the diffraction peaks is plotted in Fig. 1.41 as function of milling time. The crystallite size increased steadily with milling time, reaching about 12 nm in samples milled for 36 h. Figure 1.42 shows a TEM micrograph of the sample milled for 77 h and subsequently washed. The ZnS particles of 10–30 nm in size appear separable from each other [Tsuzuki et al. 1997].

1.2.5.4 Cadmium Sulphide CdS

Cadmium sulphide is an important semiconductor for a variety of applications including solid-state solar cell windows, laser communications, photoconductors,

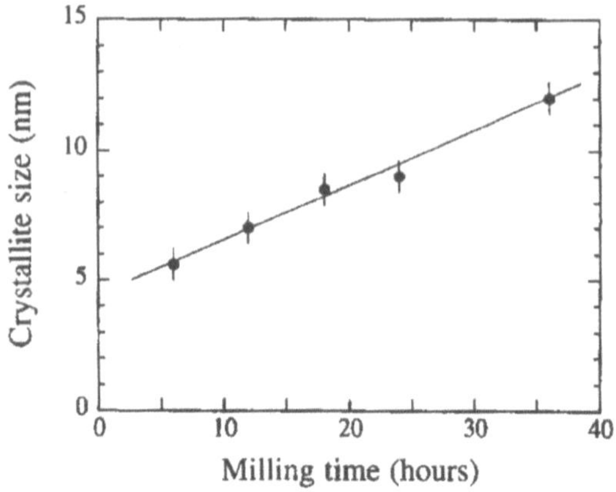

Fig. 1.41 Average crystallite size of ZnS as a function of milling according to Eq. (1.53) [Tsuzuki et al. 1997]

field effect transistors, sensors, and high-emitting applications. Therefore, much extensive studies on CdS particles and their physicochemical properties has been carried out and many methods have been used for the preparation of this material [Shao et al. 2003; Hao et al. 2003; He et al. 2003; Wang et al. 2003; Xu et al. 2004; Yan et al. 2004]. Some CdS nanoparticles were already synthesized by high-energy milling from chloride precursors [Tsuzuki and McCormick 1997, 1999, 2004].

CdS nanoparticles were prepared by high-energy milling of cadmium acetate and sodium sulphide precursors according to the reaction

$$(CH_3COO)_2Cd.2H_2O + Na_2S.9H_2O \rightarrow CdS + 2CH_3COONa + 11H_2O \quad (1.54)$$

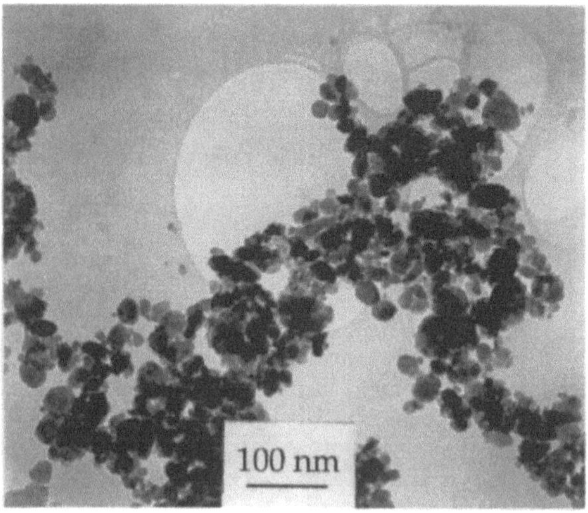

Fig. 1.42 TEM of sample milled for 77 h and subsequently washed [Tsuzuki and McCormick 1997]

The reaction is thermodynamically feasible at ambient temperature, as the enthalpy change is negative. The value $\Delta H^{\circ}_{298} = -253\,\mathrm{kJmol}^{-1}$ was calculated from thermodynamic data [Wagman et al. 1982].

The XRD pattern of the chemically synthesized sample (Fig. 1.43) appears to be composed of the hexagonal α- CdS (greenockite, JCPDS 041-1049) and cubic β-CdS (hawleyite, JCPDS 010-0454) phases. The peaks at 26.5, 43.7 and 51.9° are associated with both phases. The presence of the small peak at 31.5° also gives evidence for the existence of cubic phases. In addition, the comparison of the relative intensities of the XRD pattern of the sample with JCPDS data files (Table 1.9) reveals that the peaks associated with both phases of CdS have relatively higher intensity, suggesting once more the presence of both phases in the chemically synthesized sample. However, the hexagonal phase is dominant phase in the sample with regard to low intensity of the (200) reflection and only traces of cubic phase exist in the sample. The small peaks at around 23° are most likely due to the chemical reaction residuals.

XRD pattern of CdS nanoparticles synthesized by high-energy milling in a laboratory mill is displayed in Fig. 1.44. XRD pattern shows mainly the reflections of cubic phase. It is also supported by relative intensity as well. No reflection peaks was found at $2\theta = 36.6°$ and 47.8°, which are present only in the hexagonal form of CdS. The (022) reflection of the cubic phase at about 31° is probably masked

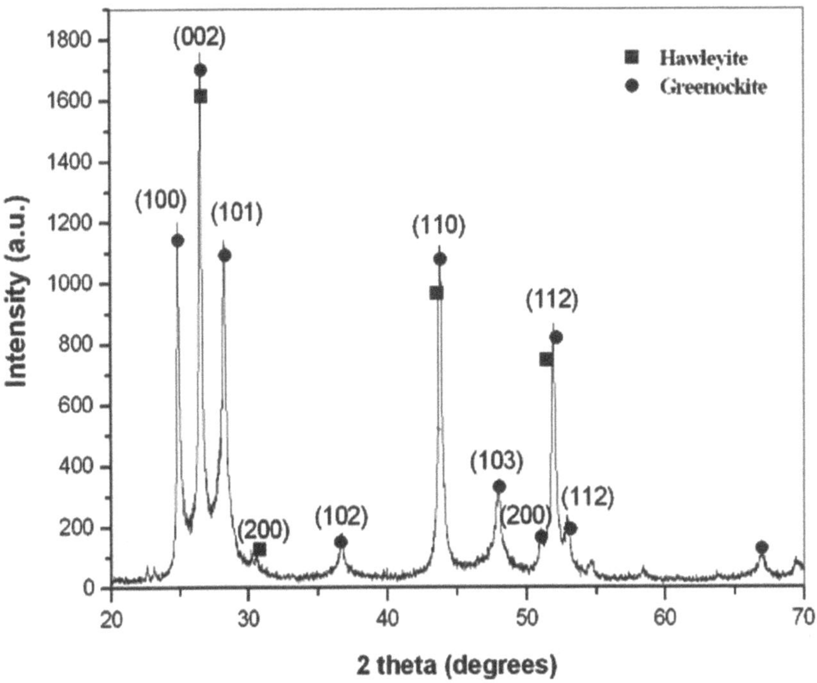

Fig. 1.43 XRD pattern of the chemically synthesized CdS

Table 1.9 Comparison of the JCPDS diffraction data and experimentally measured data for CdS

Greenockite			Hawleyite		
hkl	Relative intensity		hkl	Relative intensity	
	JCPDS	measured		JCPDS	measured
100	62	70.9	111	100	100
002	91.9	100	200	39.8	–
101	100	66.8	220	80.5	40.4
102	29.2	10.2	311	60.2	34.3
110	48.3	66.2	222	10.2	–
103	50.2	19.2	400	20	–
200	8.1	10.4	311	30	–
112	31	49.0	420	10	–
201	15	12.8	422	30	–

due to a large broadening of the (111) reflection. The peaks associated with hexagonal phase disappear with milling and only the cubic structure is evident in the diffraction pattern of the milled sample. This is in a good agreement with observations in paper where CdS was synthesized from chloride precursor [Tsuzuki and McCormick 1997]. The higher background on the XRD pattern of mechanochemically synthesized CdS implies the formation of some amorphous material. The amorphization is in fact a highly distorted periodicity of lattice elements, and it is often characterized as a short range order in contrast to the long order of a fully crystalline structure.

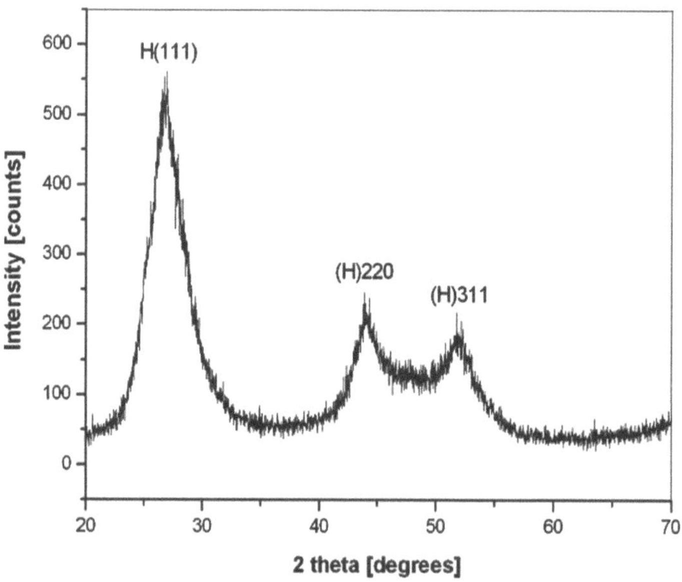

Fig. 1.44 XRD pattern of mechanochemically synthesized CdS (H-hawleyite)

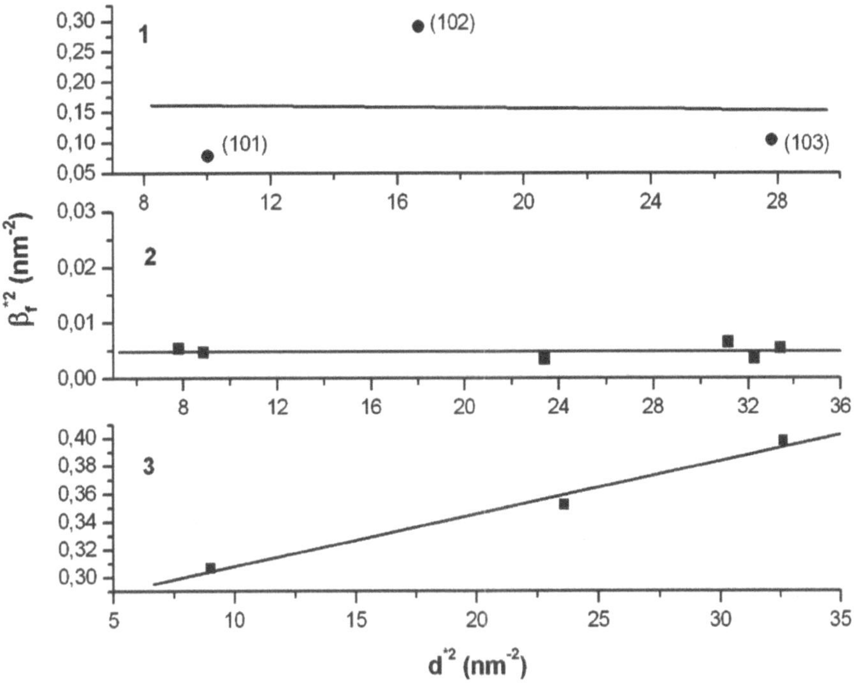

Fig. 1.45 Williamson-Hall plot of mechanochemically synthesized CdS (1), chemically synthesized CdS (2) and chemically synthesized CdS (10l), $l \geq 1$ (3)

Figure 1.45 shows the Williamson-Hall plots for mechanochemically (1) and chemically (2–3) synthesized samples. The scatter of the β_f^{*2} values for the chemically synthesized sample with low correlation coefficient indicates that the crystallite shape differs from a spherical one. In addition, the (10l), $l \geq 1$ reflections show higher deviation than other reflections. The Williamson–Hall plots for the chemically synthesized sample yields very small slopes close to zero which implies the sample is free of strain. Two main crystallite groups are expected in the chemically synthesized sample. For chemically synthesized samples 2 and 3 (10l), $l \geq 1$ the average volume weighted crystallite size are calculated using the intercept of the corresponding lines, being 14.4 nm and 2.6 nm respectively. The crystallite with 14.4 nm is dominant group in the sample. The corresponding plots yield low correlation coefficient (r <0.5).

Regarding the mechanochemically synthesized sample (1), the presence of non zero slope and intercept reveals that both size and strain components exist in the sample. The plot shows high correlation coefficient ($r = 0.991$) and negligible scatter in the β_f^{*2}, suggesting uniform crystallites. The mechanochemically synthesized sample yields larger broadening compared to the chemically synthesized sample. The volume weighted crystallite and maximum lattice strain are calculated 1.92 nm and 3.1×10^{-2} respectively.

The Warren-Averbach analysis based on Fourier analysis provides detail information regarding to crystallite size, lattice strain and their distributions. For the mechanochemically synthesized CdS (cubic phase), the Warren–Averbach method was applied as the second method to calculate both strain and crystallite size components. The three intensive reflections were used for calculation. The analysis once more revealed a narrow fraction of crystallites in the sample. The average surface weighted crystallite size was obtained 1.55 nm and the root mean square strain (RMSS) at $L = 1$ nm,$< \varepsilon_{L=1nm}^2 >^{1/2}$, accounted for 1.7×10^{-2}. As discussed in recent paper [Pourghahramani and Forssberg 2006a], the results of the Williamson–Hall plots and Warren–Averbach methods with regard to their relationships are in a good agreement.

Surface morphology of the synthesized CdS nanoparticles with estimated size from 20–30 nm is depicted in Fig. 1.46. Individual nanoparticles have tendency to form nanoparticle agglomerates during milling process and both entities can be clearly seen.

The particle morphology of CdS nanoparticles was also observed using TEM. Image of the mechanochemically synthesized CdS is presented in Fig. 1.47. Because of the extremely small dimensions and high surface energy of the as-prepared CdS nanoparticles, it is easy for them to aggregate. So, it was difficult to determine precisely the size of CdS nanoparticles by simple viewing the TEM image.

We suppose that each particle is composed of fine nanocrystallites, whose sizes were determined by XRD technique. Thus, each particle observed in the TEM micrographs was of polycrystalline nature. Selected area electron diffraction (SAED) pattern corresponding to random orientation of CdS particles is also shown in Fig. 1.47. The presence of (111), (220) and (311) planes clearly suggests presence of hawleyite phase.

Fig. 1.46 SEM of mechanochemically synthesized CdS

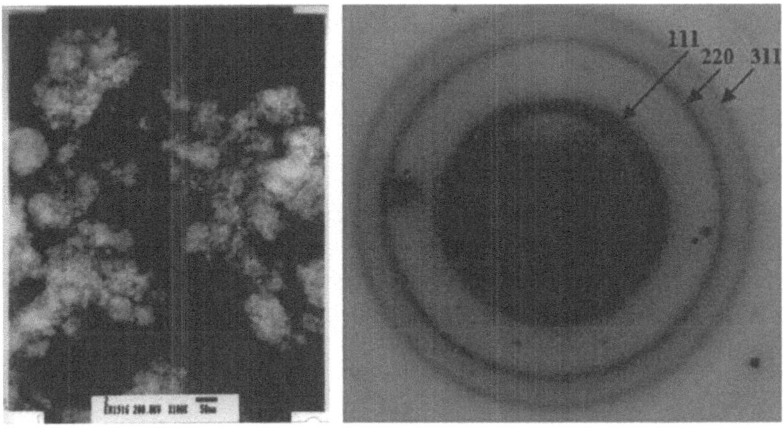

Fig. 1.47 Transmission electron micrographs (*left*) and SAED (*right*) of mechanochemically synthesized CdS

The optical properties of CdS nanoparticles can be elucidated from the UV-VIS spectrum of the as-synthesized product. UV-VIS optical absorption spectrum of the mechanochemically (1) and chemically (2) synthesized CdS are shown in Fig. 1.48. The spectrum of mechanochemically synthesized nanoparticles (1) showed that the

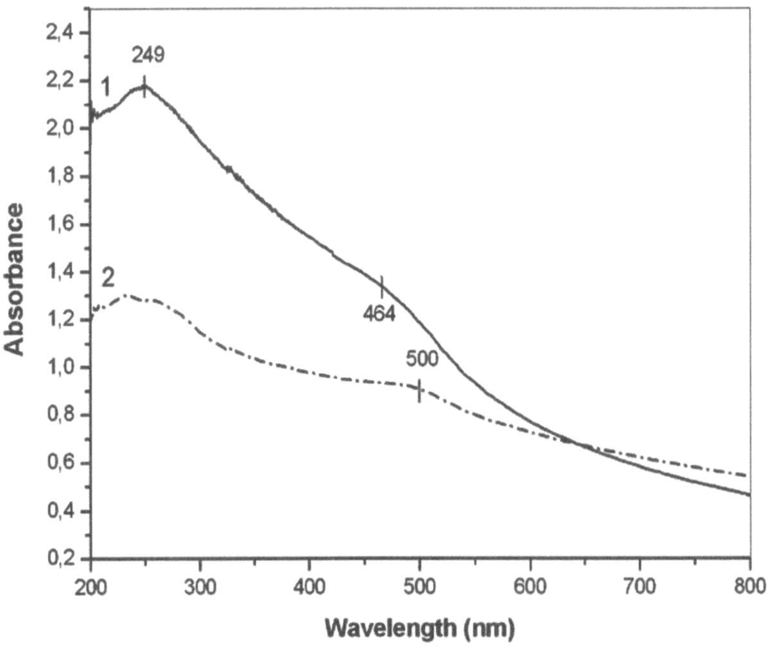

Fig. 1.48 UV-VIS spectra of mechanochemically synthesized CdS nanoparticles (1) and chemically synthesized CdS (2)

onset of the optical absorption 464 nm was blue-shifted compared with that for bulk CdS which is 512 nm. This blue shift was caused by strong quantum confinement effect, due to the decrease in particle size [Brus 1984]. The band gap of CdS nanoparticles was calculated to be about 2.67 eV, greater than that of 2.42 eV estimated for the bulk CdS [Lippens and Lanoo 1989]. The typical UV-VIS spectrum of the chemically synthesized CdS particles is shown in the same Figure, where the curve (2) has an onset at 500 nm with calculated band gap (2.48 eV).

The specific surface area of chemically synthesized CdS particles ($13\,m^2g^{-1}$) is 5-times lower in comparison with mechanochemically synthesized CdS particles. The specific surface area of mechanochemically synthesized CdS nanoparticles (54–$60\,m^2g^{-1}$) increased with the increasing milling time. It is further advantage of mechanochemical synthesis of CdS nanoparticles.

The synthesis of ultrafine CdS nanoparticles according to reaction

$$CdCl_2 + Na_2S \rightarrow CdS + 2NaCl \tag{1.55}$$

has been studied. The solid state reaction resulted in the formation of CdS particles with an average diameter of < 8 nm. The average particle size was controlled within the range of 4–8 nm by varying the size of milling media (Fig. 1.49). As shown in this Figure, the CdS particle size decreased with decreasing ball size, reaching 4.3 nm in samples milled with 4.8 nm balls. The effect of ball size on the CdS particle size appear to be due to the decrease in the energy of ball-powder collisions with decreasing ball size [Tsuzuki and McCormick 1997].

The same strategy of synthesis as presented by Eq. (1.55) has been applied in the presence of a nonionic surfactant, $C_{18}H_{37}O(CH_2CH_2O)_{10}H$ [Wang et al. 2003]. Spherical cubic-phase CdS nanoparticles with an average diameter of ca 5 nm were prepared. The surfactant solves in the free water to form a "shell" surrounding CdS particles preventing them from aggregating to larger particles.

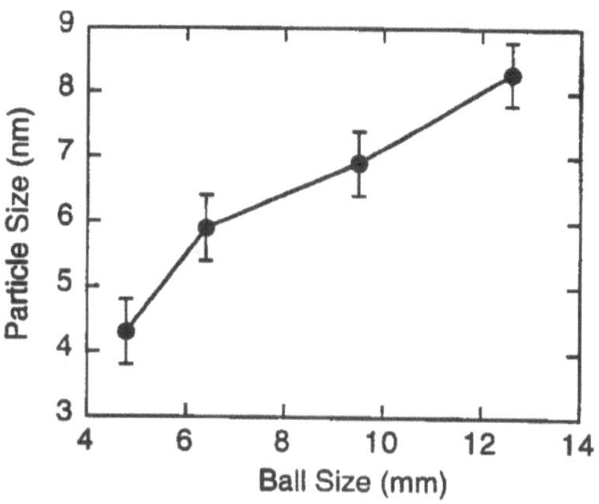

Fig. 1.49 Effect of ball size on the average CdS particle size determined from X-ray diffraction peak widths [Tsuzuki and McCormick 1997]

1.2.5.5 Zinc Cadmium Sulphides $Zn_xCd_{1-x}S$

The properties of $Zn_xCd_{1-x}S$ nanoparticles synthesized by the mechanochemical route from the corresponding metal acetates and sodium sulphide by high-energy milling according to the simplified equation

$$\begin{aligned} x(CH_3COO)_2Zn.2H_2O + (1-x)(CH_3COO)_2Cd.2H_2O + Na_2S.9H_2O \\ \rightarrow Zn_xCd_{1-x}S + 2CH_3COONa + 13H_2O \end{aligned} \tag{1.56}$$

where $x = 0.8$, 0.6, 0.5, 0.4 and 0.2 were studied.

After the completion of reaction (1.56) produced $Zn_xCd_{1-x}S$ nanoparticles have been washed, decantated and dryied according to the flowchart shown in Fig. 1.50.

Figure 1.51 gives the XRD patterns of the synthesized $Zn_xCd_{1-x}S$ nanoparticles, which are marked as samples 1–5, respectively. All XRD patterns show obvious size and strain broadenings, indicating of the finite size of these nanoparticles and strained nanoparticles. The broad spectra may be due to the presence of particles of various sizes as well as the various defects introduced during the high-energy milling. The XRD patterns have three peaks which can be indexed to the diffraction from (111), (220) and (311) lattice planes of cubic-blended structure. The mechanochemical treatment causes the complete transformation of the hexagonal phase to the cubic phase. This can be attributed to the motion of dislocation in activated solid state [Imamura and Senna 1984]. However, owing to the increasing mismatch of ZnS and CdS lattice, the crystal quality of $Zn_xCd_{1-x}S$ decreases with increasing zinc content, which is consistent with the published results [Nie et al. 2004]. As for $Zn_xCd_{1-x}S$ nanoparticles the diffraction peaks in the XRD patterns systematically shift to higher 2 theta values and a phase transition from hawleyite to sphalerite occurs with an increase of Zn content. This continuous peak shifting of nanoparticles also indicates that there is no phase separation or separated nucleation of ZnS or CdS in the $Zn_xCd_{1-x}S$ nanoparticles. The higher background on the XRD pattern of mechanochemically synthesized sample implies the formation of some amorphous material.

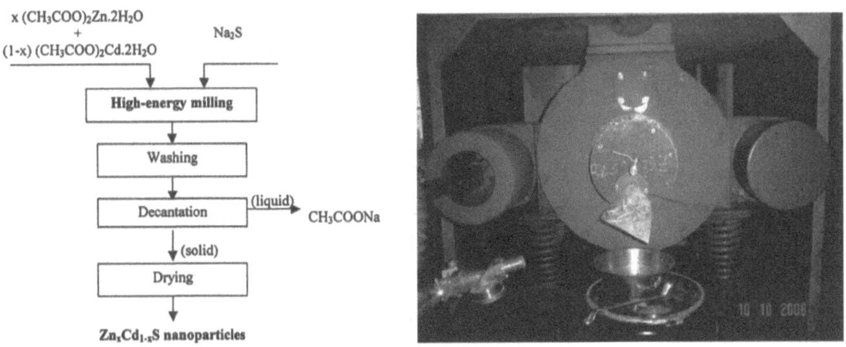

Fig. 1.50 Flowchart of the $Zn_xCd_{1-x}S$ nanoparticles synthesis in an industrial mill

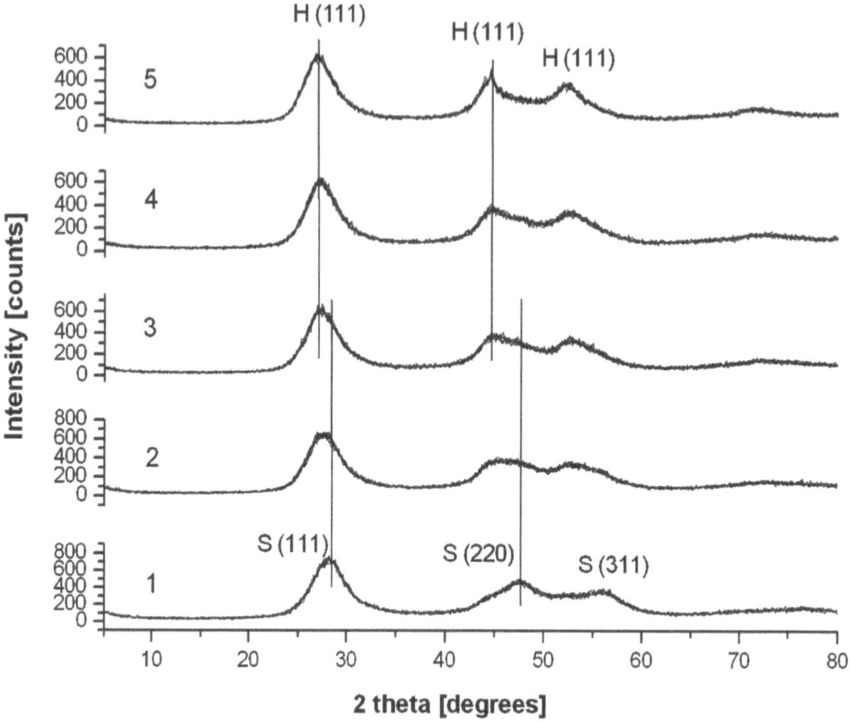

Fig. 1.51 XRD patterns of mechanochemically synthesized $Zn_xCd_{1-x}S$. (1 – $Zn_{0.8}Cd_{0.2}S$, 2 – $Zn_{0.6}Cd_{0.4}S$, 3 – $Zn_{0.5}Cd_{0.5}S$, 4 – $Zn_{0.4}Cd_{0.6}S$ and 5 – $Zn_{0.2}Cd_{0.8}S$) [Dutková et al. 2008b]

The optical properties of $Zn_xCd_{1-x}S$ nanoparticles can be elucidated from the UV-VIS spectrum of the as-synthesized products. The UV-VIS spectra of the samples are illustrated in Fig. 1.52. The as-prepared $Zn_xCd_{1-x}S$ nanocrystals allow systematic variation of their band gap from 3.7 eV to 2.4 eV which are the border values for ZnS and CdS bulk crystals, respectively.

The samples 1–5 have absorption peak at 328 nm (3.76 eV), 369 nm (3.34 eV), 413 nm (2.98 eV), 440 nm (2.80 eV) and 467 nm (2.64 eV), respectively. As for the $Zn_xCd_{1-x}S$ nanocrystals, their absorption edges gradually shift from 328 to 467 nm as the Zn content (x value) in the particles decreases from 0.80 to 0.20. From samples 1–5, a blue shift of 139 nm occurs as the composition alteration. Obviously, such a large shift in the absorption edge cannot be only attributed to quantum-size effects, but could be due to a continuous shift of the energy bandgap of the nanoparticles with a change in their composition [Zhong et al. 2003]. The profile of the UV-VIS curves of mechanochemically synthesized $Zn_xCd_{1-x}S$ can not be simulated by a superposition of the corresponding ZnS and CdS UV-VIS spectra, indicating that the $Zn_xCd_{1-x}S$ products are complex compounds rather than a simple mixture of ZnS and CdS. Absorption spectra are red shifted on increasing the particle size. At fixed size, the red shift in the absorption spectrum with increasing representation of

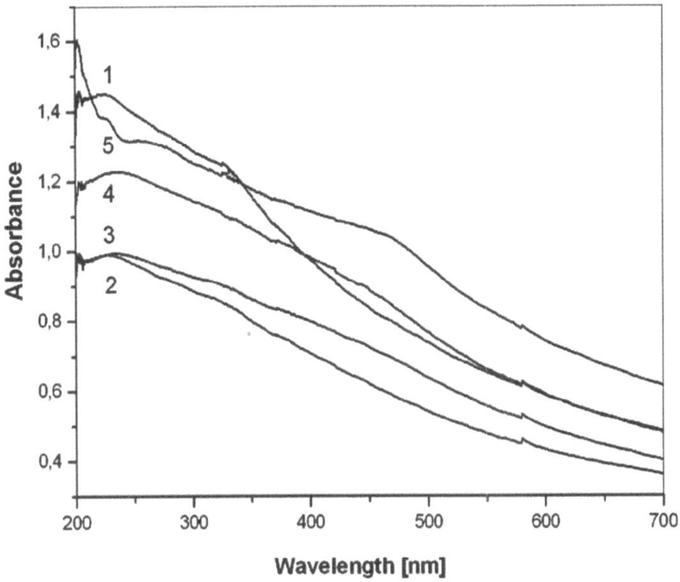

Fig. 1.52 UV-VIS patterns of mechanochemically synthesized $Zn_xCd_{1-x}S$. (1 – $Zn_{0.8}Cd_{0.2}S$, 2 – $Zn_{0.6}Cd_{0.4}S$, 3 – $Zn_{0.5}Cd_{0.5}S$, 4 – $Zn_{0.4}Cd_{0.6}S$ and 5 – $Zn_{0.2}Cd_{0.8}S$) [Dutková et al. 2008b]

Cd is observed. This is due to changes in the solid phase composition. The energy bandgap determined from the excitonic peak, smoothly increases with increasing composition from CdS to ZnS. Similar behaviour was observed for the bandgap variation of bulk $Zn_xCd_{1-x}S$.

1.2.5.6 Nickel Sulphides Ni_xS

The feasibility and kinetics of synthesizing various nickel sulphides by milling of elemental mixtures of Ni and S have been investigated [Kosmac and Courtney 1992; Kosmac et al. 1993]. According to the Ni-S phase diagram [Metals Handbook 1973], five compounds exist in this system: Ni_3S_2, Ni_7S_6, NiS, Ni_3S_4 and NiS_2. The phases Ni_3S_2 and high-temperature polymorph of NiS are formed readily via such processing. In distinction, it requires prolonged milling to obtain Ni_7S_6. NiS_2 can only be obtained as a minor reaction product and Ni_3S_4 cannot be formed by milling for the conditions of this study.

The kinetics of NiS formation is illustrated in Fig. 1.53. Sulphur coats the Ni particles and sulphide formation takes place at the interface of the elements after a certain degree of microstructural refinement due to the plastic deformation accompanying milling. Ni_3S_2 forms rapidly from elements at this stage. However, a stasis in the reaction is then observed. This is associated with NiS formation and a slight decrease in the amount of Ni_3S_2 according to the equation

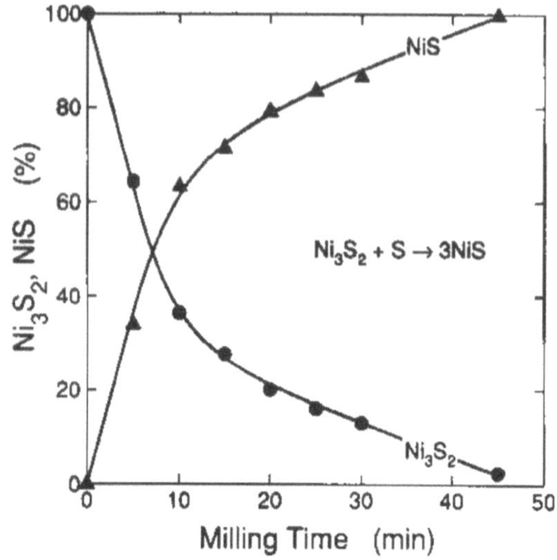

Fig. 1.53 Kinetics of NiS formation on milling an equimolar powder mixture of S and Ni_3S_2 [Kosmac et al. 1993]

$$Ni_3S_2 + S \rightarrow 3NiS \quad (1.57)$$

NiS phase is transforming into Ni_3S_2 at prolonged milling.

1.2.5.7 Tin Sulphides Sn_xS

Tin sulphides represent a complex system with a wide compositional range. They frequently show the features of low dimensional chalcogenides: polytypism, polymorphism, nonstoichiometry and the possibility of exfoliation [Imamura and Senna 1982; Jellinek 1988; Rouxel et al. 1995]. Some of them, e.g. SnS_2 consists of two layers of closely packed sulphur anions with sandwiched tin cations in octahedral coordination. The sulphides of this type have a large number of empty sites in their structure and hence they are interesting host lattices for investigation [Morales et al. 1992; Müller-Warmuth and Schöllhorn 1994].

For the mechanochemical synthesis of various tin sulphides three starting mixtures were prepared, see Table 1.10 [Baláž et al. 1999; Baláž et al. 2002].

The results of the XRD analysis show that the main products of mechanosynthesis are SnS, Sn_3S_4 (or $SnS+Sn_2S_3$), and Sn_2S_3. In case of sample Sn-S/1 the

Table 1.10 Physico-chemical characteristics of mechanochemically synthesized tin sulphides

Sample	Weight ratio S:Sn	XRD products	Crystallite size (nm)	Specific surface area (m^2g^{-1})
Sn-S/1	1:3.70	SnS	11	3.3
Sn-S/2	1:2.78	Sn_3S_4 ($SnS+Sn_2S_3$)	13	5.4
Sn-S/3	1:2.47	Sn_2S_3	27	7.3

X-ray diffraction pattern (Fig. 1.54C) is similar to that of the SnS of JCPDS card 39-0354. Almost all the peaks were indexed on the basis of an orthorhombic structure with the spacegroup Pbnm. The calculated lattice parameters ($a = 0.4318$ nm, $b = 1.1200$ nm, $c = 0.3984$ nm) were consistent with the data of JCPDS card. The X-ray diffraction pattern of sample Sn-S/2 (Fig. 1.54B) is similar to that of Sn_3S_4 of JCPDS card 27-0900. Still the JCPDS card is questionable because there is no report about the existence of the Sn_3S_4 phase in the binary phase diagram of tin-sulphur system [Sharma and Chang 1986]. The X-ray pattern seems to corresponds to a mixed SnS and Sn_2S_3 phase.

The X-ray pattern of sample Sn-S/3 (Fig. 1.54A) was consistent with that of JCPDS card 14-0619. The calculated lattice parameters ($a = 0.8889$ nm, $b = 1.4020$ nm, $c = 0.3744$ nm) were also in accordance with the data for Sn_2S_3.

In Table 1.10 the data on the crystal size and specific surface area of the mechanochemically synthesized sulphides are given. Their size varies from 11 to 27 nm and increasing with the increasing proportion of sulphur. A similar situation prevails also in case of specific surface. The specific surface of sulphides is by

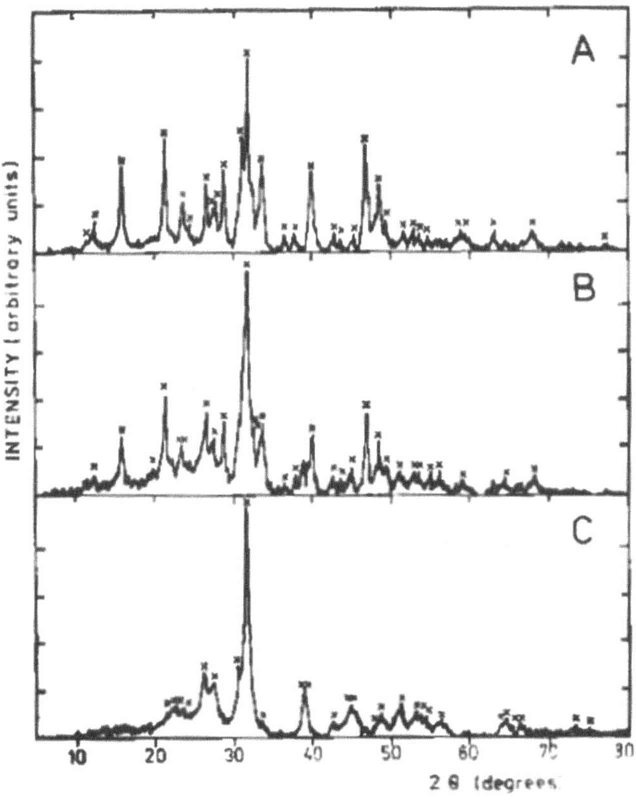

Fig. 1.54 XRD patterns in mechanochemically synthesized tin sulphides: (**A**) Sn-S/1, (**B**) Sn-S/2, (**C**) Sn-S/3 [Baláž et al. 1999]

one order of magnitude greater than in case of the precursors of the synthesis, i.e. metallic tin and elementary sulphur. The specific surface value is in the range that is achieved by intensive milling of sulphides [Baláž 2000].

The samples of mechanochemically synthesized sulphides were subjected to dissolution reaction in the alkaline environment of Na_2S. For the reaction temperature $T = 30°C$, Fig. 1.55 shows the ralationship between the degree of conversion ε_{Sn} and dissolution time t_L. The process of dissolution is characterized by great initial speed. After 5 min of reaction, more than 50% of the tin bound to a sulphidic form was dissolved.

The temperature sensitivity ($T = 30$–$60°C$) of the mechanochemically synthesized SnS is given in Fig. 1.56. Its assessment in Arrhenius coordinates shows that the apparent activation energy is $E = 27\,\mathrm{kJmol^{-1}}$. This value suggests that the factor determining the speed of the dissolution is probably the chemical reaction on phase boundaries [Habashi 1974].

Widespread use and search for technological applications of nanomaterials requires the availability of large (tonnage) quantities of well-characterized material

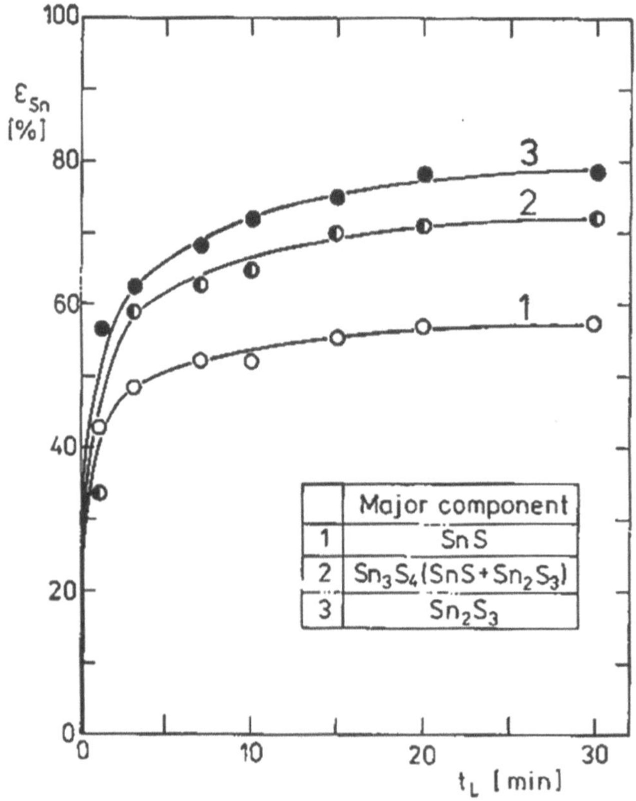

Fig. 1.55 Influence of dissolution time, t_L on the recovery of tin, ε_{Sn}, from the mechanochemically synthesized product in the tin-sulphur samples: (1) Sn-S/1; (2) Sn-S/2; (3) Sn-S/3 [Baláž et al. 1999]

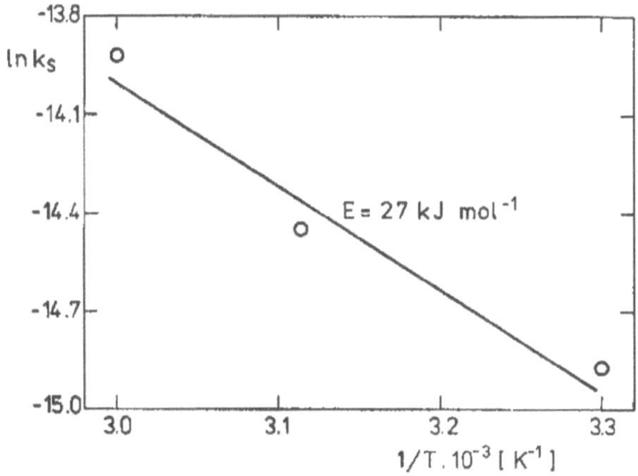

Fig. 1.56 Arrhenius plot of the specific rate constant, k_S ($k_S = k/S$; k = rate, S = specific surface area) of tin dissolution in the Sn-S/1 sample [Baláž et al. 1999]

with reproducible properties [Suryanarayana 1994]. The mechanochemical approach has the potential to fulfill such efforts.

1.2.6 Properties

Because of the extremely small size of the grains, a large fraction of the atoms in nanocrystallites is located in the grain boundaries (Fig. 1.57) and thus the

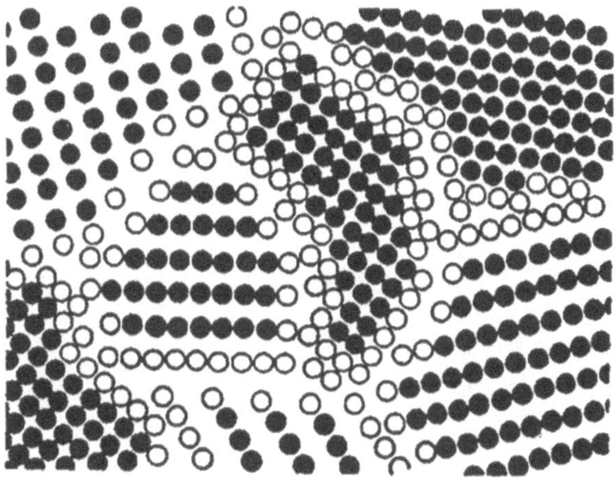

Fig. 1.57 Schematic representation of equiaxed nanocrystalline metal distinguishing between atoms associated with the individual grains (•) and those constituting grain boundary network (○) [Gleiter 1989]

nanocrystalline state exhibits enhanced combination of properties in comparison with the course-grained material.

1.2.6.1 Diffusion

Since nanocrystalline materials contain a very large fraction of atoms at the grain boundaries, the numerous interfaces provide a highly density of short-cut diffusion paths. Consequently, they are expected to exhibit an enhanced diffusivity in comparison to single crystals or conventional coarse-grained polycrystalline materials with the same chemical composition [Suryanarayana 1994].

Table 1.11 summarizes the measured diffusivities in nanocrystalline Cu 8 nm in size (A) in comparison to lattice diffusivity (B) and to diffusivity in grain boundaries (C). The large differences can be clearly seen.

1.2.6.2 Mechanical Properties

A rather considerable number and variety of experimental investigations of the mechanical properties of nanophase materials have been performed. They have included measurements of hardness, fracture toughness, stress-strain behaviour in compression and tension, strain-rate sensitivity and creep. The investigations were performed on pure metals, alloys, intermetallic compounds and ceramics [Siegel 1997].

The elastic constants of nanocrystalline materials have been found to be reduced by 30% or less. These results were interpreted as due to the large free volume of the interfacial component resulting from the increased interatomic spacing in the boundary regions [Gleiter 1989]. The most significant change resulting from a reduction in the grain size to the nanometer level is a 4–5 times increase in the strength and hardness over the coarse grained material.

Conventional polycrystalline metals and alloys show an increase in yield strength σ_y with decreasing grain size d according to the well-known Hall-Petch (H-P) equation

$$\sigma_y = \sigma_0 + kd^{-1/2} \tag{1.58}$$

Table 1.11 Self-diffusivity in nanocrystalline Cu (A), lattice self-diffusion in Cu (B) and in Cu grain boundaries [Gleiter 1989]

Temperature (K)	Self diffusivity (m^2s^{-1})		
	A	B	C
293	$2.6.10^{-20}$	4.10^{-40}	$4.8.10^{-24}$
353	$2.0.10^{-18}$	2.10^{-34}	$6.2.10^{-21}$
393	$1.7.10^{-17}$	2.10^{-31}	$2.2.10^{-19}$

where σ_0 is friction stress resisting the motion of gliding dislocation, and k is the Hall-Petch slope, which is associated with a measure of the resistance of the grain boundary to slip transfer [Hall 1951; Petch 1953].

In analogy, hardness H_v can be related to the grain size by

$$H_v = H_0 + k_H d^{-1/2} \tag{1.59}$$

where H_0 and k_H are constants.

The H-P effect in conventional coarse-grained materials is attributed to the grain boundaries acting as efficient obstacles to dislocations.

By decreasing the grain size of metals down to the order of a few tens of nanometers, the H-P slope remains positive but with smaller value. At ultra-fine grained sizes below ca 20 nm, a reverse softening effect or negative H-P relation is observed for some metals. Many investigators now accept that there exists a critical grain size in nanocrystalline materials above which the Hall-Petch slope is positive and below which it is negative; different reasons have been put forward for its presence. Several factors such as grain-boundary sliding, creep diffusion, triple junctions, pores and impurities could contribute to inverse H-P relation in metals and alloys [Gleiter 1989; Tjong and Chen 2004].

At room temperature, metal and alloys undergo a nonrecoverable plastic deformation after the elastic domain, with a gradual increase of the applied stress as a function of strain (so called work-hardening) [Ashly and Jones 1980]. It was shown that pure nanocrystalline Cu behaved differently. This form of copper displays near perfect elastoplastic behaviour characterized by Newton flow and the absence of both work-hardening and neck formation [Champion et al. 2003].

Metals are usually soft, ductile and tough while ceramics are strong, hard and brittle. However, through nanostructure processing, it appears to be possible to make metals hard and strong like ceramics and ceramics can be made ductile like metals [Suryanarayana 1994]. This behaviour can be illustrated by the qualitative framework shown in Fig. 1.58 which indicates that with decreasing grain size into the

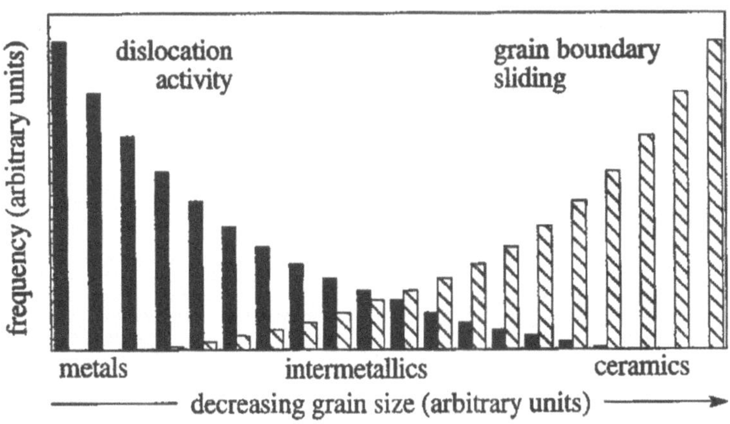

Fig. 1.58 Grain size dependance of dislocation activity and grain boundary sliding contribution for various nanophase materials [Siegel and Fougere 1995]

nanophase regime, the frequency of dislocation activity decreases and that of grain boundary sliding increases. Which of these effects dominates depends upon the grain size regime, the specific type of material, and most importantly on the nature of its interatomic bonding.

1.2.6.3 Thermal Properties

Specific heat and melting temperature are considered as the most fundamental thermal properties of any solid. Generally, nanoparticles exhibit enhanced specific heat and lowering of the melting point [Koper and Winecki 2001].

The experimentally obtained specific heats C_p/T of bulk (polycrystalline) iron and nanocrystalline iron are shown in Fig. 1.59. The specific heat for nanocrystalline Fe is larger compared to bulk iron for temperatures above approximately 10 K. The specific heat of a material is closely related to its vibrational and configurational entropy, which is significantly affected by the nearest-neighbour configurations. Thus, the increase in specific heat of nanocrystalline materials has been attributed to the small crystal size (and consequent large interfacial component) [Suryanarayana 1994]. If this is so, grain growth should reduce the specific heat of the nanocrystalline material which was indeed observed [Rupp and Birringer 1987].

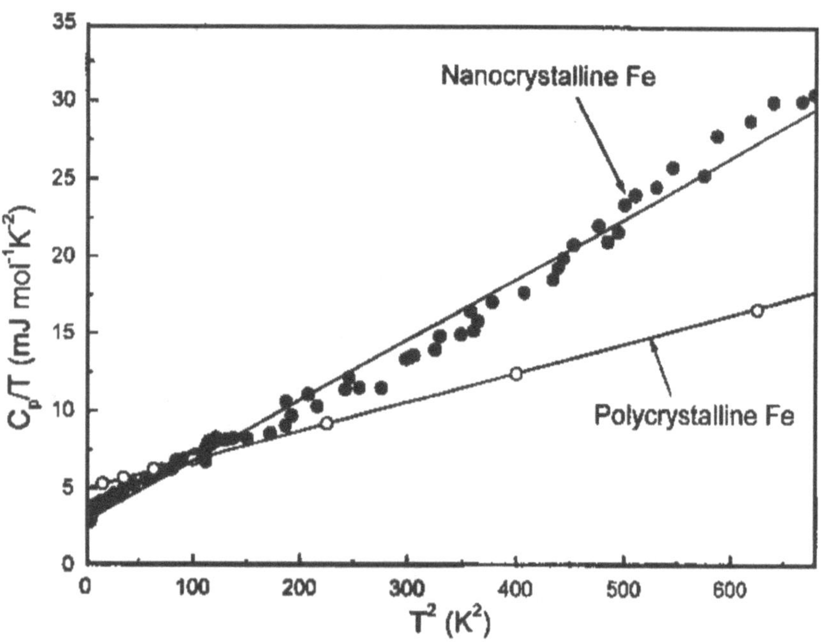

Fig. 1.59 Plot of C_p/T vs. T^2 for nanocrystalline and polycrystalline Fe [Bai et al. 1996]

The melting point, T_M is another fundamental thermal property of a nanocrystalline solid. It is the temperature above which the crystalline structure of solid disappear and is replaced by disordered atomic arrangement in the liquid [Koper and Winecki 2001].

It was recognized as early in 50s that nanosized particles melt below their corresponding bulk melting temperatures [Takagi 1954]. Theoretical model that described the lowering of melting points of nanoparticles was developed [Shi 1994]. In this model the mean-square atom displacement averaged over the entire volume of the particle, including its surface, was evaluated. As the particle decreases, the increased number of surface atoms enhances the average values of atomic displacement. According to the Lindemann criterion (a crystal will melt when the root-mean-square displacement of its atoms exceeds a certain fraction of the interatomic distance), this causes a decrease in the melting temperature [Lindemann 1910]. On the basis of these arguments Shi derived the equation

$$\frac{T_m(r)}{T_m(\infty)} = \exp\left[-(\alpha-1)\left(\frac{r}{3h}-1\right)^{-1}\right] \tag{1.60}$$

where $T_m(r)$ is the melting temperature of a nanocrystal, $T_m(\infty)$ is the melting temperature of the bulk material, h corresponds to the height of a monolayer of atoms in its crystal structure, α is the ratio of mean-square atom displacement on the surface and inside of the particle (in practise parameter α is adjusted to fit experimental data).

Figure 1.60 shows melting temperatures for Au and CdS nanoparticles as a function of the particle size. The solid lines represent predictions from Eq. (1.60). There is excellent agreement between the theory and experimental data for gold and fair agreement for CdS nanoparticles [Koper and Winecki 2001].

1.2.6.4 Electrical Properties

Because of the increased volume fraction of atoms lying at the grain boundaries, the electric resistivity of nanocrystalline materials is expected to be higher than that in the corresponding coarse-grained polycrystalline materials [Suryanarayana 1994].

The observed temperature and grain size dependance of the electric resistivity as well as the temperature coefficient may be understood in terms of electron scattering inside the crystallites as well as scattering of the electron by the boundaries by analogy to the treatment of the electric dc-resistivity of thin metallic films. If the crystal size is smaller than the electron mean free path, grain scattering dominates and hence the conductivity as well as the temperature coefficient is expected to decrease [Gleiter 1989].

1.2.6.5 Magnetic Properties

For a long time it is known, that the first great advantage of magnetic nanoparticles is that they are single-domain and therefore have large coercivities, allowing them to

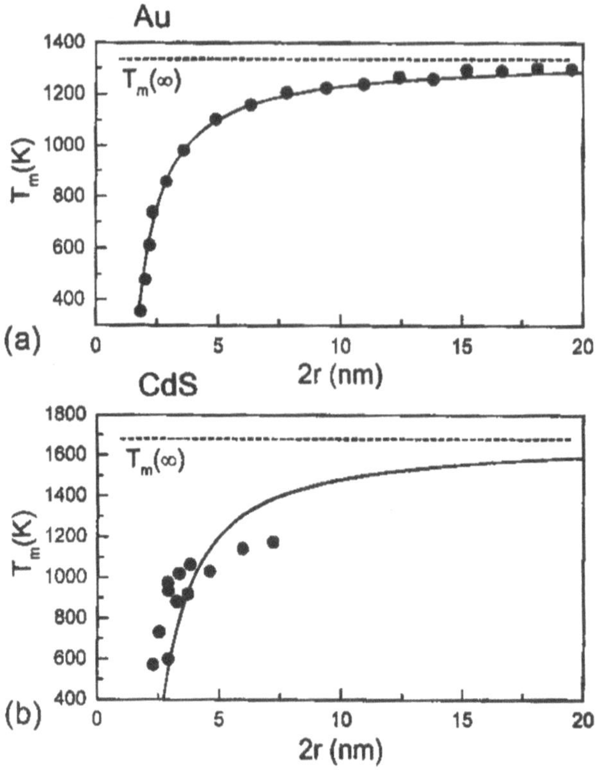

Fig. 1.60 Melting temperature T_m vs. size for Au (**a**) and CdS (**b**) nanocrystals. *Solid line* = equation (1.60) [Shi 1994]

individually retain their magnetization directions [Stoner and Wohlfarth 1948]. With single-domain particles, the magnetic and crystallographic domains are defined by the particle itself, although particles can also be polycrystalline. A typical size of a magnetic domain is of the order of 1 μm and particles much smaller than that will be single-domain. Typically critical sizes for monodomain particles are in the range 20–2000 nm and depend on the ferromagnetic material under consideration. Small magnetic particles have been of industrial importance since 50s primarily stimulated by the recording industry' desire for ever denser and more reliable recording media. The quest for smaller particles which can be used for recording purposes [Speliotis 1999; Mee 1994] continues since smaller particles implies higher data storage densities [Bertran and Zhu 1992; Lambeth et al. 1996]. There are of course material limits to how small a magnetic particle can be and still be useful. For example, not only should the particle be ferromagnetic, it shoud retain its magnetic orientation in ordinary conditions for many years [de Heer 2000].

For magnetic properties it is largely the surface/interface effects, which cause significant differences compared to the bulk. In a nanoscale particle the surface atoms/bulk atoms ratio is sufficiently large ($\sim 50\%$ for diameter ≈ 3 nm) that surface/interface effects can dominate the magnetic properties. One useful point of view is that small particles represent surface matter in macroscopic quantities. For instance, 10 mg of 3 nm cobalt particle would contain 5 mg of surface cobalt [Sorensen 2001]. At the surface, the coordination number of each surface atom is smaller than within the bulk, hence the α-band of a collection of transition metal atoms at the surface is narrower than in the bulk, leading to a huge density of states and hence enhanced magnetism [Blugel 1992].

The surface magnetic moments are enhanced by 10–30% over their bulk values in ferromagnetic Fe, Ni, and body centered cubic Co (100) and (110) surfaces [Freeman et al. 1987].

Measurement of the saturation magnetization (M_S) of nanocrystalline iron (6 nm crystallite size) revealed a reduction of M_S by about 40% relative to the saturation of bulk α-Fe. For comparison, in metallic Fe glasses (extrapolated to pure Fe), M_S is only reduced by about 2% relative to α-Fe [Birringer et al. 1986].

A reduction of the Curie temperature (T_C) of Ni by about 40°C has been reported if the crystal size was reduced to about 70 nm [Valiev et al. 1989]. This reduction was attributed to a reduction of T_C of the grain boundary regions. The interior of the crystals was assumed to have the same T_C as a single crystal of Ni.

Another useful magnetic property is called the magnetocaloric effect. When a material containing extremely small magnetic particulates in a non-magnetic or weakly magnetic matrix is placed in a magnetic field, the magnetic spins of the particulates tend to align with the field. The increase in magnetic order lowers the magnetic entropy of the spin system. If this process is performed adiabatically (i.e. no heat is exchanged with the surroundings), the reduction in spin entropy is offset by an increase in lattice entropy, and the specimen's temperature will rise. This temperature rise is reversible (the specimen cools down on removal of the magnetic field) and is known as the magnetocaloric effect. It has been shown that the magnetocaloric effect may be enhanced at low fields and high temperatures [Suryanarayana 1994].

Giant magnetoresistance is another magnetic property where nanodimension plays an important role. Magnetoresistance is a phenomenon where the application of a direct current magnetic field changes the resistance of a material. The magnetoresistance effect occurs in metals only at very high magnetic fields and low temperatures. For example, in pure copper at 4 K a field of 10 T produces a factor of 10 change in the resistance. Because of the large fields and low temperatures, magnetoresistance in metals originally had few potential application possibilities. However, that changed in 1988 with the discovery of what is now called giant magnetoresistance (GMR) in materials synthetically fabricated by depositing on a substrate alternate layers of nanometer thickness of a ferromagnetic material and a nonferromagnetic metal. In 2007, Nobel Prize was given to Albert Fert and Peter Gründberg for discovery of GMR. A schematic of the layered structure and the alternating orientation of the magnetization in the ferromagnetic layer is shown in Fig. 1.61. The effect was first observed in films made of alternating layers of Fe and Cr, but since then other layered materials composed of alternating layers of Co and Cu and AgCo and NiFe nanoparticles have been made that display much higher

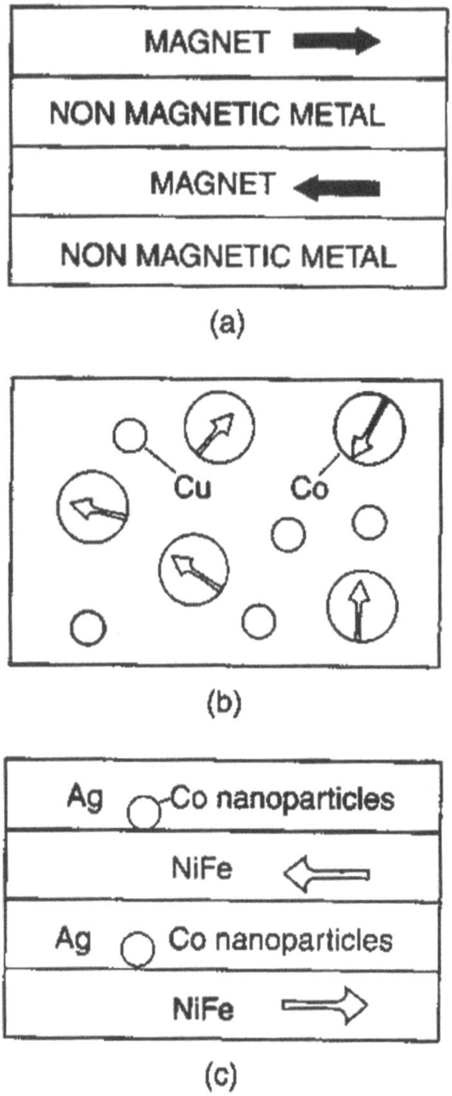

Fig. 1.61 Three arrangements for producing giant magnetoresistance [Poole and Owens 2003]

magnetoresistive effects [Poole and Owens 2003]. The giant magnetoresistance effect is utilized in reading heads that read magnetically encoded information.

1.3 Nanogeoscience

In analogy with title of the famous Feynman speech (see Sect. 1.2.1), the paper entited: "There's plenty of room at the bottom: Nanoscience in geochemistry" has been published in 2002 [Hochella, Jr. 2002]. In the same year the Nanoscience Workshop was organized in Berkeley, California [Anonymous 2002]. Three years later the nanogeoscience session was implemented into the agenda of 15th Annual Goldsmith Conference which was held at the University of Idaho, Moscow. These three events together with the book "Nanoparticles and Environment" published under umbrella of Mineralogical Society of America [Banfield and Navrotsky 2001] can be marked as the milestones in nanogeoscience.

As stated in literature, nanogeoscience is broadly defined to include the study of materials and processes at the nanoscale in their role in geological processes on the Earth and other planets. Because processes are intrinsically molecular at the nanoscale, there is an immediate symmetry and a diffuse boundary between nanogeoscience and the field of chemistry, physics, and materials science. Furthermore, geoscientists increasily recognize the major role played by microorganisms in geological phenomena. There is an equally fuzzy boundary between nanogeoscience and the life sciences in this realm because microbial processes often proceed by manipulating surface forces at the nanoscale. Though nanoparticles represent only a small fraction of the mass of material in the Earth, they represent a large number fraction of the particles in atmospheric and aqueous environments, and nanoparticles are responsible for most of the surface area (at solid-water, solid-air, and solid-liquid interfaces) of Earth materials. Their large surface-to-volume ratio ensures that surface forces exert considerable influence over the chemistry and structure of nanoparticles [Anonymous 2002].

1.3.1 Mineral Nanoparticle Surfaces

As a mineral nanoparticle shrinks to between tens of nanometer and about a nanometer, a significant and increasing fraction of the atoms are exposed on surfaces (Fig. 1.62 and previous text) rather than contained in the bulk.

Ions with non-optimal coordination geometries on surfaces give rise to excess energy. Consequently, the total energy of the system increases as size decreases because of the enhanced contribution from the energy associated with surface sites. The energy penalty caused by creation of a surface can be minimized by hydration, protonation, surface reconstruction, change in surface site coordination, displacement of atoms, and changes in bond lengths and angles [Banfield and Zhang 2001].

The impact of the surface atomic layer on adjacent layers within the solid probably decreases smoothly to zero as distance from the surface (into the particle) increases. However, in order to facilitate consideration of size-dependent characteristics and behaviour of nanoparticles, it is convenient to subdivide them into three atomic structural regions (Fig. 1.62). First, we define *a surface* atomic layer (A) that consists of ions with modified coordination geometries. Second specified as *a near-surface* region (B), this being a zone of atomic layer(s) near, but not at the surface. These layers are identified by a distribution due to proximity to the surface atomic layer. Third, a *bulk* material (C) can be specified with structure equivalent to that within macroscopic crystals [Banfield and Zhang 2001].

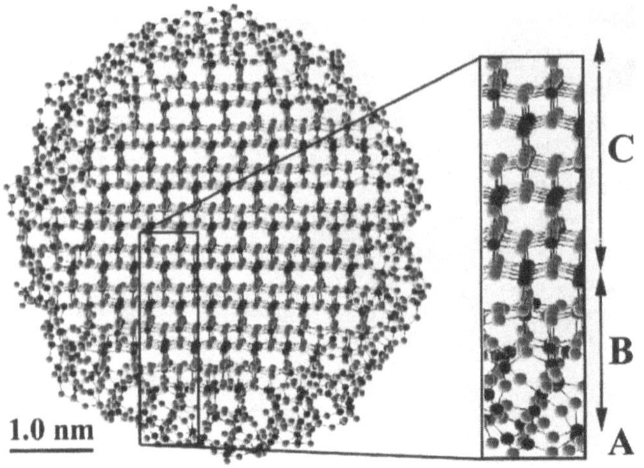

Fig. 1.62 Molecular dynamics simulation of anatase TiO_2 particle [Banfield and Navrotsky 2001]

1.3.2 Nano- Versus Mechano-Approach: Analogy

As stated in paper of Hochella [Hochella, Jr. 2002] the work of Stöber and Arnold represents one of the early study of the importance of nanogeoscience [Stöber and Arnold 1961]. The authors observed the enhanced solubility of quartz SiO_2 fragments (in fact in the nanodomain) adhering to micron-sized grains of the dry-milled mineral. This is a relative early experimental mineral dissolution study that recognized the importance of enhanced solubility of extremely fine grains relative to coarse particles. This phenomenon has been described in a modified version of the Kelvin equation which relates the solubility of a material to its surface free energy and grain size [Adamson 1982]

$$S/S_0 = \exp(2\gamma V/rRT) \tag{1.61}$$

where S is the solubility of grains with inscribed radius r, S_0 is the solubility of bulk material, γ is the surface free energy, V is the molar volume, R is the gas constant and T is the temperature. The equation says that as the grain dimension decreases, the solubility of that grain will go up exponentially relative to the measured solubility of a very large grain (S_0, where the grain size has no appreciable effect on solubility).

It is important to remember that the solubility of nanosized particles is greatly influenced by Eq. (1.61), so that thermodynamically calculated solubilities of bulk materials may be off by orders of magnitude relative to nanoscale equivalents [Hochella, Jr. 2002].

Non-homogeneity in macroscopic crystals in direction from surface to the bulk has been observed by mechanically treated solids. Structure of quartz is changing continuously by mechanical activation (Fig. 1.63).

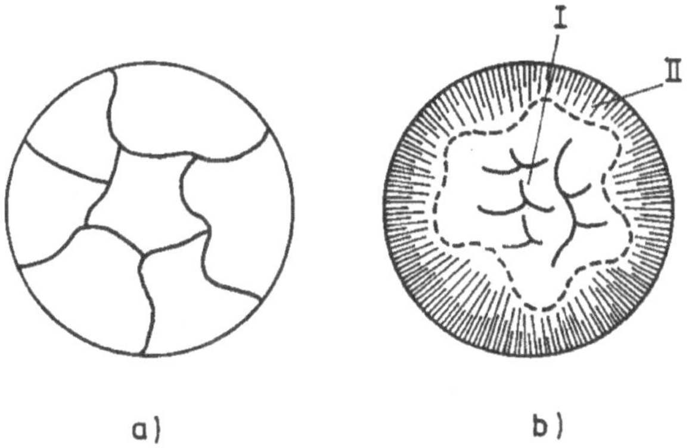

Fig. 1.63 Quartz SiO_2 particle: (**a**) non-activated, (**b**) mechanically activated [Heinicke 1984]

In non-activated mineral, primary crystallite size $> 100\,nm$ (a) can be observed. After activation, two regions can be identified (b). In the bulk (I) exists a relatively little disturbed center with regions scattering coherently and primary crystallite size $\sim 90\,nm$. At the surface (II) and in the near-surface a partial crystalline region of thickness $\sim 20\,nm$ and primary crystallite size ~ 1–$10\,nm$ can be seen. This model has been verified and proved for several oxide systems [Heinicke 1984]. Numerous investigations have shown a correlation between solubility and amorphous portion (or crystallinity degree X) for mechanically activated quartz (Fig. 1.64).

In this way, nanoparticles with the special surface properties can be produced by mechanochemical treatment. Moreover, such particles are "enriched" by defect chemistry which can further contribute to properties of nanomaterials.

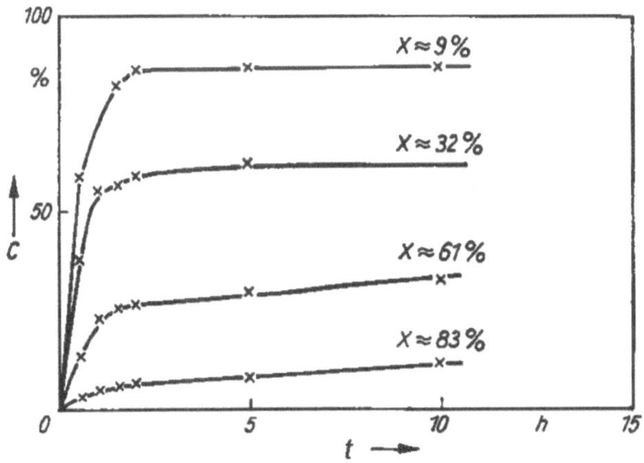

Fig. 1.64 Solubility, C vs. dissolution time, t for quartz SiO_2 mechanically activated in a vibration mill [Steinicke et al. 1978; Steinicke and Linke 1982]

1.3.3 Clusters and Aggregates Formation

Nanoscale materials can be produced in many ways and several of them including mechanochemical approach have been described in this chapter.

Formulation of multinuclear clusters from aqueous ions defines the initial step in growth of any crystal from solution. It was reported that $Zn_3S_3(H_2O)_6$ rings condense to form a neutral $Zn_6S_6(H_2O)_9$ cluster with a structure analogous to wurtzite ZnS (W). Interestingly, condensation of $Zn_3S_3(H_2O)_6$ rings in the presence of excess bisulphide results in the formation of $Zn_4S_6(H_2O)_4^{4-}$ clusters with structure found in sphalerite ZnS(S) (Fig. 1.65) [Luther III et al. 1999; Banfield and Zhang 2001]. Luther postulated that natural solutions can contain relative stable tetrameric $Zn_4S_6(H_2O)_4^{4-}$ clusters $\sim$ 1–2 nm in diameter that have structural similarity to sphalerite.

There is another way in which crystals can be assembled. This involves the solid state combination of large clusters of atoms or nanoparticles to larger crystals. The pieces are assembled in crystallographically specific ways so that the interface elimination leads to formation of a larger single crystal. The crystallographic control distinguishes this pathway from simple aggregation. Oriented assembly of nanoparticles is an effective form of crystal growth because removal of the interface eliminates the energetic contribution of two surfaces. An example of oriented aggregation in Fig. 1.66 for anatase TiO_2 nanoparticles consists of smaller anatase crystals that were oriented with respect to each other. Most particle-particle boundaries have been eliminated, but a few remain. This growth mechanism is a form of self-organization [Penn and Banfield 1999; Banfield and Zhang 2001].

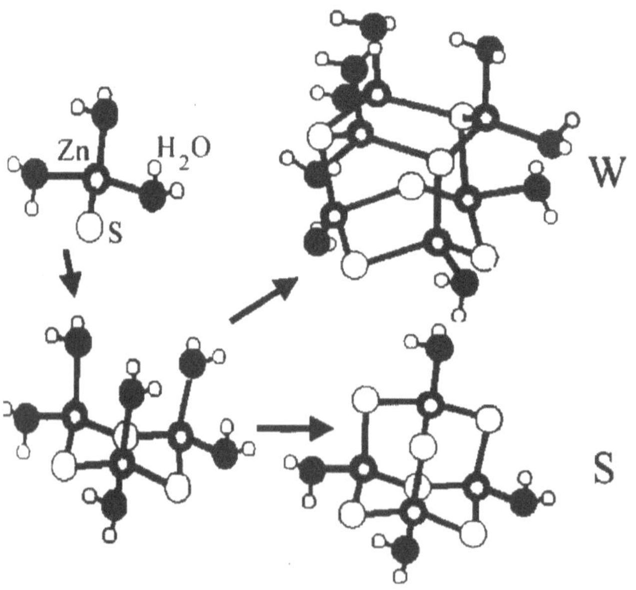

Fig. 1.65 Cluster formation in ZnS-H_2O system [Luther III et al. 1999]

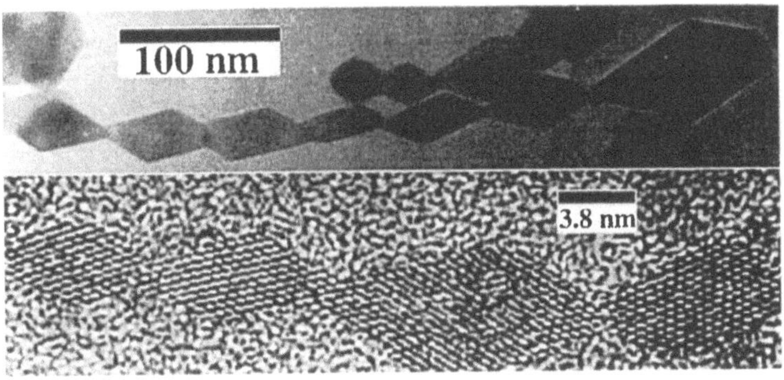

Fig. 1.66 Oriented aggregation in anatase TiO_2 nanoparticles [Banfield and Zhang 2001]

Manganese oxide MnO_2 occurs in nature as hollandite but the mineral form coexists with other phases like SiO_2, Fe_2O_3, Al_2O_3 and P_2O_5. Nanofibres of synthetic MnO_2 has been synthesized by chemical route. Originally the synthesized nanoparticles (Fig. 1.67) are in the form of spherical agglomerates, approximately (100–500 nm) in diameter, with relatively smooth surfaces (a). During aging in solution the size of the original nanoparticle agglomerates start to increase (500–3000 nm). In addition, the surface of some of the agglomerates begin to show the first signs of nanofibrous surface structure (b). With prolonging the time of aging, the size of the agglomerated particles grows to 1000–5000 nm. Further aging leads to a well-defined nanofibrous structure (c, d), so-called "bird-nest" morphology, which consists of a three-dimensional random weave of nanofibres (e). No visible remnants of the original nanoparticle agglomerates are visible. At this stage the size of nest structure is about 3000–10000 nm in size [Xiao et al. 1998].

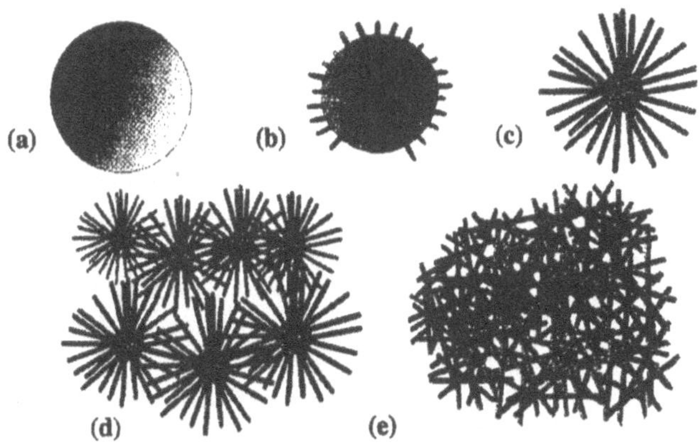

Fig. 1.67 Development of "bird's nest" morphology of synthetic MnO_2 [Xiao et al. 1998]

1.4 Nanoparticles in Medicine

Application of nanoparticles in the field of medicine is especially promising. Areas such as disease diagnosis, drug delivery and molecular imaging are being intensively researched. Nanotechnology already afford the possibility of intracellular imaging through attachment of nanocrystalline semiconductors to selected molecules, for example proteins.

In Table 1.12 several examples of nanoparticles used in biomedical applications are given.

For example gold nanoparticles are used for thermal therapy. Au nanoshells of about 150 nm are injected into the bloodstream. This size nanoshell penetrates tumor but not healthy tissue. The nanoshells are also specificially designed to absorb infrared light that passes harmlessly through body tissue. The nanoshells heat up when they absorb the near-infrared light, thereby killing the nearly tumor cells. Some examples showed that 100 tumor cells could be captured and destroyed from 50 million blood cells in less than one hour.

Table 1.12 Examples of nanosized components in biomedical applications [Kim et al. 2006]

Composition	Size (nm)	Applications
Au	2–150	Drug and gene delivery
Ag	1–80	Antimicrobial properties
Pt	1–20	Sensors and electrodes
Co	1–50	Drug targeting
CdX (X=S, Se, Te)	1–20	Fluorescence labelling
ZnX (X=S, Se, Te)	1–20	Fluorescence labelling
PbS	2–18	Photoluminescence
TiO_2	3–50	Nerve tissue monitoring
ZnO	1–30	Photoluminescence
Fe-O	6–40	Drug delivery
Fe-Pt	2–10	Drug delivery

1.4.1 Size of Biological Objects

Comparison of size of nanoparticles with objects playing important role in medicine is given in Fig. 1.68.

Note that bacteria are huge in comparison and it is helpful to realize that the volume of one Bacillus cereus bacterium could hold a million 5 nm nanoparticles [Klabunde 2001].

Table 1.13 lists the length parameters d for various biological substances and in Fig. 1.69 model of the DNA double helix is presented. Here, the indivi diameter of 2 nm, and the repeat unit containing 10 nucleotide pairs is 3.4 nm long. To accomplish the coupling of two nanostrands of DNA in an efficient manner, a small

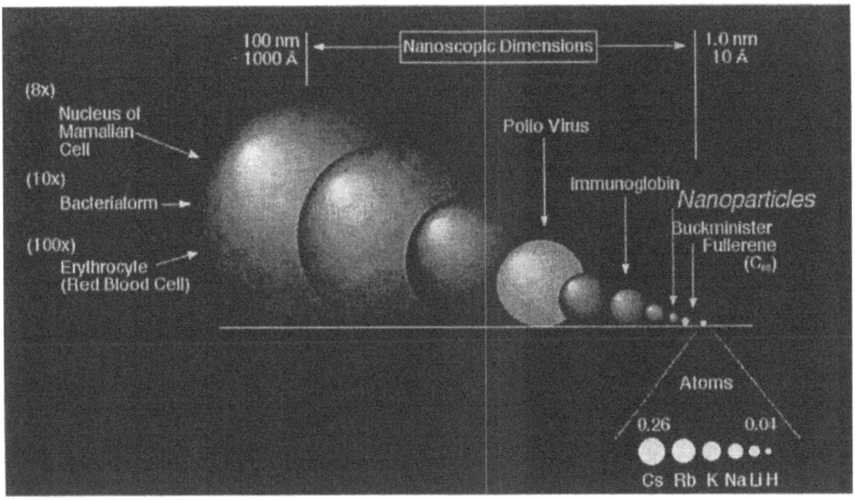

Fig. 1.68 Size comparisons of nanocrystals with bacteria, viruses, and molecules [Klabunde 2001]

Table 1.13 Typical sizes of various biological substances, adapted after [Poole and Owens 2003]

Class	Material	d (nm)	Class	Material	d (nm)
Aminoacids	Glycine	0.42	Organelles	Mitochondrion	3000
	Tryptophane	0.67		Lysosome	700
Proteins	Insulin	2.2	Cells	Bacterium (E.coli)	8000
	Hemoglobin	7.0		Human blood platelet	3000
	Albumin	9.0		Leucocytes	15000
	Lipoprotein	20		Erytrocytes	8000
Viruses	Influenza	60	Others	Chlorophyll	1.1
	Bacteriophage	140		Human chromosome	9000

single-ring pyrimidine base always pairs off with a larger two-always pairs off with a larger two-ring purine base, namely, cytosine (C) with guanine (G), and thymine (T) with adenine (A) [Poole and Owens 2003].

1.4.2 Nanomaterials for Cancer Therapy and Diagnosis

Cancer was once considered an incurable disease, but today most patients diagnosed with early stage disease will survive their illness. Advances in cancer diagnostics and therapeutics over the last few decades are largely responsible for this improvement [Gu et al. 2007]. Despite these advances, cancer continues to torment humanity as the second leading cause of death with 10.9 million newly diagnosed cases world-wide in the year 2005 alone. Therefore, there is still a strong need for a paradigm shift in the approach to cancer diagnosis and therapy. Since the biological

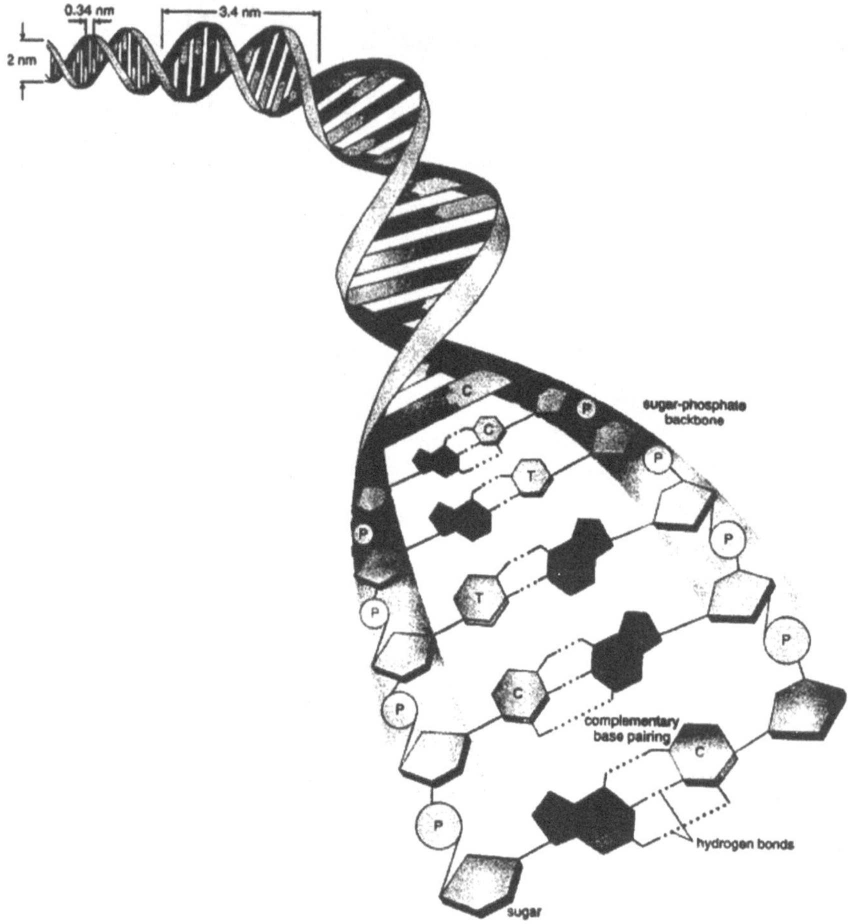

Fig. 1.69 Model of the DNA double helix, nucleotides: A – adenine, C – cytosine, G – guanine, T – thymine [Mader 2001]

processes in general are those that lead to cancer particular occur at the nanoscale, there is a great opportunity for nanotechnologies to treat cancer at an as early stage as possible [Kumar 2006].

Cancer is sa group of diseases characterized by uncontrolled growth and spread of abnormal cells. If the growth and spread of cancer cells is not controlled, the desease is fatal.

1.4.2.1 Cancer Development

In Fig. 1.70 tumor development to a steady diffusion-limited maximal size (I) and beyond diffusion-limited maximal size (II) is illustrated. Once a small tumor mass

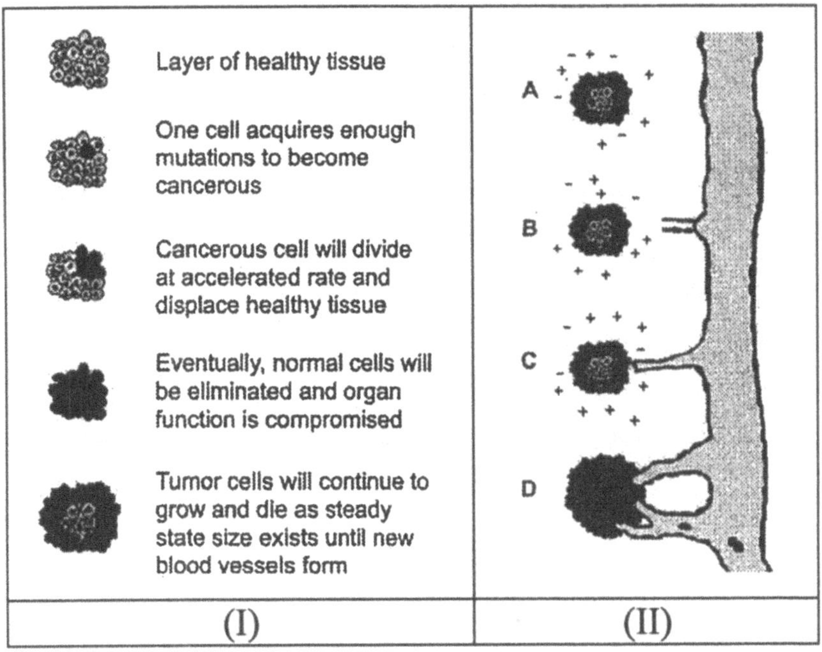

Fig. 1.70 Tumor development from initial carcinogenesis to a steady diffusion-limited maximal size (I) and to a fatal metastase formation (II) [Brannon-Peppas and Blanchette 2004]

has formed (I), the healthy tissue will not be able to compete with the cancer cells for the inadequate supply of nutrients from the blood stream. Tumor cells will displace healthy cells until the tumor reaches a diffusion-limited maximal size. This diffusion-limited maximal size of most tumors is around $2\,\text{mm}^3$ [Jones and Harris 1998].

The growth beyond this size, the tumor must recruit the formation of blood vessels to provide the nutrients necessary to fuel its continued expansion. It is thought that there could be numerous tumors at this diffusion-limited maximal size throughout the body. Until the tumor can gain that access to the circulation it will remain at this size and the process can take years. The continuous development of a tumor beyond the diffusion-limited maximal size is shown in section (II) of Fig. 1.70. Tumor cells in steady state (A) are capable of secreting molecules that initiate the angiogenic process (angiogenesis-a process vital to the continued development of a tumor mass). It is supported that this process called "the angiogenic switch" is result from a local imbalance between positive and negative regulators of angiogenesis (B). This imbalance causes endothelial cell proliferation and migration – a vessel is formed which extends towards the tumor and provides nutrients to sustain cell proliferation (C). A fully vascularized tumor (D) is capable of continued growth with metastatic potential due to the proximity to the blood stream [Brannon-Peppas and Blanchette 2004].

There are some special features of tumor cells which is important to take into account by application of nanoparticles for cancer treatment:

- leaky tumor vasculature presenting increased permeability and gaps, with pores between 350 and 800 nm and cut off between 400 and 600 nm; in comparison with normal tissue this vasculature is highly permeable to macromolecules
- a dysfunctional lymphatic drainage system, which results in enhanced fluid retention in the tumor interstitial space
- the tumor environment is oxidative and acidic, and thus oxidation destroy their anticancer properties.

It is generally accepted that nanoparticles with diameter less than 200 nm are the most effective for extravasating tumor microvasculature [Juillerat-Jeanneret 2006; Gu et al. 2007].

Cancer is treated by surgery, radiation, chemotherapy, hormones and immunotherapy. Each of these cancer treatments has advantages and disadvantages, thus the combination with other treatments is recommended to achieve the optimum outcome [Kim et al. 2006].

1.4.2.2 Nanoparticles for Cancer Therapy

Chemotherapy is a major therapeutic approach for the treatment of localized and metastasized cancer. The selective increase in tumor tissue uptake of anticancer agents would be of great interest in cancer chemotherapy since anticancer drugs are not specific to cancer cells. Routes of administration, biodistribution and elimination of available chemotherapeutic agents can be modified by drug delivery systems to optimize drug therapy [Serpe 2006].

Delivering of therapeutic agents precisely where and when they are needed in the human body is becoming realistic due to rapid tremendous progress in physiology, nanoparticle and nanobiology. Nanoparticles can improve the targeting of cytotoxic drugs to cancer only if they can be directed to cancer areas and maintain there for long time periods with their drug cargo for selective and local release of drugs [Juillerat-Jeanneret 2006]. A strategy could be to associate anticancer agents with colloidal nanoparticles, with the aim to overcome the mechanisms of resistance and to increase selectivity of drugs towards cancer cells while reducing their toxicity towards normal tissues [Brigger et al. 2002].

First generation of nanoparticles have been available for several years. Colloids are representatives of nanoparticles stabilized in solution to prevent uncontrolled growth, aggregation, and flocculation of the nanoparticles. Utilization of colloidal processing leads to attractive new concepts for the fabrication of advanced nanostructured materials [Kim et al. 2006].

According to the process used for the preparation of nanoparticles, nanospheres or nanocapsules can be obtained. Unlike nanospheres (matrix system in which the drug is dispersed throughout the particles), nanocapsules are vesicular systems in which the drug is confined to an aqueous or oily cavity surrounded by a single polymeric membrane. Nanocapsule may, thus be considered as a "reservoir" system,

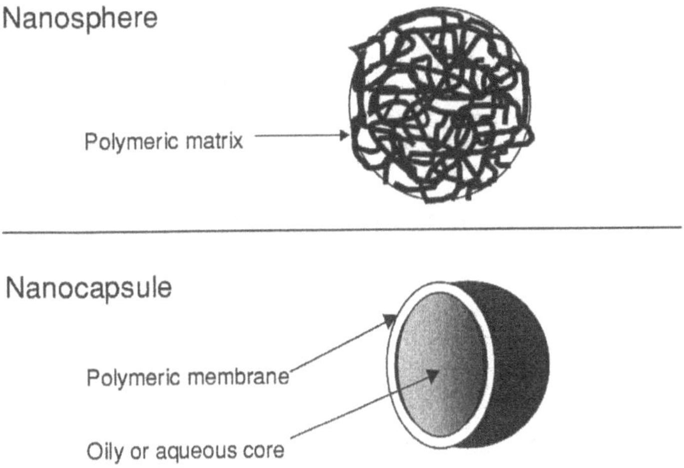

Fig. 1.71 Nanosphere and nanocapsule particles for drug delivery [Brigger et al. 2002]

Fig. 1.71 [Couvreur et al. 1996; Brigger et al. 2002]. Nanocapsules are composed of a polymeric wall containing a liquid inner core, while nanospheres are made of a polymeric matrix in which the drug can be dispersed [Zeisser-Labouèbe et al. 2006].

In *second generation of nanoparticles* a key aspect is the need for multifunctionality of these materials in which several properties are combined to achieve a specific function. For second generation nanoparticles, the surface layer (a few or several monolayers) is distinctly different from that of the core material. Such particles are categorized as core-shell structures. Figure 1.72 shows a schematic representation of different types of second generation nanoparticles with surface modification and nanoparticles with a core-shell structures (a-nanoparticle coated with surfactant to form a stable suspension, b-nanoparticle coated with a thin metallic layer (Au), c-small nanoparticle coated with a porous ceramic layer, d-dispersion of core-shell combination of (a) and (c) for stable suspension [Kim et al. 2006].

Several polymers have been tested to combine with anticancer drugs. Paclitaxel (Taxol) is a microtube-stabilizing agent that causes polymerization of tubulin and cell death. Taxol incorpolated in polyvinylpyrrolidone nanospheres with

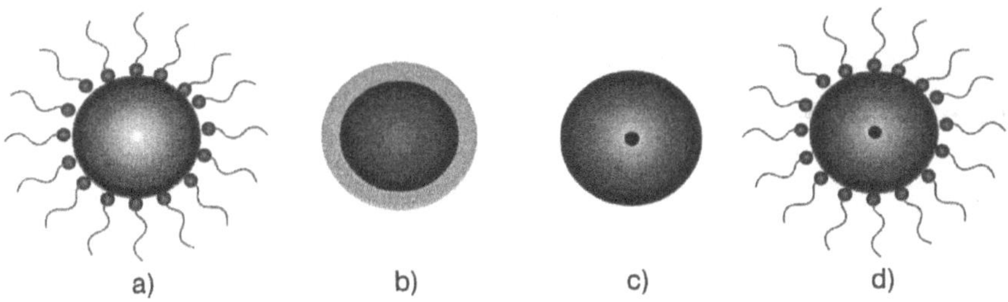

Fig. 1.72 Second generation of nanoparticles [Kim et al. 2006]

a diameter of 50–60 nm were applied to treat melanoma. Mice treated with repeated intravenous injections of taxol-loaded nanospheres showed a significant tumor regression and higher surrival rates than mice treated with free taxol [Sharma et al. 1996]. Doxorubicin is used to treat several types of cancers by inhibiting the synthesis of nucleic acids in cancer cells, but at the price of cardiotoxicity and myelosuppresion and a very narrow therapeutic index. The advantage of doxorubicin combination with polymers can be illustrated by Fig. 1.73 where (DOX-doxorubicin in saline; (DOX+Ps)-doxorubicin in saline plus polysorbate 80; (DOX+NP)-doxorubicin bound to poly (butyl cyanoacrylate) nanoparticles; DOX-NP+Ps doxorubicin bound to poly (butyl cyanoacrylate) nanoparticles coated with polysorbate 80 [Juillerat-Jeanneret 2006].

Rats treated with doxorubicin bounds to polysorbate-coated nanoparticles had significantly higher surrival times than with free doxorubicin, over 20% of the animals showed a long-term remission. All animals treated with polysorbate-containing drug also had a slight inflammatory reaction to the tumor. There was no indication of neurotoxicity [Steiniger et al. 2004]. When administered in vivo, polymeric nanoparticles are rapidly taken up by the macrophages due to the adsorption of proteins at their surface. Thus, the biodistribution of particles is mainly directed towards liver and spleen where they are captured and made unavailable to other target issues. This property has been related to the hydrophobicity of the particle surface. Recently, a great deal of work has been devoted to developing so-called stealth^{TM} particles, which are "invisible" to macrophages. The principle is based on "hydrophilization" of the surface. These stealth^{TM} nanoparticles have been shown to be characterized by a prolonged half-life in the blood compartment. This allows them to selectively extravasate in pathological sites, like tumors or inflamed regions with

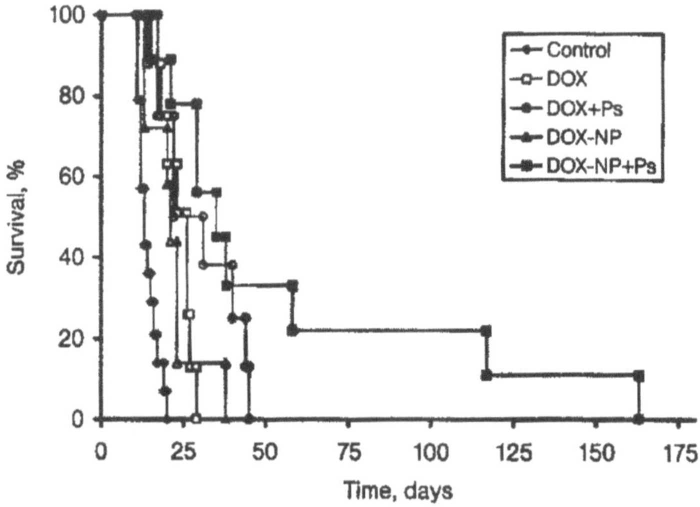

Fig. 1.73 Percentage of survival of the rats with glioblastoma tumor using various drug formulations [Serpe 2006]

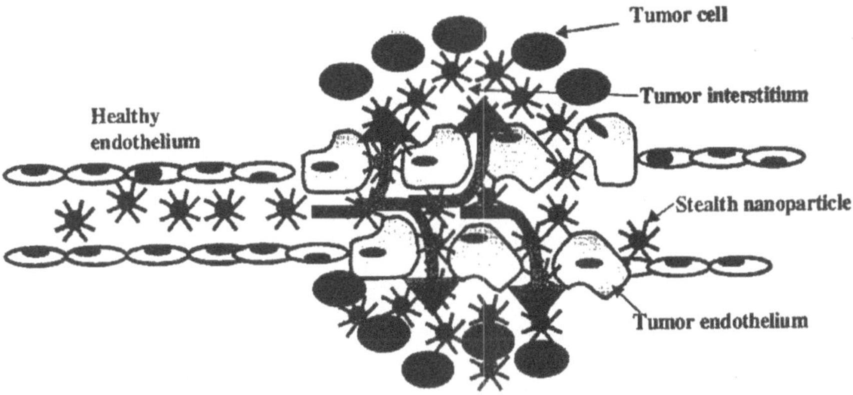

Fig. 1.74 Extravasation of long-circulating (Stealth™) nanoparticles in the tumor cells [Brigger et al. 2002]

a leaky vasculature, Fig. 1.74 [Moghimi et al. 2001; Brigger et al. 2002; Zeisser-Labouébe et al. 2006].

In conclusion, an ideal *nanovector* in cancer therapy and imaging can be defined as an effective drug consisting of

- core consistuent material,
- therapeutic and/or imaging payload and
- biological surface modifiers [Ferrari 2005].

An ideal nanovector should look as shown in Fig. 1.75. Here SPION is a reporter system, T is a targeting agent linked to the polymer; a drug is linked to the polymer via a releasing linker. Challenges include the control of drug release and bioavailability at the tumor site, enhancing selective targeting, and achieving controlled and efficient intracellular delivery, together with acceptable toxicological hazards [Juillerat-Jeanneret 2006].

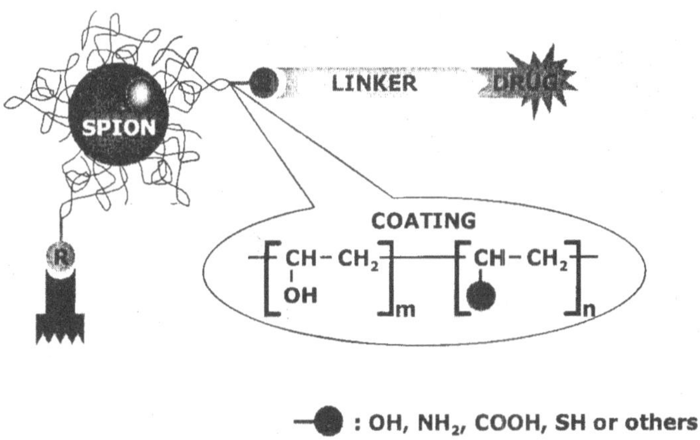

Fig. 1.75 Ideal drug delivery nanovector [Juillerat-Jeanneret 2006]

1.5 Opportunities and Other Challenges

Nanotechnology is entering many industry sections including medicine, plastics, energy, electronics, and aerospace [Klabunde 2001; Anonymous 2004; Lines 2008]. Few examples of interesting nano-applications can be mentioned for illustrations:

- *Batteries* – the ability to prepare metallic nanocrystals that can be consolidated into high surface area battery electrodes has certain inherent advances. Nanostructural materials in lithium ion batteries have proven to be very advantageous. Nickel-metal hydride batteries made of nanocrystalline Ni and metal hydrides are good candidates for lighweight, high-energy density batteries. Recently, cathode materials with three-dimensional framework structures like $LiFeO_4$ for environmentally benign, high-energy density batteries have been developed.
- *Catalysts* – in nanoform can improved activity and selectivity in processes of heterogeneous catalysis (see more in Sect. 6.4.2). The use of nanomaterials based on rare earth metal oxides allows for the preparation of thinner active layers, which can mean less precious metal usage, e.g. Pt.
- *Coatings* – long-term antimicrobial activity can be improved in many coating formulations through the incorporation of nanomaterials. It has been demonstrated, that when paints are doped with light-absorbing nanoparticles, such as TiO_2, the paints are self-cleaning.
- *Composites* – carbon fibres and bundles of multi-wall carbon nanotubes are used in polymers to control or enhance conductivity, with applications such as antistatic packaging.
- *Cutting tools* – cutting tools made of nanocrystalline materials such as WC and TiC are more wear and erosion-resistent. The surface hardness of such nanomaterials is as high as five times higher than the normal microcrystalline materials.
- *Destructive adsorbents* – nanocrystalline metal oxides exhibit high surface areas and strongly chemisorb acidic gases and polar organics. These nanomaterials are finding use in anti-chemical/biological warfare, in air purification and as an alternative to incineration of toxic substances.
- *Displays* – the color of emission is completely dependent on the semiconductor bandgap (see more in Sect. 1.2.3). Bandgap tailoring can be achieved through the size effects of nanoparticles. Through the use of successive layers of tailored particles, a palette of colour can be generated (flat-panel displays). Nanocrystalline ZnS, ZnSe, CdS, CdSe and PbTe are candidates for the next generation of light-emitting phosphors.
- *Fuel cells* – it has been reported that the use of rare earth metal oxides has utility in various aspects of fuel cell design.
- *Magnetic materials* – magnets made of nanocrystalline alloys (e.g. Y-Sm-Co) have unusual magnetic properties. Giant magnetoresistance (GMR) revealed in nanocrystalline materials enables construction of high capacity disk heads.
- *Paints* – incorporating nanoparticles in paints could improve their performance. Pigments with average particle size in the 10–20 nm range display a greatly improved printed colour that is superior to that of commercially available dye-based inks.

- *Sensors* – porous aggregates of semiconductor nanocrystals maintain high surface areas and when they undergo adsorption of various gases, their electric conductivity changes. Since more of the gas to be detected (such as SO_2) is adsorbed per unit mass compared with normal compressed powders, the electrical changes are more pronounced. Thus the use of nanoparticles yield a considerable advantage in sensor technology. Sensors for toxic waterborne materials are also possible.
- *Textile fibres* – nanoparticles have been incorporated into products such as nylon, polypropylene and other polymers to provide long-term antimicrobial character even in harsh environments and after extensive thermal cycling.

1.6 Conclusion

There is an increasing demand for nanoparticles by many companies. The research results illustrated in this chapter are a driving force in this tendency. Products produced from nanoparticles exhibit unique properties and have a wide range of high commercial applications in rapidly expanding markets. Nanoparticles are the core of these technologies. The strongest interest in nanotechnology stems from the concept

Table 1.14 Review papers combining mechano- and nano-aspects of solids

Title	Author(s)
Nanocrystalline Materials	Gleiter 1989
The Synthesis and Structure of Nanocrystalline Materials Produced by Mechanical Attrition: Review	Koch 1993
Mössbauer Effect Studies of Nanostructured Materials	Campbell and Gleiter 1993
Nanocrystalline Materials	Suryanarayana 1995
Novel Materials Synthesis by Mechanical Alloying/Milling	Murty and Ranganathan 1998
Colloid-Chemical Aspects of Mechanical Activation	Juhász 1998
Some Recent Developments in Mechanical Activation and Mechanosynthesis	Gaffet et al. 1999
Recent Development of Materials Design Through a Mechanochemical Route	Senna 2001
Mechanical Alloying and Mixing	Suryanarayana 2001
Self-sustaining Reactions Induced by Ball Milling	Takacs 2002
Mechanical Activation of Precursors for Nanocrystalline Materials	Heegn et al. 2003
Mechanosynthesis of Nanophase Powders	Miani and Maurigh 2004
Mechanical Processing for Nanomaterials	Gaffet and LeCaër 2004
Nanocrystalline Materials and Coatings	Tjong and Chen 2004
Mechanochemistry: The Mechanical Activation of Covalent Bonds	Beyer and Clausen-Schaumann 2005
Mechanochemistry and Mechanical Activation of Solids	Boldyrev 2006
Mechanochemistry of Sulphides: From Minerals to Nanocrystalline Semiconductors	Baláž et al. 2008

that structures may be designed and built that exhibit novel, superior or significantly enhanced properties [Lines 2008].

Mechanochemistry plays an important role in nanoscience and several review papers illustrate this fact (Table 1.14). Mechanochemical processing of solids as a relatively novel and simple solid-state process for the manufacture of nanoparticles is on the way to be incorporated into the world of nanoscience and nanotechnology.

References

Adamson, A.W. (1982) *Physical Chemistry of Surfaces*, 4th Edition. Wiley, New York, 664p.

Agricola, G. (1546) *De Natura Fossilicum*. Latin Edition.

Agricola, G. (1556) *De Re Metallica*. Latin Edition.

Anonymous (2002) Report of the Nanogeoscience Workshop, Berkeley, California, pp. 1–35.

Anonymous (2004) The Royal Society and The Royal Academy of Engineering, Nanoscience and Nanotechnologies. London, pp. 7–23.

Ascencio, J.A., Gutierrez-Wing, C., Espinosa-Pesqueira, M.E., Marin, M., Tehuacanero, S., Zorilla, C. and José-Yacamán, M. (1998) Structure determination of small particles by HRTEM imaging: Theory and experiment. Surface Science 396, 349–369.

Ascencio, J.A., Lin, H.B., Pal, U., Medina, A. and Wang, Z.L. (2006) Transmission electron microscopy and theoretical analysis of AuCu nanoparticles. Atomic distribution behaviour. Microscopy Research and Technique 69, 522–530.

Ashly, M.F. and Jones, D.R.H. (1980) *Engineering Materials: An Introduction to their Properties and Applications*. Pergamon Press, Oxford.

Avrami, M. (1941) Kinetics of phase change. III Granulation, phase change, and microstructure. Journal of Chemical Physics 9, 177–184.

Avvakumov, E.G., Boldyrev, V.V. and Kosobudskij, I.D. (1972) Mechanochemical activation of solid state reactions. 1. About reaction of pyrite with iron. Izvestija Sibirskovo otdelenija AN SSSR, serija chimiceskich nauk 4, 45–50.

Avvakumov, E.G. and Strugova, L.I. (1974) Mechanical activation of solid state reactions. 6. Application of non-diffusion kinetics in mechanochemical reaction in solid mixtures. Izvestija Sibirskovo otdelenija AN SSSR, serija chimireskich nauk 6, 42–45.

Avvakumov, E.G. (1986) *Mechanical Methods of the Activation of Chemical Processes*. Nauka, Novosibirsk (in Russian).

Avvakumov, E.G. and Kosova, N.V. (1991) Fast propagating solid-state mechanochemical reactions. Sibirskij Chimičeskij Zhurnal 5, 62–66.

Avvakumov, E.G., Senna, M. and Kosova, N. (2001) *Soft Mechanochemical Synthesis. A Basis for New Chemical Technologies*. Kluwer Academic Publishers, Boston, p. 201.

Bacon, F. (1658) In: *Opuscula Varia Posthuma* by Rawley (1658). See Shaw, P. *The Philosophical Works of the Hon. R. Boyle*. Ist Edition, Vol. 1, 1733, p. 149.

Bai, H.Y., Luo, J.L., Jin, D. and Sun, J.R. (1996) Particle size and interfacial effect on the specific heat of nanocrystalline Fe. Journal of Applied Physics 79, 361–364.

Baláž, P., Bálintová, M., Bastl, Z., Briančin, J. and Šepelák, V. (1997) Characterization and reactivity of zinc sulphide prepared by mechanochemical synthesis. Solid State Ionics 101–103, 45–51.

Baláž, P., Ohtani, T., Bastl, Z. and Boldižárová, E. (1999) Properties and reactivity of mechanochemically synthesized tin sulfides. Journal of Solid State Chemistry 144, 1–7.

Baláž, P. (2000) *Extractive Metallurgy of Activated Minerals*. Elsevier, Amsterdam, 290p.

Baláž, P. (2001) Mechanochemistry in extractive metallurgy: The modern science with old routes. Acta Metallurgica Slovaca, Special Issue 4, 23–28.

Baláž, P., Takacs, L., Ohtani, T., Mack, D.E., Boldižárová, E. and Achimovičová, M. (2002) Properties of a new nanosized tin sulphide phase obtained by mechanochemical route. Journal of Alloys and Compounds 337, 76–82.

Baláž, P., Boldižárová, E., Godočíková, E. and Briančin, J. (2003) Mechanochemical route for sulphide nanoparticles preparation. Materials Letters 57, 1585–1589.

Baláž, P., Dutková, E., Ficeriová, J., Pourghahramani, P. and Achimovičová, M. (2008) Mechanochemistry of sulfides: from minerals to nanocrystalline semiconductors. In: F. Delogu and G. Mulas (Eds.) *Experimental and Theoretical Approaches to Modern Mechanochemistry*. Research Signpost, Trivandrum, Kerala, India.

Balamurugan, B., Mehta, B.R. and Sharma, K. (1999) Fabrication and characterization of $Cd_xZn_{1-x}S$ nanoparticles. Nano-Structured Materials 12, 151–154.

Banfield, J.F. and Navrotsky, A. (Eds.) (2001) *Nanoparticles and the Environment, Reviews in Mineralogy & Geochemistry*. Mineralogical Society of America, Blacksburg, Virginia, Vol. 44, pp. 1–349.

Banfield, J.F. and Zhang, H. (2001) *Nanoparticles and the Environment*. In: J.F. Banfield and A. Navrotsky (Eds.) *Reviews in Mineralogy & Geochemistry*. Mineralogical Society of America, Blacksburg, Virginia, Vol. 44, pp. 1–58.

Baramboin, N.K. (1970) *Mechanochemistry of Macromolecular Substances*. Khimiya, Moscow, 357p.

Baraton, M.I. and El-Shall, M.S. (1995) Synthesis and characterization of nanoscale metal oxides and carbides. Nanostructured Materials 6, 301–304.

Barret, P. (1978) *Kinetics of Heterogeneous Reactions in Solid-Gas Systems*. Academia, Prague (in Czech).

Berlin, A.A. (1958) Mechanochimičeskoje prevraščenija i sintez polimerov. Uspechi chimiji 27, 94–112.

Bertenev, G.M. and Razumovskaya I.V. (1969) Phonone conception of solid fracture. Fizikochimičeskaja Mechanika Materialov 5, 60–64.

Bertran, H. and Zhu, J.G. (1992) Fundamental magnetization processes in thin-film recording media. Solid State Physics 46, 271–371.

Beyer, M.K. and Clausen-Schaumann, H. (2005) Mechanochemistry: The mechanical activation of covalent bonds. Chemical Reviews 105, 2921–2948.

Birringer, R., Herr, U. and Gleiter, G. (1986) Nanocrystalline materials – a first report. Grain boundary structure and related phenomena. Supplement of Transactions of Japan Institute of Metals 27, 43–52.

Blugel, S. (1992) Two-dimensional ferromagnetism of 3d, 4d, and 5d transition metal monolayers on mole metal (001) substrates. Physical Review Letters 68, 851–854.

Boldyrev, V.V. (1972) About kinetic factors which influence the mechanochemical processes in inorganic systems. Kinetika i Kataliz 13, 1411–1412 (in Russian).

Boldyrev, V.V. (1986) Mechanochemistry of inorganic solids. Proceedings of Indian Natural Sciences Academy 52, 400–417.

Boldyrev, V.V., Lyakhov, N., Pavlyuchin, Y., Boldyreva, E., Ivanov, E. and Avvakumov, E.G. (1990) Achievements and prospects in mechanochemistry. Russian Chemical Reviews 14, 105–161.

Boldyrev, V.V. (1983) Mechanochemistry and mechanical activation of solids. Solid State Ionics 63–65, 537–543.

Boldyrev, V.V. and Tkáčová, K. (2000) Mechanochemistry of solids: past, present and prospects. Journal of Materials Synthesis and Processing 8, 121–132.

Boldyrev, V.V. (2002) About history of mechanochemistry development in Siberia. Chemistry in Sustainable Development 10, 3–12.

Boldyrev, V.V. (2006) Mechanochemistry and mechanical activation of solids. Russian Chemical Reviews 75, 177–189.

Bowden, F. and Yoffe, A. (1952) *Initiation and Growth of Explosion in Liquids and Solids*. Cambridge University Press, Cambridge, 104p.

Bowden, F. and Yoffe, A. (1958) *Fast Reactions in Solids*. Butterworths, London, 163p.

Bowden, F.P. and Tabor, D. (1958) *The Friction and Lubrication of Solids*. Clarendon Press, Oxford, 372p.

Brannon-Peppas, L. and Blanchette, J.O. (2004) Nanoparticle and targeted systems for cancer therapy. Advanced Drug Delivery Reviews 56, 1649–1659.

Brigger, I., Dubernet, C. and Couvreur, P. (2002) Nanoparticles in cancer therapy and diagnosis. Advanced Drug Delivery Reviews 54, 631–651.

Brus, L.E. (1984) Electron-electron and electron-hole interaction in small semiconductor crystallites: the size dependence of the lowest excited electronic state. Journal of Chemical Physics 80, 4403–4409.

Butyagin, P.Yu (1973) Primary active centers in mechanochemical reactions. Žurnal Vsesojuznogo Chimičeskogo obščestva D. Mendelejeva 18, 90–95.

Butyagin, P.Yu (1984) Disordered structures and mechanochemical reactions in solids. Uspechi chimiji 53, 1769–1781.

Butyagin, P.Yu (1994) Problems in mechanochemistry and prospects for its developments. Russian Chemical Reviews 63, 965–976.

Campbell, S.J. and Gleiter, H. (1993) Mössbauer effect studies of nanostructured materials. In: G.J. Long and F. Grandjean (Eds.) *Mössbauer Spectroscopy Applied to Magnetism and Materials Science*. Plenum Press, New York, pp. 241–303.

Champion, Y., Langlois, C., Guerin-Mailly, S., Langlois, P., Bonnentien, J.-L. and Hytch, M.J. (2003) Near-perfect elastoplasticity in pure nanocrystalline copper. Science 300, 310–311.

Chen, S.A. and Liu, W.M. (1999) Preparation and characterization of surface coated ZnS nanoparticles. Langmuir 15, 8100–8104.

Chen, J., Li, Y., Wang, Y., Yun, I. and Cao, D. (2004) Preparation and characterization of zinc sulphide nanoparticles under high-gravity environment. Materials Research Bulletin 39, 185–194.

Chodakov, G.S. (1972) *Physics of Milling*. Nauka, Moscow (in Russian).

Christian, J.E. (1965) *The Theory of Transformation in Metals and Alloys*. Pergamon Press, New York.

Clark, G.L. and Rowan, R. (1941) Studied on lead oxides, polymorphic transition by grinding, distortion and catalytic activity in PbO. Journal of American Society 63, 1302–1305.

Couvreur, P., Couarraze, G., Devissagnet, J.-P. and Puisieux, F. (1996) Nanoparticles: preparation and characterization. In: S. Benita (Ed.) *Microencapsulation: Methods and Industrial Application*. Marcel Dekker, New York, pp. 183–211.

Dachille, F. and Roy, H. (1960) High pressure phase transformations in laboratory mechanical mixers and mortars. Nature 186, 34–71.

Dachille, F. and Roy, R. (1961) Influence of displacive-shearing stresses on the kinetics of reconstructive transformations effected by pressure in the range 0-100.000 bars. In: J.H. de-Boer et al. (Eds.) *Reactivity of Solids*. Elsevier, New York, pp. 502–511.

de Heer, W.A. (2000) Nanomagnetism. In: Z.L. Wang (Ed.) *Characterization of Nanophase Materials*. Wiley, Weinheim, pp. 289–314.

Dhas, N.A., Zaban, A. and Gedanken, A. (1999) Surface synthesis of zinc sulfide nanoparticles on silica microspheres. Sonochemical preparation, characterization and optical properties. Chemistry of Materials 11, 806–813.

Ding, J., Miao, W.F., McCormick, P.G. and Street, R. (1995) Mechanochemical synthesis of ultrafine Fe powder. Applied Physics Letters 67, 3804–3806.

Ding, J., Tsuzuki, T., McCormick, P.G. and Street, R. (1996) Ultrafine Co and Ni particles prepared by mechanochemical processing. Journal of Physics D 29, 2365–2369.

Dutková, E., Pourghahramani, P., Velumani, S., Baláž, P., Kostova, N.G. and Ascencio, J.A. (2008a) Mechanochemically synthesized ZnS nanoparticles. Journal of Nanoscience and Nanotechnology (submitted).

Dutková, E., Baláž, P., Pourghahramani, P., Nguyen, A.V., Šepelák, V., Feldhoff, A., Kováč, J. and Šatka, A. (2008b) Mechanochemical solid state synthesis and characterization of $Zn_xCd_{1-x}S$ nanocrystals. Solid State Ionics 179, 1242–1245.

Dutta, J. and Hoffmann, H. (2003) http://www.mxsg3.epfl.ch/ltp/cours/Nanomat/pdf

Eckert, J., Holzer, J.C., Krill III, C.E. and Johnson, W.L. (1992) Reversible grain size changes in ball-milled nanocrystalline Fe-Cu alloys. Journal of Materials Research 7, 1980–1983.
Efros, Al.L. and Efros, A.L. (1982) Interband absorption of light in a semiconductor sphere. Semiconductors 16, 1209–1214.
El-Shall, M.S., Slack, W., Vann, W., Kane, D. and Hanely, D. (1994) Synthesis of nanoscale metal oxide particles using laser vaporization/condensation in a diffusion cloud chamber. Journal of Physical Chemistry 98, 3067–3070.
Enzo, S., Sampoli, M., Cocco, G., Schiffini, L., and Battezzati, L. (1989) Crystal-to glass transition in the Ni-Ti system by mechanical alloying and consequent surface area. Philosophical Magazine B 59, 169–175.
Erofeev B.V. (1946) General equation of chemical kinetics and its application for solid state reactions. Doklady Akademii Nauk, USSR 52, 515–519.
Fahlman, B.D. (2007) *Materials Chemistry*. Springer, Dordrecht, 485p.
Faraday, M. (1820) The Quarterly Journal of Science, Literature and the Arts 8, 374.
Faraday, M. (1827) *Chemical Manipulations: Being Instructions to Students in Chemistry on the Methods of Performing Experiments of Demonstrations or of Research, with Accuracy and with Success*. W. Phillips, London, p. 147.
Faraday, M. (1834) Philosofical Transactions of the Royal Society, p. 55.
Faraday, M. (1857) Philosophical Transactions of the Royal Society London 7, 145–181.
Fecht, H.J. (1995) Nanostructure formation by mechanical attrition. Nano-Structured Materials 6, 33–42.
Ferrari, M. (2005) Cancer nanotechnology: Opportunities and challenges. National Review of Cancer 5, 161–171.
Fink, M. and Hofmann, U. (1932) Oxydation von Metallen unter dem Einfluss der Reibung. Zeitschrift für Metallkunde 24, 49–54.
Fink, M. and Hofmann, U. (1933) Zeitschrift für anorganische und allgemeine Chemie 210, 100.
Flavickij, F.M. (1902) About a new method of analytical study of the solids. Žurnal ruskovo fiziko-chemičeskovo obščestva 34, 8 (in Russian).
Flavickij, F.M. (1909) Special methods and reactions in solid state chemistry. *Proceedings of Mendelejev Congress*, Moscow.
Fox, P.G. and Soria-Ruiz, J. (1970) Fracture-induced decomposition in brittle crystalline solids. Proceedings of Royal Society A317, 79–90.
Fox, P.G. (1975) Mechanically initiated chemical reactions in solids. Journal of Materials Science 10, 340–360.
Freeman, A.J., Fu, C.L., Lee, J.I. and Ogushi, T. (1987) In: M. Takahashi et al. (Eds.) *Physics in Magnetic Materials*. World Scientific, Singapore, p. 221.
Gaffet, E., Bernard, F., Niepce, J.-C., Charlot, F., Gras, Ch., LeCaër, G., Guichard, J.-L., Delcroix, P., Mocellin, A. and Tillement, O. (1999) Some recent developments in mechanical activation and mechanosynthesis. Journal of Materials Chemistry 9, 305–314.
Gaffet, E. and LeCaër, G. (2004) Mechanochemical processing for nanomaterials. In: H.S. Nalwa (Ed.) *Encyclopedia of Nanoscience and Nanotechnology*. American Scientific Publishers, pp. 91–129.
Gao, L., Wang, H.Z., Hong, J.S., Miyamoto, H., Miyamoto, K., Nishikawa, Y. and Torre, S.D.D.L. (1999) SiC-ZrO_2(3Y)-Al_2O_3 nanocomposites superfast densified by plasma spark sintering. Nano-Structured Materials 11, 43–49.
Gleiter, H. (1989) Nanocrystalline materials. Progress in Materials Science 33, 223–315.
Gleiter, H. (1995) Nanostructured materials: state of the art and perspectives. Zeitschrift für Metallkunde 86, 78–83.
Grohn, H., Paudert, R. and Bisinger, H.I. (1962) Über die mechanische Anregung einiger chemischer Reaktionen anorganischer Feststoffe. Zeitschrift für Chemie 2, 88–90.
Grohn, H. and Paudert, R. (1963) Mechanochemische Reaktionen von Elementen der IV. Hauptgruppe mit einigen organischen Verbindungen. Zeitschrift für Chemie 3, 89–93.
Gu, F.X., Karnik, R., Wang, A.Z., Alexis, F., Levy-Nissenbaum, E., Hong, S., Langer, R.S. and Farokhzad, O.C. (2007) Targeted nanoparticles for cancer therapy. Nanotoday 2, 14–21.

Gutman, E.G. (1974) *Mechanochemistry of Metals and Protection against Corrosion.* Metallurgija and Protection against Corrosion, Moscow, 230p. (in Russian).

Habashi, F. (1974) *Principles of Extractive Metallurgy*, Vol. 1 – General Principles. Gordon & Breach, New York.

Hahn, H. and Averback, R.S. (2001) The production of nanocrystalline powders by magnetron sputtering. Applied Physics Letters 78, 3708–3710.

Hall, E.O. (1951) The deformation and aging of mild steel III. Discussion and results. Proceedings of the Physical Society (London) B64, 747–753.

Hannay, N.B. (1959) In: N.B. Hannay (Ed.) *Semiconductors.* Reinhold, New York.

Hao, L., You, M., Mo, X., Jiang, W., Zhu, Y., Zhou, Y., Hu, Y., Lin, X. and Chen, Z. (2003) Fabrication and characterization of ordered macroporous semiconductors CdS by colloidal crystal template. Materials Research Bulletin 38, 723–729.

He, R., Qian, X., Yin, J., Xi, H., Bian, L. and Zhu, Z. (2003) Formation of monodispersed PVC-capped ZnS and CdS nanocrystals under microwave irradiation. Colloids and Surfaces A: Physicochemical and Engineering Aspects 220, 151–157.

Heegn, H. (1989) On the connection between ultrafine grinding and mechanical activation of minerals. Aufbereitungs-Technik 30, 635–642.

Heegn, H. (1990) Mechanische Aktivierung von Festkörpern. Chemische Ingenieur Technik 63, 458–470.

Heegn, H., Birkeneder, F. and Kamptner, A. (2003) Mechanical activation of precursors for nanocrystalline materials. Crystal Research and Technology 38, 7–20.

Hedvall, J.A. (1938) *Reaktionsfähigkeit fester Stoffe.* Barth-Verlag, Leipzig.

Heinicke, G. and Sigrist, K. (1971) Zur Thermodynamik tribochemischer Reaktionen. Zeitschrift für Chemie 11, 226–235.

Heinicke, G. (1981) Recent Advances in Tribochemistry. In: *Proceedings of the International Symposium on Powder Technology 81*, Kyoto, pp. 354–364.

Heinicke, G. (1984) *Tribochemistry.* Akademie Verlag, Berlin, 495p.

Hellstern, E., Fecht, H.J., Garland, C. and Johnson, W.L. (1989) Structural and thermodynamic properties of heavily mechanically deformed Ru and AlRu. Journal of Applied Physics 65, 305–310.

Hess, K., Steurer, E. and Fromm, E. (1942) Kolloid Zeitschrift 98, 209.

Hochella, M.F., Jr. (2002) There's plenty of room at the bottom: nanoscience in geochemistry. Geochimica et Cosmochimica Acta 66, 735–743.

Hoffmann, H., Bowen, P., Jongen, N. and Lemaire, J. (2001) Nano-composite powders, a new concept for their synthesis. Scripta Materialia 44, 2197–2201.

Hoffmann, U., Horst, Ch. and Kunz, U. (2005) Reactive comminution. In: K. Sundmacher, A. Kienle, and A. Seidel-Margenstern (Eds.) *Integrated Chemical Processes.* Wiley, Weinheim, pp. 407–436.

http://www.cheaptubesinc.com

http://www.ipt.arc.nasa.gov/nanotechnology.html

Hüttig, G.F. (1943) Zwischenzustände bei Reaktionen im Fester Zustand und ihre Bedentung für die Katalyse. In: *Handbuch der Katalyse* IV. Springer Verlag, Wien, pp. 318–331.

Imamura, K. and Senna, M. (1982) Change in phase stability of zinc blende and wurtzite on grinding. Journal of Chemical Society, Faraday Transactions 78, 1131–1140.

Imamura, K. and Senna, M. (1984) Difference between mechanochemical and thermal processes of polymorphic transformation of ZnS and PbO. Materials Research Bulletin 9, 59–65.

Jellinek, F. (1988) Transition metal chalcogenides. Relationship between chemical composition, crystal structure and physical properties. Reactivity of Solids 5, 323–339.

Jones, A. and Harris, A.L. (1998) New developments in angiogenesis: a major mechanism for tumor growth and target for therapy. Cancer Journal from Scientific American 4, 209–217.

José-Yacamán, M., Ascencio, J.A. and Lin, H. (2001) The structure, shape and stability of nanometric sized particles. Journal of Vacuum Science and Technology B19, 1091–1107.

Juhász, A.Z. (1974) Mechanochemische Aktivierung von Silikatmineralen durch Trocken-Feinmahlen. Aufbereitungs-Technik 10, 558–562.

Juhász, Z. (1985) Mechanochemische Erscheinungen beim Feinmahlen von Tonmineralen. Sprechsall 118, 110–117.

Juhász, A.Z. and Kolláth, B. (1993) Mechanochemical reactions of OH-containing crystals. Acta Chimica Hungarica-Models in Chemistry 130, 725–735.

Juhász, Z.A. (1998) Colloid-chemical aspects of mechanical activation. Particulate Science and Technology 16, 145–161.

Juillerat-Jeanneret, L. (2006) Critical analysis of cancer therapy using nanomaterials. In: Challa S.S.R. Kumar (Ed.) *Nanomaterials for Cancer Therapy*. Wiley-VCH, Weinheim 2006, pp. 199–241.

Keskinen, J., Ruuskanen, P., Karttunen, M. and Hannula, S.-P. (2001) Synthesis of silver powder using a mechanochemical process. Applied Organometallic Chemistry 15, 393–395.

Khaleel, A. and Richards, R.M. (2001) Ceramics. In: K.J. Klabunde (Ed.) *Nanoscale Materials in Chemistry.* John Wiley & Sons, Hoboken, New Jersey, pp. 85–120.

Kim, D.K., Jo, Y.S., Dobson, J., El Haj, A. and Muhammed, M. (2006) Nanomaterials for controlled release of anticancer agents. In: Challa S.S.R. Kumar (Ed.) *Nanomaterials for Cancer Therapy.* Wiley-VCH, Weinheim 2006, pp. 168–198.

Klabunde, K.J., Stark, J., Koper, O., Mohs, C., Park, D., Decker, S., Jiang, Y., Lagadic, I. and Zhang, D. (1996) Nanocrystals as stoichiometric reagents with unique surface chemistry. Journal of Physical Chemistry 100, 12142–12153.

Klabunde, K.J. (Ed.) (2001) *Nanoscale Materials in Chemistry.* John Wiley & Sons, Hoboken, New Jersey, 292p.

Koch, C.C. (1993) The synthesis and structure of nanocrystalline materials produced by mechanical attrition: review. Nano-Structured Materials 2, 109–120.

Koch, C.C. (1997) Synthesis of nanostructured materials by mechanical milling: problems and opportunities. Nano-Structured Materials 9, 13–22.

Koper, O. and Winecki, S. (2001) Specific heats and melting points of nanocrystalline materials. In: K.J. Klabunde (Ed.) *Nanoscale Materials in Chemistry.* John Wiley & Sons, Hoboken, New Jersey, pp. 263–278.

Kosmac, T. and Courtney, T.H. (1992) Milling and mechanical alloying of inorganic nonmetallics. Journal of Materials Research 7, 1519–1525.

Kosmac, T., Maurice, D. and Courtney, T.H. (1993) Synthesis of nickel sulfides by mechanical alloying. Journal of American Ceramic Society 76, 2345–2352.

Kotkata, M.F. and Mahmoud, E.A. (1982) Non-isothermal crystallization kinetic studies on amorphous chalcogenide semiconductors. Materials Science Engineering 54, 163–168.

Kubaschewski, O. and Alcock, C.B. (1979) *Metallurgical Thermochemistry*. Pergamon Press, New York.

Kumar S.S.R. Challa (2006) *Nanomaterials for Cancer Therapy.* Wiley-VCH, Weinheim 2006, pp. 168–198.

Lambeth, D., Velu, E., Bellesis, G., Lee, L. and Laughlin, D. (1996) Media for $10Gb/in^2$ hard disk storage: Issues and status (invited). Journal of Applied Physics 79, 4497–4501.

Lan, Ch., Hong, K., Wang, W. and Wang, G. (2003) Synthesis of ZnS nanorods by annealing precursors ZnS nanoparticles in NaCl flux. Solid State Communications 125, 455–458.

Lea, M.C. (1866) Researches on the latent image. The British Journal of Photography 13, 84.

Lea, M.C. (1891a) On gold-coloured allotropic silver, Part I. American Journal of Science-Third Series 141, 179–190.

Lea, M.C. (1891b) On allotropic silver. Part II-Relations of allotropic silver with silver as it exists in silver compounds. American Journal of Science-Third Series 141, 259–267.

Lea, M.C. (1892) Disruption of the silver haloid molecule by mechanical force. American Journal of Science-Third Series 143, 527–531.

Lea, M.C. (1893) On endothermic decomposition obtained by pressure. Second part. Transformation of Energy by Shearing Stress. American Journal of Science-Third Series 145, 413–420.

Lea, M.C. (1894) Transformation of mechanical into chemical energy. Third part. Acting of shearing stress continued. American Journal of Science-Third Series 147, 377–381.

Lin, I.J. and Nadiv, S. (1970) Review of the phase transformation and synthesis of inorganic solids obtained by mechanical treatment (mechanochemical reactions). Materials Science and Engineering 393, 193–209.
Lin, I.J. and Somasundaran, P. (1972) Alternation in properties of samples during their preparation by grinding. Powder Technology 6, 171–179.
Lin, I.J. and Niedzwiedz, S. (1973) Kinetics of the massicot-litharge transformation during comminution. Journal of American Chemical Society 56, 62–64.
Lin, I.J, Nadiv, S. and Grodzian, D.J.M. (1975) Changes in the state of solids and mechanochemical reactions in prolonged comminution processes. Minerals Science Engineering 7, 313–336.
Lin, Y., Zapien, J.A., Shan, Y.Y., Geng, Ch.Y., Lee, Ch.S. and Lee, S.T. (2005) Wavelength-controlled lasing in $Zn_xCd_{1-x}S$ single-crystal nanoribbons. Advanced Materials 17, 1372–1377.
Lindemann, F.A. (1910) Über die Berechnung molekularer Eigenfrequenzen. Physikalische Zeitschrift 11, 609–612.
Lines, M.G. (2008) Nanomaterials for practical functional uses. Journal of Alloys and Compounds 449, 242–245.
Lippens, P.E. and Lanoo, M. (1989) Calculation of the band gap for small CdS and ZnS crystallites. Physical Review B39, 10935–10942.
Luther III, G.W., Theberge, S.M. and Richard, D.T. (1999) Evidence for aqueous clusters as intermediates during zinc sulfide formation. Geochimica et Cosmochimica Acta 63, 3159–3169.
Lyakhov, N.Z. (1984) The kinetics of mechanochemical reactions. Folia Montana, Extraordinary Number, 40–49.
Lyakhov, N.Z. (1993) Mechanical activation from the viewpoint of kinetic reaction mechanisms. In: K. Tkáčová et al. (Eds.) *Proceedings of the 1st International Conference on Mechanochemistry INCOME'93*, Vol. 1, Cambridge Interscience Publishing, Cambridge, pp. 59–65.
Mader, S.S. (2001) *Biology*. Mc-Graw-Hill, Boston.
McCormick, P.G. (1995) Application of mechanical alloying to chemical refining: Overview. Materials Transactions, JIM, 36, 161–169.
Mee, C. (1994) *The Physics of Magnetic Recording*. Amsterdam, North-Holland.
Metals Handbook. (1973) *Metallography, Structures and Phase Diagram*. American Society for Metals, Metals Park, Ohio, Vol. 8, p. 325.
Miani, F. and Maurigh, F. (2004) Mechanosynthesis of nanophase materials. In: J.A. Schwarz, C.I., Contescu, K., Putyera (Eds.) *Dekker Encyclopedia of Nanoscience and Nanotechnology*. Marcel Dekker, New York, pp. 1787–1795.
Mie, G. (1908) Beitrage zur Optik trüber Medien, speziell kolloidaler Metallösungen. Annalen der Physik 25, 377–445.
Moghimi, S.M., Hunter, A.C. and Murray, J.C. (2001) Long-circulating and target-specific nanoparticles: theory to practice. Pharmacological Reviews 53, 183–318.
Morales, J., Perez-Vicente, C. and Tirado, J.L. (1992) Chemical and electrochemical lithium intercalation and staging in 2H-SnS_2. Solid State Ionics 51, 133–138.
Murphy, C.J. and Coffer, J.L. (2002) Quantum dots: a Primer. Applied Spectroscopy 56, 16A–27A.
Müller-Warmuth, W. and Schöllhorn, R. (Eds.) (1994) *Progress in Intercalation Research*. Kluwer Academic, Dodrecht.
Murty, B.S. and Ranganathan, S. (1998) Novel materials synthesis by mechanical alloying/milling. International Materials Reviews 43, 101–141.
Nie, Q., Yuan, Q., Wang, Q. and Zhude, X. (2004) In situ synthesis of $Zn_xCd_{1-x}S$ nanorods by hydrothermal route. Journal of Materials Science 39, 5611–5612.
Oleszak, D. and Shingu, P. (1996) Nanocrystalline metals prepared by low energy ball milling. Journal of Applied Physics 79, 2975–2980.
Ostwald, W. (1887) *Lehrbuch der Allgemeinen Chemie*. 2. Auflage, 2. Band, Engelmann, Leipzig, 616p.
Ostwald, W. (1909) *Grundriss der Kolloidchemie*. Verlag von Theodor Steinkopff, Dresden, 302p.
Ostwald, W. (1927) *Die Welt der vernachlässigten Dimensionen*. Verlag von Theodor Steinkopff, Dresden, 336p.

Parker, L.H. (1914) Reactions by trituration. Journal of the Chemical Society 105, 1504–1516.

Parker, L.H. (1918) Reactions between solid substances. Journal of the Chemical Society 113, 396–409.

Pawaskar, N.R., Sathaye, S.D., Bhadhabe, M. and Patil, K.R. (2002) Applicability of liquid-liquid interface reaction technique for the preparation of zinc sulphide nanoparticulate thin films. Materials Research Bulletin 37, 1539–1546.

Penn, R.L. and Banfield, J.F. (1999) Morphology development and crystal growth in nanocrystalline aggregates under hydrothermal conditions. Insight from titania. Geochimica et Cosmochimica Acta 63, 1549–1557.

Petch, N.J. (1953) The cleavage strength of polycrystals. Journal of Iron and Steel Industry 174, 25–28.

Peters, K. (1962) Mechanochemische Reaktionen. In: H. Rumpf and D. Behrens (Eds.) *Proceedings of the 1st European Symposium on Size Reduction*. Frankfurt a.M., Dechema, Verlag Chemie, Weinheim, pp. 78–98.

Pfund, A.H. (1930) Bismuth-black and it applications. Physical Review 35, 1434–1437.

Pierce, T. (1928) Mechanochemistry and the colloid mill including the practical applications of fine dispersion. Chemical Catalog Company, New York, 191p.

Pileni, M.P. (1993) Reverse micelles as microreactors. Journal of Physical Chemistry 97, 6961–6973.

Pileni, M.P. (1997) Nanosized particles made in colloidal assemblies. Langmuir 13, 3266–3276.

Poole, Ch.P., Jr. and Owens, F.J. (2003) *Introduction to Nanotechnology*. John Wiley & Sons, Hoboken, New Jersey, 388p.

Pourghahramani, P. and Forssberg, E. (2006a) Microstructure characterization of mechanically activated hematite using XRD line broadening. International Journal of Mineral Processing 79, 106–119.

Pourghahramani, P. and Forssberg, E. (2006b) Comparative study of microstructural characteristics and stored energy of mechanically activated hematite in different grinding environments. International Journal of Mineral Processing 79, 120–139.

Rouxel, J., Meerschaut, A. and Wiegers, G.A. (1995) Chalcogenide misfit layer compounds. Journal of Alloys and Compounds 229, 144–157.

Roy, R. (1994) Accelerating the kinetics of low-temperature inorganic synthesis. Journal of Solid State Chemistry 111, 11–17.

Rupp, J. and Birringer, R. (1987) Enhanced specific heat capacity (C_p) measurements (150–300 K) of nanometric-sized crystalline materials. Physical Review B36, 7888–7890.

Schrader, R. and Hoffmann, B. (1973) Änderung der Reaktionsfähigkeit von Festkörpern durch vorhergehende mechanische Bearbeitung. In: V.V. Boldyrev and K. Meyer (Eds.) *Festkörperchemie*. Deutscher Veralg für Grundstoffindustrie, Leipzig.

Senna, M. (2001) Recent development of materials design through a mechanochemical route. International Journal of Inorganic Materials 3, 509–514.

Serpe, L. (2006) Conventional chemotherapeutic drug nanoparticles for cancer treatment. In: Challa S.S.R. Kumar (Ed.) *Nanomaterials for Cancer Therapy*. Wiley-VCH, Weinheim 2006, pp. 1–39.

Shao, M., Li, Q., Xie, B., Wu, J. and Qian, Y. (2003) The synthesis of CdS/ZnO and CdS/Pb_3O_4 composite materials via microwave irradiation. Materials Chemistry and Physics 78, 288–291.

Sharma, D., Chelvi, T.P., Kaur, J., Chakravorty, K., De, T.K., Maitra, A. and Ralhan, R. (1996) Novel Taxol^R formulation: polyvinylpyrrolidone nanoparticles-encapsulated Taxol^R for drug delivery in cancer therapy. Oncological Research 8, 281–286.

Sharma, R.C. and Chang, Y.A. (1986) In: T.B. Massalsky, H. Okamoto, P.R. Subramanian and L. Kacprzak, (Eds.) *Binary Alloy Phase Diagrams*. ASM International, Materials Park, OH, p. 3280.

Sherif El-Eskandarany, M., El-Bahnasawy, H.N., Ahmed, H.A. and Eissa, N.A. (2001) Mechanical solid-state reduction of hematite with magnesium. Journal of Alloys and Compounds 314, 286–295.

Shi, F.G. (1994) A glass transition: unified treatment. Journal of Materials Research 9, 1908–1916.

Siegel, R.W. and Fougere, G.E. (1994) In: G.C. Hadjipahayis and R.W. Siegel (Eds.) *Nanophase Materials*. Kluwer, Dordrecht.
Siegel, R.W. and Fougere, G.E. (1995) Mechanical properties of nanophase materials. Nano-Structured Materials 6, 205–216.
Siegel, R.W. (1997) Mechanical properties of nanophase materials. Materials Science Forum 235–238, 851–860.
Simionesku, K. and Oprea, K. (1971) *Mechanochemistry of Macromolecular Compounds*. Mir, Moscow, 357p.
Smalley, R. (1999) Nanotechnology: The State of Nano-Science and Its Projects for the next Decade. US Congress Hearings.
Smékal, A. (1942) Ritzvorgang und molekulare Festigkeit. Naturwissenschaften, 30, 224–225.
Smith, E.F. (1972) *Chemistry in America, Chapters from the History of the Science in the United States*, Arno Press, New York, 301p.
Smolyakov, V.K., Lapshin, O.V. and Boldyrev, V.V. (2007) Macroscopic theory of mechanochemical synthesis in heterogeneous systems. International Journal of Self-Propagating High-Temperature Synthesis 16, 1–11.
Smolyakov, V.K., Lapshin, O.V. and Boldyrev, V.V. (2008) Mechanochemical synthesis of nanosize products in heterogeneous systems: Macroscopic kinetics. Journal of Self-Propagating High-Temperature Synthesis 17, 20–29.
Sorensen, C.M. (2001) Magnetism. In: K.J. Klabunde (Ed.) *Nanoscale Materials in Chemistry*, John Wiley & Sons, Hoboken, New Jersey, pp. 169–221.
Speliotis, D. (1999) Magnetic recording beyond the first 100 years. Journal of Magnetics and Magnetic Materials 193, 29–35.
Staudinger, H. and Dreher, E. (1936) Mitteilung über hochpolymere Verbindungen. Berichte des Deutschen Chemische Gesellschaft A 69, 1901–1099.
Steigerwald, M.L. and Brus, L.E. (1990) Semiconductor crystallites: a class of large molecules. Account on Chemical Research 23, 183–188.
Steinicke, U., Ebert, I., Geissler, H., Hennig, H.-P. and Kretzschmar, U. (1978) Reaktivität on mechanisch aktivierten Quarz. Kristall und Technik 597–603.
Steinicke, U. and Linke, E. (1982) Die Rolle von Anregungszuständen und strukturdefekten be mechanisch aktivierten Festkörperreaktionen. Zwitschrift für Chemie 22, 397–401.
Steiniger, S.C., Kreuter, J., Khalansky, A.S., Skidan, I.N., Bobruskin, A.I., Smirnova, Z.S., Severin, S.E., Uhl, R., Kock, M., Geiger, K.D. and Gelperina, S.E. (2004) Chemotherapy of glioblastoma in rats using doxorubicin-loaded nanoparticles. International Journal of Cancer 109, 759–767.
Stöber, W. and Arnold, M. (1961) Anomalien by der Ablosung von Kieselsaure von der oberflache feinkorniger Siliziumdioxydpulver. Kolloid Zeitschrift 174, 20–27.
Stoner, E.C. and Wohlfarth, E.P. (1948) A mechanism of magnetic hysteresis in heterogeneous alloys. Philosophical Transactions of the Royal Society London 240, 599–644.
Suryanarayana, C. (1994) Structure and properties of nanocrystalline materials. Bulletin of Materials Science 17, 307–346.
Suryanarayana, C. (1995) Nanocrystalline materials. International Materials Reviews 40, 41–64.
Suryanarayana, C. (2001) Mechanical alloying and milling. Progress in Materials Science 46, 1–184.
Suslick, K.S. (1990) Sonochemistry. Science 247, 1439–1445.
Takacs, L. (2000) Quicksilver from cinnabar: the first documented mechanochemical reaction? Journal of Metals, January, 12–13.
Takacs, L. (2002) Self-sustaining reactions induced by ball milling. Progress in Materials Science 47, 355–414.
Takacs, L. (2003) M. Carey Lea, the father of mechanochemistry. Bulletin of Historical Chemistry 28, 26–34.
Takacs, L. (2004) M. Carey Lea, the first mechanochemist. Journal of Materials Science 39, 4987–4993.

Takacs, L. (2007) The mechanochemical reduction of AgCl with metals, revisiting experiments of Michael Faraday. Journal of Thermal Analysis and Calorimetry 90, 90–94.

Takagi, M. (1954) Electron-diffraction study of liquid-solid transition of thin metal films. Journal of Physical Society Japan 9, 359–363.

Tammann, G. (1929) Der Einfluss der Kaltbearbeitung auf die chemischen Eigenschaften insbesondere von Metallen. Zeitschrift für Elektrochemie 35, 21–28.

Teo, B.K. and Sloane, N.J. (1985) Magic numbers in polygonal and polyhedral clusters. Inorganic Chemistry 24, 4545–4558.

Thiessen, P.A., Meyer, K. and Heinicke, G. (1967) *Grundlagen der Tribochemie*. Akademie Verlag, Berlin 1967.

Thiessen, P.A., Heinicke, G. and Schober, E. (1970) Zur tribochemischen Umsetzung von Gold und CO_2 mit Hilfe radioaktiver Markierung. Zeitschrift für Anorganische und Allgemeine Chemie 377, 20–28.

Thiessen, K.P. (1979) Energetische Randbedingungen tribochemischer Prozesse. I. Stufenschema der Energie-„Zuständ", Zeitschrift für Physikalische Chemie, Leipzig 260, 403–409.

Thiessen, K.P. and Sieber, K. (1979a) Energetische Randbedingungen tribochemischer Prozesse. II. „Tribochemische Effekte" Stationäre Zustände n-ter Ordnung. Zeitschrift für Physikalische Chemie, Leipzig 260, 410–416.

Thiessen, K.P. and Sieber, K. (1979b) Energetische Randbedingungen tribochemischer. Prozesse. III. Hypothetische Geschwindigkeits/Affinitätsbeziehung tribochemischer Prozesse. Zeitschrift für Physikalische Chemie, Leipzig 260, 417–422.

Tjong, S.C. and Chen, H. (2004) Nanocrystalline materials and coatings. Materials Science and Engineering R45, 1–88.

Tkáčová, K., Paholič, G., Šepelák, V. and Sekula, F. (1988) Simple model of mechanical activation of solids. In: K. Tkáčová et al. (Eds.) *Proceedings of the 5th International Symposium "Theoretical and Technological Aspects of Disintegration and Mechanical Activation"* Part I, Tatranská Lomnica, pp. 70–79.

Tompkins, F.C. (1963) Superficial Chemistry and Solid Imperfections. Inaugural lectures, Imperial College of Science and Technology, London.

Tsuzuki, T., Ding, J. and McCormick, P.G. (1997) Mechanochemical synthesis of ultrafine zinc sulfide particles. Physica B, 378–387.

Tsuzuki, T. and McCormick, P.G. (1997) Synthesis of CdS quantum dots by mechanochemical reaction. Applied Physics A65, 607–609.

Tsuzuki, T. and McCormick, P.G. (1999) Mechanochemical synthesis of metal sulphide nanoparticles. Nanostructured Materials 12, 75–78.

Tsuzuki, T. and McCormick, P.G. (2004) Mechanochemical synthesis of nanoparticles. Journal of Materials Science 39, 5143–5146.

Valiev, R.Z., Mulyukov, R.R., Mulyukov, Kh.Ya., Novikov, V.I. and Trusov, L.I. (1989) Curie temperature and saturation (magnetic) of nickel with a submicrocrystalline structure. Pisma v Zurnal Techniceskoj Fiziki 15, 78–81.

Velumani, S. and Ascencio, J.A. (2004) Formation of ZnS nanorods by simple evaporation. Applied Physics A79, 153–156.

Wagman, D.D., Evans, W.H., Parker, V.B., Schumm, R.H., Halow, I., Bailey, S.M., Churney, K.L. and Nuttall, R.L. (1982) *The NBS Tables of Chemical Thermodynamic Properties*. Journal of Physical and Chemical Reference Data 11, Supplement No. 2.

Wanetig, P. (1921) Zur Frage der Zähflüssigkeitsänderung von Viskoselösungen. Textilforschung 3, 202–206.

Wanetig, P. (1922) Textilforschung 4, 66–69.

Wanetig, P. (1927) Zur Frage der Zähflüssigkeitsänderung von Viskoselösungen. Kolloid Zeitschrift 41, 152–158.

Wang, L.P. and Hong, G.Y. (2000) A new preparation of zinc sulfide nanoparticles by solid-state method at low temperature. Materials Research Bulletin 35, 695–701.

Wang, W., Liu, Z., Zheng, Ch., Xu, C., Liu, Y. and Wang, G. (2003) Synthesis of CdS nanoparticles by a novel and simple one-step, solid state reaction in the presence of a nonionic surfactant. Materials Letters 57, 2755–2760.

Warren, B.E. and Averbach, B.L. (1950) The effect of cold-work distortion on X-ray patterns. Journal of Applied Physics 21, 595–599.

Weichert, R. and Schönert, K. (1974) On the temperature rise at the tip of a fast running crack. Journal of Mechanics and Physics of Solids 22, 127–133.

Weichert, R. (1976) Untersuchungen zur Temperatur an der Bruchspitze, Dr.-Ing. Dissertation, Karlsruhe.

Weller, H. (1993) Colloidal semiconductor Q-particles: chemistry in the transition region between solid state and molecules. Angewandte Chemie International Edition 32, 41–53.

Wenzel, C.F. (1777) *Lehre von der chemischen Verwandtschaft.* Dresden.

Whitcomb, D. (2006) Mathew Carey Lea: chemists, photographic scientist. Chemical Herritage News Magazine 24, 1–2.

Wong, E. quoted by R.S. Boyd (1999) Knight Rider Newspapers, Kansas City Star, Monday, November 8.

Yan, B., Chen, D. and Jiao, X. (2004) Synthesis, characterization and fluorescence properties of CdS /P(N-iPAAm) nanocomposites. Materials Research Bulletin 39, 1655–1662.

Yonezawa, T. (2004) Well-dispersed bimetallic nanoparticles. In: Y. Waseda and A. Muramatsu (Eds.) *Morphology Control of Materials and Nanoparticles*. Springer, Heidelberg, pp. 85–112.

Xiao, T.D., Strutt, P.R., Benaissa, M., Chen, H. and Kear, B.H. (1998) Synthesis of high active-site density nanofibrous MnO_2-base materials with enhanced permeabilities. Nanostructured Materials 10, 1051–1061.

Xu, Ch., Zhang, Z. and Ye, Q. (2004) A novel facile method to metal sulfide (metal=Cd, Ag, Hg) nanocrystallites. Materials Letters 58, 1671–1676.

Zeisser-Labouébe, M., Vargas, A. and Delie, F. (2006) Nanoparticles for photodynamic therapy. In: Challa S.S.R. Kumar (Ed.) *Nanomaterials for Cancer Therapy.* Wiley-VCH, Weinheim 2006, pp. 40–86.

Zelikman, A.N., Voldman, G.M. and Belyayevskaya, L.V. (1975) *Theory of Hydrometallurgical Processes.* Metallurgija, Moscow (in Russian).

Zhong, X., Feng, Y.Y., Knoll, W. and Han, M.Y. (2003) Alloyed $Zn_xCd_{1-x}S$ nanocrystals with highly narrow luminiscence spectral width. Journal of American Chemical Society 125, 13559–13563.

Chapter 2
High-Energy Milling

2.1 Introduction

Particle size reduction, or comminution is an important step in many technological operations. The process itself is defined as the mechanical breakdown of solids into smaller particles without changing their state of aggregation [Bernotat and Schönert 1998]. It may be used to create particles of a certain size and shape (including nanosize), to increase the surface area and induce defects in solids which is needed for subsequent operations such as chemical reactions, sorption, etc. Milling not only increases the surface area of solids. It is likely to increase the proportion of regions of high activity in the surface. This increase is really dramatic: if a cubic solid is broken down into smaller cubes, each of side 1/n of the side of the original cube, the specific surface area is increased n-times, but the broken down mass contains n^2-times the original length of edge and n^3-times the number of corners [Welch 1953].

The size reduction of solids is an energy intensive and highly inefficient process: 5% of all electricity generated is used in size reduction [Rhodes 1998].

Conventionally, the term fine milling is used for size range below 100 μm and the ultrafine (or very fine) milling for particles size less than 10 μm. In mechanochemistry, the term high-energy milling is frequently used in order to stress the character of applied milling equipments (mills) [Boldyrev 1983; Baláž 2000].

The chapter is intended as an introduction to the topic of high-energy milling and milling equipments used for preparation of micro- and nanosized solids.

2.2 Basic Phenomena

Consider a crystal of sodium chloride NaCl as a simple and convenient model of a brittle material. Such a crystal is composed of a lattice of positively charged sodium ions and negatively charged chloride ions arranged such that each ion is surrounded by six ions of the opposite sign. Between the oppositely charged ions there is an attractive force whose magnitude is inversely proportional to the square

© Springer-Verlag Berlin Heidelberg 2008

of the separation of the ions. There is also a repulsive force between the negatively charged electron clouds of these ions which becomes important at very small interatomic distances. Therefore two oppositely charged ions have an equilibrium separation such that the attractive and repulsive forces between them are equal and opposite. Figure 2.1 shows how the sum of the attractive and repulsive forces varies with changing separation of the ions. It can be appreciated that if the separation of the ions is increased or decreased by a small amount from the equilibrium separation, there will be a resultant net force restoring the ions to the equilibrium position. The ions in the sodium chloride crystal lattice are held in equilibrium positions governed by the balance between attractive and repulsive forces. Over a small range of interatomic distances the relationship between applied tensile or compressive force and resulting change in ion separation is linear. That is, in this region (AB in Fig. 2.1) Hooke's law applies: strain is directly proportional to applied stress. The Young's modulus of the material (stress/strain) describes this proportionality. In this Hooke's law range the deformation of the crystal is elastic; i.e. the original shape of the crystal is recovered upon removal of the stress [Rhodes 1998].

In order to break the crystal it is necessary to separate adjacent layers of ions in the crystal and this involves increasing the separation of the adjacent ions beyond the region where Hooke's law applies – i.e. beyond point B in Fig. 2.1 into the plastic deformation range. The applied stress required to induce this plastic behaviour is known as the elastic limit or yield stress, and is sometimes defined as the material's strength [Rhodes 1998].

A body under stress stores energy-strain energy. This strain energy is not uniformly distributed throughout the solid but is concentrated around holes, corners and cracks.

The theory of crack formation and propagation has been well demonstrated in literature [Griffith 1920; Smekal 1922, 1936; Rumpf 1962, 1966; Schönert and Weichert 1969; Schönert and Steier 1971; Schönert 1974]. It was postulated that for

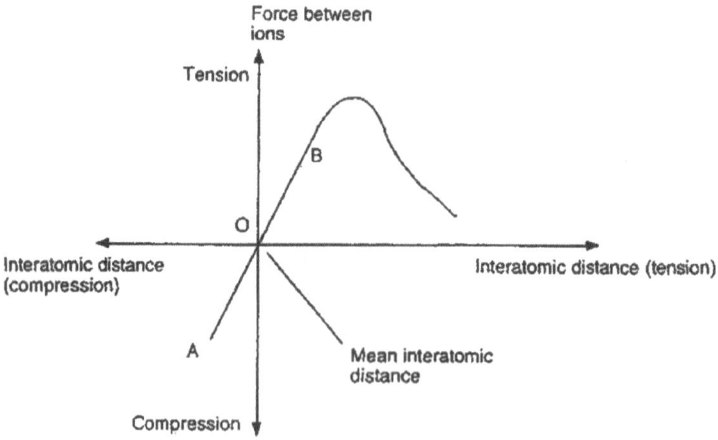

Fig. 2.1 Forces versus distance on an atomic scale [Rhodes 1998]

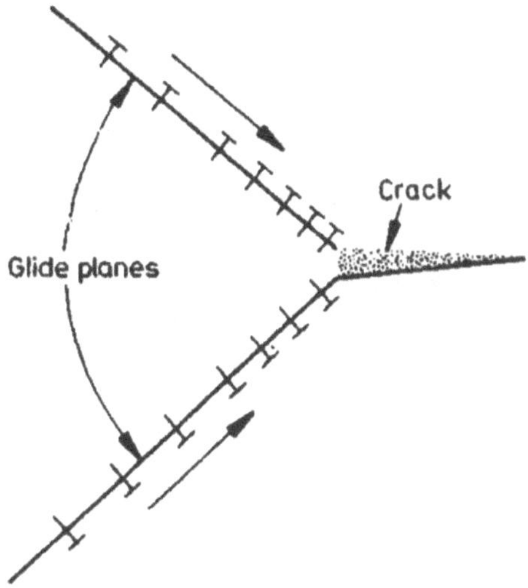

Fig. 2.2 Formation of a crack along the intersection of glide planes [Cottrell 1958]

fracture to occur the propagation of cracks has to be promoted, and the condition that should lead to a fracture is that a certain minimum length of the microcracks (defect sites) should be formed [Griffith 1920].

This concept was later modified in works of Cottrell who stated that such cracks can also be created by the coalescence of dislocations in solids (Fig. 2.2). According to this model microcracks can develop and dislocations can coalesce at the interfaces of crystals, or at the intersection of two dislocations, or in the case of mosaic structure, at the boundaries of the blocks [Cottrell 1958].

Fracture is a problem of energetics. The propagation of the initially-formed cracks requires that the energy of elastic deformation should be at least as high as the interfacial energy of the system [Griffith 1920]. It was pointed out that fracture energy was much higher than surface free energy [Rumpf 1961, 1966; Schönert and Weichert 1969; Schönert and Steier 1971; Schönert 1974]. Thus, the specific surface energy of solids is about 0.05–0.5 $\mathrm{Jm^{-2}}$, whereas the specific fracture energy for brittle solids is 5 $\mathrm{Jm^{-2}}$, for plastic solids it is 10 $\mathrm{Jm^{-2}}$, and for metals is approximately 500 $\mathrm{Jm^{-2}}$.

2.3 Energy Requirement

In milling practice the quantitative measures of energetics attracted attention for a long time. The literature on the problem is quite extensive and the theories of Rittinger, Kick-Kirpičev and Bond deserve to be mentioned.

Rittiger proposed that the energy required for particle size reduction was directly proportional to the area of new surface created

$$E = C_R \left(\frac{1}{x_2} - \frac{1}{x_1} \right) \tag{2.1}$$

where C_R is constant and x_1 is the starting and x_2 the final particle size [Rittinger 1867].

However, this law ignores the energy absorbed by elastic deformation which is several times greater than that required for creation of new surfaces.

On the basis of stress analysis theory for plastic deformation Kirpičev and independently Kick proposed the energy required for particle size reduction was directly proportional to the ratio of the volume of the feed particle to the product particle [Kirpičev 1874; Kick 1883]

$$E = C_K \ln \left(\frac{x_1}{x_2} \right) \tag{2.2}$$

The main drawback of this theory is that the same energy is required to reduce 10 μm particle to 1 μm as is required to reduce 1 m boulders to 10 cm blocks [Rhodes 1998]. This is a simplification which cannot be accepted especially in the fine and ultrafine milling.

Bond suggested a more useful formula where the total energy that is specific to any size is inversely proportional to the square root of the particle size [Bond 1952]

$$E = C_B \left(\frac{1}{\sqrt{x_2}} - \frac{1}{\sqrt{x_1}} \right) \tag{2.3}$$

In milling practice the formula (2.3) is applied in the modified form which enables determination of the Bond work index W_i. W_i is defined as the energy required to reduce the size of unit mass of material from infinity to 100 μm in size.

It is common fault of all three theories that they do not take account of particle size distribution of the feed and the product, nor do they take account of particle interactions and the energy consumed by plastic deformation [Juhász and Opoczky 1990].

Attempts have been made [Holmes 1957; Charles 1957; Hukki 1961] to find the general formula for which proposals of Rittinger, Kick-Kirpičev and Bond are special cases. These attempts concluded that these three proposals can be considered as being the integrals of the same differential equation

$$\frac{dE}{dx} = -C \frac{1}{x^N} \tag{2.4}$$

with

$N = 2$, $C = C_R$ for Rittinger
$N = 1$, $C = C_K$ for Kick-Kirpičev
$N = 1.5$, $C = C_B$ for Bond

2.4 Milling Equipments

Besides of energy considerations, especially for the energy consumed by plastic deformation, the form of mechanical activation in high-energy milling is important. The main stress types applied are compression, shear (attrition), impact (stroke) and impact (collision), Fig. 2.3.

The multi-stage character of mechanical activation requires the application of high-energy mills with different working regimes.

The principles of most frequently mills applied in high-energy milling are shown in Fig. 2.4.

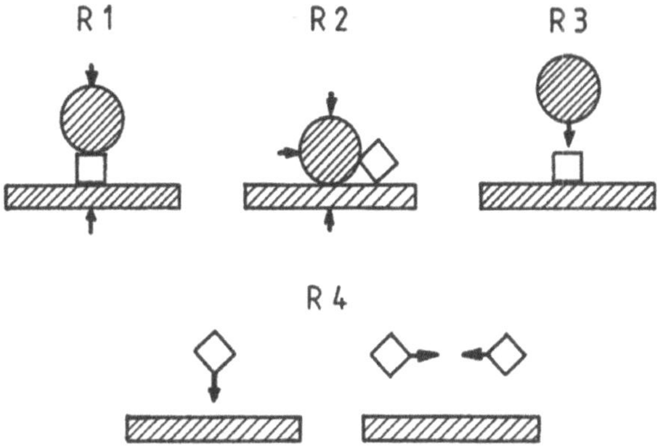

Fig. 2.3 Main stress types in mills, R1 – compression, R2 – shear (attrition), R3 – impact (stroke), R4 – impact (collision), *circle* – mass of milling media, *square* – mass of material charge, *rectangle* – mass of mill wall [Heinicke 1984]

When using mills with application of balls as milling media, different motion of the balls can be observed (Fig. 2.5). The motion can be described as cascading (A), falling or cataracting (B) or centrifugal (C).

2.4.1 Vibration Mills

A vibration mill is an reservoir or tank-like type of mill containing material and milling media, either wet or dry. The milling chamber is set in motion along a circular or elliptic trajectory by the rotation of out-of-balance weights. The three dimensional motion involves four factors: the speed of the vibration, the horizontal amplitude, the vertical amplitude and the phase angle [Klimpel 1997]. The acceleration of the vibration motion is specified by its frequency and amplitude.

The performance of a laboratory mill (Fig. 2.6) is characterized by high frequencies (usually up to 25 rev s^{-1}) but low amplitudes (2–4 mm).

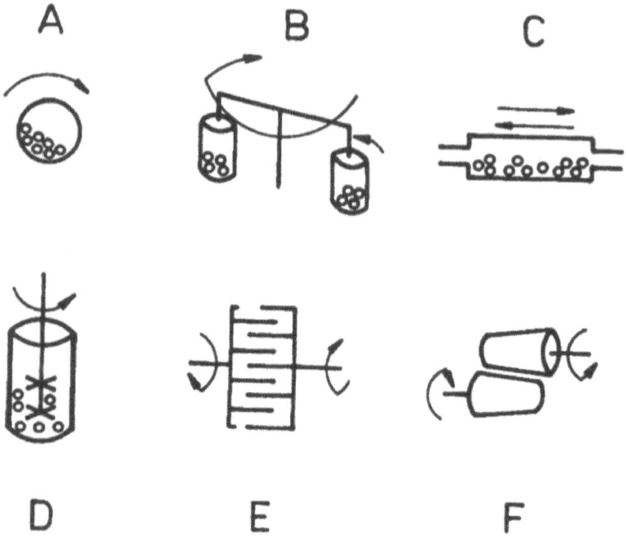

Fig. 2.4 Types of mills for high-energy milling (the stress type is given in parenthesis): (**A**) ball mill (R1–R4), (**B**) planetary mill (R1–R4), (**C**) vibration mill (R1–R4), (**D**) attritor – stirring ball mill (R1–R4), (**E**) pin mill (R4), (**F**) rolling mill (R1–R2) [Boldyrev 1986]

Industrial mills operate at frequencies 16–19 rev s^{-1} and amplitudes below 6 mm. The acceleration scarcely exceeds 10-times the gravitational acceleration (10 g). There is a tendency in the design of vibration mills in an increase the amplitude to 7–10 mm at frequencies sufficient to transport the charge. This will allow acceleration of the vibration motion to the values 12–20 g [Tkáčová 1989].

Milling media are usually balls, cylpebs or rods with degree of filling 70–80%. The application of balls predominates, rods are preferred for coarse milling, cylpebs are preferred for pulverizing [Schubert 1989].

The theory of vibration mills was elaborated [Bachmann 1940]. According to Rose high frequency vibration is to be preferred and it is essential that the

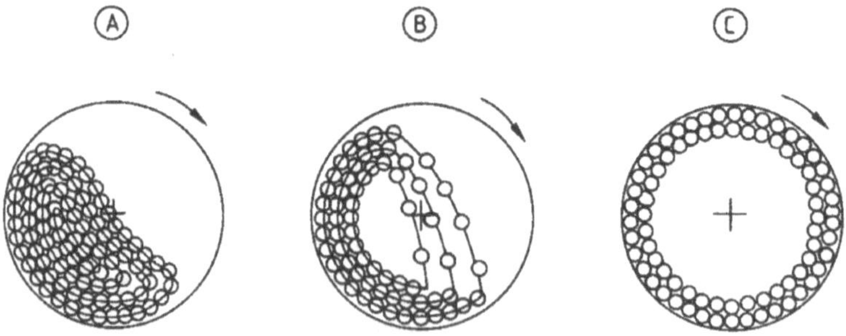

Fig. 2.5 Types of motion in a ball mill: (**A**) cascading, (**B**) falling or cataracting, (**C**) centrifugal [Bernotat and Schönert 1998]

Fig. 2.6 Laboratory vibration mill

dimensions of the milling bodies should be adjusted to the strength and particle size of the feed [Rose and Sullivan 1961; Rose 1962]. In the opinion of Bernhardt and Heegn the amplitude of the vibration mill and the quantity of the feed is prime importance in mechanical and mechanochemical activation [Bernhardt and Heegn 1976a, b; Heegn 1989].

Vibration mills are frequently applied for the preparation of very fine products and in case of induced mechanical activation the nanosized particles have been prepared [Baláž 1981]. Mechanical activation of solid, if performed in this type of mill, usually requires long milling times and is often carried out as a batch process. However, the residence time can be reduced to less than 30 minutes and mechanical activation is in principle possible also in continuously operation vibration mills [Husemann et al. 1976; Tkáčová et al. 1993].

An important disadvantage of vibration mills is their low output. Because of the high-input of energy and the low throughput of material, the high temperatures are developing in the mill. The high temperature is caused by the fact that the mill is almost completely filled and by the complex trajectory of the grinding bodies within the mill [Schubert 1989].

The new concept of vibration mills working in "eccentric" mode has been introduced by Gock et al. [Beenken et al. 1996; Gock and Kurrer 1996, 1998, 1999; Kurrer and Gock 1997] and applied in milling practice (Fig. 2.8).

Unlike conventional vibratory mills with circular vibrations, this machine performs elliptical, circular and linear vibrations (Fig. 2.7). As a result of dismissing the homogeneous circular vibrations, the motion process drastically changes: the amplitude of the vibrations of the individual milling media increases, the rotation speed of the grinding media filling increases and the direction of motion is irreversibly set. Apart from the surface area enlargement, eccentric vibratory milling with 30 mm diameter balls have led to a mechanical activation of the material.

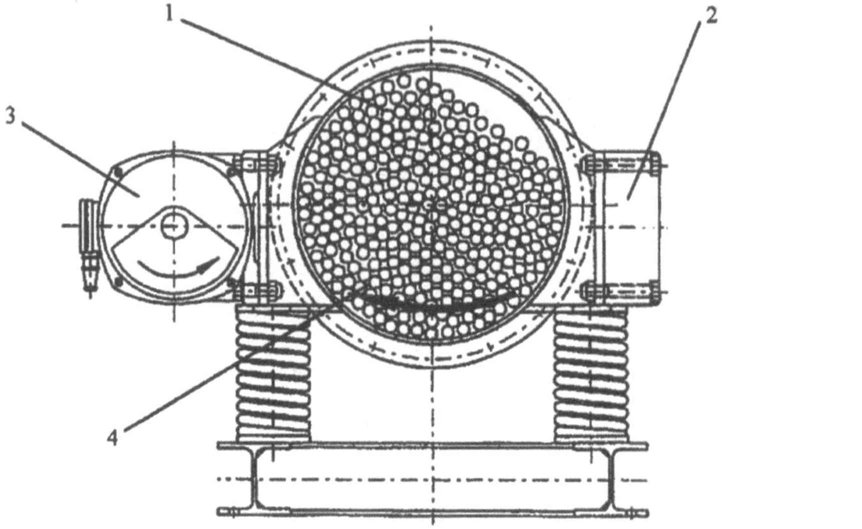

Fig. 2.7 Diagrammatic view of the eccentric vibration mill [Gock and Kurrer 1999]

Fig. 2.8 Industrial eccentric vibration mill ESM 656 (Siebtechnik, Germany) [http://www.siebtechnik.com]

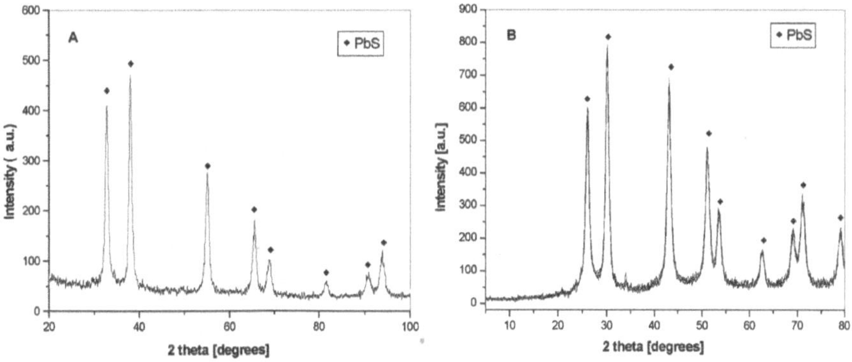

Fig. 2.9 XRD patterns of PbS nanoparticles mechanochemically synthesized in a laboratory (**A**) and in an industrial mill (**B**) [Baláž et al. 2004]

Recently, various mechanochemical synthesis of the new nanosized materials have been performed with this type of mill [Baláž et al. 2004, 2005; Godočíková et al. 2006]. XRD patterns of synthesized PbS semiconductor particles are given in Fig. 2.9. The values of 18 nm and 13 nm were estimated for PbS nanocrystallites prepared in laboratory mill and industrial mill, respectively.

The grain size of the ball-milled solids are found to decrease with milling time down to a constant value. An advantage of the use of high-energy milling is the possibility to introduce large strains and produce nanocrystalline structures. The product of milling is powder, typically a few micrometers in diameter, with the internal nanocrystalline structure [Koch 1993].

2.4.2 Planetary Mills

The planetary ball mill (Fig. 2.10) owes its name to the planet-like movement of its vial (s) (Fig. 2.4). Since the vials and the supporting disc rotate in opposite directions, the centrifugal forces alternatively act in like and opposite directions. This causes the milling balls to run down the inside wall of the vial – the friction effect, followed by the material being milled and milling balls lifting by of and travelling freely through the inner chamber of the vial and colliding against the opposite inside wall [Suryanarayana 2001].

Planetary mills exploit the principle of centrifugal acceleration instead of gravitational acceleration. The enhancement of the forces acting on the balls in relation to the conventional ball mill is achieved by the combined action of two centrifugal fields [Golosov 1971; Molčanov et al. 1988].

The charge inside vials performs two relative motions: a rotary motion around the mill axis and a planetary motion around the vial axis (Fig. 2.11).

The application of laboratory-scale planetary mills for mechanical activation has been studied intensively by Russian scientists [Golosov 1971; Molčanov et al. 1988;

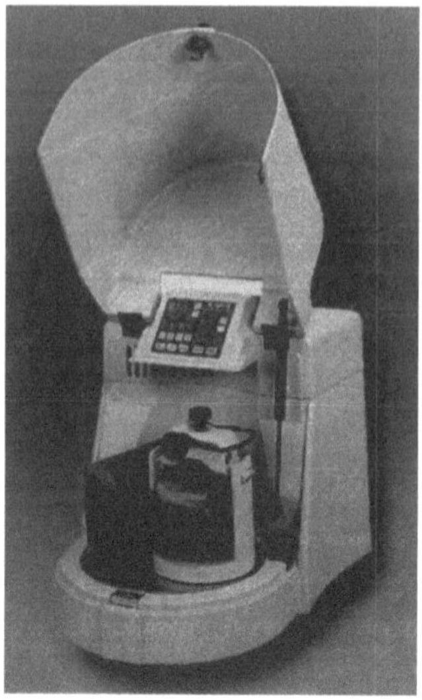

Fig. 2.10 Laboratory planetary mill Pulverisette 6 (Fritsch, Germany) [http://www.fritsch.com]

Molčanov and Jusupov 1981; Avvakumov 1986]. Their general conclusion was that, by observing certain conditions, these mills produce high mechanical activation after a relatively short milling time. In principle, it is possible to obtain the gravitational accelerations in values 50–100 g. The energy density in these mills is 100–1000 higher than the energy density used earlier in conventional milling equipments [Fokina et al. 2004].

However, it was noted that manufacturing of high throughput rate encounters design difficulties. During the past decades the problem of continuous feeding of

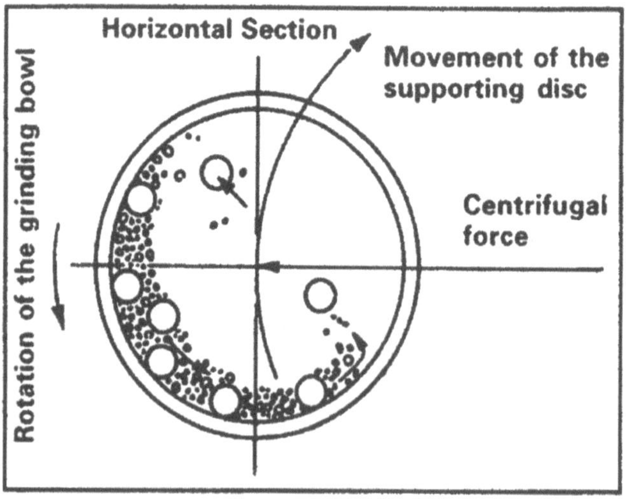

Fig. 2.11 Movements of working parts and balls in a planetary mill

a material into a planetary mill has been successfully solved and novel planetary mills of the industrial scale operating in a continuous mode are now manufactured [Kochnev 1992; Kochnev and Simakin 1994]. In this mode the initial material is continuously fed into the mill, with coarser powder returned for another cycle of milling and the final fine or nanoscale powder product continuously provided as a result of the milling process. The productivity per unit volume of the working chamber for these mills is significantly higher (at least ten times) than that of conventional ball mills. Industrial planetary mills of continuous action, characterized up to 20 g acceleration and by productivity up to 3–5 tons of powder (of minus 10 micrometers fraction) per hour, are now-commercially available [http://www.ttd.spb.ru].

2.4.3 Mixer Mills

Special mills have been developed in the USA under the trade name Spex mills (Fig. 2.12). The common variety of the mill has one vial containing the sample and milling balls, secured in the clump and swung energetically back and forth several thousand times a minute. The back- and- fourth shaking motion is combined with lateral movements of the end of the vial, so that the vial appears to be describing a figure eight or infinity sign as it moves. The vial is vibrated with an amplitude 50 mm and a frequency of 20 Hz. These mills can be considered as high-energy variety [Koch 1993].

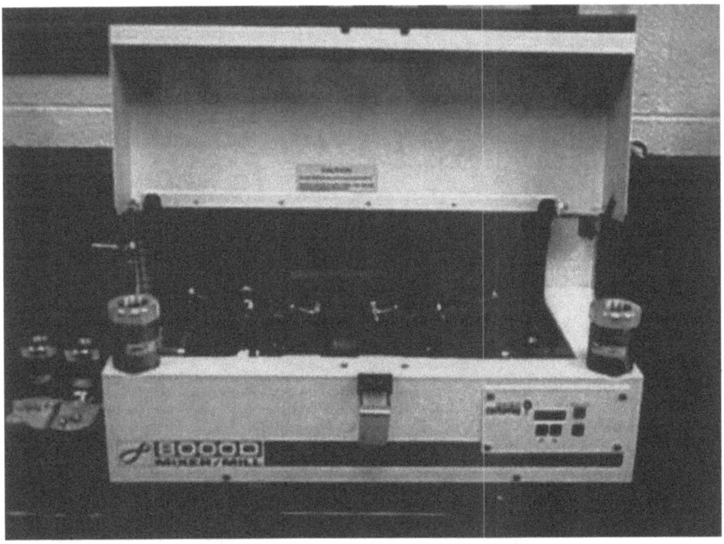

Fig. 2.12 Laboratory mixer mill Spex 8000 (CertiPrep, USA) [http://www.utk.edu]

2.4.4 Attritors (Stirred Ball Mills)

The attritor was invented in 1922 by Szegvari in the USA for a quick dispersion of fine sulphur particles needed to complete the vulcanization of rubber. After patenting the principle of attritor in the fifties in the USA, in 1956 the license for their manufacturing was transferred to Germany. This type of mill originally used for applications in chemical and pharmaceutical industry was later applied for powder and extractive metallurgy. An attritor was the first high-energy mill used for preparation of materials by mechanical alloying [Benjamin 1970].

Attritors (attrition mills, stirred ball mills) consist of a cylindrical grinding chamber with a drive shaft having multiple impellers sticking out from the rotating shafts (Fig. 2.13). There is a wide variety of impeller types including flat discs, discs with various geometric openings, and concentric rings. This rotating shaft puts the fine media, particles and, possibly a liquid into a stirring motion. The impellers energize the balls charge, causing powder size reduction because of impact between balls, between balls and container wall, and between balls, agitator shaft, and impellers [Baláž 2000]. Common media materials are glass, sand, alumina, zirconia, silicon carbide, steel, rubber, polyurethane and various plastics. Typical media sizes can range from 0.5 to 4 mm in diameter. The density of the media used is important because the media need to move through the mix of particles, media and fluids. Cover seals can be provided for processing under an atmosphere [Klimpel 1997].

Attritors use the comminution intensity between the contact surfaces of moving balls, similar to the operation of conventional ball mills, but without the disadvantages of the latter [Klimpel 1997; Püpke 1971; Netzsch 1974]. The most important

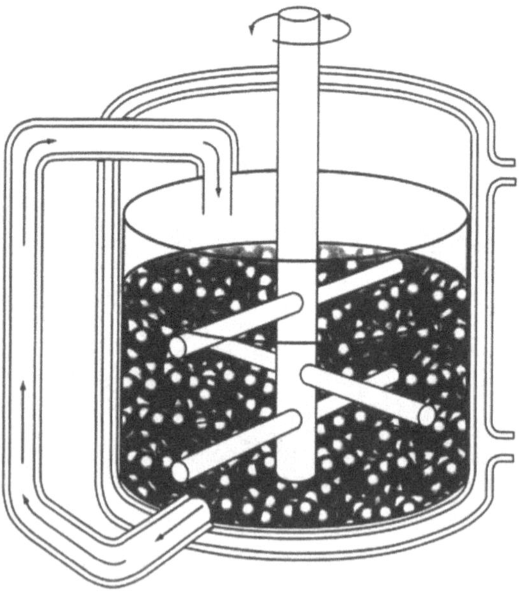

Fig. 2.13 Diagrammatic view of the attritor [Klimpel 1997]

Fig. 2.14 Laboratory attritor PC 2043 of batch mode of operation (Union Process, USA) [http://www.unionprocess.com]

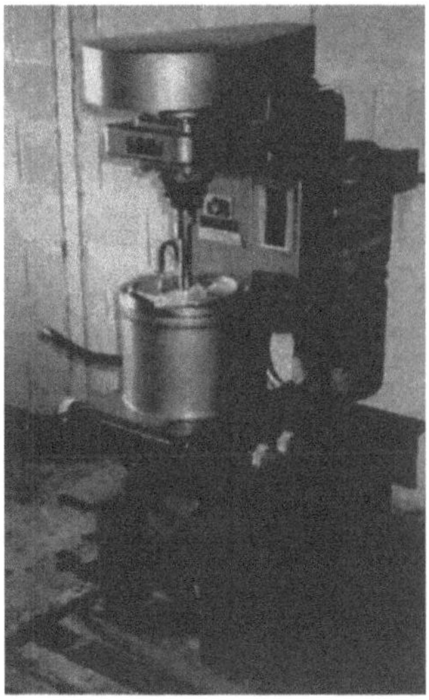

concept in the attritor is that the power input is used directly to agitating media to achieve milling. It is not used for rotating or vibrating a heavy grinding chamber in addition to the media.

Attritors can be operated wet or dry, in batch, circulation a continuous mode. Batch attritors (Fig. 2.14) are used to process hard-to-mill materials, such as tungsten carbide, silicon carbide and various metals. The circulation attritor is a combination of an attritor and a holding tank.

The high circulation rate is achieved in this system. One advantage of the circulation is that large quantities of material can be handled. Continuous attritors are best suited for the continuous production of large quantities of material (Fig. 2.15).

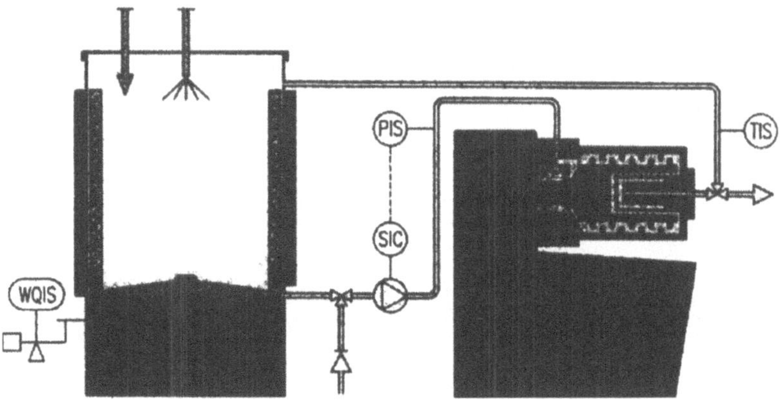

Fig. 2.15 Industrial attritor LMZ 25 of continuous mode of operation (Netzsch, Germany) [http://www.netzsch.com]

It was demonstrated that this type of mill may, in some instances, offer certain advantages in comparison to other mills, e.g. vibration mills [Heegn et al. 1974, 1980]. If strontium ferrate is activated in a vibration mill then extensive amorphisation takes place even after brief milling. This result was detrimental to the later use of this material because of deterioration of it magnetic properties. In contrast, when milling was carried out in an attritor, the specific surface increased monotonously while crystallinity was either preserved or it decreased much less than on milling in a vibration mill.

Baláž documented the differences in vibratory and attrition milling on changes of chalcopyrite milling (Fig. 2.16). From this Figure it is evident, that the products of milling in an attritor and vibration mill differ in specific structural disordering. According to the published data, these differences are due to the differences in milling environment and ball dimensions [Boldyrev 1983]. It is known that milling in aqueous environment and/or the use of small balls is more favourable for new surface formation whereas dry milling and/or the use of larger balls favour amorphization.

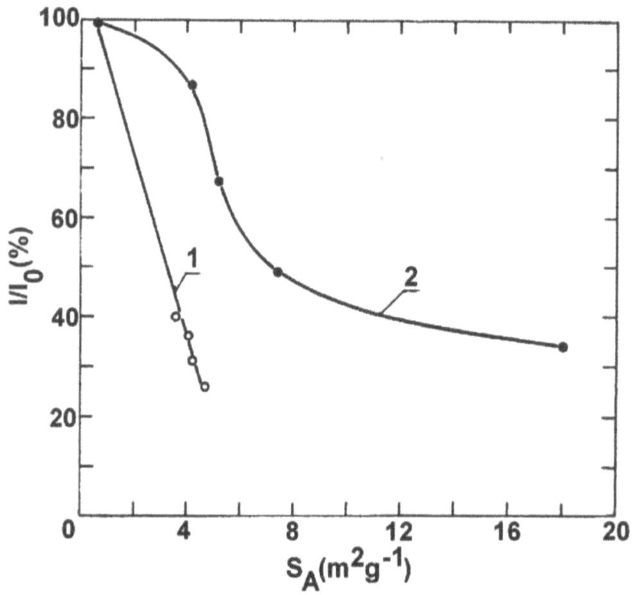

Fig. 2.16 The relative intensity of the selected diffraction line, I/I_0 vs. specific surface, S_A of chalcopyrite $CuFeS_2$ mechanically activated in a vibration mill (1) and in an attritor (2) [Baláž 2000]

2.4.5 New Mill Design

The brief review presented obviously could not cover all features of high-energy mills. Special new designs of mills have been developed in recent years for specialised purposes. The selected examples are given bellow and this list is, of course, not complete.

Nishin Giken (Tokyo, Japan) has developed the mill with the ability to control the temperature of milling from very low temperatures by spraying liquid nitrogen up to a high temperature of 300°C by electrical heating [http://www.nishineng.com]. In Australia concepts of magnetic mechanical milling and electric-discharge-assisted mechanical milling have been developed [Calka and Wexler 2002]. This concept of milling enhancement can result in both faster reactions and new synthesis and processing routes. In Italy a new high-energy mill for the synthesis of nanophase materials in large quantities has been designed and developed as described in [Miani and Maurigh 2004]. ZOZ company designed several new concepts of attrition mill scalable for production of nanosized materials [http://www.zoz.com].

2.4.6 Process Variables

There are several variables which influence the milling process (see e.g. Fig. 2.17).

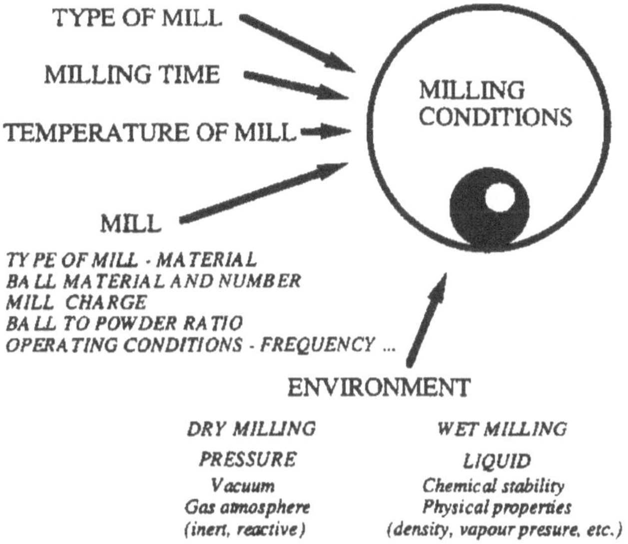

Fig. 2.17 An overview of the main factors to be considered in the milling process [Campbell and Kaczmarek 1996]

2.4.6.1 Type of the Mill

Usually specially designed mills are applied for special purposes. Sometimes the stress mechanism is decisive for particle size reduction in the mill, sometimes it is the size of the feed and the product size distribution which are most important [Rhodes 1998].

2.4.6.2 Material of Milling Media

The material used for the milling media (milling chamber, vial, balls) is important due to impact of the milling balls on the inner walls of the milling chamber. Some specific materials are used for specialized purposes [Suryanarayana 2001]. Commonly materials include various steels (stainless, tool, hardened chromium, tempered, bearing steel and tungsten carbide lined), tungsten carbide, zirconia, agate, alumina and silicon nitride. The specific gravity plays an important role. In general, high-density and larger balls give better results because of high impact forces on the powders. The balls should be more dense than the material to be milled.

Even though most of the investigators generally use only one size of the milling medium, there have been instances where different size balls have been used in the same investigation. It has been predicted that the highest collision energy can be obtained if balls with different diameters are used [Suryanarayana 2001].

Use of milling balls of the same size has been shown to produce tracks. Consequently, the balls roll along a well-defined trajectory instead of hitting the end surfaces randomly. Therefore, it is necessary to use several balls, generally a combination of smaller and larger balls to "randomize" the motion of the balls [Rhodes 1998; Suryanarayana 2001; Takacs and Pardavi-Horvath 1994].

The properties of materials used as milling media are given in Table 2.1.

Table 2.1 Milling media in high-energy milling [http://www.fritsch.com]

Material	Main composition	Density (g cm^{-3})	Abrasion resistance
Agate	SiO_2	2.65	Good
Corundum	Al_2O_3	> 3.8	Fairly good
Zirconium oxide	ZrO_2	5.7	Very good
Stainless steel	Fe, Cr, Ni	7.8	Fairly good
Tempered steel	Fe, Cr	7.9	Good
Tungsten carbide	WC, Co	14.7–14.9	Very good

2.4.6.3 Ball-to-Powder Ratio

The ball to powder ratio (BPR) or charge ratio is another milling parameter and represents the weight ratio of the milling balls to the powder charge. A wide range of BPR have been used in a variety of investigations from values as low 1:1 to as high 220:1 [Suryanarayana 2001].

2.4.6.4 Extent of Filling of the Milling Chamber

It is necessary that there is enough space for the balls and the powder particles to move around freely in the milling chamber. The filling depends on type of mill. Generally about 50% of the space is left empty [Suryanarayana 2001] but in attritors this value is 20–30%.

2.4.6.5 Milling Atmosphere

The milling atmosphere can contaminate the powder and as a results milling is frequently carried out in evacuated, argon, or helium charged milling chambers [Suryanarayana 2001]. Other atmospheres can be used if particular effects are desired. Nitrogen and ammonia have been used to produce nitrides and hydrogen atmospheres to produce hydrides [Calka 1991; Calka and Wexler 2002; Calka and Radlinski 1991; Ogino et al. 1993; Chen et al. 1990, 1995; Schulz et al. 1995; Murty and Ranganathan 1998]. The intention application of hydrogen in milling chamber was a real breakthrough in hydrogen storage technology for synthesis of nanocrystalline hydrides. The presence of air in the milling chamber can result in the formation of oxides, oxisulphates or sulphates [Baláž 2000]. Contamination can be avoided by milling the powders with a milling media made up of the same material as that of the powders being milled.

2.4.6.6 Milling Speed

The milling speed can have an important influence but this varies with the type of mill. Above a certain critical speed, the balls will be pinned to the walls of the milling chamber and not exert any impact force on the powder. Below this critical speed, however, the higher the milling speed, the higher the milling intensity will be. For the higher speeds the temperature of the milling chamber may reach a high value [Suryanarayana 2001].

2.4.6.7 Milling Time

According to many researchers the milling time is the most important milling parameter. However, the level of contamination will increase with milling time and some undesirable phases may form if a powder is milled too long. Substantial reduction in particle size usually occurs with milling time and typically takes the form of exponential decay.

2.5 Accompanying Phenomena

2.5.1 Temperature Effects

Since diffusion processes are involved in the formation of a nanostructure, it is expected that the temperature of milling will have a significant effect. Two kinds of temperature effects during milling are usually taken into account: local temperature pulses due to ball collisions and the overall temperature in a vial. During the milling process part of the milled powder is attached to the vial walls and, other part is attached to the balls. The remaining part is in a free state moving in the volume of the milling chamber. All these parts have different temperatures [Suryanarayana 2001].

The temperature of the powder influences the diffusivity and defect concentration in the powder thus influencing the phase transformations induced by milling. The results of measurement and calculation of the milling temperatures show that there is only a moderate temperature rise during milling and that local melting can be ruled out [Murty and Ranganathan 1998]. Bi powder was milled and the temperature of miling was calculated. The value was below its melting point 271°C [Koch 1997].

The temperature increase of the milling balls in two laboratory mills was studied recently [Takacs and McHenry 2006]. The ball temperature remains below 100°C in a SPEX mixer mill and it is cooler when flat-ended rather than round-ended vial is used. Temperatures over 200°C are typical in a planetary mill operating at similar milling intensities. The results presented in Fig. 2.18 were obtained using a Fritsch P-5 mill with two vial sizes and rotational speeds. The final temperatures are indeed much higher than the temperatures obtained with the SPEX 800 mill. The temperature could be increased further via increasing the speed of the mill. In fact, the highest temperature obtained with the smaller bowl was 327°C at 318 rpm. A decrease of the temperature is expected at very high speeds, as the balls stay attached to the container wall for too long, reducing both heating and efficiency of milling [Abdellaoui and Gaffet 1996]. It is suggested that the higher ball temperatures result from more oblique collisions and friction, while the lower temperature but higher intensity of the mixer mill with flat-ended vial is due to the larger portion of frontal impacts.

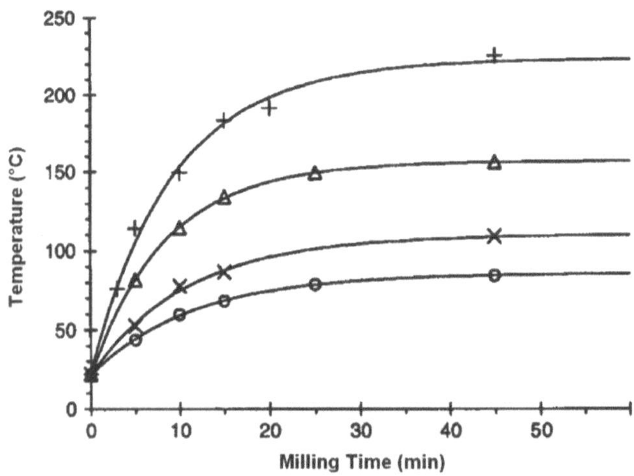

Fig. 2.18 Ball temperature-milling time curves obtained using a Fritsch P-5 planetary mill. The (+) and (x) symbols represent data with a 10-cm diameter milling bowl and 100 balls and 280 rpm (+) and 200 rpm (x) speeds and the (Δ) and (○) symbols correspond to a 7.5-cm bowl, 50 balls and 280 rpm (Δ) and 200 rpm (○) speeds [Takacs and McHenry 2006]

2.5.2 Contamination

The serious problem which is usually cited in mechanochemical research is contamination. The small size of milled particles, availability of large surface area, and formation of new surfaces during milling all contribute the contamination of the powder [Suryanarayana et al. 2001]. As mentioned above, the milling atmosphere can be one source. However, if steel balls and milling chambers are used iron contamination can contribute as a further factor. It is most serious for the high-energy mills. Low-energy mills result in much less, often negligible Fe contamination. Other milling media, such as tungsten carbide or ceramics, can be used but contamination from such media is also possible. Surfactants (process control agents) may also be used to minimize contamination. Using a "seasoned" milling chamber – i.e. media coated with the product powder-resulted in very low values for Fe contamination [Davis and Koch 1987; Koch 1997]. Application of shortest milling times and self-coating of the balls with milled material can be also helpful [Suryanarayana et al. 2001].

2.5.3 Amorphization

During high-energy milling, the size of crystals decreases to some critical values. Further energy supply to these crystals of limiting size causes further deformation of crystals, energy accumulation in the volume or at the surface of crystals, and subsequently amorphization [Boldyrev and Tkáčová 2001].

The occurrence of an amorphous phase is generally interferred by observing the presence of broad peaks in the X-ray diffraction patterns. It is not possible to distinguish among solids which are truly amorphous, extremely fine grained or a solid in which very small crystals are embedded in an amorphous matrix. Hence, it has been the practice to recognize such observations as "X-ray amorphous", suggesting that the identification was done only by X-ray diffraction methods. It is desirable that the X-ray diffraction observations are confirmed by other techniques as well [Suryanarayana 2001].

2.5.4 Mechanochemical Equilibrium

The objective of high-energy milling is the preparation of a milled material having the finess suitable for further application. However, this particle size reduction is, in many cases, complicated by particle size enlargement where smaller particles are put together to form larger entities in which the original particles can still be identified. Experience shows that after achievement of a certain degree of dispersion further comminution is reduced, or there may even be an increase of particle size, due to the interaction of particles [Juhász and Opoczky 1990].

In a typical experiment, copper was milled in a mill loaded with steel balls for 110 hours followed by separation of milled material in a 60 μm sieve [Hüttig 1943]. Both the coarse and the fine fractions were milled for another 48 hours and a product of approximately the same sieve analysis was obtained because of the comminution of the coarse fraction and aggregation of the fines. Hüttig called the particle size distribution and he demonstrated the same phenomenon with other brittle materials. On the basis of these experiments Hüttig postulated the existence of an "equilibrium state of comminution".

Tanaka suggested the following equation for characterizing the kinetics of milling

$$S_\infty = S_\infty \left(1 - e^{-kW_z}\right) \tag{2.5}$$

where W_z is the energy input, S_∞ the specific surface at equilibrium and k a constant [Tanaka 1958].

Later the process of new surface area formation was described by equation

$$S = S_m \left(1 - e^{-k_2 t}\right) \tag{2.6}$$

where S is specific surface after time t and S_m is maximum specific surface [Chodakov 1972]. Constant k_2 implies the significance of rate constant of new surface formation. Equation (2.6) describes also processes in which the formation of new surface is limited by milling equilibrium after a certain time of milling.

Jimbo described the phenomenon characterized by the increase of the size, the decrease of the specific surface area and the specific total pore volume with the increase of milling time as "the negative grinding phenomenon" [Jimbo et al. 1990]. This phenomenon is in a very close relation to the aggregation and agglomeration effects discussed in the next paragraph.

2.5.5 Agglomeration and Aggregation

In the course of the milling process a gradual deterioration of effectivity is to be observed. The energy input proportional surface increase gets impossible to obtain. A first explanation in mechanochemistry was given in the year 1937: with decrease of particle size, the number of defect locations runs out hence the increase of milling resistance, at extreme dispersity is approached [Smekal 1936]. Equations (2.5) and (2.6) can be applied for the mathematical description of such process. As it is well known in mechanochemistry, attaining a certain value, the surface area begins to diminish. In Fig. 2.19 the characteristic plots are given for several sulphides after application of high-energy milling.

Thorough investigation of this process on several solids has shown [Opoczky 1977; Beke 1984; Juhász and Opoczky 1990] that three stages can be clearly distinguished:

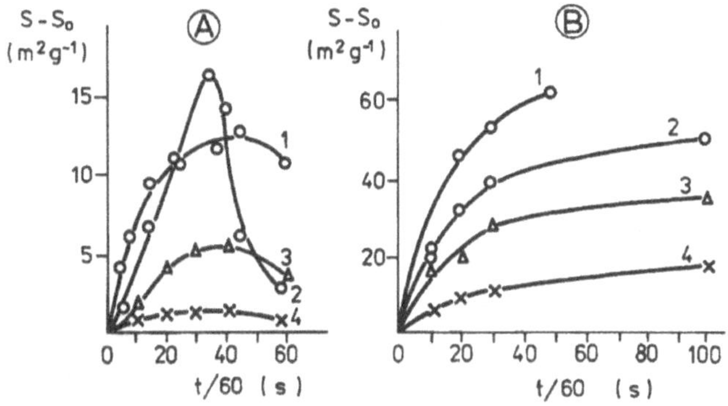

Fig. 2.19 Specific surface area, S–S_0 vs. time of mechanical activation, t: 1 – FeS_2, 2 – ZnS, 3 – FeS, 4 – PbS, A – milling in air, B – milling in H_2O, S_0 – specific surface area of non-activated solid [Avvakumov 1986]

- *the Rittinger stage*, in which the interaction of particles can be neglected and the energy input is approximately proportional to the new surface area formation (Fig. 2.20a);
- *the aggregation stage* in which the new surface area produced is not proportional to the energy input because of particle interaction (aggregation). However, the degree of dispersion is still increasing significantly (Fig. 2.20b). The particles adhere on each other. This adherence takes place without structural changes in consequence of van der Waals forces of the magnitude 0.04–4 kJmol^{-1}. These aggregates can be dissolved by slight mechanical intervention;
- *the agglomeration stage* in which the increase in dispersion first drops to a negligible value then stops altogether; it may even give way to a decrease of surface area because of particle interaction (agglomeration). Here particles are grown

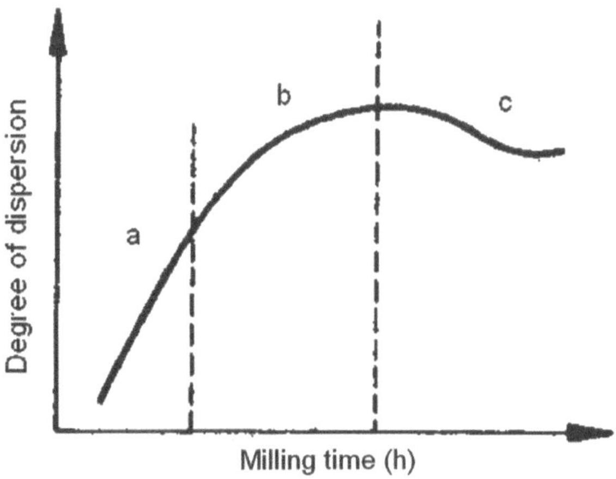

Fig. 2.20 Three stages during high-energy milling [Opoczky 1977]

together by chemical bonds of the magnitude 40–400 kJmol^{-1} and separation becomes impossible (Fig. 2.20c). Mechanochemical reactions and changes in the crystal structures mainly occur at this stage.

To obtain agglomerates, binding forces must act between the particles. The possible mechanism can be divided into the following major groups: solid bridges (1), interfacial forces and capillary pressure at freely movable liquid surfaces (2), adhesion and cohesion forces at not freely movable binder bridges (3), attraction forces between solid particles (4) and form-closed bonds (5) [Rumpf 1962]. Another classification into only two groups distinguishes between the presence of material bridges between the primary particles in the agglomerate and attraction forces [Pietsch 1972, 1984].

The most important characteristic of all forms of agglomerates is their strength. Rumpf defined the tensile strength as the tensile force at failure divided by the cross section of the agglomerate. Because with high probability failure occurs as the result of the highest tensile strength in all stressing situations, this proposal is justified.

Figure 2.21 illustrates the relative magnitude of the different bonds discussed above as a function of particle size. We see that van der Waals forces become important only for particles below 1 μm in size, adsorbed vapour forces are relevant below 80 μm and liquid bridge forces are active below about 500 μm [Rhodes 1998].

Van der Waals forces act between atoms and molecules without the transfer of electrons from one partner to the other. Adhesion forces between solid particles can be classified as long-distance forces (Coulombian-electrostatic and van der Waals forces) or a short-distance attractive forces (chemical and intermediate bonds, e.g. hydrogen bonds). Experiments and calculations support the view that the major role in adhesion of solids is played by attractive electrostatic and van der Waals forces [Juhász and Opoczky 1990].

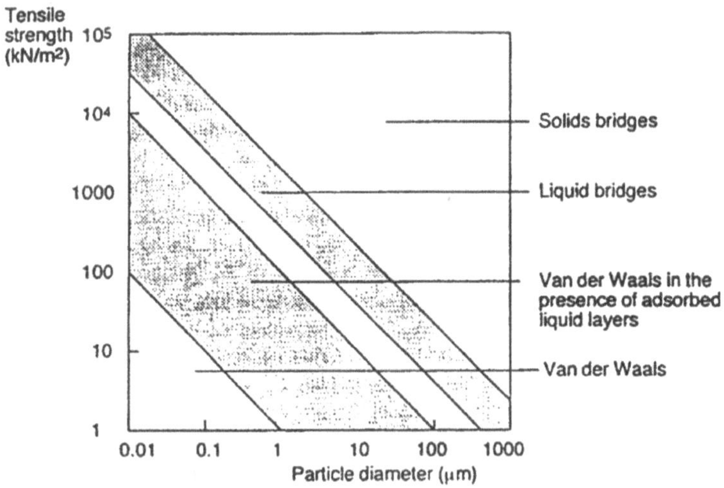

Fig. 2.21 Theoretical tensile strength of agglomerates with different bonding mechanism [Rumpf 1962]

According to many authors, the formation, rupture and severance of surface bonds that are characteristic for particle interactions in high-energy milling can be summarized as follows. After a relatively short period of milling and at relatively low fineness the adherence of particles sets in as a consequence of van der Waals adhesion forces. The distance between particles and the number of contacts plays an important role in the appearance of adhesion. In real polydisperse systems the presence of very fine particles along with relative coarse particles will greatly promote the formation of aggregates, and this accounts for the marked tendency for aggregation in milled products that have a broad range of particle sizes. Plastic deformation at contact points greatly increases both the area of the adhesion forces and, thereby increases the strength, compactness, and resistance to mechanical effects of particle agglomerates [Rumpf 1962, 1966, 1972–1974; Rumpf and Herrmann 1970; Krupp 1967; Juhász and Opoczky 1990].

In the course of adhesion of particles in mechanochemically activated systems the chemical interactions may also proceed at the contacting surfaces. These interactions result in a very compact, irreversible adhesion of the particles and forming of agglomerates. Agglomerates can be formed by various processes, e.g. by welding, coalescence of crystals, by direct initiation at the contacting surfaces, or by means of "melt-bridges" [Juhász and Opoczky 1990]. A typical agglomerate formed by mechanical activation is given in Fig. 2.22.

In nanoparticles, van der Waals forces are extremely important. Particles in the nanometer size range have a strong tendency to agglomerate owing to their relatively large specific surface area and other properties, which in turn accentuates their van der Waals interactions. Nanoclusters are only kinetically stable. They must be stabilized against aggregation into larger particles. This is frequently performed by adding further substances into the system (see next paragraph).

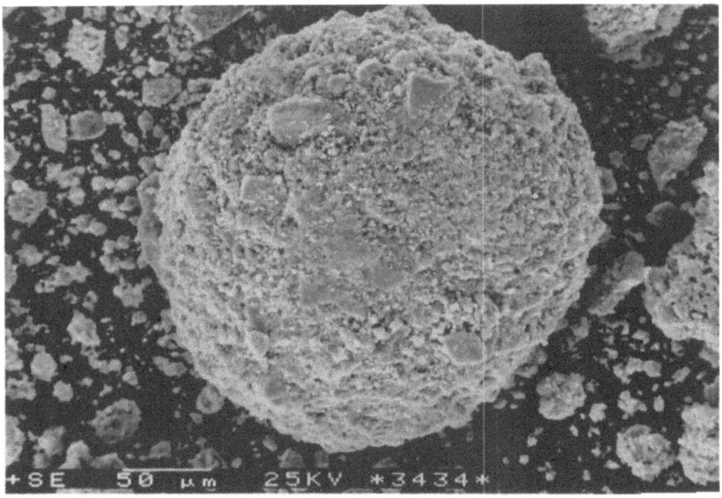

Fig. 2.22 A typical agglomerate of chalcopyrite $CuFeS_2$ particles formed by mechanical activation

2.6 Surfactants

There are several expressions used in literature for substances which preferentially adsorb onto the interfaces, prevent aggregation and stabilize the separated nanoparticles. The terms such as *surfactants, surface active agents, process control agents* and *lubricants* are most frequently applied.

In principle, stabilization can be accomplished by *electrostatic* (charge, or "inorganic") stabilization or *steric* ("organic") *stabilization* [Overbeeck 1981].

Steric stabilization is achieved by surrounding particles with a layer of substances which are sterically bulky such as polymers. These large adsorbates provide a steric barrier that prevents close contact of the fine particles (Fig. 2.23). The coil dimensions of surfactants are usually larger than the range over which the attraction forces between nanoparticles are active [Dutta and Hofmann 2004].

The water molecule plays here a special role. If a surfactant is added to water, its hydrocarbon chain tend to self-associate to minimize contact with water molecules. If the water is able to solubilize the polar head and the alkyl chains equally, no aggregates are formed.

The shape of the surfactant (Fig. 2.24) plays an important role in forming the assembly [Pileni 2003]. If the surfactant molecules have a very large polar head and a small chain forming a cone shape (Fig. 2.24a), the chains tend to self-associate to form a spherical aggregate called a normal micelle (Fig. 2.24b). If the surfactant has a shape of a champagne cork (Fig. 2.24c) spherical reverse micelles are formed (Fig. 2.24d). This is termed a water-in-oil droplet. In surfactant systems containing large amounts of both oil and water, the shape and dimension of the aggregates changes (Fig. 2.24e). Adding more water induces a new phase transition, with the surfactant molecules reorganizing as a planar or lamellar film (Fig. 2.24f). The other types of surfactant shapes are also possible as an onion shape on Fig. 2.24g.

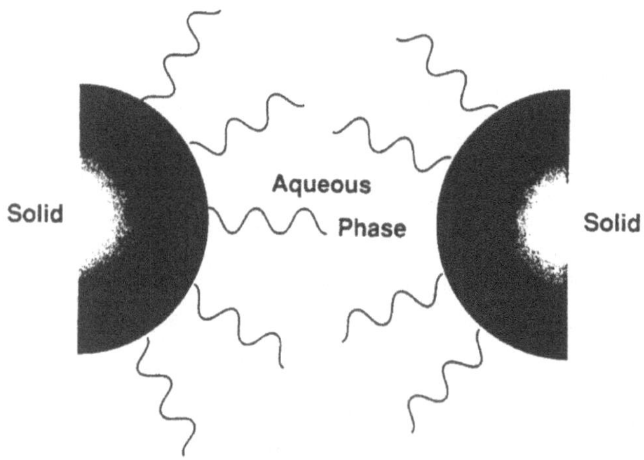

Fig. 2.23 Steric stabilization of nanoparticles by polymeric surfactants [Klimpel 1997]

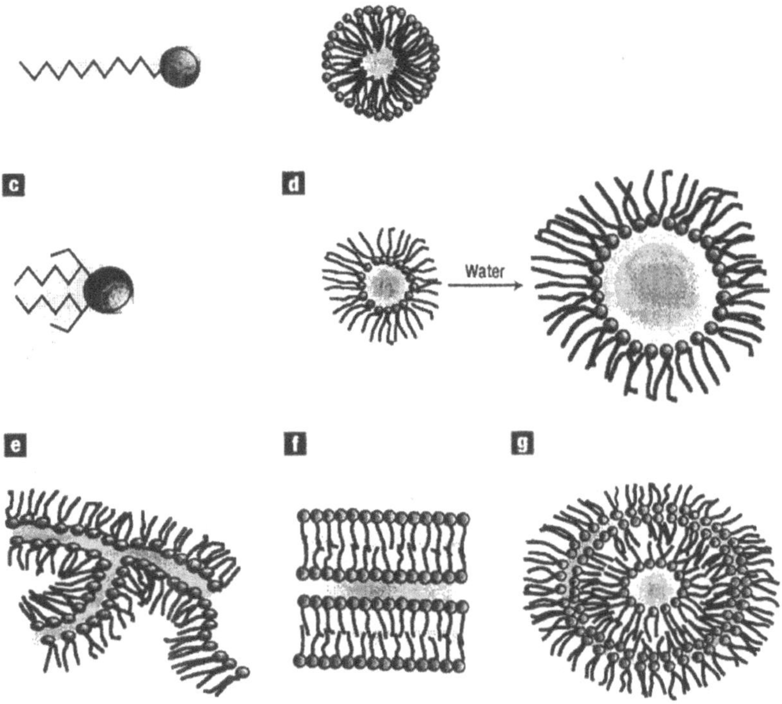

Fig. 2.24 Surfactant shapes and various self-assemblies in colloidal solution [Pileni 2003]

Electrostatic stabilization (Fig. 2.25) occurs by the adsorption of ions to the electrophilic surface [Hunter 1987]. Electrostatic stabilization involves the creation of an electrical double layer arising from ions adsorbed on the surface and associated counterious which surrounds the particle [Dutta and Hofmann 2004].

Surfactant chemicals as a group have some unusual chemical and physical properties. In addition to the tendency for their adsorption at phase interfaces, they

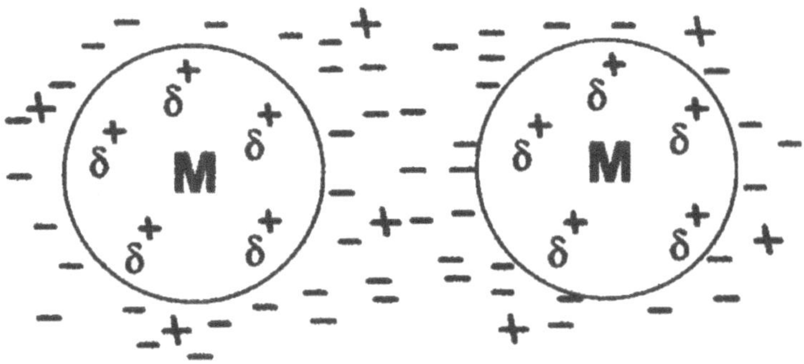

Fig. 2.25 Electrostatic stabilization of nanoparticles by polymers [Aiken III et al. 1996]

also demonstrate several other important properties. One of these properties is self-association between surfactant molecules of exactly the same structure leading to the formation of micelles [Klimpel 1997].

In high-energy milling, surfactants have been used at a level of about 1–5 wt.% of the total powder charge into the mill [Suryanarayana 2001]. The nature and quantity of surfactants used and the type of powder milled would determine the final size, shape, and purity of the powder particles. Use of a larger quantity of surfactants normally reduces the particle size by 2–3 orders by magnitude. For example, it was reported that milling aluminium for 5 hours produced a particle size of about 500 μm when 1% stearic acid was used as a surfactant. However, when 3% of stearic acid was used, the particle size was only 10 μm [Lai and Lu 1998]. Several surfactants are given in Table 2.2.

The choice of polymer surfactants varies from case to case. They must be mobile enough to provide access for the addition of monomer units, while stabile enough to prevent the aggregation of nanocrystals [Mana et al. 2002]. By using surfactants (or mixtures of surfactants) that bind differently to the different crystallographic faces, we can control the shape of the nanocrystals [Puntes et al. 2001]. However, it should be realized that there is no universal surfactant.

There have been several reports in literature that milling can be influenced by the addition of surfactants into the mill [El Shall and Somasundaram 1984]. The most of the results have been interpreted in terms of two major mechanisms.

The first mechanism is called *Rebinder effect*. According to this concept, the adsorption of surfactants results in a decrease of the surface energy of solids where by their strength and resistance to mechanical effects decrease [Rebinder and Kalinovskaja 1932]. The classical explanation of the chemomechanical effect is based on the Griffith's theory of fracture of solids weakened by a priori cracks. Molecules and ions of the surfactant intrude into the microcracks. The concept of Rebinder effect comprises the whole complex of phenomena promoting deformation and fracture. The essential feature is that, because of the reversible interaction of the solid with the surfactant, less work is required for the generation of new surface. After Juhász and Opoczky, Rebinder concept about the reduction of surface energy and the consequent decrease of fracture energy is untenable, since it is known that fracture energy is primarily determined by plastic deformation at the crack front and that stresses arising on fracture surfaces can only marginally, if at all, influence the propagation of cracks [Juhász and Opoczky 1990].

Table 2.2 Surfactants and the quantity used (adapted after Suryanarayana 2001)

Benzene	–	Octane	1%
Ethanol	4%	Polyethyleneglycol	–
Methanol	–	Sodium chloride	2%
Graphite	0.5%	Stearic acid	1%
Heptane	0.5%	Toluene	5 ml

The second mechanism states that adsorption of surfactants influences the properties of fragments created by milling. This mechanism is based on the empirical observation that surfactants alter the conditions of energy and mass transport by high-energy milling.

References

Abdellaoui, M. and Gaffet, E. (1996) The physics of mechanical alloying in a modified horizontal rod mill: mathematical treatment. Acta Materialia 44, 725–734.

Aiken III, J.D., Lin, Y. and Finke, R.G. (1996) A perspective on nanocluster catalysis: polyoxoanion and $(n\text{-}C_4H_9)_4N^+$ stabilized Ir(0) ~300 SD nanocluster soluble heterogeneous catalysts. Journal of Molecular Catalysis A: Chemical 114, 29–51.

Avvakumov, E.G. (1986) *Mechanical Methods of Chemical Processes Activation.* Nauka, Novosibirsk (in Russian).

Bachmann, D. (1940) Über den Einfluss des Mahlvorganges auf die Eigenschaften von Hochpolymeren Zellulose und Polystyrol. Verfahrenstechnik Zeitschrift, VDI-Beiheft 2, 43–55.

Baláž, P. (1981) *Intensification of Chalcopyrite Oxidative Leaching.* PhD. Thesis, Mining Institute of Slovak Academy of Sciences, Košice.

Baláž, P. (2000) *Extractive Metallurgy of Activated Minerals.* Elsevier, Amsterdam.

Baláž, P., Godočíková, E., Kril'ová, L., Lobotka, P. and Gock, E. (2004) Preparation of nanocrystalline materials by high-energy milling. Materials Science Engineering A386, 442–446.

Baláž, P., Godočíková, E., Takacs. L. and Gock, E. (2005) Mechanochemical preparation of metal/sulphide nanocomposite particles. International Journal of Materials Products Technology 23, 26–41.

Beenken, W., Gock, E. and Kurrer, K.E. (1996) The outer mechanics of the eccentric vibration mill. International Journal of Minerals Processing 44–45, 437–446.

Beke, B. (1984) Considerations about the energetic effectivity of fine grinding. In: K. Iinoya, J.K. Beddow and G. Gimbo (Eds.) *Proc. Int. Symp. on Powder Technology*. Kyoto, 1981, Hemisphere Publishing Corporation, Washington, pp. 373–379.

Benjamin, J.S. (1970) Dispersion strengthened superalloys by mechanical alloying. Metallurgical Transactions 1, 2943–2951.

Bernhardt, C. and Heegn, H.P. (1976a) Contribution to the investigation of the mechanical activation in fine grinding mills. In: H. Rumpf and K. Schönert (Eds.) *Proc. IVth European Symp. on Comminution*. Nürnberg 1975, Dechema Monographien, Bd. 79, Verlag Chemie, Frankfurt am Main, pp. 213–225 (in German).

Bernhardt, C. and Heegn, H.P. (1976b) Zur mechanischen Aktivierung von Ton. Folia Montana, extraordinary number 296–307.

Bernotat, S. and Schönert, K. (1998) Size reduction. In: *Ullmann's Encyclopedia of Industrial Chemistry.* VCH Verlagsgesellschaft, Weinheim, Vol. B2, pp. 5.1–5.39.

Boldyrev, V.V. (1983) *Experimental Methods in Mechanochemistry of Solid Inorganic Materials.* Nauka, Novosibirsk (in Russian).

Boldyrev, V.V. (1986) Mechanochemistry of inorganic solids. Proceedings of Indian National Science Academy 52, 400–417.

Boldyrev, V.V. and Tkáčová, K. (2001) Mechanochemistry of solids: past, present and prospects. Journal of Materials Synthesis 8, 121–132.

Bond, F.C. (1952) The third theory of comminution. Transactions of AIME, Minerals Engineering 193, 484–494.

Calka, A. (1991) Formation of titanium and zirconium nitrides by mechanical alloying. Applied Physics Letters 59, 1568–1569.

Calka, A. and Radlinski, A.P. (1991) Universal high performance ball milling device and its application for mechanical alloying. Materials Science and Engineering A, 134, 1350–1353.

Calka, A. and Wexler, D. (2002) Mechanical milling assisted by electrical discharge. Nature 419, 147–151.

Campbell, S.J. and Kaczmarek, W.A. (1996) Mössbauer effect studies of materials prepared by mechanochemical methods. In: G.J. Long and F. Grandjean (Eds.) *Mössbauer Spectroscopy Applied to Materials and Magnetism.* Plenum Press, New York Vol. 2, pp. 273–330.

Charles, R.J. (1957) Energy-size reduction relationships in comminution. Transactions of AIME 208, 80–88.

Chen, Y., Fitzgerald, J., Williams, J.S. and Willis, P. (1990) Mechanochemical synthesis of boron nitride nanotubes. Journal of Metastable and Nanocrystalline Materials 2, 375–380.

Chen, Y., Li, Z.L. and Williams, J.S. (1995) The evolution of hydriding and nitriding reactions during ball milling of titanium in ammonia. Journal of Materials Science Letters 14, 542–544.

Chodakov, G.S. (1972) *Physics of Milling.* Nauka, Moscow (in Russian).

Cottrell, A.M. (1958) Intercrystalline creep fracture. Transactions of AIME 212, 192–203.

Davis, R.M. and Koch, C.C. (1987) Mechanical alloying of brittle components: silicon and germanium. Scripta Metallurgica 21, 305–310.

Dutta, J. and Hofmann, H. (2004) Self organization of colloidal nanoparticles. In: H.S. Salva (Ed.) *Encyclopedia of Nanoscience and Nanotechnology*, American Scientific Publishers, California, Vol. 9, pp. 617–640.

El-Shall, H. and Somasundaran, P. (1984) Physico-chemical aspects of grinding: a review of use of additives. Powder Technology 38, 275–293.

Fokina, E.L., Budim, N.I., Kochnev, V.G. and Chernik, G.G. (2004) Planetary ball mills of periodic and continuous action. Journal of Materials Science 39, 5217–5222.

Gock, E. and. Kurrer, K.E. (1996) Eccentric vibratory mill-a new energy-efficient way for pulverisation. Erzmetall 49, 434–442.

Gock, E. and Kurrer, K.E. (1998) Increased efficiency of the vibratory milling process with the eccentric vibratory mill. Aufbereitung-Technik 39, 103–111.

Gock, E. and Kurrer, K.E. (1999) Eccentric vibratory mills-theory and practice. Powder Technology 105, 302–310.

Godočíková, E., Baláž, P., Criado, J.M., Real, C. and Gock, E. (2006) Thermal behaviour of mechanochemically synthesized nanocrystalline CuS. Thermochimica Acta 440, 19–22.

Golosov, S.J. (1971) *Introduction to Ultrafine Milling in Planetary Mills.* Nauka, Novosibirsk (in Russian).

Griffith, A.A. (1920) The phenomena of rupture and flow in solids. Philosophical Transactions of the Royal Society (London) A 221, 163–198.

Heegn, H.P., Bernhardt, C., Gottschalk, J. and Hussemann, K. (1974) Activation effects on the comminution of quartz and calcite in various laboratory mills. Chemische Technik (Leipzig) 26, 696–701.

Heegn, H.P, Bernhardt, C. and Ludwig, G. (1980) Zur Feinstmahlung von Strontiumferrit. Folia Montana, extraordinary number, 195–200.

Heegn, H.P. (1989) On the connection between ultrafine grinding and mechanical activation of minerals. Chemie Ingenieur Technik 62, 458–464.

Heinicke, G. (1984) *Tribochemistry.* Akademie-Verlag, Berlin.

Holmes, J.A. (1957) A contribution to the study of comminution: a modified form of Kick's law. Transactions of Institute Chemical Engineers (London) 35, 125–138.

http://www.fritsch.com

http://www.netzsch.com

http://www.nishineng.com

http://www.siebtechnik.com

http://www.ttd.spb.ru

http://www.unionprocess.com

http://www.utk.edu

http://www.zoz.com

Hukki, R.T. (1961) Proposal for a Solomonic settlement between the theories of von Rittinger, Kick and Bond. Transactions of AIME 220, 403–408.

Hunter, R.J. (1987) *Foundations of Colloid Science*, Vol. 1. Oxford University Press, New York.

Husemann, K., Wolf, R., Hermann, R. and Hoffmann, B. (1976) Erhöhung der Effektivität trockener Feinstmahlprozesse durch grenzflächenaktive Zusätze. Aufbereitungstechnik 35, 393–403.

Hüttig, G. (1943) Intermediate steps at solid-state reactions and their significance in catalysis. In: G.M. Schwab (Ed.) *Handbook of Catalysis*. Springer Verlag, Wien, Vol. 4, pp. 318–577 (in German).

Jimbo, G., Zhao, Q.Q., Yokoyana, T. and Taniyana, Y. (1990) The grinding limit and the negative grinding phenomenon. In: *Proc. IInd World Congress Particle Technology, Society of Powder Technology*, Kyoto, Part II, pp. 305–312.

Juhász, A.Z. and Opoczky, L. (1990) *Mechanical Activation of Minerals by Grinding: Pulverizing and Morphology of Particles*. Ellis Horwood, Chichester.

Kick, F. (1883) A critique of industrial methods of measurement of strength. Dinglers Polytechnik Journal 247, 1–15.

Kirpičev, V.L. (1874) Similarity in elastic phenomena.Žurnal Russkogo Fiziko-chimičeskogo Obščestva, čast' fizičeskaja IX, 152–158 (in Russian).

Klimpel, R.R. (1997) *Introduction to the Principles of Size Reduction of Particles by Mechanical Means*. Engineering Research Center at the University of Florida, Gainesville, pp. 1–41.

Koch, C.C. (1993) The synthesis and structure of nanocrystalline materials produced by mechanical attrition: a review. Nanostructured Materials 2, 109–129.

Koch, C.C. (1997) Synthesis of nanostructured materials by mechanical milling: problems and opportunities. Nanostructured Materials 9, 13–22.

Kochnev, V.G. (1992) Planetary mill. Russian Patent 1358990.

Kochnev, V.G. and Simakin, S.A. (1994) Planetary mill feeder. Russian Patent 2094120.

Krupp, H. (1967) Particle adhesion theory and experiment. Journal of Colloid Interface Science 1, 111–239.

Kurrer, K.E. and Gock, E. (1997) Eccentric vibratory mills for ultrafine comminution. Zement-Kalk-Gips International 50, 362–373.

Lai, M.O. and Lu, L. (1998) *Mechanical Alloying*. Boston, Kluwer Academic Publishers.

Mana, L., Scher, E.C. and Alivisatos, A.P. (2002) Shape control of colloidal semiconductor nanocrystals. Journal of Cluster Science 13, 521–532.

Miani, F. and Maurigh, F. (2004) Mechanosynthesis of nanophase powders. In: J.A. Schwarz, C.J. Contescu and K. Putyera (Eds.) *Dekker Encyclopedia of Nanoscience and Nanotechnology*. Marcel Dekker, New York, pp. 1787–1795.

Molčanov, V.I. and Jusupov, T.S. (1981) *Physical and Chemical Properties of Fine Milled Minerals*. Nedra, Moscow (in Russian).

Molčanov, V.I., Selezneva, O.G. and Žirnov, E.N. (1988) *Activation of Minerals by Milling*. Nedra, Moscow (in Russian).

Murty, B.S. and Ranganathan, S. (1998) Novel materials synthesis by mechanical alloying/mixing. International Materials Review 43, 101–143.

Netzsch Feinmahltechnik GmbH-Technical Information FT 002, 1974. Zirkulationsmühle System ZETA.

Ogino, Y., Yamasaki, T., Atzumi, N. and Yoshioka, K. (1993) Nitriding of transition metal powders by ball milling in nitrogen gas. Materials Transactions JIM 34, 1212–1216.

Opoczky, L. (1977) Fine grinding and agglomeration of silicates. Powder Technology 17, 1–7.

Overbeeck, J.T.G. (1981) *Colloidal Dispersions*. Royal Society of Chemistry, London.

Pietsch, W.B. (1972) Über Grenzflächenvorgänge in der Agglomerationstechnik. CZ Chemie-Technik 1, 116–119.

Pietsch, W.B. (1984) Agglomerate bonding and strength. In: N.E. Fayed and L. Otten (Eds.) *Handbook of Powder Science and Technology*. Van Nostrand, New York, pp. 231–251.

Pileni, M.P. (2003) The role of soft colloidal templates in controlling the size and shape of inorganic nanocrystals. Nature Materials 2, 145–150.

Puntes, F., Krishnan, K.M. and Alivisatos, A.P. (2001) Colloidal nanocrystal shape and size control: the case of cobalt. Science 291, 2115–2117.
Püpke, I. (1971) Construction and operating method of an attrition mill in powder metallurgy. Powder Metallurgy International 3, 94–96.
Rebinder, P.A. and Kalinovskaja, N.A. (1932) Lowering of strength of solid surface layers at surfactants adsorption. Žurnal techničeskoj fiziki 2, 286–302.
Rhodes, M. (1998) *Introduction to Particle Technology.* John Wiley and Sons, Chichester.
Rittinger, P.R. (1867) *Lehrbuch der Aufbereitungskunde.* Ernst und Korn, Berlin.
Rose, H.E. and Sullivan, R.M. (1961) *Vibration Mills and Vibration Milling.* Constable and Company Ltd., London.
Rose, H.E. (1962) Hochleistungsschwingmühlen. Chemie Ingenieur Technik 34, 414–417.
Rumpf, H. (1961) Problemstellungen und neuere Ergebnisse der Bruchtheorie. Materialprüfung 3, 253–265.
Rumpf, H. (1962) The Strength of granules and agglomerates. In: W.A. Knepper (Ed.) *Agglomeration.* John Wiley, New York, pp. 379–418.
Rumpf, H. (1966) Struktur der Zerkleinerungswissenschaft. Aufbereitungstechnik 7, 421–435.
Rumpf, H. and Herrmann, H. (1970) Eigenschaften Bindugsmechanismen und Festigkeit von Agglomeraten. Aufbereitungstechnik 11, 117–127.
Rumpf, H. (1972) Haftung und Festigkeit von Agglomeraten Vergleich zwischen Modellrechnung und Experiment. Die Pharmaceutische Industrie 34, 270–281.
Rumpf, H. (1973) Physical aspects of comminution-a new formulation of a law of comminution. Powder Technology 7, 148–159.
Rumpf, H. (1974) Die Wissenschaft des Agglomerierens. Chemie Ingenieur Technik 46, 1–11.
Smekal, A. (1922) Technische Festigkeit und molekuläre Festigkeit. Naturwissenschaften 10, 799–804.
Smekal, A. (1936) Bruchtheorie spröder Körper. Zeitschrift für Physik 103, 495–525.
Schönert, K. and Weichert, A. (1969) Die Wärmetönung des Bruches in Eisen und ihre Abhängigkeit von der Ausbreitungsgeschwindigkeit. Chemie Ingenieur Technik 41, 295–300.
Schönert, K. and Steier, K. (1971) Die Grenze der Zerkleinerung bei kleinen Korngrössen. Chemie Ingenieur Technik 43, 773–777.
Schönert, K. (1974) Über die Eigenschaften von Bruchflächen. Chemie Ingenieur Technik 46, 711–715.
Schubert, H. (1989) *Aufbereitung fester mineralischer Rohstoffe.* VEB Deutscher Verlag für Grundstoffindustrie, Leipzig, Band I.
Schulz, R., Boily, S., Zaluski, L., Zaluska, A. and Ström-Olsen, J.O. (1995) Nanocrystalline Mg-based hydrides: hydrogen storage for the zero-emission vehicle. Innovations in Metallic Materials, 529–535.
Suryanarayana, C. (2001) Mechanical alloying and milling. Progress in Materials Science 46, 1–184.
Suryanarayana, C., Ivanov, E. and Boldyrev, V.V. (2001) The science and technology of mechanical alloying. Materials Science and Engineering A, 304–306, 151–158.
Takacs, L. and Pardavi-Horvath, M. (1994) Magnetic properties of nanocomposites prepared by mechanical milling. In: R.D. Shull and J.M. Sanchez (Eds.) *Nanophases and Nanocrystalline Structures.* Warrendale, Pa, pp. 135–144.
Takacs, L. and McHenry, J.S. (2006). Temperature of the milling balls in shaker and planetary mills. Journal of Materials Science 41, 5246–5249.
Tanaka, T. (1958) Staub 18, 157–168.
Tkáčová, K. (1989) *Mechanical Activation of Minerals.* Elsevier, Amsterdam.
Tkáčová, K., Heegn, H.P. and Števulová, N. (1993) Energy transfer and conversion during comminution and mechanical activation. International Journal of Mineral Processing 40, 17–31.
Welch, A.J.E. (1953) The reaction of crystal lattice discontinuities to mineral dressing. In: *Developments in Mineral Dressing.* The Institution of Mining and Metallurgy, London, pp. 387–392.

Chapter 3
Selected Identification Methods

3.1 Introduction

During the history of mechanochemistry development the number of applied identification methods gradually increased. At the beginning, these methods were directed to at obtaining the characteristics of mechanically activated solids, mostly in integral form. At present, the number of methods applied in the investigation of mechanochemical phenomena comprises a few tens [Baláž 2000]. Moreover, with the intergrowth of mechanochemistry into nanoscience, special techniques such as TEM, HRTEM, AFM, etc. have been applied in different fields. This chapter is concerned with principle of several representative identification methods which are currently used for characterization of nanosized solids prepared by mechanochemical processes.

3.2 Infrared Spectroscopy

If a molecule is placed in an electromagnetic field (IR light), a transfer of energy from the field to the molecule will occur only when Bohr's frequency condition is satisfied

$$\Delta E = h\nu \qquad (3.1)$$

where ΔE is the difference in the energy between two energy levels, h is Planck's constant ($6.6262.10^{-34}$ Js) and ν is the frequency of light. If $\Delta E = E_1 - E_2$, where E_1 is a vibrational state of higher energy than E_2, the molecule absorbs radiation when it is promoted from E_2 to E_1 and emits radiation when it reverts from E_1 to E_2 [Berčík 1977].

According to its wave nature, electromagnetic radiation is characterized by two quantities: the wavelength λ and the frequency ν. In the IR spectral region the wavelength is usually given in units of [μm], the frequency is given in units of [s^{-1}]. The product of λ and ν is the velocity of light ($c = \lambda\nu$). An additional parameter which is commonly used in vibrational spectroscopy instead of the frequency is the wavenumber $\bar{\nu}$, defined as the reciprocal of the wavelength in cm [cm^{-1}]. The following relation exists between the wavelength λ, the frequency ν, the wavenumber

© Springer-Verlag Berlin Heidelberg 2008

$\bar{\nu}$ and the velocity of light c: $\bar{\nu}$ [cm^{-1}] $= 10^4/\lambda$ [μm] $= \nu$ [s^{-1}]$/c$ [cm s^{-1}]. Then the relation

$$\Delta E = h\nu = hc\bar{\nu} = \frac{hc}{\lambda} \tag{3.2}$$

can be written.

The method of infrared spectroscopy is based on absorption of IR light by molecules of gaseous, liquid or solid substances. The ability of molecules to absorb radiation is due to changes in the energetic state of electrons, vibrational motion of atoms and rotation of whole molecules. The difference in the energy ΔE is equal to the sum of energy differences between individual states

$$\Delta E = \Delta E_\nu + \Delta E_e + \Delta E_r \tag{3.3}$$

where ν denotes vibration state, e electron state and r rotation state.

In diatomic molecules, the vibration of point masses occurs only along the line connecting two atoms. In polyatomic groups (e.g. CO_3, SiO_4, PO_4 groups), the situation is much more complicated because all the atoms perform their own harmonic oscillations.

The normal modes of vibration are usually classified by type and symmetry. Two main types can be distinguished: (1) valence or stretching vibrations (symbol ν), characterised by changing bond lengths, and (2) planar bending vibrations (symbol δ), where one or more bond angles change, while bond lengths remain constant. Additionally, in out-of-plane bending vibrations (symbol γ), one atom oscillates through a plane defined by at least three neighbouring atoms (e.g. CO_3 group). Stretching frequencies are usually higher than bending frequencies. There are also two symmetry modes: (1) symmetric vibrations (index s), where the symmetry of the atomic group is retained, and (2) asymmetric vibrations (index as), where the symmetry changes during the vibrations. In highly symmetric atomic groups degenerate vibrations occur. In this case two or more vibrations, depending on the degree of degeneracy, have different coordinates but the same energy, giving cause for absorptions at the same wavenumber [Beran et al. 2004].

The comparison of normal vibrations of water and carbon dioxide illustrates the various vibrations of a non-linear and a linear molecule (Fig. 3.1). Both molecules have two stretching vibrations, one symmetric $\nu_s(\nu_1)$, where both bonds are stretched simultaneously, and one asymmetric $\nu_{as}(\nu_3)$, where one bond is compressed, while the other is elongated. The H_2O molecule has one bending vibration $\delta(\nu_2)$. The CO_2 molecule, as a consequence of its linearity, has two bending vibrations which are perpendicular to each other (ν_{2a} and ν_{2b}). Both bending vibrations of CO_2 have the same energy, they are doubly degenerate [Beran et al. 2004; Nakamoto 1978].

Changes in the number or position of IR absorption bands are mostly analysed in terms of structural changes. Characteristics "group frequencies" can be observed in the spectra of all compounds that contain the respective functional groups and can be used for their identification. Frequencies of atomic groups relevant for inorganic compounds are compiled in Table 3.1.

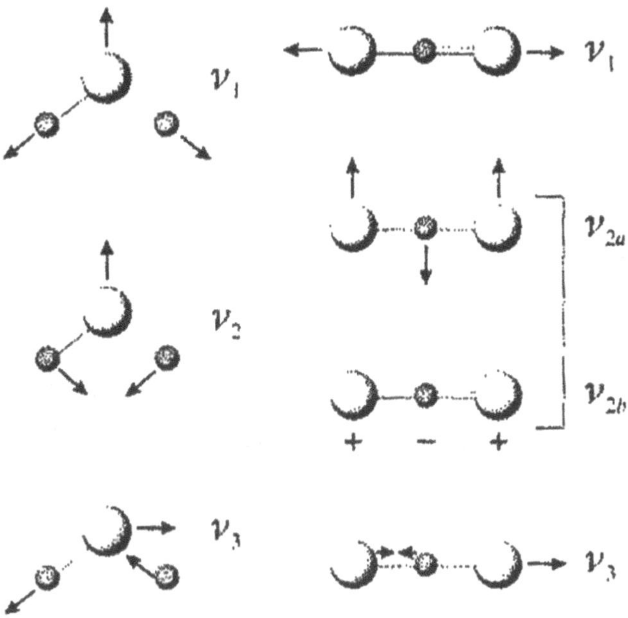

Fig. 3.1 Normal vibrations of the H_2O and CO_2 molecules. ν_1 – symetric stretching, ν_2 – asymetric stretching, ν_3 – bending, ν_{2a}, ν_{2b} – doubly degenerated bending vibrations of CO_2

Table 3.1 Characteristic group frequencies in cm^{-1} of functional atomic groups relevant for common minerals [Nakamoto 1978; Liese 1974]

Group	Stretching vibrations	Bending vibrations	Group	Stretching vibrations	Bending vibrations
MOH	3700–2900	1300–400	PO_4^{3-}	1100–950	600–550
H_2O	3700–2900	1650–1600	SiO_4^{4-}	1000–800	550–400
CO_3^{2-}	1600–1300	950–650	$Si_xO_y^{2-}$	1200–900	800–400
NO_3^-	1500–1250	900–700	AsO_4^{3-}	900–750	400
BO_3^{3-}	1300–1200	800–600	VO_4^{3-}	900–750	400
SO_4^{2-}	1200–1050	700–600	WO_4^{2-}	850–750	350–300

In an infrared spectrometer the sample is gradually exposed to radiation of different frequencies and it is determined at which frequencies the absorption occurs (Fig. 3.2). The radiation of source Z goes through sample S and impinges on prism P where it is resolved in monochromatic components.

Fig. 3.2 The scheme of IR spectrometer. Z – source of radiation, S – sample, P – prism, D – detector, R – registration

The gradual rotation of the prism results in radiation of different frequencies falling on the detector D where both the frequency and radiant flux is registered. The obtained spectrum expresses the dependence of absorption on radiation frequency. As to a double-beam instrument, absorbance $A = -\log \Phi/\Phi_0$ is recorded instead of radiant flux Φ.

The classical way to carry out infrared spectroscopy is to scan the frequency of the incoming light to enable the detector to record changes in the light intensity for those frequencies at which the sample absorbs energy. A major disadvantage of this method is that the detector records meaningful information only while the scan is passing through absorption lines, while most of the time is spent scanning between lines when the detector has nothing to record. To overcome this deficiency, modern infrared spectrometers irradiate the sample with a broad band of frequencies simultaneously, and then carry out a mathematical analysis of the resulting signal, called a Fourier transformation, to convert the detected signal back into the classical form of the spectrum. The resulting signal is called a Fourier transform infrared (FTIR) spectrum.

Figure 3.3 shows the IR spectrum of nanocrystalline ZnS prepared by high-energy milling of zinc and sulfur in a planetary mill in a presence of air and scanned by classical IR. The spectrum consists of a series of bands with frequencies corresponding to sulphates (Table 3.1). In comparison with the standard spectrum of $ZnSO_4.7H_2O$ it was concluded the formation of zinc sulphate as the product of zinc sulphide mechanochemical surface oxidation.

The application of infrared spectroscopy in examining sulphides is associated with problems caused by the general application of solids and, in particular, with the structure of sulphides. These problems are reflected both in data collection and interpretation and include the special properties of nanophase sulphides.

Sulphides are typically characterised by the low strength of interatomic bonds and these are excited in the low wavenumber range $\overline{\upsilon} < 1000$–$400\,\text{cm}^{-1}$ where IR

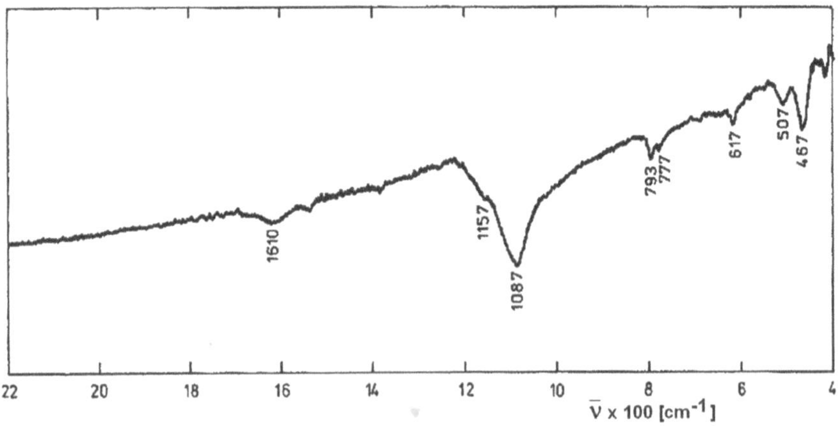

Fig. 3.3 IR spectrum of $ZnSO_4.7H_2O$ [Baláž et al. 1997]

spectroscopy is least efficient. These solids have the main absorption bands in the region $\overline{\upsilon} < 400\,\text{cm}^{-1}$, and only pyrite FeS_2 and arsenopyrite FeAsS have high peaks at $\overline{\upsilon} \sim 400\,\text{cm}^{-1}$ [Liese 1974; Kössler 1970; Abramov et al. 1982; Farmer 1974; Boldyrev 1976]. The spectra of sulphides are characterized by broadened peaks due to the specific features of the crystalline structure and the nature of Me-S bonds. The form of the spectrum is greatly affected by the proportion of the metallic bond. The scatter and absorption of infrared radiation by free electrons causes smoothing and partial "broadening" of the peaks which overlap and the spectra becomes indistinc and complex to interpret.

The majority of currently available IR spectrometers work in the regions above $400\,\text{cm}^{-1}$ with a few operating above $200\,\text{cm}^{-1}$. Due to this restriction the IR method in solids is limited by the method of preparing samples for actual measurements. Satisfactory results have been obtained with the method of preparing tablets in a mixture with potassium bromide which uses the effect of plasticity of halides of alkali metals at elevated pressure [Kössler 1970]. The KBr method solves the problems associated with low transmission of sulphide powders but it is essential to overcome other procedural problems such as, for example, purity of KBr, evacuation of tablets, application of optimum pressure in tablet pressing, possible interaction of KBr with sulphides, etc.

Despite these restrictions, the IR spectroscopy method has been used in examining the effects resulting from mechanical activation of solids [Juhász 1978; Vlasov and Kakazej 1979; Boldyrev 1983; Heinicke 1984; Molčanov et al. 1988; Chodakov 1972; Boldyrev and Meyer 1973; Molčanov and Jusupov 1981; Juhász and Opoczky 1990; Kulebakin 1983; Avvakumov 1986]. The application possibilities of the method are utilised mainly for

- evaluating the particle size and disordering of their structure and
- identifying new compounds formed on the surface of solids.

The first case is concerned with the solids of the oxide and silicate type [Duyckaerts 1959; Hlavay 1978; Hlavay and Inczédy 1979] and is based on Rayleigh's equation which links the scatter of light S with the particle diameter d

$$S = kd^3\lambda^{-4} \tag{3.4}$$

where λ is the wavelength of incident light. The effect of the particle size is then reflected in the situation in which the absorption band does not change frequently but becomes more intense and narrower with the reduction in particle size. Since the shape of the absorption band also depends on the orientation of the crystal lattice, polymorphism etc., is also possible to evaluate the disordering of the structure of the solids which takes place during their mechanical activation [Hlavay 1978]. The methods of infrared spectroscopy and X-ray diffraction analysis have been compared in evaluating the degree of disordering of a series of oxide-type solids. Certain correlations have been found between the particle size and the quantitative data obtained by means of these methods [Hlavay and Inczédy 1979].

Since the low strength of bonds of sulphides may, during mechanical activation, cause their disordering or even fracture, it may be expected that the dry high-energy

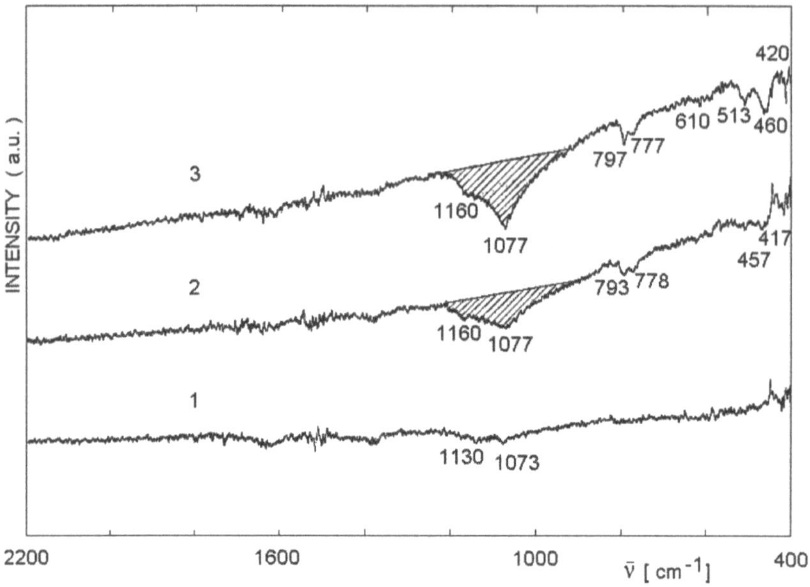

Fig. 3.4 IR spectrum of PbS, mechanical activation: 1–0 min, 2–2 min, 3–15 min [Godočíková et al. 2002]

milling (in air) will be accompanied by the formation of oxide compounds on their surfaces. Sulphates, oxysulphates, carbonates and other species can be formed. For such species the IR spectroscopy is a suitable identification tool because of their typical wavelengths (see Table 3.1). The typical example can be seen in Figs. 3.3 and 3.4.

The IR spectrum of non-activated lead sulphide PbS contains weak bands at 1130 and 1073 cm^{-1} which are characteristic for the sulphate group. The spectra of mechanically activated samples are more rich for the occurence of the bands. The bands at 1160, 1077 and 610 cm^{-1} belong to $PbSO_4$, the band at 457 cm^{-1} corresponds to the sulphate group and bands at 793–797, 778–777 and 417 cm^{-1} is due to admixed SiO_2.

3.3 X-Ray Photoelectron Spectroscopy

X-ray photoelectron spectroscopy (XPS) is a surface analytical technique based on the photoelectric effect which was described by Einstein in 1905. Each atom in the surface has core electron with the characteristic binding energy is approximately equal to the ionization of that electron. If monochromatic X-rays of $h\upsilon$ energy impinge the sample surface, the energy of the photon is adsorbed completely by the core electron of an surface atom. If the photon energy is large enough, the core electron will then escape from the atom and emit out of the surface. In XPS, these photoelectrons are collected and their kinetic energy E_k is determined resulting in a recorded spectrum of electron intensity as a function of the measured energy. Using Einstein's equation for photoelectric effect

$$E_{\text{kin}} = h\upsilon - E_B \tag{3.5}$$

where E_B – the binding energy and $h\upsilon$ – the X-ray photon energy, it is possible to calculate the binding energy E_B for a certain intensity peak since the incoming X-rays have a known frequency. This binding energy is material dependent. By comparing the XPS spectrum with tables of known elements and bindings, the chemical composition on the surface is determined. The variation of binding energy results in the shift of the corresponding XPS peaks ranging from 0.1 eV to 10 eV. This effect is termed *chemical shift* which can be used for studying the chemical status of element in the surface. Therefore, XPS was sometimes called Electron Spectroscopy for Chemical Analysis (ESCA).

Since the number of photoelectrons of an element is dependent upon the atomic concentration of that element in the sample, XPS is used not only to identify the elements but also to quantify the chemical composition. After the value of peak intensity (the peak area after background removal) is obtained, the atomic concentration of an element, c_i can be expressed as

$$c_i = \frac{I_i/S_i}{\sum I_i/S_i} \tag{3.6}$$

where I_i is the peak intensity for element i, and S_i is the sensitivity factor for the peak i [O'Connor et al. 1992]. Quantitative accuracy is 5%.

The theoretical elaboration of XPS method was developed in Sweeden by a group of Siegbahn, who in 1981 won the Nobel Prize for his work in developing the technique [Siegbahn et al. 1969].

The scheme of spectrometer is represented in Fig. 3.5. A sample is irradiated with monochromatic X-rays AlK_α where $h\nu = 1486.6$ eV and the energy of the emitted electrons is measured. The energy analyzer is, in principle, a monochromator

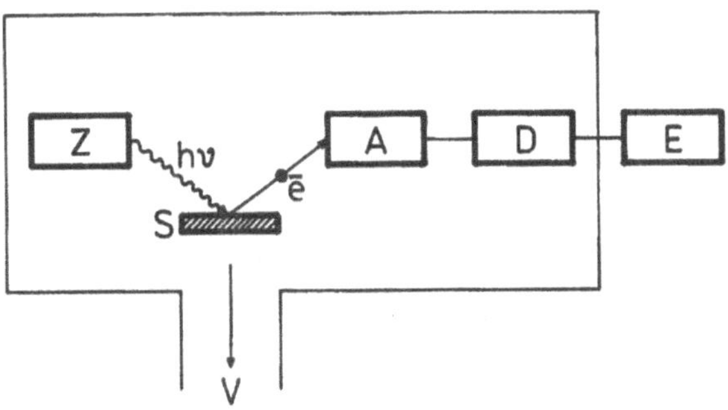

Fig. 3.5 The scheme of XPS spectrometer. Z – X-ray source, V – vacuum, S – sample, A – energy analyzer, D – detector, E – electronic control

of electrons which transmits only electrons of a certain kinetic energy. The beam is monochromatized using Bragg reflection on a crystal and is then directed towards the sample. This whole part of the instrument is kept in an ultra-high vacuum environment to avoid sample and instrument contamination due to adsorption of molecules.

Equation (3.5) adapted for measurement on a spectrometer assumes

$$E_{\mathrm{kin}} = h\upsilon - (E_k - \phi_s) \tag{3.7}$$

where ϕ_s is the work function of spectrometer which is constant for a given spectrometer. For practical purposes, the values of $h\upsilon$ and φ_s are combined within the spectrometer constant and the value of E_B can be immediately calculated from the measured value of E_{kin}.

The result of measurement is a XPS spectrum which gives the dependence of intensity of the detected signal on the kinetic energy of electron (Fig. 3.6). The spectral peaks correspond to different bonding states of electrons. The penetration depth of the exciting X-rays amounts to micrometers by decimal order. However, the escape pathlength of the emitted electrons and thus the thickness of the analyzed surface layer is substantially smaller due to interactions between the electrons and the components of the layer. The escape pathlength depends on kinetic energy of the electrons and is equal to 0.5–8 nm making XPS highly surface selective. Thus for studying the bulk characteristics of samples, identical chemical composition in surface and bulk is a necessary (but not sufficient) postulate [Nefedov 1984].

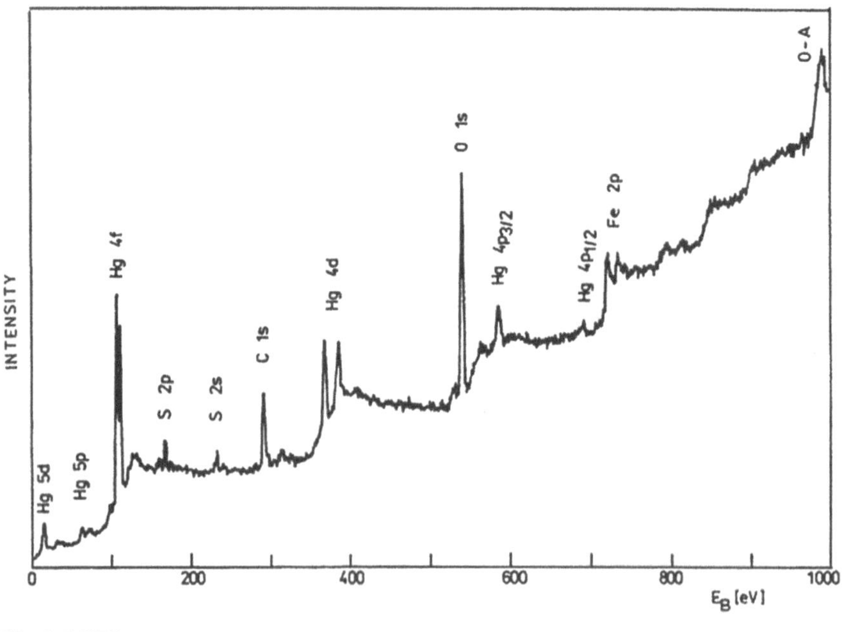

Fig. 3.6 XPS survey spectrum of HgS

Every line of a photoelectron spectrum is characterized by its binding energy, shape, width and, as the case may be, splitting. These data are a source of information about electron structure (e.g. oxidation state of the present elements) and chemical composition of the surface layer of the investigated sample. Thus, quantitative analysis is feasible, but owing to the microscopically heterogeneous structure of powdered sample it cannot be recommended as a method of estimating the bulk composition [Boiteux 1986]. The typical depth resolution is 5–30 nm.

If the values of binding energy are to be more precisely measured in the regime of high resolution, the spectrometer is usually calibrated for two main lines: Au $4f_{7/2}$ ($E_B = 84.0$ eV) and Cu $2p_{3/2}$ ($E_B = 932.7$ eV). Other than hydrogen and helium, all elements can be determined, most of them to the level of 0.1 atomic percentage. The amount of sample necessary for analysis is usually equal to a few milligrammes.

A potential complication in measuring the values of binding energy consists in accumulation of positive charge on the surface of sample. This cause the shift of the whole spectrum by the value of this charge towards higher values of binding energies. In this case, the binding energy has to be calibrated with an internal calibrating peak. The C 1s peak from the adventitious carbon-based contaminant with binding energy of 284.8 eV, is commonly used as the reference for calibration. This effect is most problematical in electrically non-conductive samples. In order to ascertain the value of charging or to eliminate this phenomenon, a great number of methods have been put forward [Briggs and Seah 1983]. However, none of them is of universal use and corrections need to be made for each sample.

In the case of unknown samples, the first spectrum is usually taken over a wide energy interval ($E_B = 0$–1000 eV) as a survey spectrum (Fig. 3.6) which serves for identification of the elements present. Then special spectra are taken over the close energy interval to identify special species (Fig. 3.7).

The finding that non-equivalent atoms of a certain element give spectra with measurable difference between binding energies is of outstanding analytical importance. This difference is called *chemical shift* ΔE_{ch} by analogy with a similar effect appearing in Mössbauer spectroscopy [Briggs and Seah 1983]. The non-equivalence of atoms may be due to the differences in their chemical environment or to different location of atoms in the lattice.

The structure of individual lines is examined at high resolution over a narrow energy interval. Besides the fundamental electron lines we may observe [Briggs and Seah 1983]

- spin-orbital splitting which results in a doublet for all levels except s-electrons,
- multiplet splitting which may appear if there are unpaired electrons on valence levels of the system,
- satellite lines and bands of different origin and some peaks of low intensity which are due to non-monochromaticity and
- impurity of X-ray sources.

These structures frequently help us to interpret the measured results and in the case of multiplet splitting and satellites they are also a valuable source of additional information about the state of nanocrystalline solids.

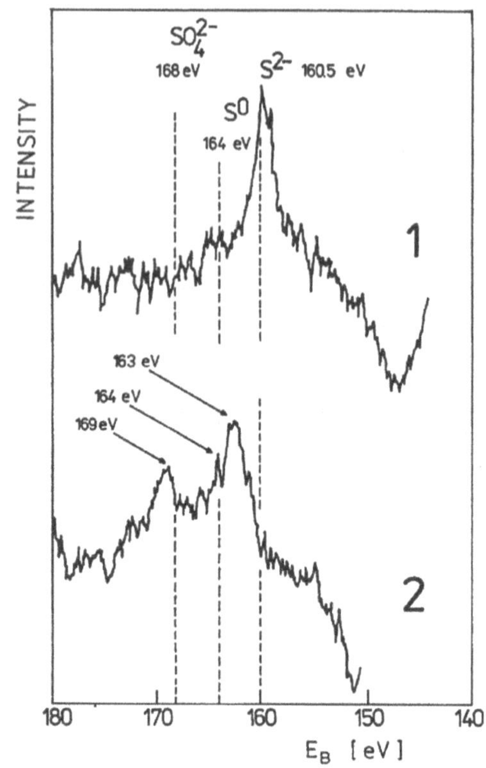

Fig. 3.7 XPS S 2p spectrum of PbS. 1 – PbS mineral, 2 – PbS synthetic [Pugh and Bergström 1986]

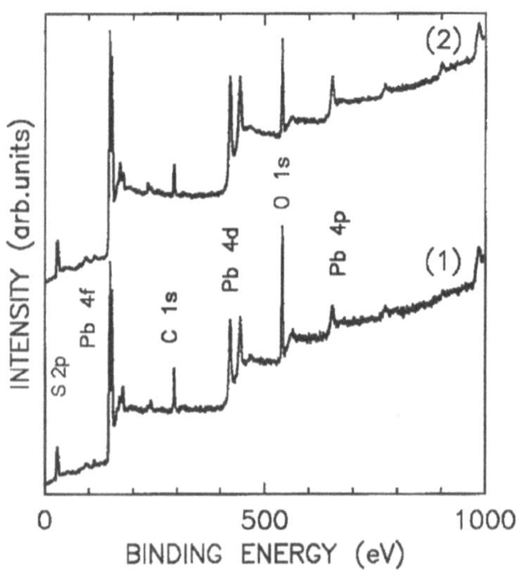

Fig. 3.8 XPS survey spectrum of PbS, mechanical activation: 1 – 0 min, 2 – 15 min [Godočíková et al. 2002]

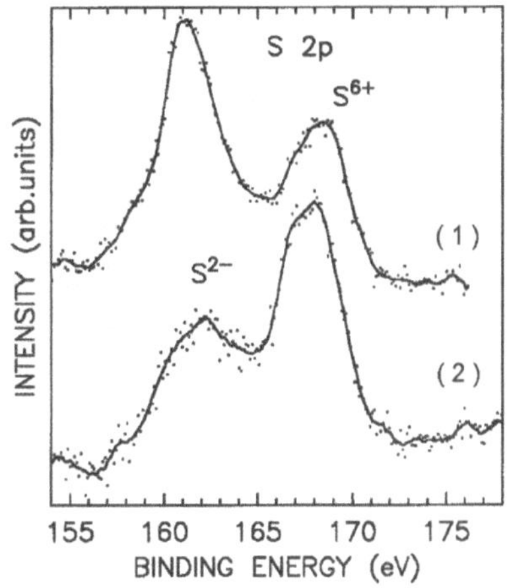

Fig. 3.9 XPS S2p spectrum of PbS, mechanical activation: 1 – 0 min, 2 – 15 min [Godočíková et al. 2002]

The typical example for nanocrystalline lead sulfide PbS prepared by mechanical activation is given in Figs. 3.8 and 3.9. Pb, S and contaminating C and O are seen in the survey XPS spectrum. The sulphate sulphur enrichment on the surface is illustrated by Fig. 3.9, where spectra of S 2p electrons for both are given. While about 42% of hexavalent sulfur is present in the surface layer of the non-activated PbS, it increased to 70% in the nanocrystalline PbS after milling.

3.4 Scanning Electron Microscopy

The scanning electron microscope (SEM) is a type of electron microscope capable of producing high resolution image of a sample surface. The first electron microscope was built in 1931 by Ruska and Knoll in Berlin. Ruska won in 1986 the Nobel Prize for his works in scanning electron microscopy. The technique was greatly developed through the 50s and has allowed great advances in different fields of science and technology.

The advantage of SEM in comparison with the optical microscope which utilizes light is its depth of focus and resolving capability. For the optical microscope, the depth of focus is the distance above and below the image plane over which the image appears in focus. As the magnification increases in the optical microscope, the depth of focus decreases. The three-dimensional appearance of the specimen image is a direct result of the large depth of field of the SEM. It is this large depth of fields that is the most attractive feature of the scanning electron microscope. The steep surface and edges (cliffs) tend to be brighter than flat surfaces resulting in images with good three-dimensional contrast (Fig. 3.10).

The other feature of SEM is resolving power, which is orders of magnitude better than that of an optical microscope because the wavelength of the probing beam is orders of magnitude smaller. The limit on what size can be resolved is set by the wavelength. The wavelength of the visible light used in optical microscopes is

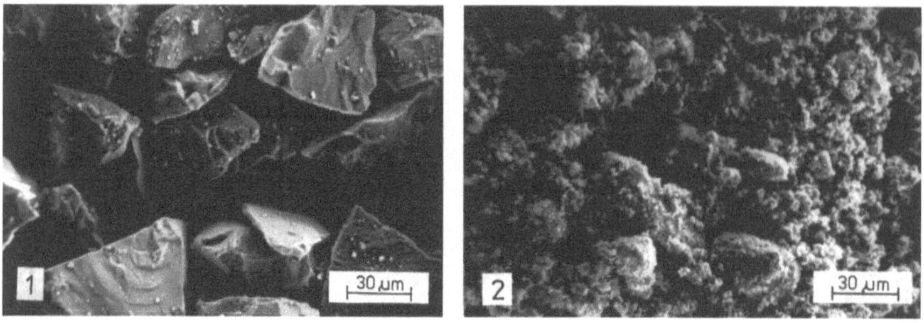

Fig. 3.10 SEM micrographs of $CuFeS_2$, mechanical activation: 1 – 0 min, 2 – 60 min [Baláž 2000]

between 400 and 700 nanometers. The resolving powers of high-quality light microscopes are limited by the wavelength of imaging light to about 200 nanometers. Scanning electron microscopy uses electrons with energies of a few thousand electron volts, energies a thousand times greater than that of visible light (2–3 eV). However, even electron microscopes with higher resolution are used. Thus the method provides valuable information regarding the structure arrangement, density and geometrical features of nanomaterials.

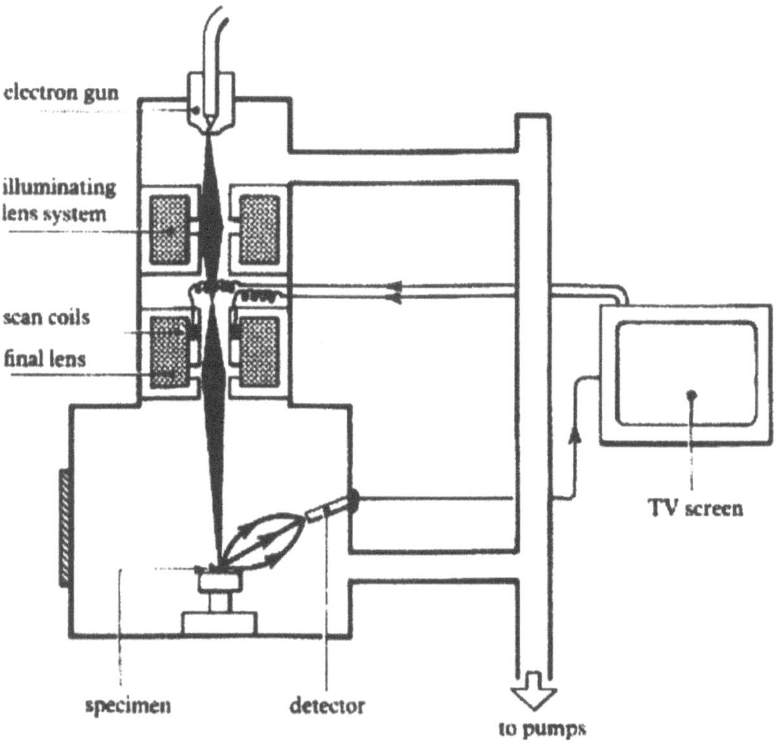

Fig. 3.11 The scheme of scanning electron microscope

The scheme of scanning electron microscope is given in Fig. 3.11. A beam of electrons is generated in the electron gun by heating a filament in a vacuum tube. This beam is accelerated and focused by magnetic lenses on the specimen. The electrons are collected by detectors, converted to a voltage, and amplified. This amplification generates a point of brightness on a cathode ray tube (or TV screen). An image is built up simply by scanning the electron beam across the specimen in exact synchromy with the scan of the electron beam in the cathode ray tube.

Accelerated voltage in microscope range from 10 to 40 kV, and this means that the incident electrons have sufficient energy to eject core electrons from atoms in the targeted specimen. Some of the electrons do not cause electrons to be ejected but are instead backscattered in the direction of the incident beam. In addition, a finite number of secondary electrons are emitted from the specimen via the photoelectron and Auger processes. Thus, three types of signals can be monitored. Backscattered electrons will have an energy and probability of backscater that is directly proportional to the average atomic number (Z) of the sample target. Thus, by rastering the beam over the sample surface and using a position sensitive detector set to energies just below the incident electron energy, SEM can produce a backscattered image of the sample where bright areas correspond to high – Z regions and dark areas correspond to low Z – regions. In contrast, the secondary electrons have an extremely short mean free path within the solid and are thus only detected if they are produced within the upper nanometric layer of the sample surface. The third type of signal is

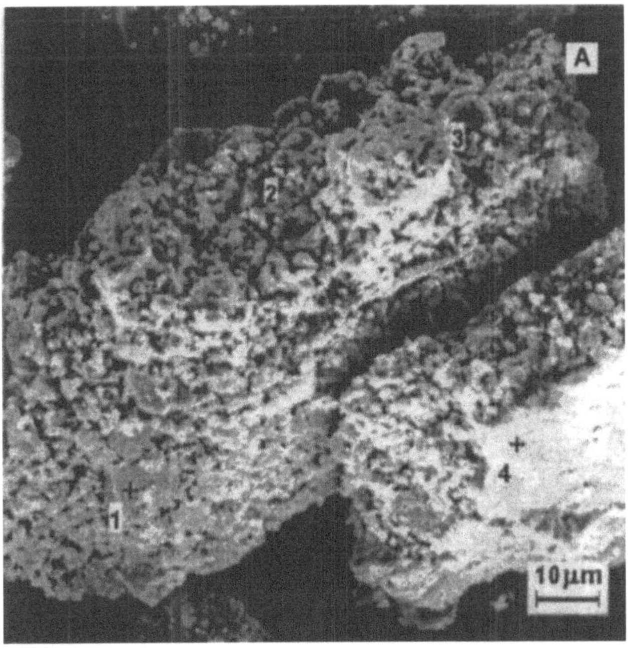

Fig. 3.12 SEM micrograph of Cu/FeS nanocomposite [Baláž et al. 2002a]

characteristic X-ray emission. If rastering of the incident beam is switched off and an EDS detector is used to collect a spectrum of the characteristic X-rays emitted from the sample, a spatially resolved chemical analysis can be acquired [Wogelius and Vaughan 2000].

The combination of SEM 2nd EDS techniques for analysis of nanocrystalline composite particles is illustrated by Figs. 3.12A and 3.13B.

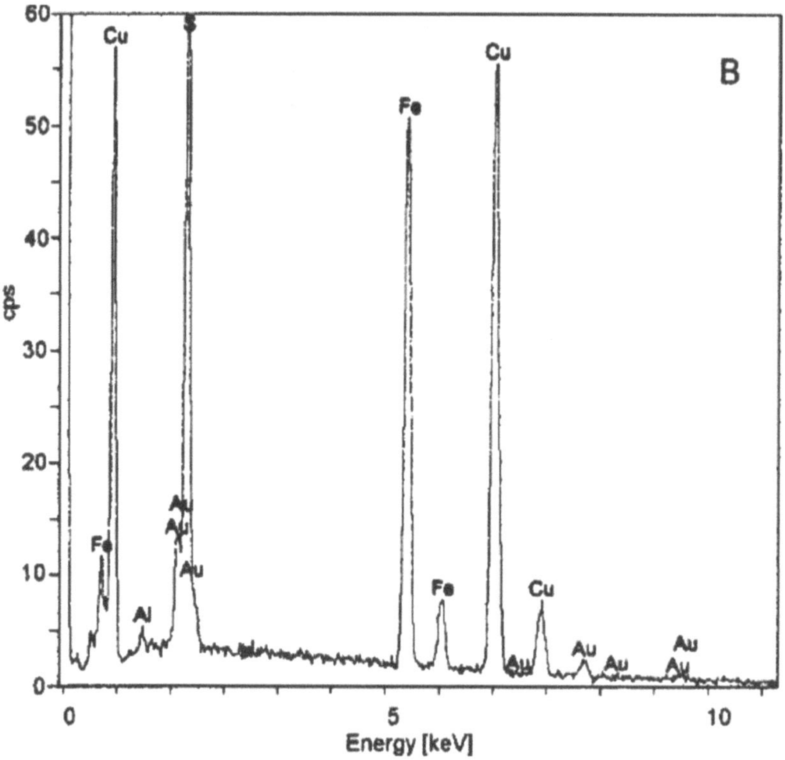

Fig. 3.13 EDS of Cu/FeS nanocomposite from Fig. 3.12 [Baláž et al. 2002a]

3.5 Scanning Tunnelling Microscopy

The scanning tunneling microscope (STM) is the most powerful type of microscope ever built. It was invented in 1981 by Binning and Rohrer, who went to collect the Nobel Prize in 1986.

In the STM method, which is a non-optical one, a sharp metallic tip is brought to within a few nanometers of a conducting surface and a small potential difference (usually in the mV range) is applied between the tip and the sample. If the tip is biased positively relative to the sample, electrons will flow from the sample to the tip in a process known as electron tunneling which produces a small, yet measurable, current. The magnitude of this current is exponentially dependent on tip – surface separation, the larger the distance between the tip and the surface, the smaller the current. Thus, by measuring the magnitude of the tunnelling current as the tip is moved across the surface, a topographic image of the surface can be obtained. In practice, this can be achieved by mounting a tip on a piezoelectric tube scanner. Because the piezoelectric material employed has the useful property of expanding or contracting when a voltage is applied across it, the tip can be manoeuvred with ultra-high precision (to within 0.01 nm). The experimental system, as illustrated in Fig. 3.14, gives the possibility of such high precision movement in x, y and z directions and hence producing a 3-dimensional image of the surface at nanometer resolution [Wogelius and Vaughan 2000].

Using STM it is possible to image surfaces of conducting and semiconducting materials such as metals, most metal sulfides and many metal oxide minerals, down to the atomic level. STM is particularly useful in nanoscale chemistry labs, where the study of defects and physical structure of synthetic chemical compounds is of vital importance.

STM images are sensitive to chemisorptive bonding at the surface. Several studies have used this effect to great advantage in the study of sorbed and defective overlayers of semiconductors. In semiconductors, electronic wavefunctions (states) are localized on particular atoms and we should be able to distinguish between different atomic sites.

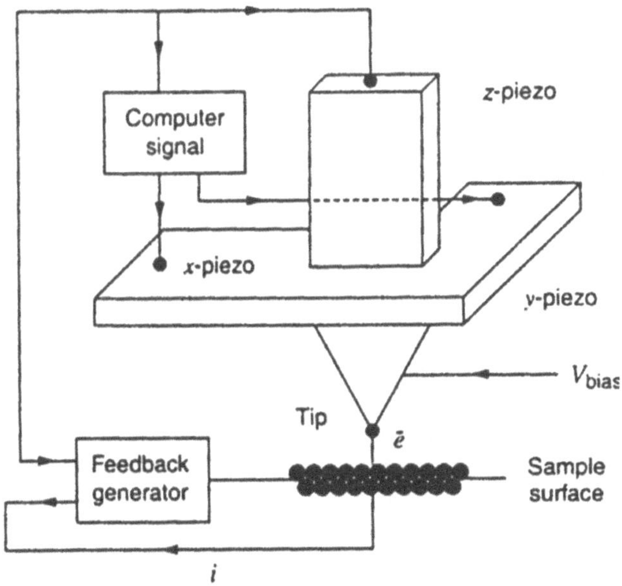

Fig. 3.14 The scheme of scanning tunneling microscope [Wogelius and Vaughan 2000]

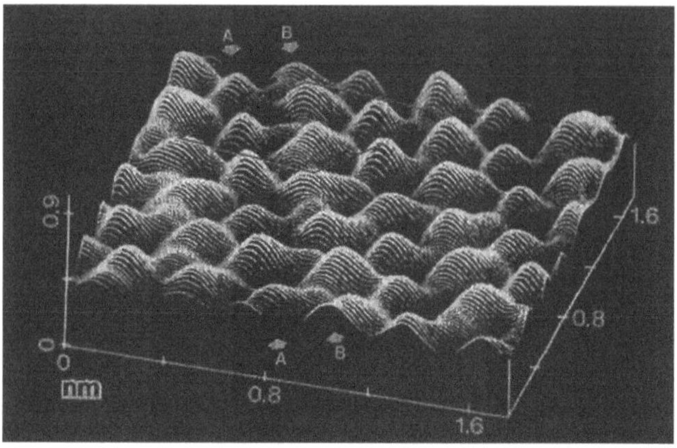

Fig. 3.15 STM image of PbS [Eggleston and Hochella, Jr. 1990]

Figure 3.15 shows STM image of lead sulphide PbS. The symmetry of a surface unit cell is slightly distorted due to drift but two different sites are distinguishable. The sites in the row labeled "A" are narrower than those in the row labeled "B". The surface cell is drawn so that wider (B) sites occur at the cell corners and narrower (A) sites occur at cell edge midpoints. Based on the interpretation of this and other images, we suggest that the B sites correspond to sulphur and the A sites correspond to lead.

In nanotechnology scanning tunneling microscope can be used to alter the observed material by manipulating individual atoms, creating ions by removing individual electrons from atoms and then reverting them to atoms by replacing the electrons.

3.6 Atomic Force Microscopy

The atomic force microscope (AFM) was invented by Binning, Quate and Gerber in 1986. The principles involved in this technique as illustrated in Fig. 3.16a, involve using a piezoelectric scanner to move a tip across the surface of the sample. The tip is mounted on a cantilever of force constant between approximately 0.001 and $0.2\,\mathrm{Nm}^{-1}$. When this tip is brought just into contact with the surface, it experiences a very small force (of the order of nanonewtons) as a result of interaction with the surface atoms. In this kind of operation in AFM, the tip is scanned across the surface at the tip-sample separation corresponding to a chemical bond length of the tip-sample combination. As it is scanned across the surface, the tip will be subject to varying attractive and repulsive forces of the kind associated with van der Waals bonds, and the movements of the tip will be registered by deflection of cantilever.

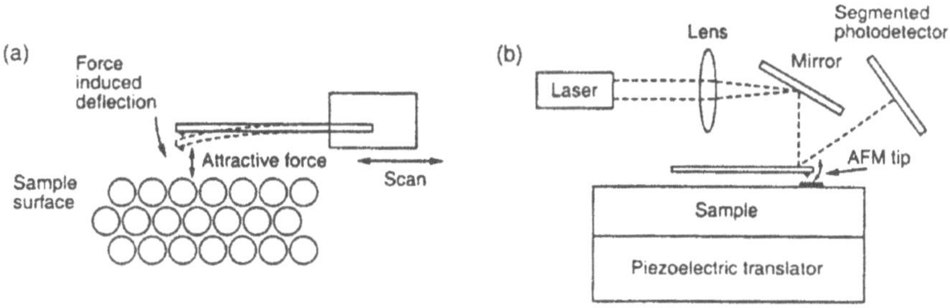

Fig. 3.16 The scheme of atomic force microscope [Wogelius and Vaughan 2000]

In a typical AFM instrument, as illustrated in Fig. 3.16b, deflection of the cantilever is monitored by reflecting a laser beam from the back of the cantilever on to a segmented photodetector [Wogelius and Vaughan 2000].

The AFM has several advantages over the electron microscope. Unlike the electron microscope which provides a two – dimensional image of sample, the AFM provides a true three – dimensional surface profile (Fig. 3.17). While an electron microscope needs an expensive vacuum – environment for proper operation, the AFM can work perfectly well in an ambient or even liquid environment.

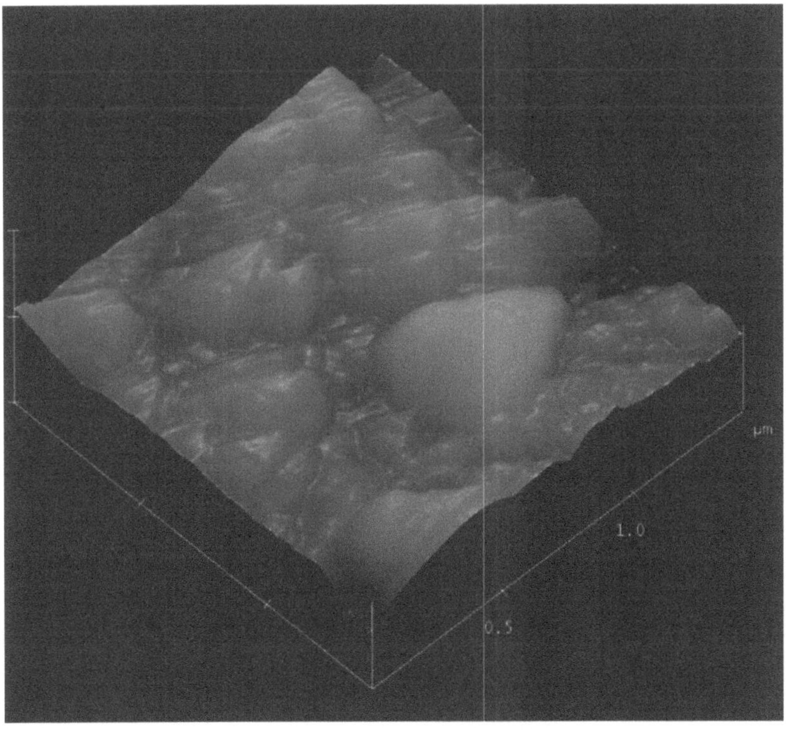

Fig. 3.17 AFM image of nanosized ZnS prepared by mechanochemical synthesis

While STM method is suitable for studying of conducting and semiconducting materials, the AFM offers the same atomic resolution capabilities but also for insulating materials. Besides imaging, it is also one of the foremost tools for the manipulation of matter at the nanoscale.

3.7 Transmission Electron Microscopy

Transmission electron microscopy (TEM) is an imaging technique whereby a beam of electrons is focused onto a specimen causing an enlarged version to appear on a fluorescent screen or layer of photographic film, or can be detected by a charge couple device (CCD) camera. TEM is a powerful technique for direct imaging the projected shapes of nanoparticles particularly when the particle size is small (Fig. 3.18).

A modern transmission electron microscope can be schematically shown in Fig. 3.19. The whole measuring system is composed of a illumination system, a specimen stage, an objective lens system, a magnification system, data recording system(s) and a chemical analysis system. The electron gun is the heart of the illumination system, which typically uses LaB_6 thermionic emission source or a field emission source. The LaB_6 gun gives a high illumination current but the current density and the beam coherence are not as high as those of a field emission source. Field emission source is unique for performing high coherence lattice imaging, electron holography and high spatial resolution microanalysis. The illumination system also includes the condenser lenses that are vitally important for forming a fine electron

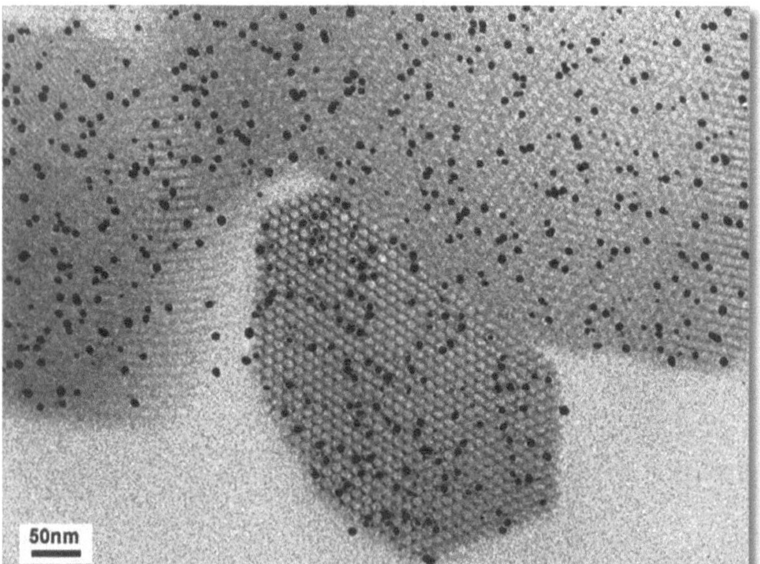

Fig. 3.18 TEM image of Ag-nanoparticles in the channels of zeolite [http://www.rz-berlin.mpg.de]

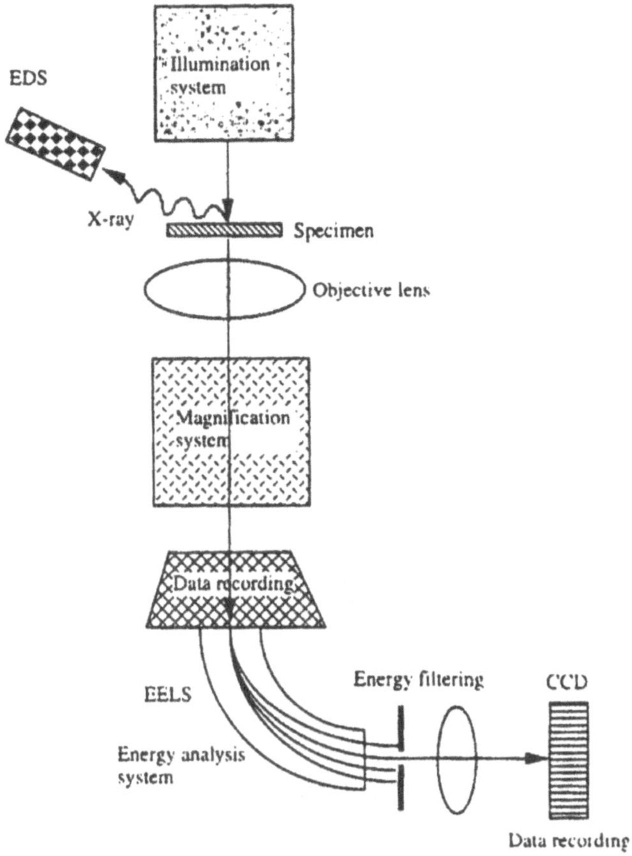

Fig. 3.19 The scheme of transmission electron microscope [Wang 2000]

probe. Specimen stage is a key for carrying out structure analysis, because it can be used to perform in-situ observations of phenomena induced by annealing, electric field, or mechanical stress, giving the possibility to characterize the physical properties of individual nanostructures. The objective lens is the heart of a TEM, which determines the limit of image resolution. The magnification system consists of intermediate lenses and projection lenses and it gives a magnification up to 1.5 million. The data recording system tends to be digital with the use of a CCD, allowing quantitative data processing and quantification. Finally, the chemical analysis system is the energy dispersive X-ray spectroscopy (EDS) and electron energy-loss spectroscopy (EELS); both can be used complimentary to quantify the chemical composition of the specimen. EELS can also provide information about the electronic structure of the specimen [Wang 2000].

In the most powerful diffraction contrast TEM instruments it is possible to produce a diffraction pattern image which is directly analogous to the planes of atoms in crystal. TEM is certainly the most direct method, providing real images of the particles. However, lack of contrast or overlap of particles frequently complicates the analysis of TEM images, because the particle boundaries cannot always be seen preciously. The degree of agglomeration is critical when bulk nanostructures or nanometer dispersions are desired. If crystallites are weekly agglomerated, forming small necks between the crystallites, and the crystallites are equiaxed, nearly spherical in shape, then TEM and BET average particle sizes should be nearly identical. The average TEM crystallite size, and the average BET particle size should be within 30% of each other in order to have weak enough agglomeration to form a nanostructured material.

High – resolution transmission electron microscopy (HRTEM) is a technique to obtain interference patterns from very thin specimen (thickness smaller than 15 nm). To interpret the images in terms of atomic coordinates simulations are usually necessary. HRTEM is used on magnifications above 400 000 and is the common tool to image grain boundaries, interfaces, nanocrystals, and lattice defects in atomic scale (Fig. 3.20).

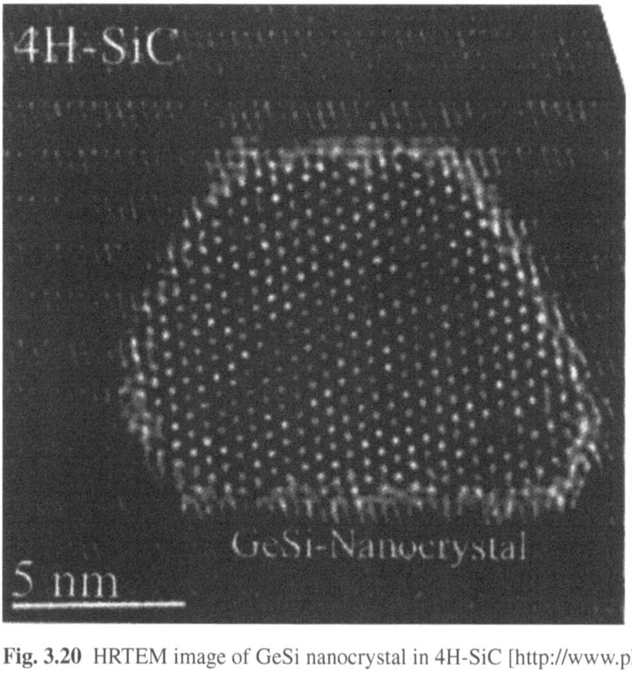

Fig. 3.20 HRTEM image of GeSi nanocrystal in 4H-SiC [http://www.physik.uni-jena.de]

3.8 X-Ray Diffraction

The X-ray beams are electromagnetic radiation of wavelength about 10^{-10} m. They occur in that portion of the electromagnetic spectrum between γ-rays and the ultraviolet. The discovery of X-rays in 1895 enabled to probe crystallite structure of solids at the atomic level. The inner atomic distances in solids are in the same area as the wavelength of X-rays. This is the reason why beams are reflected from the planes of atoms within the crystal. The reflection occurs according to Bragg law

$$n\lambda = 2d \sin\Theta \tag{3.8}$$

where n is the order of reflection, λ is the wavelength, d is the spacing between crystal planes and Θ is the angle between transmitted beam and sample (Bragg angle). The wave character of X-rays was revealed by Max von Laue in 1912. Owing to the small wavelength they show high penetrating power.

The X-ray diffraction experiments requires an X-ray source, the sample under investigation and a detector to register the diffracted X-rays. The scheme of a powder X-ray diffractometer is given in Fig. 3.21.

The result of measurement is the X-ray diffraction pattern which represents a set of diffraction lines (or peaks) of different intensities. The position of a diffraction line is fixed and the overall pattern is characteristics for a given substance (Fig. 3.22).

The ideal powder diffraction pattern consists of narrow, symmetrical, delta-function like peaks, positioned according to a well-defined unit cell. The aberrations from the ideal powder pattern can be conceived as [Ungar 2004]

- peak shift
- peak broadening
- peak asymmetries
- anisotropic peak broadening and
- peak shape.

In principle, the most common changes manifest themselves by peak shift and/or peak broadening (Fig. 3.23).

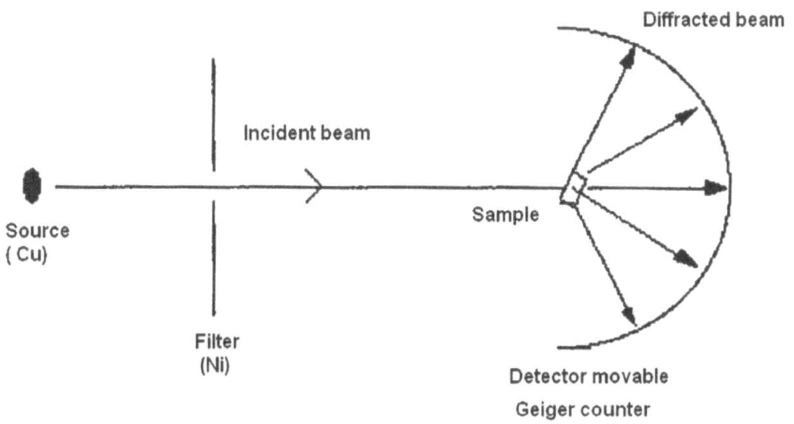

Fig. 3.21 The scheme of powder X-ray diffractometer

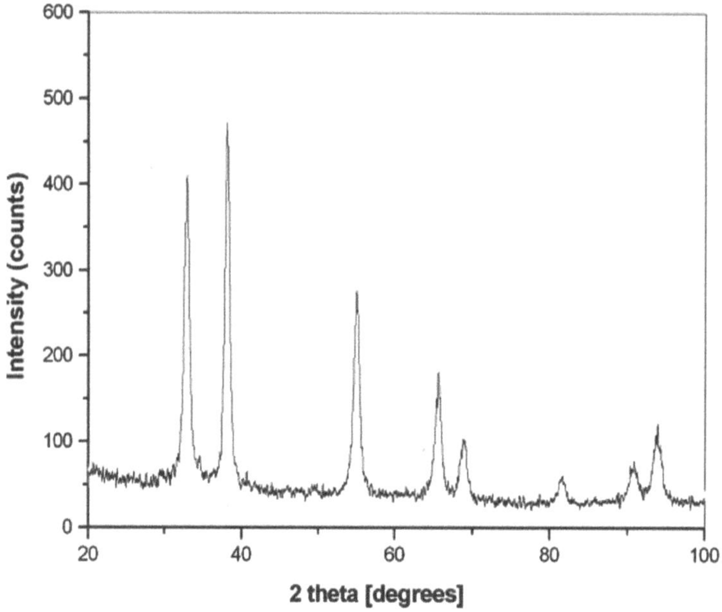

Fig. 3.22 X-ray diffraction pattern of nanocrystalline PbS [Baláž et al. 2004b]

The shift in a diffraction line is the result of uniform strain or macrostrain while the broadening of a diffraction line is due to non-uniform strain or microstrain. The most common sources of strain can be given as: dislocations, stacking faults, twinning, microstresses, grain boundaries, internal stresses, coherency strains, chemical heterogeneities, point defects or precipitates and inclusions. If a crystal is broken into smaller incoherently diffracting domains by dislocation arrays (small-angle boundaries), stacking faults, twins, or other extended imperfections, then crystallite size broadening occurs. A particle may consist of one or more crystals (primary grains). The different crystals may be separated by large angle boundaries as well as amorpous or crystalline interfaces. It is worth mentioning that the particle size itself can not be detected and measured by the XRD diffraction method. Similarly, the size of crystal is not accessible by XRD diffraction. The size of a crystal is, in general, equal to or less than the particle size. Each crystal may consist of one or more crystallites and the size of a crystallite is in general equal to or less than the crystal size. The crystallite size can be determined by means of powder diffraction, but not directly. Each crystallite may consist of one or more coherently reflecting domains.

X-ray diffraction lines broaden when the crystal lattice becomes imperfect. According the theory of kinematical scattering, X-ray diffraction peaks broaden either when crystallite become smaller than about a micrometer or if lattice defects are present in large enough abundance [Ungar 2004]. The whole picture is illustrated in Fig. 3.24.

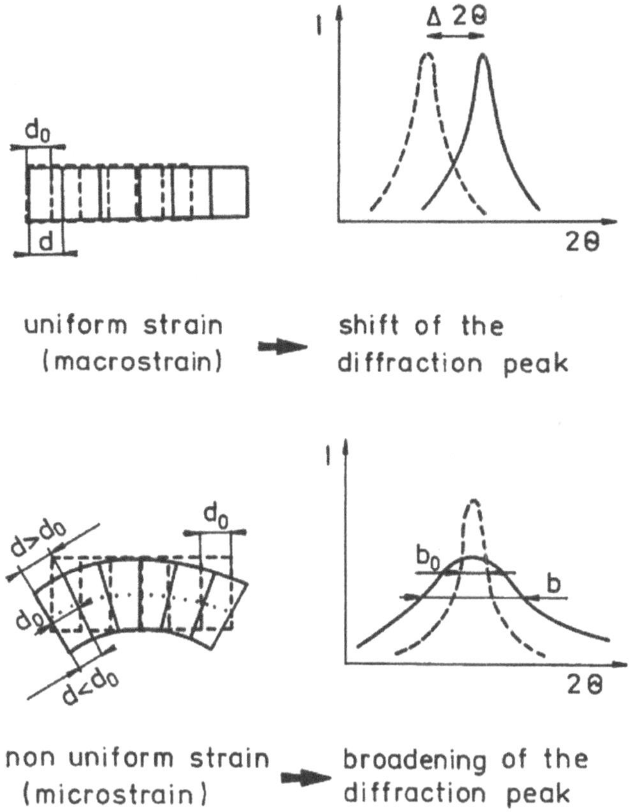

Fig. 3.23 Peak shift and broadening of a diffraction line [Maeder 1986]

Several theoretical models have been developed to obtain structural information through mathematical analysis of the XRD profile [Scherrer 1918; Warren and Averbach 1950; Williamson and Hall 1953]. The leading idea is to find a way how to substract the effect of instrumental broadening of X-ray diffraction profile and how to approximate peak shape by proper mathematical model. In Williamson Hall method it is assumed that the size and strain profile components have Cauchy shape and a method for deconvolution of size and strain broadening by looking at peak widths as a function of 2Θ is applied. The corresponding integral breadths are linearly combined and the integral breadth of the total physical broadening β_f can be written as (known also as Cauchy-Cauchy or Williamson-Hall plot)

$$\beta_f^* = \frac{\beta_f \cos\Theta}{\lambda} = \frac{1}{D_v} + 2\varepsilon d^*; \frac{2\sin\Theta}{\lambda} = \frac{1}{d} = d^* \tag{3.9}$$

To make a Williamson Hall plot, a plot of β_f^{*2} should result in a straight line, and the values for size and strain can then be obtained from the intercept, equal to D_v, and the slope of strain line respectively. Others variants of the Williamson-Hall

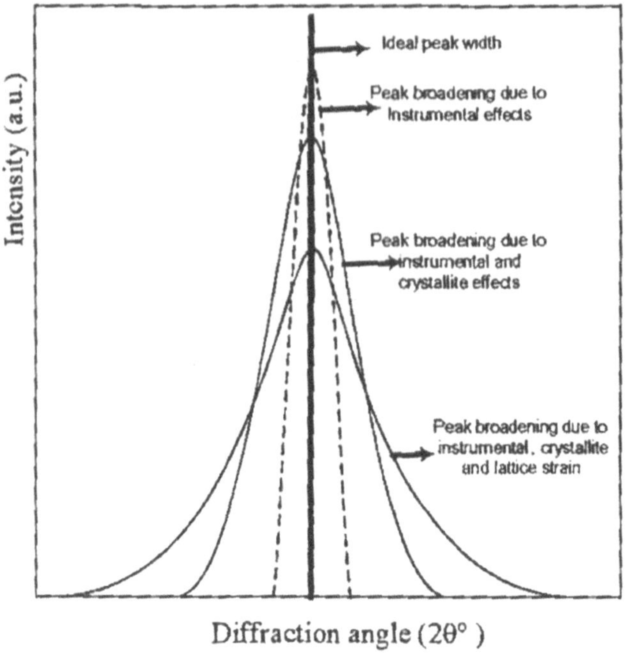

Fig. 3.24 The peak broadening of a diffraction line [Pourgahramani and Baláž 2008]

method exist, adapting Gaussian shaped functions. The Cauchy-Cauchy (or often called the Williamson-Gaussian) and Gaussian-Gaussian peak shape functions are used most frequently.

In practice, various methods of determination of diffraction line broadening are used for investigating the microstrain in mechanically activated solids (Fig. 3.25).

The HWB (diffraction line broadening) can be found out by measuring the width of diffraction line at half of its maximum intensity (I_{max}) whereas ILB (integral width) has to be calculated from

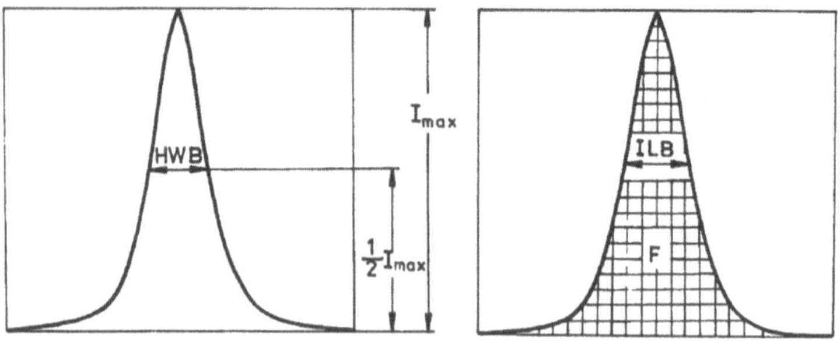

Fig. 3.25 Diffraction line broadening (HWB) and integral width (ILB) of X-ray line [Ludwig 1978]

$$ILB = \frac{F}{I_{max}} \tag{3.10}$$

where F stands for the area under the diffraction line whose maximum intensity is I_{max}.

The diffraction line broadening depends on the experimental conditions used in the measurement of the diffraction peak (*instrumental line broadening*) and on the properties of the measured solid phase (*physical line broadening*). For obtaining further information from physical line broadening it is necessary to eliminate instrumental line broadening. The correction can be achieved by comparing the obtained result with the resulting diffraction of the sample that has not been disordered by mechanical activation [Klugg and Alexander 1974] or by eliminating instrumental effects by means of correction coefficients [Köchendorfer 1944].

The physical line broadening is dependent on two principal factors

- *crystallite size* (called coherently diffracting domains) and
- *lattice strain* which is created by microstrain.

In mechanical activation these factors can be effective either jointly or separately. A serious problem that faces us is to distinguish the size effect from the strain effect. The possibility of determining the both coincident effects has been described by [Köchendorfer 1944] and is represented in Fig. 3.25. According to this procedure the quantities $b\cos\theta$ and $b/\tan\theta$ are plotted against Bragg angle θ. The change in $b\cos\theta$ characterises the lattice strain while the crystallite size is responsible for the change in $b/\tan\theta$.

Provided $b\cos\theta$ is independent of θ, then the physical line broadening b depends only on crystallite size D. Provided $b/\tan\theta$ is independent of θ, then the physical line broadening b is due only to lattice strain and the relative lattice strain $\Delta\varepsilon/\varepsilon$ can be calculated from formula

$$\frac{\Delta\varepsilon}{\varepsilon} = \frac{b}{4r\tan\theta}1000 \tag{3.11}$$

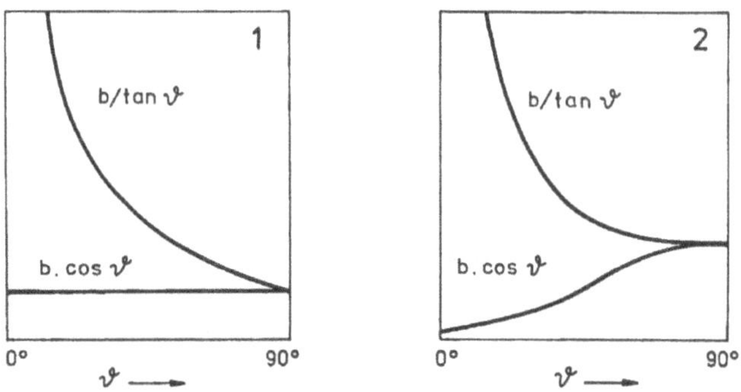

Fig. 3.26 Effect of crystallite size and lattice strain on diffraction line broadening: 1 – crystallite size, 2 – lattice strain and crystallite size [Ludwig 1978]

If the crystallite size and the lattice strain are simultaneously present in mechanically activated solids, both values can be assessed on the basis of physical broadening of two selected diffraction lines by using the equations

$$D = \frac{Kr\lambda\,(b_2 \cos\theta_2 \sin\theta_2 - b_1 \cos\theta_1 \sin\theta_1)}{b_2 \cos\theta_2 b_1 \cos\theta_1\,(\sin\theta_2 - \sin\theta_1)} \tag{3.12}$$

$$\frac{\Delta\varepsilon}{\varepsilon} = \frac{\lambda\,(b_2 \cos\theta_2 - b_1 \cos\theta_1)}{D\,(b_2 \cos\theta_2 \sin\theta_2 - b_1 \cos\theta_1 \sin\theta_2)} \tag{3.13}$$

where the symbol K, r, and λ stands for the shape factor, the radius of goniometer and the wavelength of the X-radiation used, respectively. Subscripts 1 and 2 correspond to two diffraction lines for which $\theta_2 > \theta_1$.

An advantage of XRD method under discussion in nanoscience is that this method provides a very simple possibility for estimating the particle size from broadening of the XRD lines by using the Scherrer formula

$$D = \frac{K\lambda}{b\cos\theta} \tag{3.14}$$

where D is the particle size, λ is the wavelength of the X-ray radiation used, b is the line width at half of the maximum peak intensity on a 2θ scale (HWB), θ is the angle of the considered Bragg reflection and K is a constant close to unity [Scherrer 1918].

Values for the coefficient K depend on such factors such as the geometry of crystallites and are unfortunately, not always consistently used in the literature. Moreover, XRD is sensitive to the size of coherent scattering domains which significantly differ from the particle size in the case present lattice defects or amorphous surface layers which is frequent case in mechanochemistry.

Frequently the formula (3.14) is simply used with $K = 1$ to give a rough estimation of particle size. For more precise calculations, the formula should reflect the geometrical properties of the sample; that is, one should take into account factors such as the shape of crystallites and the size distribution because the number of parallel lattice planes responsible for the line broadening may locally vary in the sample. Initially, Scherrer developed his formula for cubic crystallites and gave a value $K = 0.94$. In nanoscience a nearly spherical particles are present and value for $K = 1.2$ was derived and used for such type of particles including semiconductor nanocrystals [Borchert et al. 2005; Nanda et al. 1999, 2000].

In Fig. 3.27 a powder X-ray diffractogram for nanocrystalline $CuPt_3$ sample is given with values of d calculated after formula (3.13) with $K = 1.2$.

In Table 3.2 comparison of particle diameter estimation by XRD and TEM methods for the sample is given.

Several methods have been compared for calculation of the crystallite size and strain values for mechanically activated hematite α-Fe_2O_3 [Pourgahramani and Forssberg, 2006; Pourgahramani and Baláž 2008]. The values obtained by application of Scherrer, Williamson-Hall and Warren-Averbach methods are given in Fig. 3.28 as a function of specific energy input by high-energy milling. The crystallite sizes decrease and strain values increase drastically as the specific energy

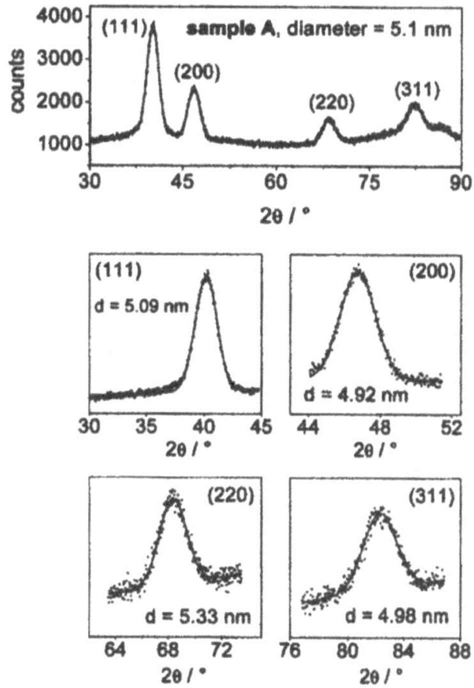

Fig. 3.27 Powder X-ray diffraction patterns of $CuPt_3$. The four available reflections have been fitted with Gaussian profiles [Borchert et al. 2005]

input grows. Among all of methods, the Scherrer equation gives the smallest crystallites and the highest values of strain.

In literature there are also other relationships which estimate the change in real structure of mechanically activated crystalline substances [Ludwig 1978].

The degree of structure disorder F by formula has been defined

$$F = \frac{(\mathrm{ILB})_x}{(\mathrm{ILB})_0} \tag{3.15}$$

where $(\mathrm{ILB})_x$ is integral width corresponding to the mechanically activated sample and $(\mathrm{ILB})_0$ is integral width corresponding to non-activated (reference) sample

Table 3.2 Nanosized particle size estimation by XRD and TEM methods [Borchert et al. 2005]

Method	Diameter (nm) and standard deviation
XRD (111)	5.09
XRD (200)	4.92
XRD (220)	5.33
XRD (311)	4.98
XRD average	5.1±0.2
TEM average	4.86±0.45

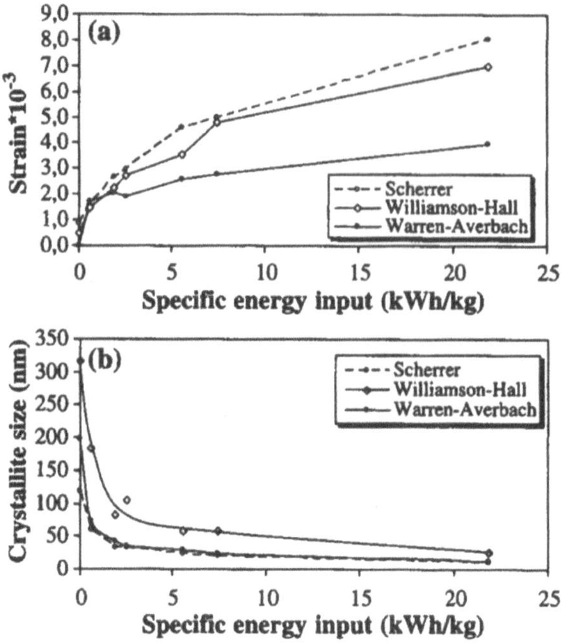

Fig. 3.28 Variation of crystallite size and strain obtained using different methods with specific energy input by mechanical activation of α-Fe_2O_3 [Pourgahramani and Forssberg, 2006]

[Patzak 1966]. Parameter F is a summary quantity for all factors which bring about a broadening of diffraction line (after elimination of instrumental line broadening). Thus we can obtain relative values in which the effect of activation on a given substance is reflected.

Another relative method serving for estimation of the degree of crystallinity (content of crystalline phase) X is the method put forward by Ohlberg and Strickler. The effect of mechanical activation can be evaluated by a mass fraction of the crystalline phase in the activated sample, X compared with the reference substance (non-activated substance) which is assumed to correspond to 100% crystallinity. Thus it holds that

$$X = \frac{U_0}{I_0}\frac{I_x}{U_x}100 \tag{3.16}$$

where U_0 and U_x denote the backgrounds of non-activated (reference) sample and activated sample while I_0 and I_x are integral intensities of diffraction lines of non-activated (reference) sample and activated sample, respectively [Ohlberg and Strickler 1962].

Equation (3.16) is based on the assumption that the mechanical activation is not accompanied by a texture error (e.g. due to preferential orientation where there was none in the non-activated sample). Sometimes, the complementary value of amorphization A (also called degree of amorphization or content of X-ray amorphous phase) is used. It is defined by equation

$$A = 100 - X \tag{3.17}$$

The precision of determination of individual X-ray – quantities varies and is dependent on a great number of factors. An important factor is elimination of the effect of primary extinction by removing the particles exceeding 40 μm [Johan et al. 1970]. The primary particle size can be determined with precision of 5–10% up to the upper limit of 200 nm size of nanoparticles and in the case of lattice strain up to any small value. The precision of determination of the content of crystalline phase also varies within the range of a few percent.

The problem of precise size determination is multiplied in diffraction patterns of nanometer-sized particles. Single crystal nanoparticles exhibit features in diffraction that are size-dependent, including slight shifts in the position of Bragg peaks, anomalous peak heights and widths. In fact, the familiar concept of a diffraction peak begins to loose meaning when considering diffraction from such small particles. There are some important consequences, which have been known for some time [Grigson and Barton 1967]

- not all peaks associated with a particular structure are resolved in small crystalline particles
- those peaks that are resolved may have maxima that do not align with expected bulk peak positions
- peak shapes, peak intensities and peak widths may differ from extrapolated bulk estimates
- few minima in intensity between peaks actually reach zero and
- small, size-related features appear in the diffraction pattern.

The very small grain size of clusters in nanometer-sized particles gives their diffraction pattern the appearance of an amorphous material. Of-course they are not amorphous. In short, the common assumption that there exist some kind of underlying long-range order in the system under study does not apply to nanometer-sized particles [Gleiter 1989]. It must also be remembered that the entire diffraction pattern contains structural information about the nanoparticles and therefore that good quality data should be collected for the whole profile, not just in the more intense regions (peaks) of the diffraction pattern [Zanchet et al. 2000].

In this case techniques applying the concept of short-range order can be applied. The most proper is the technique of X-ray Absorption Spectroscopy (XAS) which explores variations in the absorption coefficient of matter with phonon energy. When a monochromatic X-ray beam passes through a material, its intensity is reduced by various interaction processes (scattering, absorption, etc.) For hard X-rays (more than 1000 eV), the photoelectric effect dominates [Koningsberger and Prins 1988]. Based on the energy of ejected electrons it is possible to roughly divide the absorption spectrum in two regions, according to different interaction regimes with the surrounding atoms

- XANES (X-ray Absorption Near Edge Structure): ≈ 0–40 eV above E_0 (E_0– the threshold photon energy for core electron excitation), yielding information about symmetries and chemical state

- EXAFS (Extended X-ray Absorption Fine Structure): ≈ 40–1000 eV above E_0, yielding information about coordination numbers and inter-atomic distances.

A large number of EXAFS studies have been done on nanocrystalline semiconductors. For example, Rockenberger et al. examined nanocrystals of CdS in the range 1.3–12 nm in diameter with different types of surface stabilizing agents [Rockenberger et al. 1997].

For diffraction, it would be desirable to include a variety of techniques in future studies of nanometer sized particles in order to obtain a coherent picture of their structure.

3.9 Mössbauer Spectroscopy

Mössbauer spectroscopy is a versatile technique that can be used to provide information in many areas of mechanochemistry and nanoscience. It can give very precise information about chemical, structural, magnetic and time-dependent properties of a solid.

In 1957 Mössbauer showed that nuclear radiation can be emitted or absorbed recoilless if the atoms a placed in a solid state. This new phenomena originally called recoil-free nuclear resonance fluorescence later become commonly known as the "Mössbauer effect" and author received the Nobel Prize in 1961 for his work [Mössbauer 1958].

A schematic illustration of the principle of the Mössbauer effect is given in Fig. 3.29. The γ-photons emitted from a special radioactive source are resonantly absorbed by the corresponding nuclei of a suitable absorber and registered by a detector.

The Mössbauer effect occurs when atoms are in a solid lattice or matrix. The chemical binding energies in solids (1–10 eV) are much greater than free atom recoil energies, E_R. For entire crystal, which can be order of 10^{15} greater than a single atom

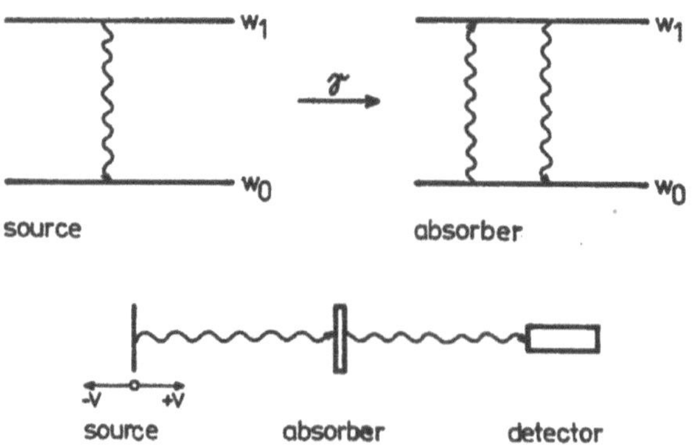

Fig. 3.29 Principle of the Mössbauer effect

E_R is negligible. However, although the nucleus is bound within the lattice it is still free to vibrate. The recoil energy can still be transferred to the lattice as a quantised lattice vibration, or phonon. If the recoil energy is less than the lowest quantised vibration mode, then a recoil free event occur.

Suitable Mössbauer isotopes are those with a relatively low energy of the γ-radiation and therefore a high recoil free fraction as well as a comparatively long mean lifetime. The most appropriate isotope is ^{57}Fe which has been used in more experiments than all the other 40–50 possible isotopes. ^{57}Fe Mössbauer spectroscopy uses ^{57}Co source. The half-life of ^{57}Co is 271.7 days and this isotope decays by electron capture to the $I = 5/2$ excited state of ^{57}Fe. After several steps the excited iron isotope release and energy quantum of 14.41 keV with suitable properties for Mössbauer effect. Usually ^{57}Co atoms are applied in a Rh foil matrix which provides a solid environment for the ^{57}Co atoms with a high recoil-free fraction and a cubic, non-magnetic site environment to produce a high recoil-free fraction and a cubic, non-magnetic site environment to produce mono-energetic γ-rays.

Figure 3.30 shows a scheme of a Mössbauer spectrometer. The source velocity is controlled by a transducer which is oscillated with constant acceleration. A waveform generator sends a reference waveform to the drive amplifier. This signal is then sent to the vibrator where it is converted to a mechanical oscillation of the drive shaft and source. The detector is a proportional counter. The pulse magnitude from the detector is processed and amplified. The Mössbauer spectrum is then obtained on a recorder after accumulating the count-rates by means of a multichannel analyzer which gives the dependence of the measured number of impulses on the velocity of moving source.

Experiments are usually performed with stationary absorber whereas the moving source is kept at ambient conditions.

The interaction between a nucleus and its surrounding environment is known as a hyperfine interaction. In practice, only three kinds of hyperfine interactions

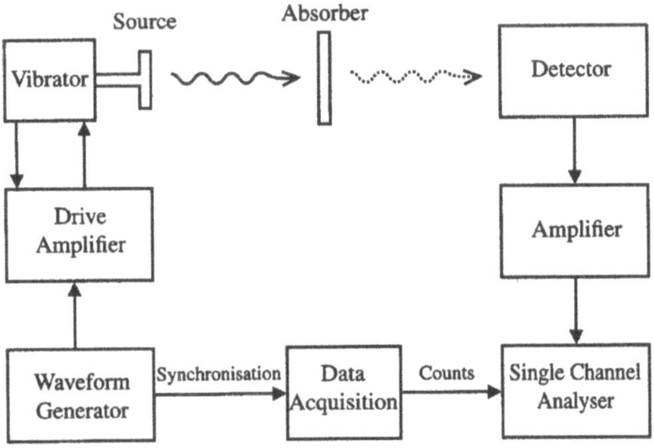

Fig. 3.30 The scheme of Mössbauer spectrometer [Bland 2002]

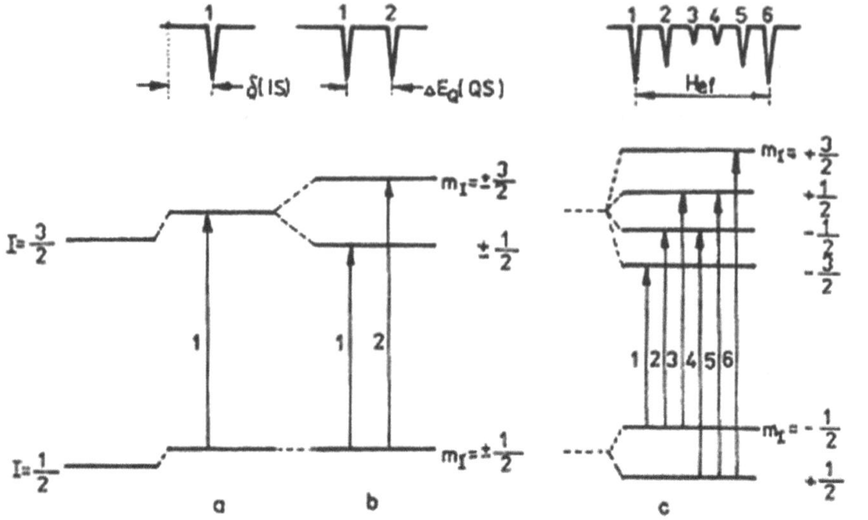

Fig. 3.31 Hyperfine interactions in Mössbauer spectroscopy: a – isomer shift (δ or IS), b – quadrupole splitting ΔE_0 (or QS), c – magnetic splitting (H_{ef})

are considered in Mössbauer spectroscopy: *isomer shift, quadrupole splitting* and *magnetic splitting* (Fig. 3.31a). The values of these parameters are dependent on kind and magnitude of the hyperfine interactions between charge distribution and magnetic moment of ^{57}Fe and electric and magnetic field acting on nucleus.

Isomer shift (δ or IS) arises due to the non-zero volume of the nucleus and the electron charge density due to s-electrons within it, leading to an electric monopole (Coulomb) interaction which alters the nuclear energy levels. The isomer shift (IS) provides information about the local chemical bonding. For this reason, it is frequently called *chemical shift* (δ). The magnitude of isomer shift increases with the decreasing density of the s-electrons surrounding the nucleus and with increasing density of the d-electrons (note: 3d-electrons screen 4s-electrons and thus reduce their density on nucleus).

The isomer shift is a measure of the deviation of Mössbauer spectrum from the zero point on the scale of velocity v of the source (Fig. 3.29), the zero point is referred to a standard (usually elemental Fe). On the basis of the values of the isomer shift we can estimate the type of chemical bond, oxidation state and coordination number of iron. For example, Fe^{2+} and Fe^{3+} have configurations of $(3d)^6$ and $(3d)^5$ respectively. The ferrous ions have less s-electron density at the nucleus due to the greater screening of the d-electrons. This produces a positive IS greater in Fe^{2+} ion than in Fe^{3+} ion (Table 3.3).

As the shift cannot be measured directly is it quoted relatively to a known absorber. For example ^{57}Fe Mössbauer spectra will be often quoted relative to α-Fe at room temperature.

Quadrupole splitting ΔE_0 (or QS) is a consequence of interaction between the nuclear quadrupole moment and the gradient of the electric field at the site of nucleus.

Table 3.3 Values of isomer shift IS for bivalent and trivalent iron in sulfides [Marfunin and Mkrtčjan 1967]

Oxidation state of Fe	+2	+3
IS ($mm\,s^{-1}$)	0.38–1.20	0.34–0.47

The value of the quadrupole splitting is proportional to the deviation of nuclear surroundings from cubic symmetry (note: if the symmetry of surroundings of the nucleus is cubic, then QS = 0). The neighbouring ions affect the wave functions around the nucleus and thus QS gives information about the size and symmetry of the surrounding crystal field. In the case of an isotope with $I = 3/2$ excited state, such as ^{57}Fe or ^{119}Sn, the excited state is split into two substances $m_I = \pm 1/2$ and $m_I = \pm 3/2$. This is shown in Fig. 3.31b, giving a two line spectrum or "doublet". The distance between the centroid of both lines and the zero point of the velocity scale corresponds to the isomer shift. Both lines should have equal intensities except for the occurrence of an anisotropic recoil-free fraction in a polycrystalline sample or in the presence of texture or relaxation. The quadrupole splitting ΔE_0 is another important parameter for the application of Mössbauer spectroscopy in mechanochemistry and nanoscience, because ΔE_0 provides a wealth of information on the electronic and crystallographic structure.

Magnetic splitting (H_{ef}) is a consequence of the interaction between the nuclear magnetic dipole moment and internal magnetic field at the nucleus. Owing to splitting of the ground level of ^{57}Fe into two sublevels and of the excited first level into four sublevels and taking the selection rules into account, six possible γ-energies arise. The allowed transitions between these produce a six-line spectrum (Fig. 3.31c). The distance between the first and sixth line of spectrum characterize the effective magnetic field of nucleus H_{ef} with QS = 0. The magnetic hyperfine splitting gives information about magnetic moments and their magnetic arrangement relative to the electric field gradient and is also important for qualitative analysis.

In mineralogy the application of Mössbauer spectroscopy is conditioned mainly by the presence of iron though there are records of the spectra of minerals with tin and tungsten as Mössbauer isotopes [Greenwood and Gibb 1971; Bancroft 1973; Malševa 1975; Vaughan and Craig 1978]. The outstanding sensitivity of Mössbauer spectroscopy enables the resolution of Fe^{2+} from Fe^{3+} with a relative precision of $\pm 1\%$ which is a great advantage for determining the oxidation state of iron. The method is more sensitive when compared with chemical analysis and the results of measurements are not affected by the presence of accessory ions of transition metals which coincide, e.g. in the course of dissolution preceeding a chemical analysis. The methods of XRD analysis as well as other methods, such as EDS analysis or electron microprobe analysis, however, are not able to distinguish between the above-mentioned oxidative states. For iron it is also possible to determine electronic configurations (low or high spin), coordination symmetry (tetra-or octahedral positions) and the pertinent deviations.

Table 3.4 Mössbauer data for selected sulfides

Mineral	Temperature (°C)	IS* ($mm\,s^{-1}$)	QS ($mm\,s^{-1}$)	H_{ef} (kgauss)	Reference
Sphalerite	27	0.66	0.80	–	Marfunin and Mkrtčjan 1967
	−193	0.86	2.04		Scott 1971
	27	0.64–0.67	0.62–0.67		
Wurtzite	27	0.69	0.80		Marfunin and Mkrtčjan 1967
	−193	0.91	2.05		
Pyrite	27	0.30	0.60		Imbert et al. 1963
	−196	0.31	0.61		Temperley and Lefevre 1966
	27	0.31	0.61		
	−192	0.40	0.62		
Marcasite	27	0.27	0.51		Temperley and Lefevre 1966
	−192	0.37	0.50		
Arsenopyrite	27	0.25	1.05		Imbert et al. 1963
	−196	0.26	1.07		
Chalcopyrite	27	0.20	0	350	Vaughan and Burns 1972
	−193	0.27	0	365	Marfunin and Mkrtčjan 1967
	27	0.20	0	246	Cabri and Goodman 1970
	−196	0.37	0.17	206	Greenwood and Whitfield 1968
Bornite	27	0.38	0		Marfunin and Mkrtčjan 1967
	−193	0.48	0.20		
	27	0.39	0.16		Vaughan and Burns 1972
	−196	0.52	0.25		Greenwood and Whitfield 1968

*relative to metallic iron

The Mössbauer spectrum characteristic of a certain mineral can be used for the identification of individual minerals in a mixture (the so-called "fingerprint" technique). The spectra of sulphides, oxides, hydroxides, jarosites, silicates etc. are currently identified and quantitatively evaluated. The sensitivity of the hyperfine structure to deviations caused by surroundings of the Mössbauer atom enables the study of changes in the stoichiometry of minerals. In Table 3.4 the Mössbauer parameters for different sulfides are summarized.

However, Mössbauer spectroscopy alone rarely provides sufficient insight into a nanosized solid for a full and proper understanding of its properties [Campbell and Kaczmarek 1969]. The information obtained by other techniques such as mentioned in previous paragraphs is also needed. In Fig. 3.32 such approach is illustrated.

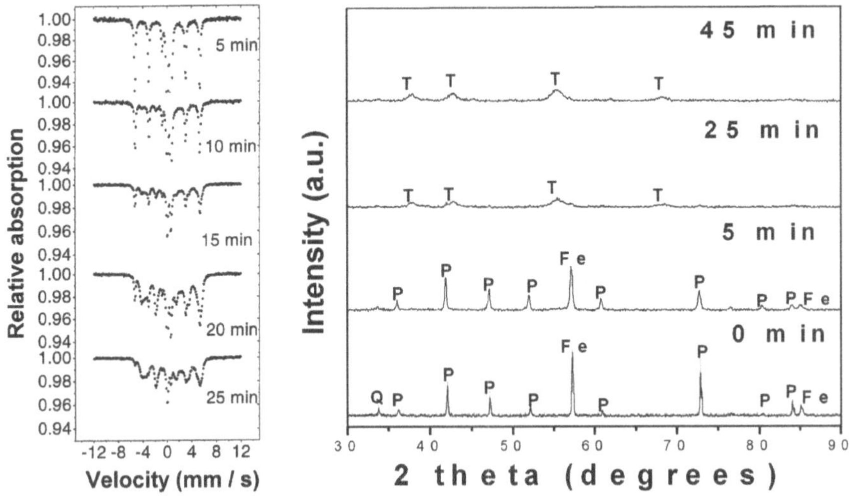

Fig. 3.32 Mössbauer spectra (*left*) and XRD patterns (*right*) for mechanochemical synthesis of nanosized troilite FeS according to reaction $FeS_2 + Fe \rightarrow 2FeS$, T-troilite FeS, P-pyrite FeS_2, Q-quartz SiO_2 [Baláž et al. 2004a]

3.10 Electron Paramagnetic Resonance

Electron paramagnetic resonance (EPR) also known as electron spin resonance (ESR) and electron magnetic resonance (EMR) deals with paramagnetic compounds to study the magnetic properties of solids. The method was discovered by Zavoisky in 1944 in USSR who published the first EPR spectrum of $CuCl_2.2H_2O$.

The electron paramagnetism is a result of the existence of unpaired electrons which have magnetic moments. If a substance containing unpaired electrons is put into strong magnetic field two energy levels arise. These levels correspond to conformable and nonconformable orientation of the electron magnetic moment to the external magnetic field. The difference between energies of both levels satisfies the so-called resonance condition of EPR and is equal to

$$\Delta E = g\mu_B \overrightarrow{B_o} \tag{3.18}$$

where g, μ_B and $\overrightarrow{B_o}$ stand for g-factor of spin-orbital interaction, magnetic moment of electron (Bohr magneton) and induction of magnetic field, respectively. The Eq. (3.18) is the fundamental equation in EPR theory. Knowledge of g-parameter gives us information about paramagnetic center electron structure. Due to several reasons g parameter of EPR signal for many paramagnetic atoms in solid state has different value. In the case of free ion, the g parameter has isotropic properties and its value being equal to 2.0023. In practice single paramagnetic probe never occurs but only population of probes with many paramagnetic centers. If this configuration of probes is in thermal equilibrium, statistical placing is described by Boltzmann distribution. In general, the g parameter is an anisotropic quantity with deviation from the mentioned value of g arising as a consequence of orbital contributions to the total magnetic moment of electron [Hedvig and Zentai 1969].

The magnetic moment of electron (Bohr magneton) is coupled with the nuclei that it is associated with. Most nuclei and their isotopes have a magnetic moments characteristic of them. The magnetic moment of electron interacts with the nuclear magnetic moment. This results in multiple lines in EPR spectrum. When the system has more than one unpaired electron, the energy levels are split even in the absence of applied magnetic field due to the interaction between the electrons. It is known as the zero-field splitting. The splitting of the energy levels can be given by (2S+1), where S is the total spin the electron. The nuclear hyperfine splitting is given by (2I+1), where I is the nuclear quantum number.

The experimental equipment for taking EPR spectra is represented in Fig. 3.33. The source of microwave radiation is a clystron K. The radiation is passed through a metallic waveguide T into the resonance hole R containing the sample V and subsequently to the detector D. This detector transforms the microwave radiation in direct

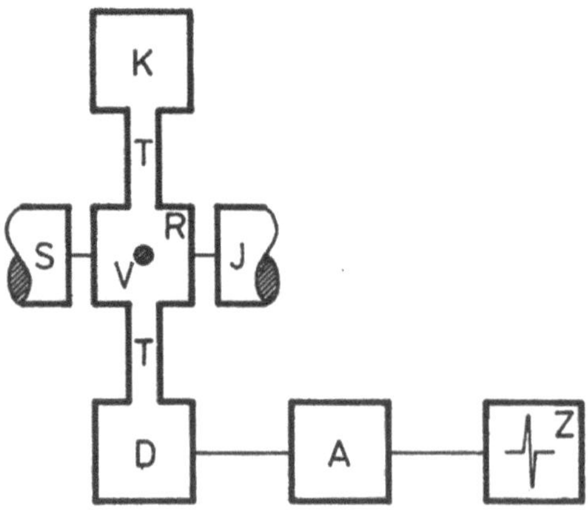

Fig. 3.33 The scheme of EPR spectrometer. K – klystron, T – waveguide, R – resonance hole, D – detector, SJ – magnet, V – sample, A – amplifier, Z – registration

current. The splitting of magnetic energy is achieved in the field of strong electromagnet SJ. After reaching the condition of resonance, i.e. when the microwave radiation is absorbed by the sample, the energy in the resonance hole decreases. The obtained signal is amplified in an amplifier A and recorded. The record of signal Z expresses the dependence of absorption of radiation on variable magnetic field and for better resolution it is given in the form of the first derivative of absorption signal (Fig. 3.34).

The most commonly used EPR spectrometers are in the range of 9–10 GHz. However, advances in electronics have facilitated the development of spectrometers working at frequencies ranging from several hundred MHz to several hundred GHz.

Useful information about the structure and magnetic properties of the solids like hyperfine couplings, intermolecular interactions, exchange coupling, etc. can be obtained by EPR method [Wertz and Bolton 1972; Pilbrow 1990].

The technique is very sensitive and it is possible to detect sub-micrometer quantities of paramagnetic species. The most frequent paramagnetic ions are ions of transition metals of the 3d-group, ions of rare earth metals of the 4f-group and donor admixtures. These ions have unpaired electrons in incomplete orbitals. Natural substances often contain these ions which are either within their crystal structure, or are present in the form of admixtures.

However, the paramagnetic centres present are not always of genetic origin and can also originate from the breakage of chemical bonds giving rise to free radicals or from the trapping of electrons in defects of solids (colour centres). The colour centres were originally attributed to halides of alkali metals but were subsequently also identified in other solids. We distinguish centres F, V, N, Z_1 and others [Henderson 1972]. F-centre and a V-centre are represented in Fig. 3.35. In principle, the electrons in F-centres are not bound and migrate between the cations surrounding the hole (vacancy) due to removal of an anion from the crystal lattice site. For instance, the formation of V-centre can be interpreted by the idea that the electrons are bound to a neutral atom formed by tearing an electron away from an anion. The neutral atoms fall into holes and thus form the V-centres [Rákoš 1988].

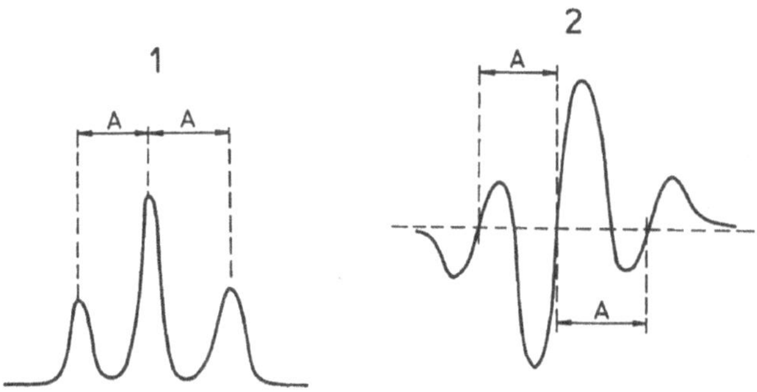

Fig. 3.34 EPR spectrum: 1 – normal spectrum, 2 – 1st-derivative

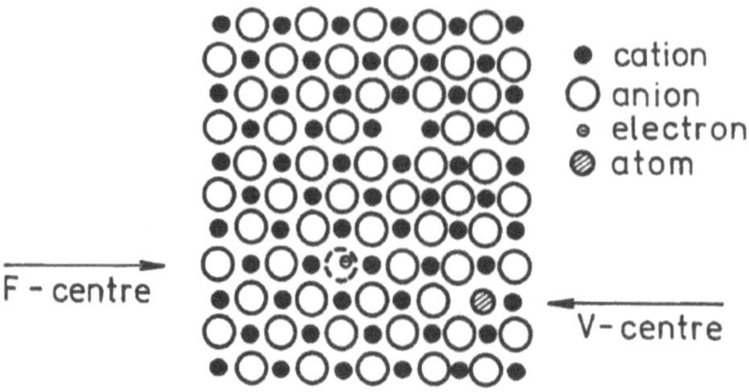

Fig. 3.35 F-centre and V-centre in the structure of a crystalline solid

In the field of study of mechanical activation the EPR method has proved to be very useful [Vlasov and Kakazej 1979; Boldyrev 1983; Avvakumov 1986; Tkáčová 1989; Baláž 2000]. In the course of high-energy milling the surface area of solids is disordered, the bonds between atoms are interrupted and radicals, atoms or molecules come into existence. These structures often cannot react rapidly with other and are therefore detectable by the EPR method. The point, linear and volume defects generated in the process of milling represent a notable source of paramagnetism.

For instance, the vacancies which are due to breakage of chemical bonds belong among point defects. As a matter of fact, the F-centre represented in Fig. 3.35 is a new defect due to an electron trapped in an anion vacancy. The admixture centres which bring about considerable distortion of interatomic bonds also belong among point defects. They are frequently used as paramagnetic markers which are either introduced in the form of admixtures into the investigated object or are formed in this object in the process of technological operation. The addition of manganese is frequently employed for this purpose. The disturbances produced in this element are relatively small which brings about long times of relaxation and a good detectability [Ljudvig and Vudberi 1964]. The EPR spectrum is fairly distinguishable and its identification is facilitated by the fact that manganese exhibits a characteristic spectrum of hyperfine interaction comprising 6 lines. Each line of hyperfine structure contains 5 components of fine structure owing to which the total number of lines in the spectrum is equal to 30. In crystals, subject to strain and in polycrystalline materials only one resonance line of manganese can usually be observed. In literature, we can find information about the correlation of parameters of the EPR spectrum of bivalent manganese with particle size of mechanically activated periclase [Kakazej 1974] or with concentration of defects in crystalline lattice of this solid [Koloskova 1962; Hennig et al. 1984].

EPR has been applied in nanoscience to study conduction electrons in metal nanoparticles, and to detect the presence of conduction electrons in nanotubes to determine where in the tubes are metals or very narrowband semiconductors [Poole and Owens 2003]. (ZnS:Mn) semiconductor nanoparticles have been prepared and EPR method was applied for characterization of Mn^{2+} ions distribution in ZnS matrix (Fig. 3.36).

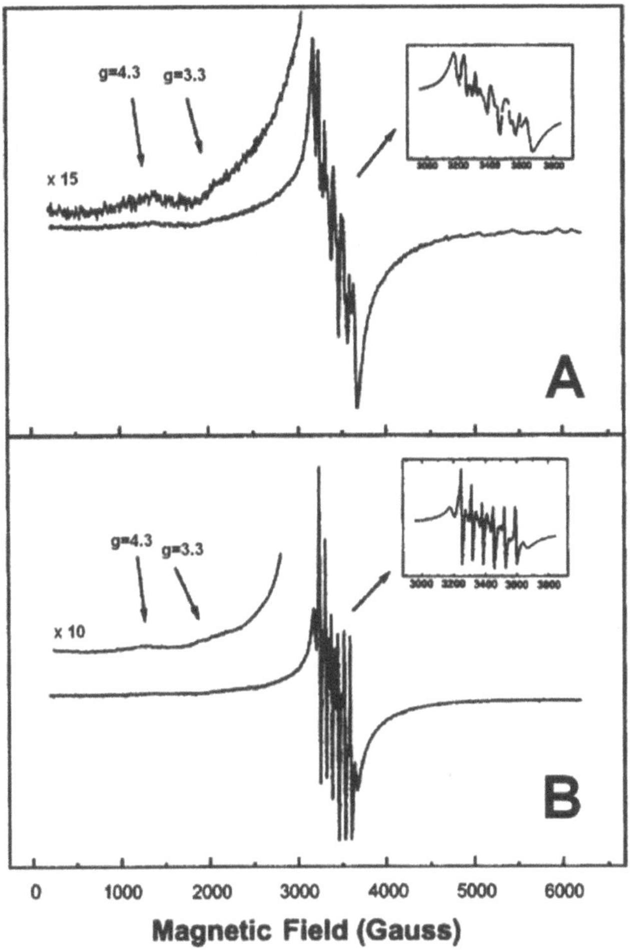

Fig. 3.36 EPR spectra of (ZnS: Mn) nanocrystals, A – classical precipitation, B – precipitation with addition of metacrylic acid [Baláž et al. 2002b]

3.11 Conclusion

The advanced methods of nanocrystalline solids identification described in previous paragraphs do not cover the whole spectrum of the possible methods related to the fine particles behaviour. Characterization of degree of dispersion, particle size distribution and specific surface area are not covered here. There is a great number of publications which cover this field [Chodakov 1972; Juhász 1978; Boldyrev 1983; Heinicke 1984; Avvakumov 1986; Molčanov et al. 1988; Tkáčová 1989; Juhász and Opoczky 1990]. Examples of the application of the advanced identification methods will be presented in further chapters of this monograph.

References

Abramov, A.A., Leonov, S.B. and Sorokin, M.M. (1982) *Chemistry of Flotation Systems*. Nedra, Moscow (in Russian).

Avvakumov, E.G. (1986) *Mechanical Methods of Chemical Processes Activation*. Nauka, Novosibirsk (in Russian).

Baláž, P., Bálintová, M., Bastl, Z., Briančin, J. and Šepelák, V. (1997) Characterization and reactivity of zinc sulphide prepared by mechanochemical synthesis. Solid State Ionics 101–103, 45–51.

Baláž, P. (2000) *Extractive Metallurgy of Activated Minerals*. Elsevier, Amsterdam.

Baláž, P., Takacs, L., Jiang, J.Z., Godočíková, E., Luxová, M. and Bastl, Z. (2002a) Preparation of Cu/FeS nanoparticles by mechanochemical reduction of copper sulphide. Metallic Materials 40, 268–280.

Baláž, P., Valko, M., Boldižárová, E. and Briančin, J. (2002b) Properties and reactivity of Mn-doped ZnS nanoparticles. Materials Letters 52, 188–191.

Baláž, P., Aláčová, A., Godočíková, E., Kováč, J., Škorvánek, I. and Jiang, J.Z. (2004a) Study of magnetic properties of nanopowders prepared by pyrite-troilite transformation via high-energy milling. Czechoslovak Journal of Physics, Supplement D 54, D197–D200.

Baláž, P., Godočíková, E., Kril'ová, L., Lobotka, P. and Gock, E. (2004b) Preparation of nanocrystalline materials by high-energy milling. Materials Science and Engineering A 386, 442–446.

Bancroft, G.M. (1973) *Mössbauer Spectroscopy, an Introduction for Inorganic Chemists and Geochemists*. McGraw Hill, London.

Beran, A., Voll, D. and Schneider, D. (2004) In: A. Beran, E. Libowitzky (Eds.) *EMU Notes in Mineralogy*, Vol. 6, pp. 189–226.

Berčík, J. (1977) *Physical and Physico-Analytical Methods*. Alfa, Bratislava (in Slovak).

Bland, J. (2002) *A Mössbauer Spectroscopy and Magnetometry Study of Magnetic Multilayers and Oxides*. PhD. Thesis, University of Liverpool.

Boiteux, Y.P. (1986) *Surface Characterization and Manipulation of Si_3N_4 and SiC Powders*. M.Sc. Thesis, Berkeley.

Boldyrev, V.V. and Meyer, K. (Eds.) (1973) *Festkörperchemie*. VEB Deutscher Verlag für Grundstoffindustrie, Leipzig (in German).

Boldyrev, A.I. (1976) *Infrared Spectra of Minerals*. Nedra, Moscow (in Russian).

Boldyrev, V.V. (1983) *Experimental Methods in Mechanochemistry of Inorganic Solids*. Nauka, Novosibirsk (in Russian).

Borchert, H., Shevchenko, E.V., Robert, A., Mekis, I., Kornowski, A., Grübel, G. and Weller, H. (2005) Determination of nanocrystalline sizes: a comparison of TEM, SAXS and XRD studies of highly monodisperse $CoPt_3$ particles. Langmuir 21, 1931–1936.

Briggs, D. and Seah, M.P. (1983) *Practical Surface Analysis by Auger and X-ray Photoelectron Spectroscopy*. Wiley, Chichester.

Cabri, L.J. and Goodman, R.H. (1970) Mössbauer spectra of some copper-iron sulfides. Geochimija 5, 636–639.

Campbell, S.J. and Kaczmarek, W.A. (1969) Mössbauer effect studies of materials prepared by mechanochemical methods. In: G.J. Long and F. Grandjean (Eds.) *Mössbauer Spectroscopy Applied to Magnetism and Materials Science*. Plenum Press, New York, Vol. 2, pp. 273–330.

Chodakov, G.S. (1972) *Physics of Grinding*. Nauka, Moscow (in Russian).

Duyckaerts, G. (1959) The infrared analysis of solid substances-a review. Analyst 84, 201–214.

Eggleston, C.M. and Hochella Jr., M.F. (1990) Scanning tunnelling microscopy of sulfide surfaces. Geochimica et Cosmochimica Acta 54, 1511–1517.

Farmer, V.C. Ed. (1974) *The Infrared Spectra of Minerals*. Monograph 4, Mineralogical Society, London.

Gleiter, H. (1989) Nanocrystalline materials. Progress in Materials Science 33, 223–315.

Godočíková, E., Baláž, P., Bastl, Z. and Brabec, L. (2002) Spectroscopic study of the surface oxidation of mechanically activated sulphides. Applied Surface Science 200, 36–47.

Greenwood, N.N. and Whitfield, H.J. (1968) Mössbauer effect studies on cubanite ($CuFe_2S_3$) and related iron sulfides. Journal of Chemical Society A 1697–1699.
Greenwood, N.N. and Gibb, T.C. (1971) *Mössbauer Spectroscopy.* Chapman and Hall, London.
Grigson, C.W.B. and Barton, E. (1967) The development of the face-centred cubic interface functions as crystals grow. Journal of Applied Physics 18, 175–184.
Hedvig, G. and Zentai, G. (1969) *Microwave Study of Chemical Structures and Reactions.* Akadémiai Kiadó, Budapest.
Heinicke, G. (1984) *Tribochemistry.* Akademie-Verlag, Berlin.
Henderson, B. (1972) *Defects in Crystalline Solids.* Edward Arnold, London.
Hennig, H.P., Ebert, I., Tkáčová, K., Jost, H., Pielert, L. and Števulová, N. (1984) Mechanisch induzierte Strukturdefekte und deren Einfluss auf die Reaktivität von MgO. Folia Montana, extraordinary number, 380–388.
Hlavay, J. (1978) Die Untersuchung fester pulverförmiger Materialien mittels Infrarotspektroskopie. In: Z. Juhász (Ed.) *Untersuchungsmethoden zur Charakterisierung mechanisch aktivierter Festkörper.* Közlekedési Dokumentációs Vállalat, Budapest, pp. 128–137 (in German).
Hlavay, J. and Inczédy, I. (1979) Sources of the error of quantitative determination of the solid crystalline minerals by infrared spectroscopy. Acta Chimica Academiae Scientiarum Hungarical 102, 11–18.
http://www.physik.uni-jena.de http://www.physik.uni-jena.de
http://www.rz-berlin.mpg.de
Imbert, P., Gerard, A. and Wintenberger, M. (1963) Étude de sulfure, arséniosulfure et arseniure de fer naturels par effect Mössbauer Comptes Rendus de l'Academie des sciences 256, 4391–4393.
Johan, Z., Rotter, R. and Slánský, E. (1970) *X-ray Analysis of Materials.* SNTL, Prague (in Czech).
Juhász, Z. (Ed.) (1978) *Untersuchungsmethoden zur Charakterisierung mechanisch aktivierter Festkörper.* Közlekedési Dokumentációs Vállalat, Budapest (in German).
Juhász, A.Z. and Opoczky, L. (1990) *Mechanical Activation of Minerals by Grinding: Pulverizing and Morphology of Particles.* Ellis Horwood Limited, Chichester.
Kakazej, N.G. (1974) Defective structure of fine periclase particles. Soviet Powder Metallurgy and Metal Ceramics 13, 322–325.
Klugg, H.P. and Alexander, L.E. (1974) *X-ray Diffraction Procedures for Polycrystalline and Amorphous Materials.* John Willey and Sons, New York.
Koloskova, N.G. (1962) Fizika Tverdogo Tela 4, 3129.
Koningsberger, D.C. and Prins, R. (1988) *X-ray Absorption: Principles, Applications, Techniques of EXAFS, SEXAFS and XANES.* John Wiley and Sons, New York.
Kössler, I. (1970) *Quantitative Infrared Spectroscopy.* SNTL, Prague (in Czech).
Köchendorfer, A. (1944) Zeitschrift für Kristallographie 105, 393–398.
Kulebakin, V.G. (1983) *Transformations of Sulphides by Activation.* Nauka, Novosibirsk (in Russian).
Liese, H. (1974) Infrared absorption (435 to 250 cm^{-1}) and ultraviolet emission analyses of selected sulfides and sulfosalts: a correlative study. Applied Spectroscopy 28, 135–139.
Ljudvig, D. and Vudberi, G. (1964) *Electron Spin Resonance.* Mir, Moscow (in Russian).
Ludwig, G. (1978) Bestimmung von Kristalitgrössen, Gitterfehlordnung und Amorphisierungserscheinungen durch Röntgenuntersuchungen. In: Z. Juhász (Ed.) *Untersuchungsmethoden zur Characterisierung mechanisch aktivierter Festkörper.* Közlekedesi Dokumentációs Vállalat, Budapest, pp. 113–127 (in German).
Maeder, G. (1986) X-ray diffraction and stress measurements. Chemica Scripta A 26, 23–31.
Malševa, T.V. (1975) *Mössbauer Effect in Geochemistry and Cosmochemistry.* Nauka, Moscow (in Russian).
Marfunin, A.S. and Mkrtčjan, A.R. (1967) Mössbauer spectra Fe^{57} of sulphidic minerals. Geochimija 10, 1094–1103.
Molčanov, V.I. and Jusupov, T.S. (1981) *Physical and Chemical Properties of Fine Ground Minerals.* Nedra, Moscow (in Russian).

Molčanov, V.I., Selezneva, O.G. and Žirnov, E.N. (1988) *Activation of Minerals by Grinding*. Nedra, Moscow (in Russian).

Mössbauer, R.L. (1958) Kernresonanz fluoreszenz von Gammastrahlung in Ir^{191}. Zeitschrift für Physik 151, 124–143.

Nakamoto, K. (1978) *Infrared and Raman Spectra of Inorganic and Coordination Compounds*. Wiley, New York.

Nanda, J., Kuruvilla, B.A. and Sarma, D.D. (1999) Photoelectron spectroscopic study of CdS nanocrystallites. Physical Reviews B 59, 7473–7479.

Nanda, J., Sapra, S. and Sarma, D.D. (2000) Size selected zinc sulfide nanocrystallites: synthesis, structure and optical studies. Chemistry of Materials 12, 1018–1024.

Nefedov, V.I. (1984) *XPS Spectroscopy of Chemical Compounds*. Chimija, Moscow (in Russian).

O'Connor, D.J., Sexton, B.A. and Smart, R. (Eds.) (1992) *Surface Analysis Methods in Materials Science*. Springer, Heidlberg.

Ohlberg, S.M. and Strickler, D.W. (1962) Determination of percent crystallinity of partly devitrified glass by X-ray diffraction. Journal of American Ceramic Society 45, 170–171.

Patzak, I. (1966) Berichte des Deutsches keramische Gesselschaft 43, 77–80.

Pilbrow, J.R. (1990) *Transition Ion Electron Paramagnetic Resonance*. Clarendon Press, Oxford.

Poole, CH.P. and Owens, F.J. (2003) *Introduction to Nanotechnology*. Wiley-Interscience, New Jersey.

Pourgahramani, P. and Baláž, P. (2008) Structural changes and characterizations in the powder materials obtained from milling processes. In: M. Yekeler (Ed.) *Powder Technology and Characterization*. Trivandrum, Kerala, pp. 217–250.

Pourgahramani, P. and Forssberg, E. (2006) Microstructure characterization of mechanically activated hematite using XRD line broadening. International Journal of Mineral Processing 79, 106–119.

Pugh, R.J. and Bergström, L. (1986) Surface and solution chemistry studies on galena suspensions. Colloids and Surfaces 19, 1–20.

Rákoš, M. (1988) *Radiospectroscopical Methods*. Alfa, Bratislava (in Slovak).

Rockenberger, J., Tröger, L., Kornowski, A., Vossmeyer, T., Eychmüller, A., Feldhaus, J. and Weller, H. (1997) EXAFS studies on the size dependence of structural and dynamic properties of CdS particles. Journal of Physical Chemistry B 10, 2691–2701.

Scott, S.D. (1971) Mössbauer spectra of synthesis iron-bearing sphalerite. Canadian Mineralogist 10, 882–885.

Scherrer, P. (1918) Bestimmung der Grösse und der inneren Struktur von Kolloidteilchen mittels Röntgenstrahlen. Nachrichten von der Gesellschaft der Wissenschaften zu Götingen, Mathematisch-Physikalische Klasse, 2, 96–100.

Siegbahn, K., Nordling, C. and Johansson, G. (1969) *ESCA: Applied to Free Molecules*. North-Holland, Amsterdam.

Temperley, A.A. and Lefevre, H.W. (1966) The Mössbauer effect in marcasite structure iron compounds. Journal of Physics and Chemistry of Solids 27, 85–92.

Tkáčová, K. (1989) *Mechanical Activation of Minerals*. Elsevier, Amsterdam.

Ungar, T. (2004) Microstructural parameters from X-ray diffraction peak broadening. Scripta Materialia 51, 777–781.

Vaughan, D.J. and Burns, R.G. (1972) Mössbauer spectroscopy and bonding in sulfide minerals containing four-coordinated iron. In: *Proc. 24th International Geological Congress*, Section 14, Montreal, pp. 156–167.

Vaughan, D.J. and Craig, J.R. (1978) *Mineral Chemistry of Metal Sulfides*. Cambridge University Press, Cambridge.

Vlasov, M.V. and Kakazej, N.G. (1979) *Electron Spin Resonance in Mechanically Disordered Solids*. Naukova dumka, Kijev (in Russian).

Wang, Z.L. (2000) Transmission electron microscopy and spectroscopy of nanoparticles. In: Z.L. Wang (Ed.) *Characterization of Nanophase Materials*. Wiley-VCh, Weinheim, pp. 37–80.

Warren, B.E. and Averbach, B.L. (1950) The effect of cold-work distortion on X-ray patterns. Journal of Applied Physics 21, 595–599.

Wertz, J.E. and Bolton, J.R. (1972) *Electron Spin Resonance*. McGraw Hill, New York.
Williamson, G.K. and Hall, W.H. (1953) X-ray broadening from filled aluminium and wolfram. Acta Metallurgica 1, 22–31.
Wogelius, R.A. and Vaughan, D.J. (2000) Analytical, experimental and computational methods in environmental mineralogy. In: D.J. Vaughan and R.A. Wogelius (Eds.) *EMU Notes in Mineralogy*. Eötwös University Press, Budapest, Vol. 2, pp. 7–87.
Zanchet, D., Hall, B.D. and Ugarte, D. (2000) X-ray characterization of nanoparticles. In: Z.L. Wang (Ed.) *Characterization of Nanophase Materials*. Wiley-VCH, Weinheim, pp. 13–36.

Chapter 4
From Minerals to Nanoparticles

4.1 Introduction

In a typical technological process, the minerals are transformed by liquid and/or gas phase reactions at high temperatures and pressures. The final products are then separated from the by-products, from the remaining starting materials and/or from the solvent. Very often, the final product is solid again [Boldyrev 1998; Boldyrev and Tkáčová 2000].

Mechanochemical processing may offer the possibility to simplify the entire technological flowchart by avoiding operations in the gaseous and liquid states and to design the process according to the scheme [Boldyrev 1996a, b]:

$$\text{Mineral} \rightarrow \text{Solid state reaction} \rightarrow \text{Desired product.}$$

4.2 Solid–Gas Reactions Stimulated by Mechanical Activation

Reactivity of solids in heterogeneous reactions can be affected by different preparative treatments prior to the reactions. The possibilities for forming different solid state defects by various pretreatment techniques are presented in Table 4.1. Mechanical activation belongs among the effective processes enabling to control and regulate the course of solid-gas reactions via formation of different defects.

It has been shown the mechanical treatment influences the sorption processes by solid-gas reactions as well as the following reactions of adsorbed gases with solids. In several cases the strengthening of sorption processes by mechanical activation depends on the penetration depth of gas into disordered solids [Thiessen et al. 1970]. Butyagin illustrated the process by studying the oxidation of carbon monoxide and hydrogen on mechanically activated quartz mineral SiO_2 [Butyagin 1971, 1984].

As for gas phase selection in solid-gas reactions, methane, carbon monoxide and lately hydrogen have been used. Hydrogen has special position among gas reactants. In contrast to oxidizing processing where SO_2 is envolved and has to be utilized in the form of problematically soleble H_2SO_4, the utilization of H_2S as gaseous reduction product of hydrogen application is more hopeful. H_2S can be further utilized in

© Springer-Verlag Berlin Heidelberg 2008

Table 4.1 Relationship between different types of defects and preparative treatment of the solid phase [Boldyrev 1979]

Preparative treatment	Crystal defects				Lattice defects	
	Surface	Habit	Heterophase inclusions	Dislocations	Doping ions	Point defects
Control of crystal growth	+	+		+	+	+
Doping					+	+
Physical aging				+		
Chemical aging			+		+	
Chemical pretreatment			+			
Radiative pretreatment			+	+	+	+
Mechanical pretreatment	+		+	+		+

the form of elemental sulphur and hydrogen which is more prospective and ecologically less objectionable route. The arisen hydrogen can be recycled into the primary process [Prasad and Mankhand 1983].

However, hydrogen alone is mostly useless as a reducing agent from thermodynamical point of view (Table 4.2) but some reactions were studied and mechanical activation has been applied as an intensification step [Baláž 2000].

For the reaction

$$MeS + H_2 \leftrightarrow Me + H_2S \tag{4.1}$$

the equilibrium constant $K = P_{H_2S}/P_{H_2}$ at 1073 K is typically 2.10^{-1} to 6.10^{-3} for Cu, Ni, Co and Fe sulphides. Dynamic regime is important where hydrogen is

Table 4.2 Heats of reaction, ΔH° for the direct and lime-enhanced reduction by H_2, CO of selected MeS at 1200 K [Prasad and Mankhand 1983]

$MeS + H_2 \rightarrow Me + H_2S$ (A)
$MeS + H_2 + CaO \rightarrow Me + CaS + H_2O$ (B)
$MeS + CO \rightarrow Me + COS$ (C)
$MeS + CO + CaO \rightarrow Me + CaS + CO_2$ (D)

MeS	ΔH°, kJ per g-atom of sulfur			
	(A)	(B)	(C)	(D)
Sb_2S_3	−5.9	70.7	70.7	−105.9
Ag_2S	−15.9	−81.6	−81.6	−114.6
Bi_2S_3	26.8	−37.7	−37.7	−72.8
Cu_2S	39.9	−31.8	−31.8	−64.9
Ni_3S_2	51.0	−13.8	−13.8	−49.0
PbS	70.7	4.6	4.6	−28.5
Co_9S_8	77.4	12.6	12.6	−22.2
MoS_2	105.4	39.3	39.3	6.3
CdS	215.9	150.2	150.2	117.2
ZnS	297.1	231.4	231.4	198.3

continously added and H_2S is continously removed from the reaction system as soon as it is formed.

The enhancement of solid-gas reactions by mechanical activation has been studied for reduction of antimony, mercury, lead and zinc sulphides [Baláž and Godočíková 2001; Baláž and Briančin 2001].

4.2.1 Stibnite Sb_2S_3

The direct reduction of Sb_2S_3 by hydrogen is represented by the equation

$$Sb_2S_3 + 3H_2 \rightarrow 2Sb + 3H_2S \quad (4.2)$$

The reaction (4.2) has been investigated at atmospheric pressure and at temperatures of 473–873 K [Torma and Inal 1979]. The activation energy of this process has been found to be $72\,\mathrm{kJ\,mol^{-1}}$. The H_2S liberated can be converted into elemental sulphur or H_2S. Consequently this method will not pollute the environment since all by-products are recovered and are saleable. On the basis of the experimental data a flow-sheet has been developed for the production of metallic antimony from high grade sulphide-bearing concentrates. However, only 60% of Sb_2S_3 was converted to metallic antimony.

The results of mechanically activated Sb_2S_3 reduction by hydrogen under dynamic conditions where hydrogen sulfide is carried out from the product are illustrated by the kinetic relationships in Fig. 4.1.

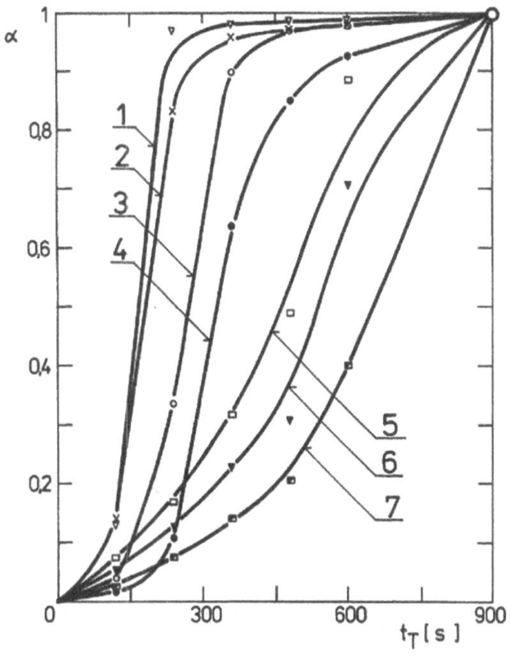

Fig. 4.1 The influence of the reaction time, t_T on the conversion degree, α of mechanically activated Sb_2S_3 for 15 min. Reaction temperatures: 1–954 K, 2–929 K, 3–903 K, 4–877 K, 5–851 K, 6–825 K, 7–800 K [Baláž and Godočíková 2001]

These plots show that the conversion reaches the value 100% and the temporal course of kinetic curves is dependent on temperature. X-ray phase analysis indicated the presence of metallic antimony (JCPDS 05-0562) together with small amount of monoclinic sulphur (JCPDS 13-0141). Mechanical activation brought about a decrease in activation energy from 130 kJ mol^{-1} for non-activated sample to 58 kJ mol^{-1} for a sample activated for 15 min. Both values indicate that the surface chemical reaction of the Sb_2S_3 particles is the rate determining step of reduction. The value 121 kJ mol^{-1} was found for both reactions performed under dynamic conditions [Chunpeng et al. 1988].

4.2.2 Cinnabar HgS

At temperatures exceeding 613 K the reaction between HgS and hydrogen takes place according to the equation [Mills 1974]

$$HgS + H_2 \rightarrow Hg + H_2S \quad (4.3)$$

The dependence of the degree of conversion α on the time of thermal decomposition t_T for mechanically activated HgS is given in Fig. 4.2.

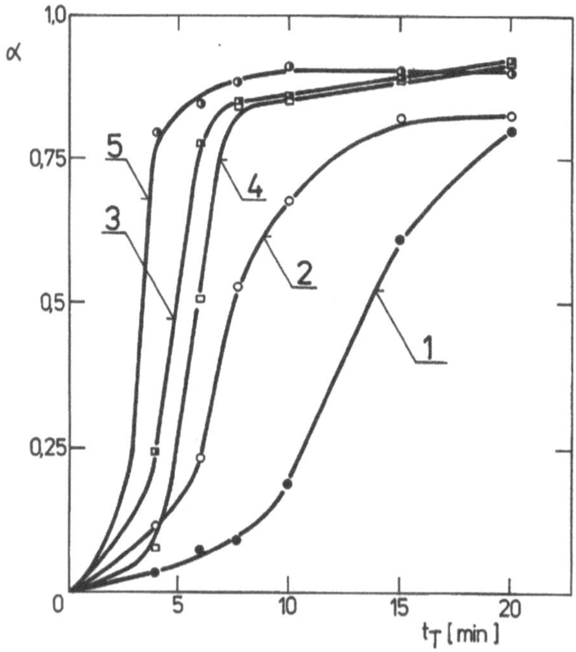

Fig. 4.2 The influence of reaction time, t_T on the conversion degree, α of mechanically activated HgS for 15 min. Reaction temperatures: 1–636 K, 2–679 K, 3–743 K, 4–722 K, 5–835 K [Baláž and Godočíková 2001]

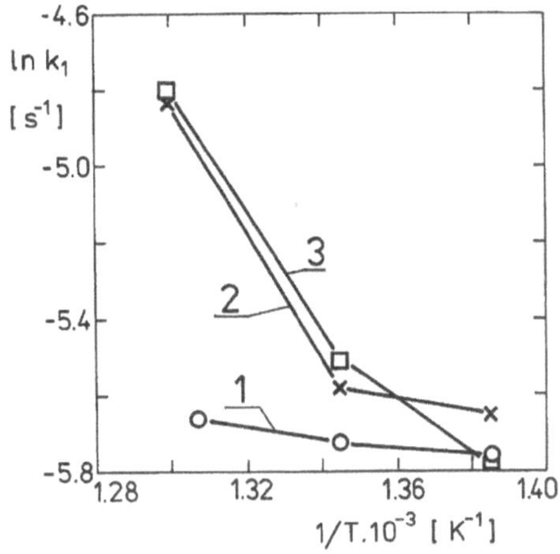

Fig. 4.3 The influence of mechanical activation of HgS on Arrhenius plot, T = 722–765 K. Time of mechanical activation: 1–0 min, 2–5 min, 3–15 min [Baláž and Godočíková 2001]

The Arrhenius plots in Fig. 4.3 give evidence of a change in reaction mechanism at T = 744 K (1/T = $1.35.10^{-3}$ K) which is manifested by the change slope for the non-activated samples and mechanically activated samples for 5 and 15 minutes.

The change in mechanism can be related with the process of dissociative sublimation which begin just at this temperature [Baláž et al. 1992]. At temperatures above 744 K, a process involving simultaneous dissociative sublimation and reductive decomposition of cinnabar proceeds. The elemental sulphur formed in the first process immediately reacts with hydrogen under dynamic conditions to give H_2S, owing to which the reaction surface is set free and the overall process is accelerated. For the activated samples the values of apparent activation energy in the temperature region 744–751 K are equal to 155–162 kJ mol^{-1}, which points out that the chemical reaction is the rate determining step of the whole process.

4.2.3 Galena PbS

In the eighties several papers dealing with the kinetics of reduction of galena with hydrogen appeared [Onajev and Spitčenko 1988; Zviadadze et al. 1985, 1986; Jovanovic et al. 1986a, b]. It was found by thermogravimetric investigations that mass loss from galena heated in flowing hydrogen occurs at temperatures over 771 K. At temperature over 1023 K the reduced lead vaporizes. The stoichiometry of the process can be expressed by the equation

$$PbS + H_2 \rightarrow Pb + H_2S \tag{4.4}$$

The Arrhenius plots for the reduction of mechanically activated as well as for non-activated PbS by hydrogen are represented in Fig. 4.4.

Their character indicates that no change in mechanism due to temperature or structure disordering of mechanically activated sample takes place in the investigated

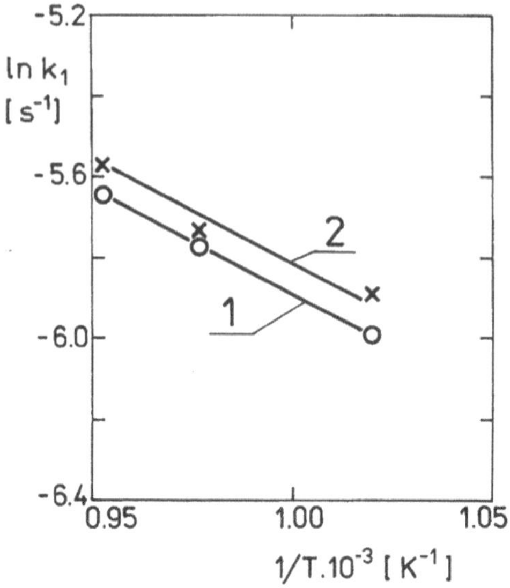

Fig. 4.4 The influence of mechanical activation of PbS on Arrhenius plot, T = 980–1048 K. Time of mechanical activation: 1–0 min, 2–5 min, 3–15 min [Baláž and Briančin 2001]

Table 4.3 Arrhenius data for PbS reduction by hydrogen [Baláž and Briančin 2001]

Mechanical activation (min)	Activation energy (kJ mol^{-1})
–	42 (r = 0.999)
15	38 (r = 0.979)

temperature interval. The values of activation energy for both samples (Table 4.3) indicate that chemical reaction is the rate determining step of PbS reduction.

4.2.4 Sphalerite ZnS

In comparison with galena the reducibility of ZnS is worse [Onajev and Spitčenko 1988]. The degree of reduction reaches only 35% for the temperature of 1070 K [Jovanovic et al. 1986]. This process is complicated by the fact that zinc is evaporated in the hydrogen flow at high temperatures. At the same time H_2S, originating in the reaction of hydrogen with the sulphur atoms of sphalerite leaves the surface. At the temperatures between 1173 and 1223 K hydrogen sulfide can react with Zn vapour to form secondary ZnS [Onajev and Spitčenko 1988; Čižikov 1976]. In temperature range 673–1023 K the reduction can be described by the equation

$$ZnS + H_2 \rightarrow Zn + H_2S \qquad (4.5)$$

Elemental zinc (JCPDS 04-0831) and sulphur (JCPDS 08-0247) were detected by X-ray phase analysis of the sample mechanically activated for 15 min and

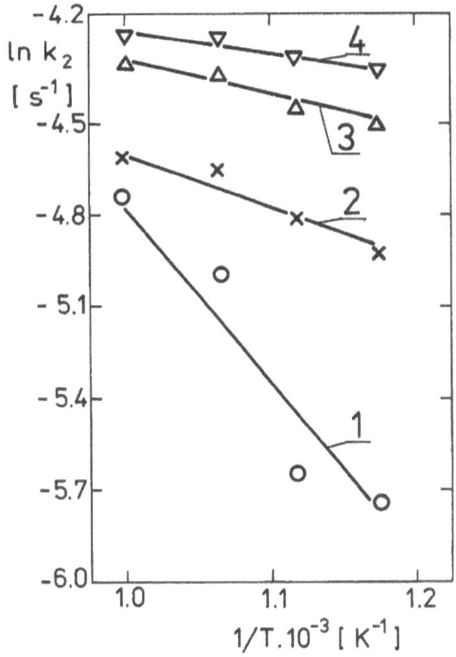

Fig. 4.5 The influence of mechanical activation of ZnS on Arrhenius plot, T = 851–1023 K. Time of mechanical activation: 1–0 min, 2–5 min, 3–10 min, 4–30 min [Baláž and Briančin 2001]

subsequently reduced by hydrogen for 20 min at 937 and 1023 K. The presence of elemental sulphur in reaction products may be a result of H_2S decomposition.

The values of activation energy calculated from plots in Fig. 4.5 are 49, 12, 7 and 4 kJ mol^{-1} for non-activated sample and samples mechanically activated for 5, 10 and 30 min, respectively. The disordering of sphalerite by milling brings about a reduction in the values of activation energy. These values show that the diffusion regime, probably involving the secondary ZnS originating from the above-mentioned recombination of Zn and H_2S determines the rate of reaction (4.5).

Generally, the direct reduction of metal sulphides by gaseous hydrogen is thermodynamically unfavorable (with exception of Sb_2S_3 and Ag_2S, see Table 4.2) and a shift of reaction regime into dynamic one and/or disordering of solids by mechanical activation is needed to improve the kinetics of the process [Baláž and Godočíková 2001]. A good way of driving such reactions forward is to decrease the in-situ partial pressure of gas products (H_2S in case of hydrogen reactant) and this may readily achieved by the incorporation of a powerful sulphur receptor [Prasad and Mankhand 1983].

Calcium oxide CaO is known to be one of the most potent absorbents for sulfur bearing gases and the reduction of metal sulphides by H_2 or CO in its presence may be represented by the overall reactions

$$MeS + H_2 + CaO \rightarrow Me + CaS + H_2O \tag{4.6}$$

$$MeS + CO + CaO \rightarrow Me + CaS + CO_2 \tag{4.7}$$

Except of MoS_2, CdS and ZnS, all other reduction reactions in CaO presence are highly feasible thermodynamically as it is evident from Table 4.2.

A major objective of the lime-enhanced reduction of metal sulphides is to exploit the sulphur fixation ability of lime (as solid CaS) to reduce atmospheric pollution. From a kinetic point of view, lime can be used effectively as a scavenger for sulphur bearing gases like H_2S only above 773 K [Prasad and Mankhand 1983]. In 1969, the lime-enhanced reduction of metal sulphides as means of preparing filamentary products of Cu, Ni and Co under suitable conditions was studied [Cech and Tiemann 1969].

The outstanding capacity of lime to enhance the reduction rates of copper sulphide Cu_2S by hydrogen is clearly demonstrated in Fig. 4.6 and in Table 4.4.

It is clear that the addition of even theoretical amount of lime enhances the reduction rate of the sulphide by a factor 24, which can be increased still further to 49 by employing 200% excess CaO in the reduction charge [Mankhand et al. 1978].

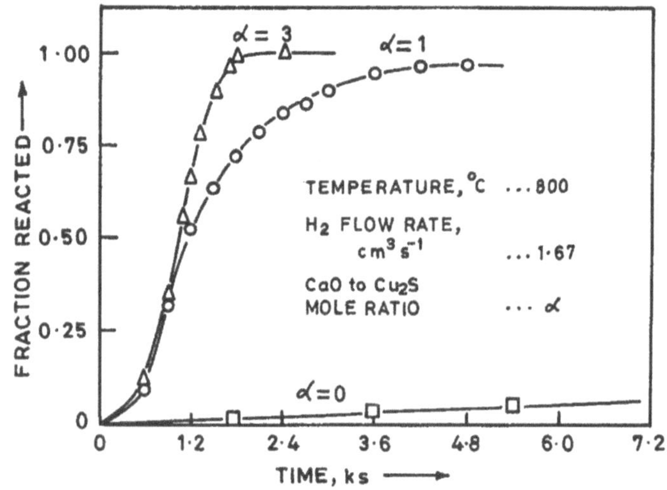

Fig. 4.6 Effect of CaO addition on the reduction of Cu_2S by hydrogen [Mankhand et al. 1978]

Table 4.4 Effect of CaO on the reduction of Cu_2S [Prasad and Mankhand 1983]

Reducing agent	CaO/Cu_2S (mole ratio)	Temperature (°C)	Fraction Cu_2S reacted
H_2	0	800	0.38
H_2	1	800	0.73
H_2	3	800	0.93
C	0	900	*
C	0.5	900	0.58
C	1	900	0.73
C	2	900	0.88

*not significant

An interesting observation from Table 4.4 is that whereas no significant reduction of Cu_2S was observed by carbon alone, the presence of CaO was found to drastically enhance the percent Cu_2S reduced.

It has been reported by different investigators that the time for lime-enhanced reduction may be shortened by the use of fine sized reactant sulphides. This can be a new challenge for application of mechanical activation. This beneficial effect, however, had to be viewed from the point of achieving separation of the reduced metal from the only dry slag (CaS+unused CaO).

Chalcopyrite $CuFeS_2$ reduction in the presence of CaO for sulphur elimination increases with the increasing temperatures and increasing amounts of CaO and is practically complete at 900°C at high CaO addition [Habashi 1986]. Depending on the amount of CaO added, the reduction follows equations

$$2CuFeS_2 + H_2 + CaO \rightarrow Cu_2S + 2FeS + CaS + H_2O \tag{4.8}$$

$$2CuFeS_2 + 3H_2 + 3CaO \rightarrow Cu_2S + 2Fe + 3CaS + 3H_2O \tag{4.9}$$

$$2CuFeS_2 + 4H_2 + 4CaO \rightarrow Cu_2S + 2Fe + 4CaS + 4H_2O \tag{4.10}$$

Due to the presence of metallic Fe the product after reduction is magnetic and can be separated by a magnet. Iron can be leached away by HCl leaving behind Cu powder, a typical analysis of which is shown in Table 4.5. In this process, ZnS initially present in the ore is also reduced and is alloyed with copper.

CaS can be best disposed of by heating in air at 500°C so that it is oxidized to $CaSO_4$

$$CaS + 2O_2 \rightarrow CaSO_4 \tag{4.11}$$

while $FeCl_2$ solution obtained by leaching the metallic iron can be decomposed according to the reaction

$$2FeCl_2 + 2H_2O + \frac{1}{2}O_2 \rightarrow Fe_2O_3 + 4HCl \tag{4.12}$$

Hydrogen chloride generated is recycled for leaching [Habashi 1986].

Table 4.5 Analysis of two chalcopyrite concentrates and copper prepared therefrom by reduction followed by leaching [Habashi 1986]

	Noranda concentrate (%)	Crude Cu (%)	Gaspe concentrate (%)	Crude Cu (%)
Cu	25.4	78.6	25.0	81.7
Fe	24.2	2.1	31.3	0.9
Zn	6.6	12.3	0.7	1.0
S	29.9	6.1	27.2	2.8
Gangue (by difference)	13.9	0.9	15.8	13.6

4.3 Mechanochemical Reduction via Solid–Solid Reactions

During the early history of the chemistry of solid state reactions, the question whether solid substances could react with each other without the aid of liquids or gases was strongly debated. The original view was that reactions could not occur directly between solids. The works of Tamman and Hedvall established in the 1930s that such reactions do occur in the solid state. By the end of the 1960s, it was generally accepted that the rate of a reaction between two solids could be substantial if the compounds were intimately mixed and heated. The reaction is initiated as lattice vibrations with increasing amplitude result in more frequent exchange of atomic positions in the crystal lattice, leading to faster macroscopic diffusion. The exchange of atomic positions is especially easy in disordered structures [Habashi 1969].

When intimate mixing of the solid reactants was carried out by using high-energy milling, the potential for initiating mechanochemical reactions was established. Until very recently, it was universally assumed that chemical reactions took place during the mechanical treatment of solids due to the evolution of heat. The increase of the reactivity was attributed to the increase of the specific surface of the solid upon treatment [Boldyrev 1993]. The importance of structural disorder, already well established in solid state chemistry [Hedvall 1938; Tamman 1932], was recognized by mechanochemists as the reason for enhanced reactivity. In early suggestions, it was accepted that the reactivity of solids is determined by the total content of defects. Later it was concluded that only those defects should be taken into account which directly or indirectly take part in the chemical reaction [Schrader and Hoffman 1973].

The preparation of metals and alloys by reduction of their salts with metals has been described in pyrometallurgy [Habashi 1969]. The method is characterized by the fact that the reducing metal is converted by high temperature to a solid or a liquid product and not to a gas in other reducing processes, e.g. by carbon and hydrogen where $CO+CO_2$ and H_2O are formed, respectively. In general, such processes are called *metallothermic*. In special case, when Al is used as a reducing agent the process is called *aluminothermic* reduction, when Ca is used it is called *calciothermic* reduction [Habashi 1969]. General scheme of metallothermic processes is given in Fig. 4.7.

As described by Habashi the different substances like *fluxes, thermal boosters* and *alloying metals* are added to improve the process of metallothermic reduction.

Fluxes form a fluid slag so that separation from the metal can be easily achieved. For example, in the production of metal Mg by ferosilicon process $CaO{\cdot}MgO$ is preferred to MgO as a raw material because CaO reacts with SiO_2 liberated to form a low-melting slag

$$2MgO.CaO + FeSi \rightarrow 2Mg + Ca_2SiO_4 + Fe \tag{4.13}$$

Thermal boosters are used to improve the thermal balance of reduction. These react with an excess reducing agent liberating appreciable amount of heat. For example, sulphur is added during the reduction of V_2O_5 by Ca; the formation of CaS is highly exothermic and raises the temperature of reduction.

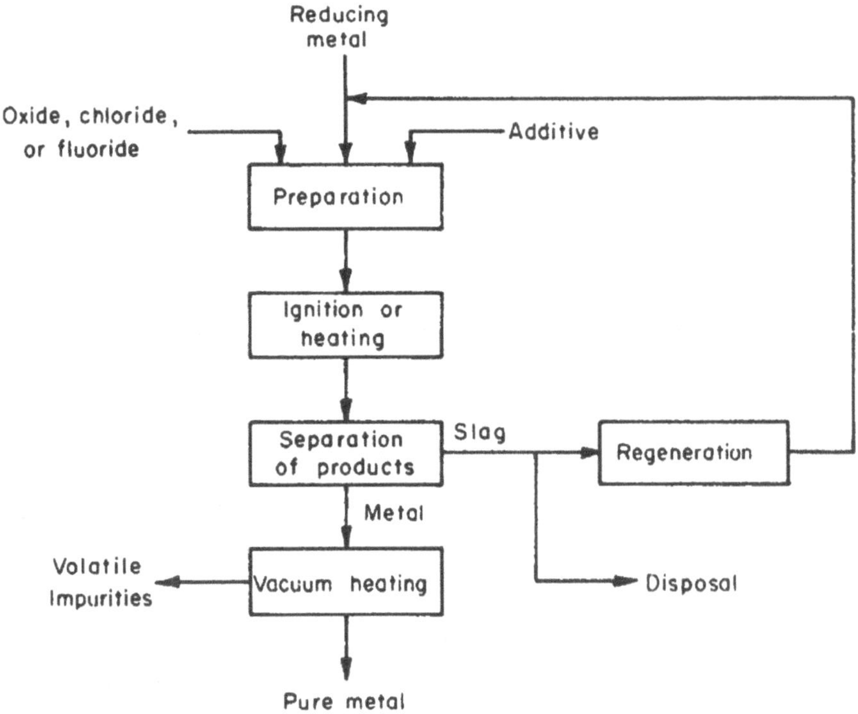

Fig. 4.7 General scheme of metallothermic reduction [Habashi 1969]

Alloying metals lower the melting point of the metal to be recovered. These metals can be removed from the alloy by vacuum distillation, e.g. Zn.

Solid-state reactions usually start at the contact points of reacting particles [Budnikov and Ginstling 1971]. It follows the number of contact points and their area has influence on solid-solid reactions. By applying mechanical activation these parameters can be increased. It has been shown that the coefficient of diffusivity of solids can be also enhanced: in case of Cr, diffusion into Fe is $10^{-12}\,cm^2\,s^{-1}$, at 1400–1600 K is equal to $2.10^{-8}\,cm^2\,s^{-1}$ but by mechanical activation of the system, the values 10^{-5}–$10^{-7}\,cm^2\,s^{-1}$ have been obtained [Butyagin and Yuščenko 1987].

4.3.1 Redox Reactions

These reactions can be expressed in general by equation

$$Me_1X + Me_2 \rightarrow Me_1 + Me_2X \tag{4.14}$$

where Me_1-reduced metal, Me_2-reducing metal, X-oxide, chloride, fluoride, sulphide, etc. Redox reactions involve simultaneous oxidation and reduction processes.

According to present views a redox reaction is a reaction in which there is a transfer of electrons from reducing metal to reduced metal. Sometimes they are called *exchange* or *metathesis reactions* [Treece et al. 1995; Gillan and Kaner 1996].

Many well-known industrial solid-state reactions are driven by these reactions. The most commonly used reducing metals are Al, Ca, Mg, Na and Si (as FeSi). The other metals such as Cu, Mn and Fe are also applied. The selected examples of application are given in Table 4.6.

Table 4.6 Commonly used reducing metals and their application [Habashi 1986]

Reducing metal	Metal produced	
	Large scale	Small scale
Al	V, Nb, Ca	Sr, Ba, Ta
Ca	V, U	rare earths, Ti, Cr
Mg	Be, Ti, Zn	Sc, Y, V, Nb, Ta
Na	Ta, Ti	Zr, Hf, V, Nb, Th
FeSi	Mg	

In choosing a metal for reduction, the following points should be observed [Habashi 1986]:

- the metal should have a strong affinity, based on thermodynamic consideration
- to ensure complete reduction, the reducing metal is usually added in slight excess
- the metal can be easily obtained in high purity and at low-cost
- the metal does not form intermetallic compounds and
- the metal should be easily handled, i.e. does not need inert atmosphere in a post-reduction operation.

Separation of reaction products play the important role in the process of reduction. The method of separation varies from one reduction process to the other. Sometimes small amounts of salts adhering to the metal have to be removed. The method of heating of the product under high vacuum is applied in this case. In mechanically activated solid–solid state reactions the problem of the reaction products separation is more underlined. Mechanical activation increases significantly kinetics in such reactions by the dynamic maintenance of high reaction volumes and rapid diffusion rates related to short diffusion paths, enhanced defect density and high reaction interface areas. In such cases where various nanostructures are formed the formation of nanocomposites is very probable.

4.3.2 Nanocomposites

In nanocomposites constituents are mixed on a nanometer-length scale. They have been generally classified based on the matrix (such as metal-matrix composites

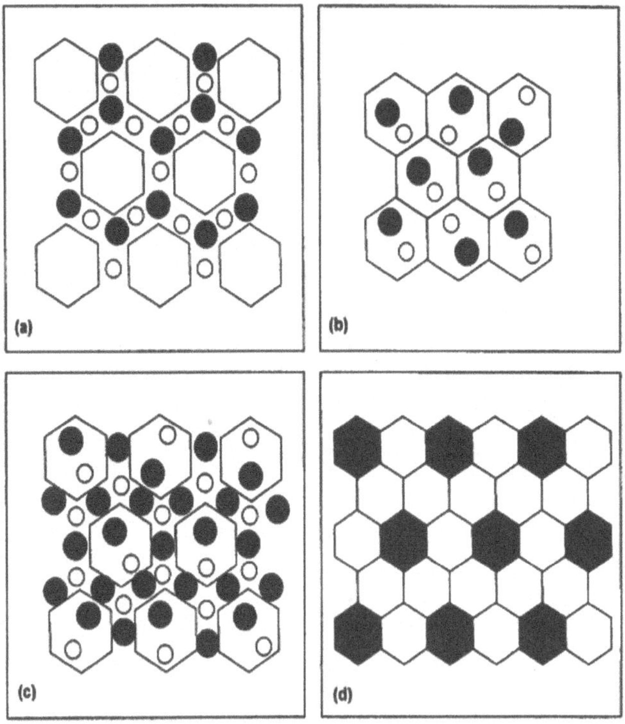

Fig. 4.8 Classification of nanocomposites: (**a**) intergranular, (**b**) intragranular, (**c**) hybrid, (**d**) nano/nano [Niihara 1991]

(MMC), ceramic-matrix composites (CMC) or polymer-matrix composites (PMC). Nanocomposites can be divided into four categories: intragranular, intergranular, hybrid and nano/nano composites (Fig. 4.8).

In the first three types, the reinforcement phase is in the nanometer level whereas the matrix is not. In the last cathegory, however, both the matrix and reinforcement are of nanometer division and these two components are randomly distributed. They often have properties that are superior to conventional microscale composites and can be synthesized using surprisingly simple and inexpensive techniques such as a mechanochemical method described above. Nanocomposites as systems with high density of interface boundaries provide the large surface of initial components. Moreover, concentration of atoms near boundaries reaches 15–30% of their total number. The mixing of components is performed at the nanometer level when the contact between the components is performed by atomically pure planes and the effect of the surrounding atmosphere is reduced to a minimum [Niihara 1991].

Nanocomposites are either prepared in a host matrix of inorganic materials or by using conventional polymer as one component of the nanocomposites. The second type of nanocomposites which are a special class of hybrid materials are termed *polymeric nanocomposites*. These materials are intimate combination (up to almost the molecular level) of one or more inorganic material (nanoparticles, e.g.) with a polymer so that unique properties of the former can be mixed with the existing qualities of the polymer to result in a totally new material suitable for novel applications [Dutta and Hofman). Resulting nanocomposites have found successful applications in various fields of material application such as baterry cathodes [Nazar et al. 1992], microelectronics [Vassilion et al. 1990], nonlinear optics [Beecroft and Ober 1997], sensors [Cao et al. 1992], etc.

Ceramic nanocomposites bring new light into properties of brittle ceramics which play crucial role in technology. Brittle ceramic can be strengthened and toughened by the incorporation of various reinforcements. Ductile metallic reinforcement appear to be one of the most promising toughening mechanism. It has already been investigated and the enhancement is mainly contributed by plastic inclusions which bridge the advancing crack. They are stretched as the cracks opens until they fracture or separate from the matrix [Gaffet and Bernard 2002].

High-energy milling is a very effective process for synthesizing *metal-ceramic nanocomposites* as it allows incorporation of the metal and the ceramic phases into each powder particle, as shown schematically in Fig. 4.9.

In the initial stage, the microstructure of the composite powder particles is at the micrometer level. With further milling, the metal phase is deformed and fractured, while the ceramic phase is mainly fractured. If it is a metal matrix composite, the ceramic particles are continually fractured into small particles, so the ceramic particle size will keep decreasing until such a point that the fracture strength of the small particles will be equal to or grater than the stress caused by collision. Often this balance point corresponds to a ceramic particle size in the range of a few nanometers to 100 nm [Zhang 2004].

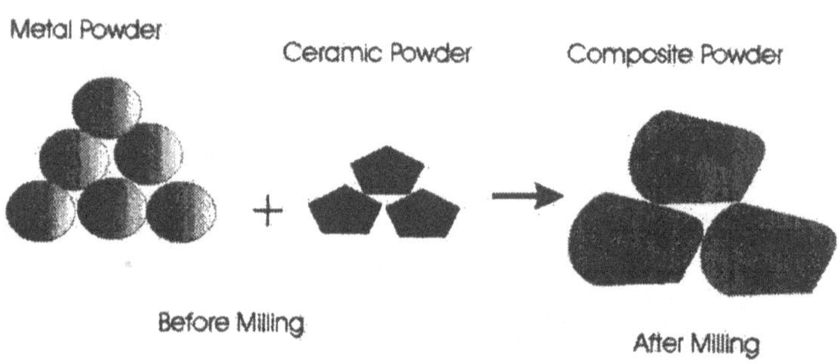

Fig. 4.9 Schematic diagram showing the formation of composite powder after high-energy milling [Zhang 2004]

4.3.3 Oxides

Mechanochemical solid-state reduction of oxides can be schematically expressed by general equation

$$Me_1O + Me_2 \rightarrow Me_1 + Me_2O \tag{4.15}$$

where Me_1-reduced metal, Me_2-reducing metal. A large variety of Me_1 and Me_2 combinations has been used, e.g. Me_1 = Fe, V, Cr, Mn, Co, Ni, Ti, Sn, Pb, Cu, Zn, Nb, Mo, W, Si and Me_2 = Al,C, Ti, B, Cr, Si, Mg, Zr, Ca, Zn, Ni and Mn (see Table 4.7).

One of the primary driving forces behind the design of successful solid-state reactions is a careful consideration of thermodynamics. The contribution of product formation on the overal heat of reaction (ΔH) varies greatly depending on such factors as product structure, type of bonding and lattice energy [Gillan and Kaner 1996]. Several mechanically induced systems will be studied in this Chapter where ΔH values summarized from literature are given in Table 4.7.

Mechanochemical reduction of oxides is an extention of the mechanical alloying process developed by Benjamin [Benjamin 1970]. At the laboratories of INCO Alloys International he applied this process as a means to overcome traditional problems associated with producing an oxide dispersion and γ'-strengthened nickel base superalloy. The scope and application of mechanical alloying have been considerably widened since the process was first announced. As with rapid solidification, with which it is often compared, mechanical alloying finds application in those high-value added materials which cannot be produced by alternative means [Schaffer and McCormick 1992a].

Mechanical alloying is a high-energy ball milling process where powder particles undergo repeated fracturing and cold welding during ball-powder-ball and ball-powder container collisions. Both processes can occur during a simple collision. When particles being impacted overlap, atomically clean interfaces are brought into contact. Cold weld form and layered composite powder particles are built up. Concurrently, work-hardened particles fracture, decreasing the particle size and creating new surfaces. This interplay between welding and fracture effectively kneads the internal structure so that the powders are continually refined [Schaffer and McCormick 1989a].

In these highly exothermic reactions, it has been shown that mechanical alloying can cause an unstable reaction which proceeds by the propagation of a combustion wave through the partly reacted powders. The combustion conditions is reached after a critical ignition time, t_{ig}, which has been shown to be the milling time required for the ignition temperature of the powder mixture, T_{ig}, to be reduced to a level equal to that achieved locally during collision event [Schaffer and McCormick 1989b, 1991].

The examination of powders immediately after combustion has revealed a rich variety of structures which reflect the high temperatures reached during the combustion event [McCormick 1995]. In general the as-combustioned mixtures of various size from nodules as large as 1–5 mm down to particles 10 nm in diameter are formed. A sudden conversion of reagents into products is observed in such processes, which is characteristic of systems displaying a high and negative reaction enthalpy.

The process also called *mechanically induced self-sustaining reactions* (MSR) was broadly studied [Tschakarov et al. 1982; Schaffer and McCormick 1992b; Yang and McCormick 1993, 1994; Takacs 1993, 1996a, b, c; Ye et al. 1995;

Table 4.7 Mechanically induced redox reactions of oxides

Reaction	ΔH° (kJ mol^{-1})	ΔG° (kJ mol^{-1})	Reference
$2CuO + Ti \rightarrow 2Cu + TiO_2$	−620		Mulas et al. 2001
$2CuO + Zr \rightarrow 2Cu + ZrO_2$	−736		Mulas et al. 2001
$2CuO + Hf \rightarrow 2Cu + HfO_2$	−796		Mulas et al. 2001
$3CuO + 2Al \rightarrow 3Cu + Al_2O_3$	−273		Schaffer 1990; Matteazzi and LeCaër 1992a
$CuO + Ca \rightarrow Cu + CaO$	−475		Schaffer 1989b; McCormick 1995
$2CuO + C \rightarrow 2Cu + CO_2$	−138		Tschakarov 1982
$4CuO + 3Fe \rightarrow 4Cu + Fe_3O_4$	−488		McCormick 1995; Schaffer 1990
$CuO + Mg \rightarrow Cu + MgO$	−445		McCormick 1995; Schaffer 1990
$CuO + Mn \rightarrow Cu + MnO$	−231		McCormick 1995; Schaffer 1990
$CuO + Ni \rightarrow Cu + NiO$	−382		Schaffer 1990
$2CuO + Si \rightarrow 2Cu + SiO_2$	−596		Xi 1996
$2Cu_2O + Ti \rightarrow 4Cu + TiO_2$	−598		Mulas et al. 2001
$2Cu_2O + Zr \rightarrow 4Cu + ZrO_2$	−754		Mulas et al. 2001
$2Cu_2O + Hf \rightarrow 4Cu + HfO_2$	−771		Mulas et al. 2001
$Cu_2O + Mg \rightarrow 2Cu + MgO$	−434		Banza and Gock 2003
$Cu_2O + Zn \rightarrow 2CuO + ZnO$	−183		Banza and Gock 2003
$3Ag_2O + 2Al \rightarrow 6Ag + Al_2O_3$	−532		Takacs 1993
$CdO + Ca \rightarrow Cd + CaO$	−377		McCormick 1995
$Cr_2O_3 + 2Al \rightarrow 2Cr + Al_2O_3$	−273		Matteazzi and LeCaër 1992a; Takacs 1993
$Cr_2O_3 + 3Zn \rightarrow 2Cr + 3ZnO$	+49		Takacs 1993
$3V_2O_5 + 10Al \rightarrow 6V + 5Al_2O_3$	−3727		Yang 1994
$V_2O_5 + 5Mg \rightarrow 2V + 5MgO$	−1457		Matteazzi and LeCaër 1992a; Yang 1994
$2V_2O_5 + 5Ti \rightarrow 4V + 5TiO_2$	−1623		Yang 1994
$WO_3 + 3Mg \rightarrow W + 3MgO$			Mukopadhyay et al. 1996
$WO_3 + 2Al \rightarrow W + Al_2O_3$	−833		Torosyan and Takacs 2004
$ZnO + Ca \rightarrow Zn + CaO$	−285		Schaffer 1989a; McCormick 1995
$ZnO + Mg \rightarrow Zn + MgO$	−253		Suryanarayana 2001
$2ZnO + Ti \rightarrow 2Zn + TiO_2$	−249		McCormick 1995
$2ZnO + 2Al \rightarrow 3Zn + Al_2O_3$			Matteazzi and LeCaër 1992a
$3CoO + 2Al \rightarrow 3Co + Al_2O_3$			Matteazzi and LeCaër 1992a
$3MnO_2 + 2Al \rightarrow 3Mn + 2Al_2O_3$			Matteazzi and LeCaër 1992a
$MoO_3 + 2Al \rightarrow Mo + Al_2O_3$			Matteazzi and LeCaër 1992a
$6Nb_2O_5 + 10Al \rightarrow 12Nb + 5Al_2O_3$			Matteazzi and LeCaër 1992a
$3NiO + 2Al \rightarrow 3Ni + Al_2O_3$			Matteazzi and LeCaër 1992a
$2NiO + Si \rightarrow 2Ni + SiO_2$			Suryanarayana 2001

Table 4.7 (continued)

Reaction	ΔH° (kJ mol^{-1})	ΔG° (kJ mol^{-1})	Reference
$3SiO_2 + 2Al \rightarrow 3Si + 2Al_2O_3$			Matteazzi and LeCaër 1992a
$WO_3 + 2Al \rightarrow W + Al_2O_3$			Matteazzi and LeCaër 1992a
$2WO_3 + 3Ti \rightarrow 2W + 3TiO_2$			Matteazzi and LeCaër 1992a
$3SnO_2 + 2W \rightarrow 3Sn + 2WO_3$		+25	Boldyrev and Avvakumov 1971; Avvakumov et al. 1974; Varnek et al. 1974; Avvakumov and Strugova 1974; Butyagin et al. 1974
$SnO_2 + Si \rightarrow Sn + SiO_2$		−276	Boldyrev and Avvakumov 1971; Avvakumov et al. 1974; Varnek et al. 1974; Avvakumov and Strugova 1974; Butyagin et al. 1974
$3SnO_2 + 4Al \rightarrow 3Sn + 2Al_2O_3$		−535	Boldyrev and Avvakumov 1971; Avvakumov et al. 1974; Varnek et al. 1974; Avvakumov and Strugova 1974; Butyagin et al. 1974
$2SnO + Si \rightarrow 2Sn + SiO_2$		−954	Patel et al. 2004
$2Sb_2O_3 + 3Si \rightarrow 4Sb + 3SiO_2$		−2823	Patel et al. 2004
$4CdO + 4S + 3Fe \rightarrow Fe_3O_4 + 4CdS$		−673	Zhang et al. 2001; Saito et al. 2004
$4PbO + 4S + 3Fe \rightarrow Fe_3O_4 + 4PbS$		−649	Zhang et al. 2001; Saito et al. 2004
$4ZnO + 4S + 3Fe \rightarrow Fe_3O_4 + 4ZnS$		−535	Zhang et al. 2001; Saito et al. 2004
$2Fe_2O_3 + Al \rightarrow 2Fe + Al_2O_3$			Nasu et al. 1999; Basset et al. 1994; Matteazzi and LeCaër 1992a; Matteazzi and LeCaër 1991
$Fe_2O_3 + 2B \rightarrow 2Fe + B_2O_3$			Matteazzi and LeCaër 1991
$Fe_2O_3 + 2Cr \rightarrow 2Fe + Cr_2O_3$			Matteazzi and LeCaër 1991
$2Fe_2O_3 + 3Si \rightarrow 4Fe + 3SiO_2$			Matteazzi and LeCaër 1991
$Fe_2O_3 + Ca \rightarrow 2Fe + 3CaO$	−1080		McCormick 1995
$Fe_2O_3 + 3Mg \rightarrow 2Fe + 3MgO$			El-Eskandarany et al. 2001
$Fe_2O_3 + 2Al + Cr_2O_3 \rightarrow FeCr + 2Al_2O_3$			Matteazzi and LeCaër 1992a
$Fe_2O_3 + 3Al + Cr_2O_3 + 3NiO \rightarrow FeCrNi + 3Al_2O_3$			Matteazzi and LeCaër 1992a
$3Fe_3O_4 + 8Al \rightarrow 9Fe + 4Al_2O_3$	−1716		Pardavi-Horvath and Takacs 1992; Takacs and Pardavi-Horvath 1994
$Fe_3O_4 + 2Ti \rightarrow 3Fe + 2TiO_2$	−395		Takacs 1993
$Fe_3O_4 + 2Zr \rightarrow 3Fe + 2ZrO_2$			Takacs 2002
$Fe_3O_4 + 4Zn \rightarrow 3Fe + 4ZnO$	−143		McCormick 1995; Takacs 1993; Takacs and Pardavi-Horvath 1994

Ma et al. 1993; Mulas et al. 2001]. MSR process has been recently reviewed [Takacs 2002]. High-energy milling induces self-sustaining reactions in many sufficiently exothermic powder mixtures. The process begins with an activation period, during which size reduction, mixing and defect formation take place. The MSR is ignited when the powder reaches a well defined critical state. Once started, the reaction propagates through the powder charge as a combustion process.

The first self-propagating solid state reaction was the so-called *thermite reaction*, in which a metal oxide is reduced with aluminium (this process is also called aluminothermic reduction, see Sect. 4.3). The process takes place in a mixture of the powders, initiated by either a high-temperature reaction or an electrically heated wire. A combustion front develops and propagates across the sample.

The first systematic investigations on ball milling induced self-propagating reactions-explosive mechanochemical reactions, according to the terminology of the authors-were performed in Bulgaria [Tschakarov et al. 1982]. The authors studied the formation of metal chalcogenides from mixtures of the elemental powders. The ignition of a self-sustaining reaction was detected by measuring an abrupt temperature increase inside the mill.

4.3.3.1 Copper Oxides

To test whether the mechanical alloying technique can be used as a vehicle for the chemical reduction of oxides, the powders of copper oxide and calcium metal were milled [Schaffer and McCormick 1989a, b, 1991]. The historically first mechanochemical reduction of oxides was reaction

$$CuO + Ca \rightarrow Cu + CaO. \tag{4.16}$$

The reaction (4.16) is associated with the high negative enthalpy ($\Delta H = -475\,kJ$). Figure 4.10 shows X-ray diffraction patterns of samples milled for various times. Reducing metal such as Ca results in the reduction of the cupric oxide to form Cu metal.

The presented process effectively combines conventional mechanical alloying with the reduction-diffusion process [Schaffer and McCormick 1990].

The thermal reduction of copper oxide with the other solid reducing agent carbon

$$2CuO + C \rightarrow 2Cu + CO_2 \tag{4.17}$$

is a widely used process for the production of copper. This reaction is thermodynamically favoured at room temperature ($\Delta H = -138\,kJ$), but will not occur for kinetic reasons. However, it has been shown that CuO can be mechanochemically reduced with graphite [Cech 1974]. The reduction was found to occur via a two-stage, $CuO \rightarrow Cu_2O \rightarrow Cu$, process. The efect of milling on the phase present as determined from quantitative analysis of X-ray diffraction patterns is shown in Fig. 4.11.

Examination of the powder using TEM method showed the as-milled particles contained copper crystallites of 15–30 nm in size.

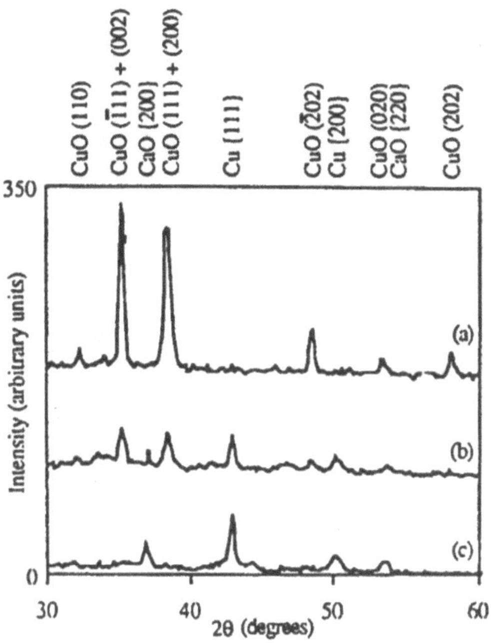

Fig. 4.10 X-ray diffraction patterns of the mixture of CuO+Ca as a function of milling time, (**a**) 7.5 min, (**b**) 60 min, (**c**) 1440 min [Schaffer and McCormick 1989a]

Italian researchers dealt with redox reactions in the Cu_2O+Me, and CuO+Me systems under self propagating regimes [Mulas et al. 2001]. Possible reaction schemes for the oxygen displacement in the studied systems are reported by equations

$$2Cu_2O + Me \rightarrow 4Cu + MeO_2 \tag{4.18}$$

$$2CuO + Me \rightarrow 2Cu + MeO_2 \tag{4.19}$$

where Me = Ti, Zr, Hf.

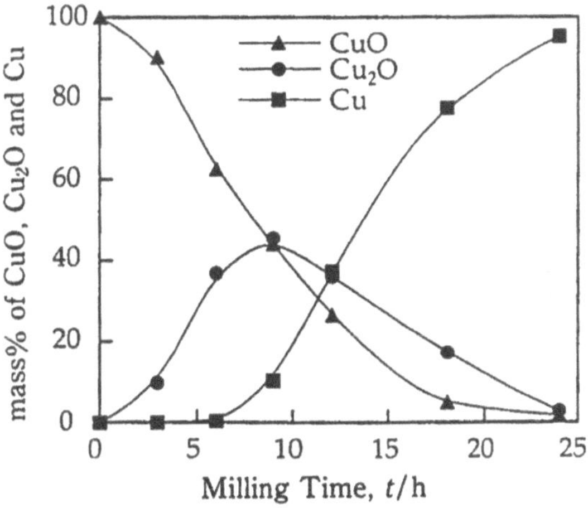

Fig. 4.11 Efect of milling time on the two stage reduction of CuO by carbon [Schaffer and McCormick 1989b]

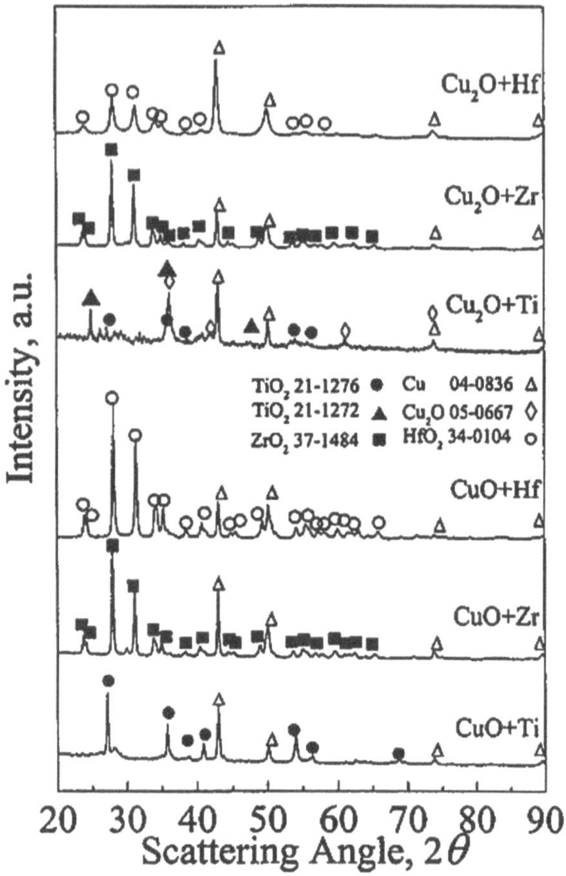

Fig. 4.12 X-ray diffraction patterns of copper oxides reacted systems [Mulas et al. 2001]

All these reactions are strongly exothermic and the values of reaction enthalpy ΔH are given in Table 4.7.

The almost complete transformation into products was definitely confirmed by XRD analysis of the combusted powders presented in Fig. 4.12. Almost in all cases, pure Cu and relative oxides are formed, while neither signals due to reaction precursors, nor intermediate phases are present.

In Fig. 4.13 ignition time, t_{ig} is plotted versus $A_{S.}$ Value A_s is the work necessary to open a new surface area unit [Streleckij et al. 1996]. This correlation suggests that an active role is played by the metal partner in reaching a critical state of reactants dispersion before the ignition. Furthermore, a greater extention of fresh interface area is required to ignite the reaction in passing from Zr to Ti and to Hf, and therefore a proper amount of mechanical work, typical of the mechanical and elastic characteristic of each metal, must be done on the system to this end. Only when these conditions are met, the reaction spreads spontaneously, and the propagation can involve all the powder dispersed in the vial [Mulas et al. 2001].

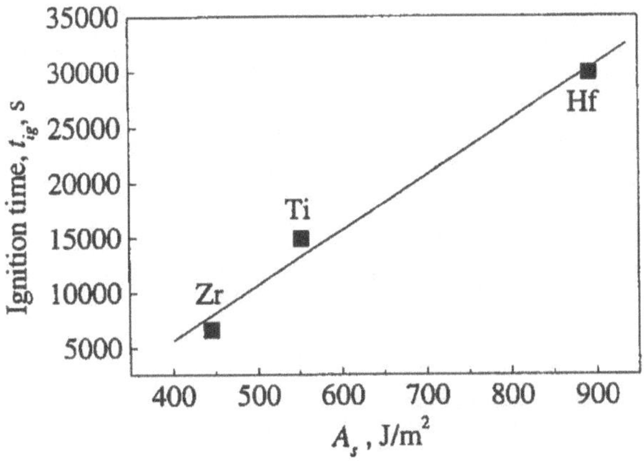

Fig. 4.13 Ignition time, t_{ig} as a function of the energy required to open a fresh interfacial surface, A_s [Mulas et al. 2001]

4.3.3.2 Iron Oxides

The reduction of hematite α-Fe_2O_3 by Al and Mg powder has been studied [Nasu et al. 1999]

$$Fe_2O_3 + 2Al \rightarrow 2Fe + Al_2O_3 \tag{4.20}$$

$$Fe_2O_3 + 3Mg \rightarrow 2Fe + 3MgO \tag{4.21}$$

The XRD patterns of hematite and Al powder are shown in Fig. 4.14 as a function of milling time.

The peak of α-Fe appeared suddenly at 1.8 ks of milling and, simultaneously, the peaks of starting materials disappeared. The same picture can be seen at Fig. 4.15

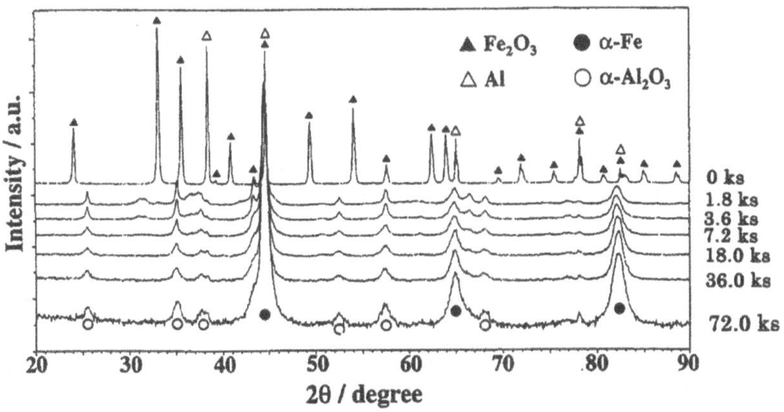

Fig. 4.14 X-ray diffraction patterns of the mixture Fe_2O_3+Al as a function of milling time in k_s [Nasu et al. 1999]

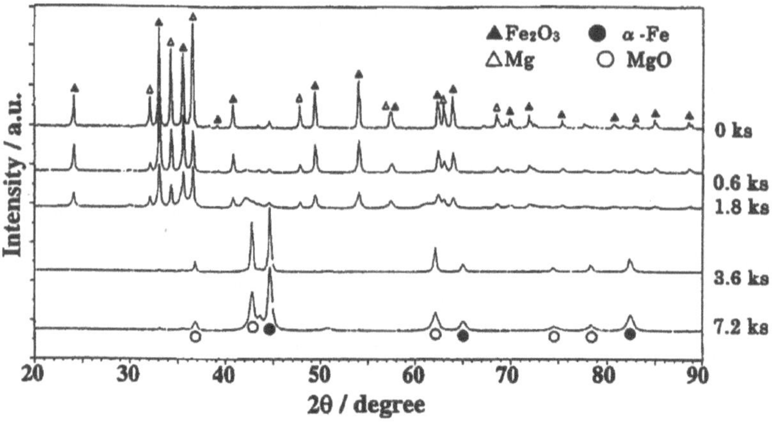

Fig. 4.15 X-ray diffraction patterns of the mixture Fe_2O_3+Mg as a function of milling time in k_s [Nasu et al. 1999]

where, together with the appearing of α-Fe the peaks for MgO can be seen. By the Mössbauer spectroscopy wustite-like phase, $Fe_{1-x}Mg_xO$ was also proved among the reaction products.

On the contrary in other paper more products in reaction (4.20) have been identified [Basset et al. 1994]. After the authors, the reduction of Fe_2O_3 with Al proceeds gradually with milling time and the reduction of hematite is accompanied by the formation of magnetite Fe_3O_4. The formation of Fe_3O_4 accompanies the reduction of hematite. The milling products are hercynite, $FeAl_2O_4$, an Fe-Al alloy (2–3% Al), α-Al_2O_3 and small iron clusters in alumina. The crystalline sizes of α-Al_2O_3 and of the Fe-Al alloy are below 15 nm after 6 h milling.

The mechanism and kinetics of reaction (4.21) has been studied [El-Eskandarany et al. 2001]. During the early stage of milling the Fe_2O_3 particles are embedded into soft matrix of Mg (reducing agent) particles to form coarse composite powders of the reactant materials. Increasing the milling time leads to the formation of fresh active surfaces of Mg which have high reducing potential and reacts with Fe_2O_3. After 120 hours the Mg powders are completely oxidized to MgO, whereas the $Fe_2O_3$is reduced successfully to metallic Fe.

The XRD patterns of the ball-milled hematite and magnesium powders are displayed in Figs. 4.16 and 4.17.

The kinetics of reaction (4.21) have been studied assuming the nucleation and crystal growth according to the Avrami equation

$$\sqrt[n]{-\ln(1-x)} = kt \tag{4.22}$$

where n is the order of the reaction and k is the reaction rate constant. From the calculated value of n and the behaviour of the experimental data is clear that reaction (4.21) is a first-order reaction. The reaction proceeds with one-dimensional growth and is diffusion controlled.

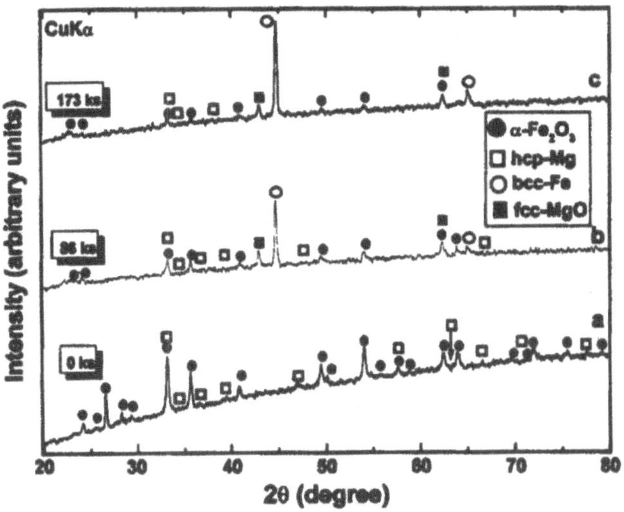

Fig. 4.16 X-ray diffraction patterns of the mixture Fe_2O_3+Mg as a function of milling time in k_s [El-Eskandarany et al. 2001]

The interest for new methods of preparation and properties of iron oxides stems from the fact that high-energy milling can lead to the preparation of new materials in magnetic materials technology.

A schematic representation of the iron oxides transformation is presented in Fig. 4.18 [Burkin 1966; Campbell et al. 1995]. Of concern here are the phases

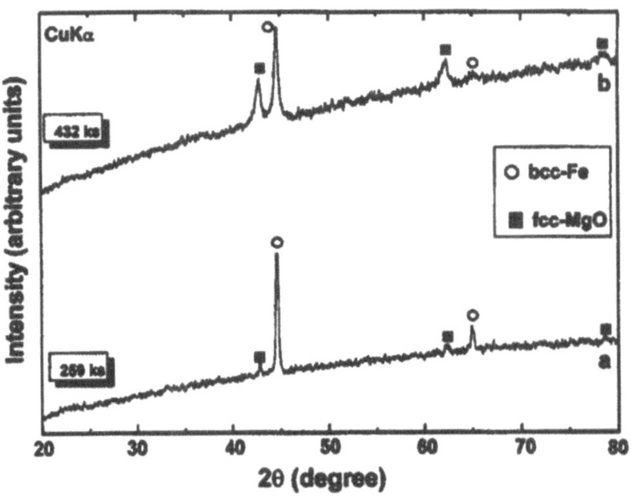

Fig. 4.17 X-ray diffraction patterns of the mixture Fe_2O_3+Mg as a function of milling time in k_s [El-Eskandarany et al. 2001]

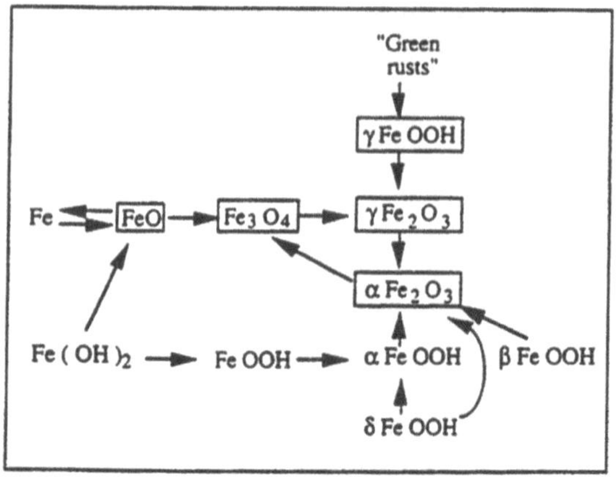

Fig. 4.18 A schematic diagram of the iron oxides transformations

magnetite Fe_3O_4, maghemite γ-Fe_2O_3 and hematite α-Fe_2O_3 that are "central" in this diagram. Phase transformations and reactions of Fe_2O_3 during milling were extensively studied [Lin et al. 1975; Senna and Kuno 1973; Nakatani et al. 1983; Kosmac and Courtney 1992; Matteazzi et al. 1993].

In the comprehensive investigation of the effects of a variety of dry and wet environments was concluded that the total phase transformation of α-Fe_2O_3 to Fe_3O_4 was best obtained by wet milling α-Fe_2O_3 in vacuum [Kaczmarek and Ninham 1994]. The mechanochemical reduction can be described by the equation

$$6Fe_2O_3 \rightarrow 4Fe_3O_4 + O_2 \tag{4.23}$$

Wet milling in low vacuum ($\sim 10^{-3}$ Pa) and by room temperature can be characterized as the fastest and total transformation hematite into magnetite. No other phases or contamination can be detected. It was suggested that oxygen bonds on the cleaved α-Fe_2O_3 oxide surface were broken during the milling and oxygen is released to the dispersing water and/or to the vial. It was also found that the oxygen pressure during the process as well the polar dispersing liquid or milling energy have a major influence on successful and fast phase transformation. All preparations performed in air, dry conditions or with hydrocarbons (benzene, anthracene) show the process of hematite reduction is non existent, or at last very slow. Normal air pressure and/or use of hydrocarbons suppress the oxygen release.

Mössbauer effect measurements have confirmed the transformation of α-Fe_2O_3 to Fe_3O_4on milling with no interdissolution of the phases. The milled products, of crystal block size $\sim$30 nm, exhibit hyperfine interaction parameters characteristic of the tetrahedral and octahedral sites of bulk Fe_3O_4.

The process of hematite-magnetite transformation has been verified with mechanically activated steelmaking dust where hematite was the main component (Fig. 4.19) [Baláž et al. 2001c].

Magnetite Fe_3O_4 shows mechanically induced self-sustaining reactions (MSR) [Takacs 2002]. The reaction of $Fe_3O_4$with zirconium

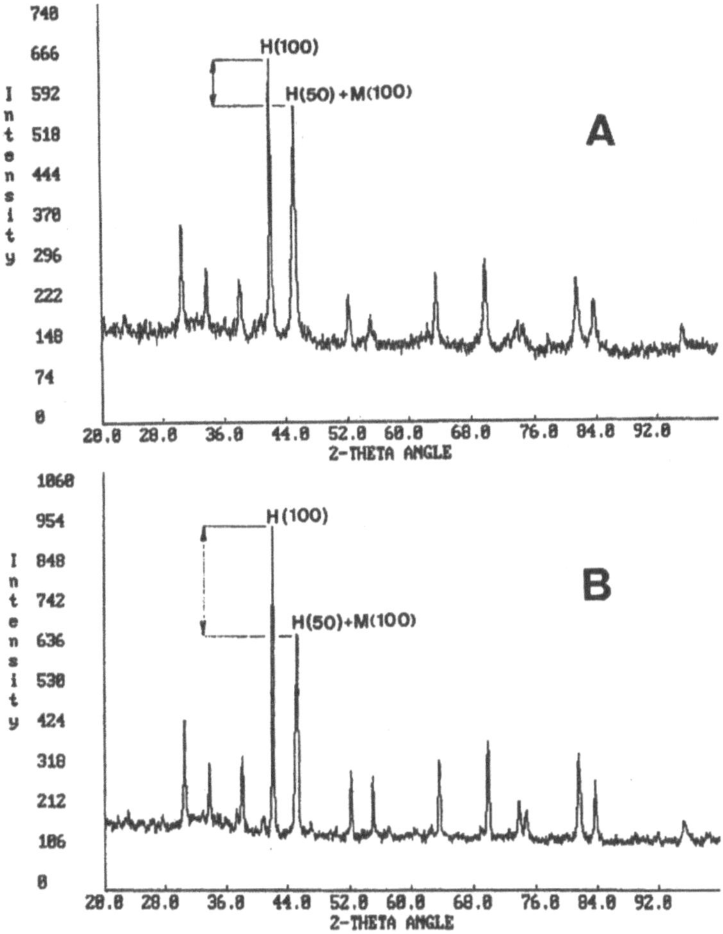

Fig. 4.19 The increase of magnetite Fe_3O_4 (M) in comparison with hematite α-Fe_2O_3 (H) in X-ray diffraction patterns for mechanically activated steelmaking dust, (**A**) sample milled in a planetary mill, (**B**) as received sample [Baláž et al. 2001c]

$$Fe_3O_4 + 2Zr \rightarrow 3Fe + 2ZrO_2 \tag{4.24}$$

has a typical temperature jump (Fig. 4.20) with three reaction steps

- the first 110 sec in the activation period, during which size reduction and mixing take place, chemically active sites are created, but very little product is formed. By the end of the activation period, the powder reaches a critical state
- ingnition occurs between two colliding balls or a ball and the wall of the vial and a combustion front propagates through the powder charge. Compared to the rest of the process, this step is almost instantaneous. The release of the reaction heat results in an abrupt increase of the temperature and

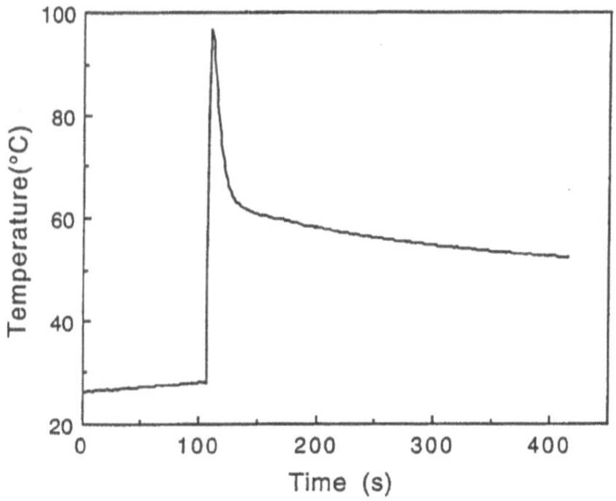

Fig. 4.20 Temperature of the milling vial during the mechanochemical reduction of Fe_3O_4 with Zr [Takacs 2002]

- the reaction is often incomplete immediately after the self-sustaining process. Continued milling is needed to obtain fully reacted product with uniform structure and properties [Takacs 2002].

4.3.3.3 Non-Ferrous Oxides

There is considerable interest in the synthesis of composites of non-ferrous metals for various applications such as conductive coatings, catalysis and energy sources (batteries).

One of the major components in alkaline batteries is γ-MnO_2. This phase is also present in Li-batteries as the intercalation host. The major commercial route to γ-MnO_2 is by electro-oxidation of purified $MnSO_4$ solution on Ti anodes. The crude solution is commercially prepared from natural Mn ores via thermal reduction to soluble MnO using carbon as reductant. In general, carbon reduction has been examined at >1100°C with resultant formation of carbide phases. Lower temperatures have been examined, but these have used hydrogen, carbon monoxide or methane as reductant [Welham 2002].

It has been shown previously [Welham 1996, 1997, 1998a, 2000] that the rate of carbothermic reduction reactions can be greatly increased by premilling the mineral and carbon together when compared with powders milled separately and then mixed. This rate increase leads to a larger amount of material reduced per unit time and therefore increases the throughput of the thermal reduction stage by the simple addition of a mill prior to the kiln.

Premilling of manganese ore, comprised of cryptomelane KMn_8O_{16}, braunite Mn_7SiO_{12} and an unidentified Mn-phase with graphite led to enhanced reduction at decreased temperatures. The longer premilled powder showed complete reduction to MnO within 30 min at 600°C, the unmilled powder showed Mn_3O_4 as the major phase after 30 min at 800°C [Welham 2002].

Tin and antimony compounds are attractive candidates for advanced materials because of their unique electrical and electrochemical properties. For example, antimony compounds have been investigated as possible superconducting materials due to their semiconductor nature. Likewise, in an experiment on tin oxide anodes, it was discovered that, during lithium alloying, the oxide is chemically reduced resulting in the formation of Li_2O clusters surrounding nanosized clusters of tin [Courtney and Dahn 1997; Parson et al. 2000; Patel et al. 2004].

The mechanochemical reduction of tin and antimony oxides by silicon proceeds according to equations

$$2SnO_2 + Si \rightarrow 2Sn + SiO_2 \tag{4.25}$$

$$2Sb_2O_3 + 3Si \rightarrow 4Sb + 3SiO_2 \tag{4.26}$$

After 170 min of milling tin oxide, and 230 min of milling antimony oxide, the above reactions progressed to completion. The non-ferrous metal/insulator composites were analyzed by X-ray diffraction (Fig. 4.21).

The X-ray patterns of the fully reduced metal oxides indicate only peaks corresponding to tin or antimony metals. The results indicate the metal oxides are chemically reduced. However, the formation of silica (SiO_2) is not obvious from XRD results, although it is possible that the silica is amorphous. FTIR analysis clearly indicates that the silicon fully consumes the oxygen with typical asymmetric stretching of the silicon-oxygen bond occuring at 1060–1080 cm^{-1}. The particle size 156 nm and 173 nm have been estimated using Scherrer equation for tin and antimony metal, respectively. Based on SEM analysis it was shown that the homogeneous metal/insulator composite was formed and the composite particles were less than a micron [Patel et al. 2004].

The mechanochemical reduction of tin oxide has also been studied [Boldyrev and Avvakumov 1971; Avvakumov et al. 1974; Avvakumov and Strugova 1974; Varnek et al. 1974; Butyagin et al. 1974; Avvakumov et al. 1975]. Tungsten, silicium and aluminium have been applied as reducing metals

$$2SnO_2 + 2W \rightarrow 3Sn + 2WO_3 \tag{4.27}$$

$$SnO_2 + Si \rightarrow Sn + SiO_2 \tag{4.28}$$

$$3SnO_2 + 4Al \rightarrow 3Sn + 2Al_2O_3 \tag{4.29}$$

Tin in form of β-Sn has been registered as reduced metal in products of reactions (4.27–4.29). The kinetics of reactions is shown in Fig. 4.22.

A new geometry was proposed and explored for the investigation of mechanochemical reduction of lead, copper and tungsten oxides with Al as the reducing metal [Torosyan and Takacs 2004]. Instead of starting with a mixture of two powder reactants, aluminium is introduced in the form of plate attached to the inside wall of the milling vial and only the oxide is milled in a powder form.

The suggested "powder-plate" geometry was applied to oxi-reduction reactions between a 1-mm thick aluminium plate and the oxide powders PbO, CuO and WO_3. The reactions are

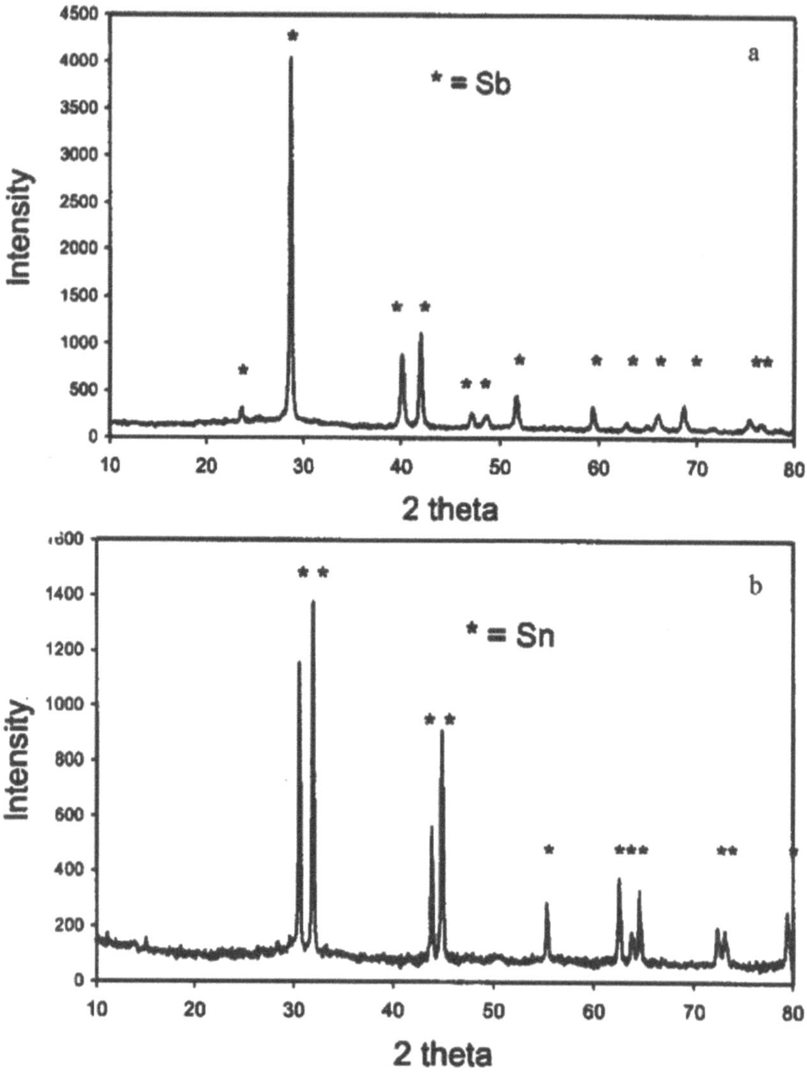

Fig. 4.21 X-ray diffraction patterns of: (**a**) the as-milled powder obtained after milling Sb_2O_3 and Si for 230 min and (**b**) the as-milled powder obtained after milling SnO and Si for 170 min [Patel et al. 2004]

$$3PbO + 2Al \rightarrow 3Pb + Al_2O_3 \tag{4.30}$$

$$3CuO + 2Al \rightarrow 3Cu + Al_2O_3 \tag{4.31}$$

$$WO_3 + 2Al \rightarrow W + Al_2O_3 \tag{4.32}$$

XRD patterns reflect the deposition of the oxide powder onto the Al plate as well as its chemical interaction with the plate. The products of reactions (4.30–4.32) are clearly visible [Torosyan and Takacs 2004].

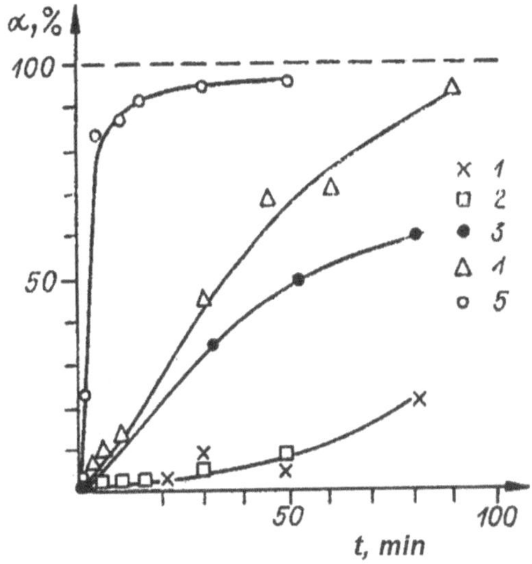

Fig. 4.22 Kinetics of mechanochemical reduction of SnO_2 by various reducing elements: 1 – without reducing element, 2 – carbon, 3 – W, 4 – Si, 5 – Al [Avvakumov et al. 1975]

The above method can be extended from the investigation of metal oxide mechanochemical redox reactions to other systems such as the formation of a carbide layer when milling graphite in the presence of a metal plate. This arrangement may also lead to useful technologies for the mechanochemical preparation of coatings [Takacs 1996a; Torosyan et al. 2002].

A different approach has been applied by Urakajev et al. who used for performing the mechanochemical reactions a reactive wear from milling balls instead of static reducing metal plate [Urakajev et al. 2001, 2003, 2004].

Precursors of redox reaction (4.33) have been mechanically activated and then the onset temperature has been studied

$$3TiO_2 + 13Al \rightarrow 3TiAl_3 + 2Al_2O_3 \qquad (4.33)$$

This temperature was observed to decrease from 1050 to 660°C after 5 hours of milling. Increasing the milling time to 100 h decreased the onset to 560°C [Welham 1998b, c].

The XRD traces are shown in Fig. 4.23.

Clearly, after 100 hours of milling, there is no evidence of reaction with only peaks for Al and rutil TiO_2 present. The longer milling times caused a broadening of all the peaks indicating that a decrease in the crystallite size and an increase in the lattice strain was occuring. The crystallite size for Al decreased more rapidly than for rutile, as it has been stated in literature [Koch 1991; Gaffet and Harmelin 1990; Weeber and Bakker 1998] that in two phase milling systems the softer material amorphises more rapidly. Elemental mapping of the powders showed that milling for 5 hours caused significant intermixing between the original phases which was not evident at shorter milling times. The reaction became net exothermic after 5 hours

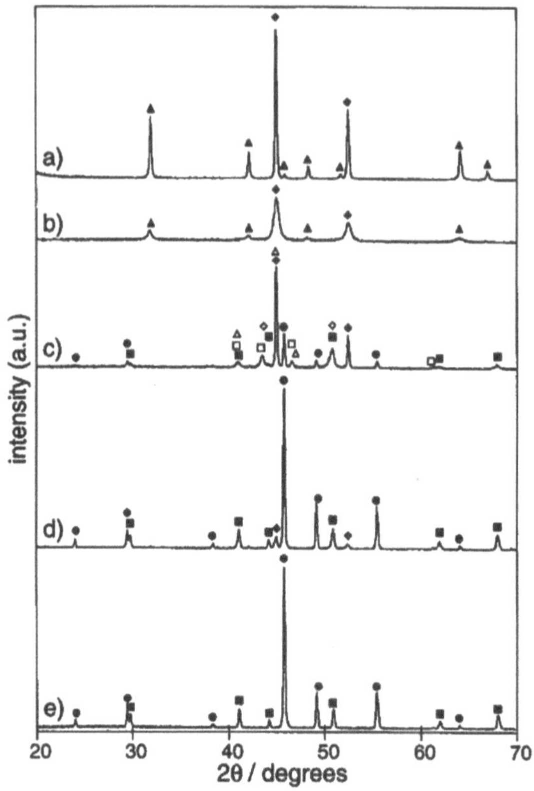

Fig. 4.23 X-ray diffraction patterns of 3:13 molar ratio TiO_2+Al, milling time 1 h (**a**) and 100 h (**b**) respectively; (**c**) and (**d**) 1 h milled powder heated to 900°C and 1200°C, respectively; (**e**) 100 h milled powder heated to 1200°C. ♦ Al, ■ Al_2O_3, • $TiAl_3$, □ Ti, ❖ TiO, Δ Ti_2O [Welham 1998a]

of milling turning from an endothermic process into a potentially self-propagating reaction once the initial thermal reaction has started.

Overview of mechanically induced redox reactions performed with oxides is given in Table 4.7.

The new method has been developed for preparation of nanocrystalline compounds by redox reaction of corresponding oxides with elements [Zhang et al. 2001; Saito et al. 2004]. The method is based on co-milling of corresponding oxide with sulphur and iron and stimulation of solid-state reaction to form nonferrous metal sulfides and iron oxides

$$4MeO + 4S + 3Fe \rightarrow 4MeS + Fe_3O_4 \quad (4.34)$$

where Me = Zn, Pb, Cd.

Comparison of ΔG^0 changes for reactions of type (4.34) and possible iron sulfide formation is summarized in Table 4.7. The thermodynamic data indicate not only the negative changes in ΔG^0 values but also higher thermodynamical probability for metal sulfidization in comparison with the iron sulfide formation.

4.3.3.4 Mechanism of Redox Reactions in Oxides

In most solid-state reactions, the reaction volume continually diminishes as the reactants become spatially separated by products. The chemical kinetics of these reactions are therefore determined by the diffusion rates of the reactants through the product phases, and the activation energy for the reaction is that for the diffusion of the reactant through the products [Carter 1991]. The reaction is therefore generally controlled by factors which influence diffusion rates. These include defect structures, defect densities and local temperatures.

Apart from diffusion, reaction rates are influenced by initial contact areas. However, for most solid-state process, the initial contact area is fixed and diffusion is the rate-limiting step. It is for this reason that solid-state reactions are so temperature dependent and that the reactions are thermally induced. The reaction interfacial area is not only dependent on initial contact areas but increases during milling [Schaffer and McCormick 1992b]. Furthermore, this increased reaction interface area is dynamically maintained during milling as the precursors are continuously broken up during fracture events. Moreover, diffusion rates are substantially increased in mechanically treated substances. This arises from the large number of crystal defects introduced through plastic deformation, which allows significant diffusion to occur at room temperature [Schaffer and McCormick 1991]. Diffusivities in nanocrystalline particles can be many orders of magnitude higher than both lattice and grain boundary diffusion in conventional polycrystalline materials [Karch et al. 1987).

The redox mechanochemical reactions are typical example of ductile (metal)/ brittle (metal oxide) systems. In these systems the brittle constituent becomes evenly distributed through the ductile metal reductant powders at the beginning of redox reaction. However, the reaction will produce two new phases, one brittle and one ductile. Both reactants and products are evenly dispersed through the reaction volume at later stage of redox reaction. Particles are reduced to nanometer-sized crystallites and brought into intimate contact, thus greatly increasing reacting areas. The high defect densities induced in the powder further increase reaction rates by providing short circuit diffusion path [Schaffer and McCormick 1990].

In special cases, the combination of small particle size, high defect densities, and the effective of product from between the reactants caused the material to be brought to a critical precombustion condition when ignition can occur. This is the case of several redox reactions covered in this Chapter.

4.3.3.5 Ferrite-Type Complex Oxides

Spinel ferrites $MeFe_2O_4$ (Me is divalent cation) posess great potential for many applications. Owing to the flexibility of their spinel structure, these compounds provide a wide range of physical and chemical behaviour which can be tuned by mechanical activation [Pavlyuchin et al. 1981, 1982, 1983, 1984, 1988; Jermakov et al. 1982]. The spinel structure provides interstices of tetrahedral and octahedral coordination, part of which is occupied by Me^{2+} and Fe^{3+} cations. In general, cations can reside on both types of sites, thus giving rise to a characteristic cation distribution [O'Neill and Navrotsky 1983, 1984]. The crystal structure of spinel ferrites can be formulated in greater detail as $(Me_{1-\lambda}Fe_{\lambda})$ $[Me_{\lambda}Fe_{2-\lambda}]O_4$, where round and square bracket enclose ions on sites of tetrahedral and octahedral coordination, respectively, and where γ denotes the so-called degree of inversion. The most important changes occur during mechanical activation in the cation sublattice, which are manifested in the redistribution of bi- and trivalent cations over the tetrahedral and octahedral cavities. An annealing of spinel ferrites at 500–600°C results in the return of the cations to their typical positions so that the normal properties of these substances are restored [Boldyrev 2006].

The results on several mechanically treated spinel ferrites like $ZnFe_2O_4$, $MgFe_2O_4$, $CuFe_2O_4$, $MnFe_2O_4$, $NiFe_2O_4$ and $CoFe_2O_4$ have been recently summarized [Šepelák et al. 2005]. Several interesting properties were determined, e.g. superparamagnetism, an enhanced chemical reactivity, an irreversible magnetic behaviour and an altered saturation magnetization in comparison with the corresponding bulk materials. As for reactivity, the hot-gas desulphurization tests have revealed that mechanochemically synthesized $ZnFe_2O_4$ posesses the highest sulfur absorption capacity when compared with the reactivity of the unmilled ZnO/Fe_2O_3 mixture as well as with the reactivity of milled $ZnFe_2O_4$ that was prepared by the conventional thermal method [Šepelák et al. 1997].

The application of high-energy milling to the chemical reduction of spinel ferrites has been reported in several papers [Takacs 1993; Takacs and Pardavi-Horvath 1994; Sorescu 1988; Botta et al. 2000; Šepelák and Becker 2000; Menzel et al. 2001; Shi and Ding 2001; Goya et al. 1998].

The high-energy milling of $NiFe_2O_4$ in a steel vial using steel balls was found to induce a mechanochemical reduction process of the material leading in a redox process to the formation of a disordered solid solution of FeO and NiO with wüstite structure [Menzel et al. 2001; Shi and Ding 2001]. To determine the phase evolution of $NiFe_2O_4$ during high-energy milling the mechanochemical reduction was followed by X-ray powder diffraction. The XRD-pattern of the non-activated sample (Fig. 4.24) is characterized by the sharp crystalline peaks corresponding to $NiFe_2O_4$ (JCPDS 10-0325). During the early stage of milling, XRD reveals only a decrease of the intensity and an associated broadening of the Bragg peaks of the spinel. This reflects the formation of the disordered state with a small crystallite size and with internal strain introduced during the mechanical treatment. With increasing milling time, changes in diffraction intensities are observed. This can be attributed to the formation of a solid solution of $Ni_{1-x}Fe_xO$ and of crystalline iron (JCPDS 06-0696) which are formed in addition to the formation of nanoscale ferrite particles. The fraction of the reduced phases increases with increasing milling time.

Mechanochemical reduction of magnesium ferrite $MgFe_2O_4$ has been studied [Šepelák et al. 2002]. The XRD pattern of the non-activated sample is characterized by the sharp peaks corresponding to $MgFe_2O_4$ (Fig. 4.25).

During the early stage of milling, XRD merely reveals a decrease of intensity and an associated broadening of the Bragg peaks of the ferrite. This reflects the formation of the disordered state with a small crystallite size and with internal strain

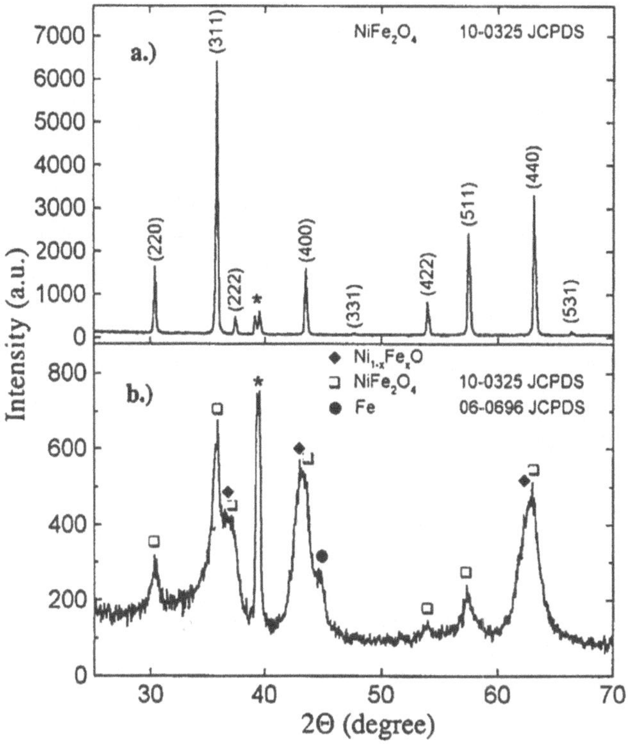

Fig. 4.24 X-ray diffraction patterns of (**a**) non-activated $NiFe_2O_4$ and of (**b**) $NiFe_2O_4$ activated for 20 min. Asterisk indicates a diffraction peak of the sample holder [Menzel et al. 2001]

introduced during the mechanical treatment. With increasing milling time, qualitative changes are observed in diffraction patterns of the samples. This can be attributed to the formation of a solid solution of $Mg_{1-x}Fe_xO$ which possesses the wüstite structure and of crystalline iron. It was concluded that both the solid solution of $Mg_{1-x}Fe_xO$ and the ferromagnetic Fe are predominantly obtained as the result of the mechanochemical reduction of $MgFe_2O_4$.

In the contact of a mechanistic analysis, it was concluded that the metallic nature of the vial and of the balls plays an essential role in the mechanochemical reduction process of $MgFe_2O_4$. It appears that also in case of ferrite-type complex oxides the event of mechanically induced redox reactions present novel opportunities for the nonthermal manipulation of such type of materials [Šepelák et al. 2002].

4.3.4 Wolframates

Tungsten carbide WC is probably the most common hard material in use today for cutting edges and other high strength applications [Welham 1999]. It is usually

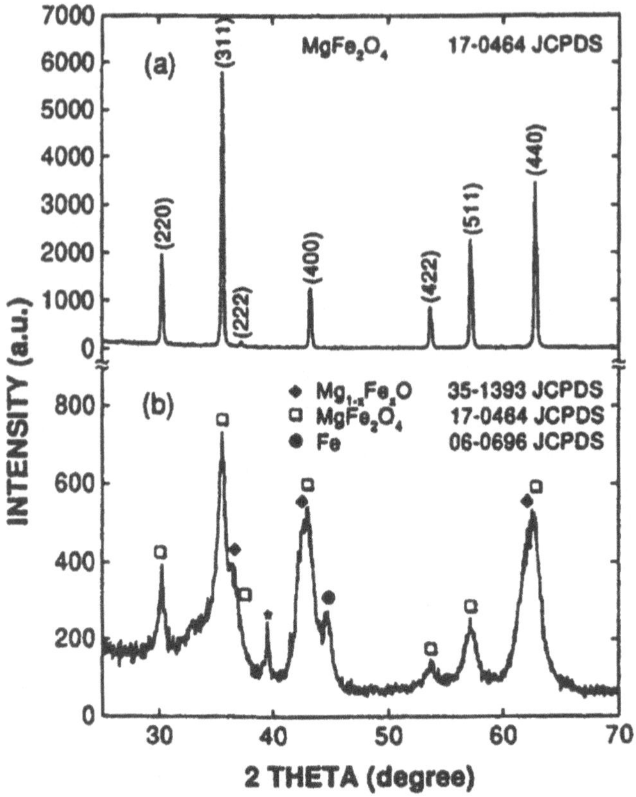

Fig. 4.25 X-ray diffraction patterns of non-activated $MgFe_2O_4$ (**a**) and of material mechanically treated for 20 min (**b**). [Šepelák et al. 2002]

made by a somewhat complicated route from either wolframite $FeWO_4$ or from scheelite $CaWO_4$ minerals. Previous attempts to produce WC directly have shown that the reduction is possible using excess carbon at temperatures >1200°C [Terry et al. 1994].

Scheelite reduction has been tested by milling in a laboratory stainless steel ball mill. The XRD trace of the as-milled carbon bearing powder is presented in Fig. 4.26b.

Clearly, there has been a significant change from the original $CaWO_4$, shown in Fig. 4.26a. The peaks present seemed to be for MgO and W_2C. No calcium-bearing phase was apparently present, presumably due to a small crystallite size. After annealing for 1 h at 1000°C, trace (c) was obtained. There has been significant crystallite grows with narrower, more intense peaks now evident for both W_2C and MgO. After milling in nitrogen the XRD trace is much clearer. Figure 4.26d shows that the as-milled powder contains elemental W and MgO. The crystallite size of the tungsten was estimated to be 9.5 nm. Only the two main MgO peaks are evident.

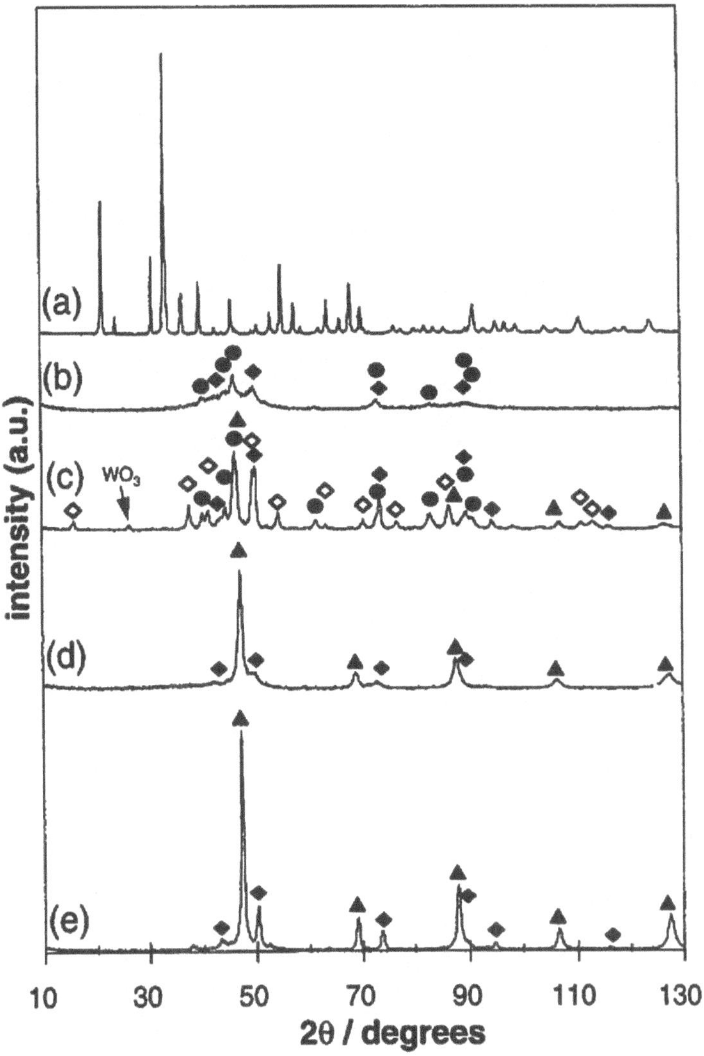

Fig. 4.26 X-ray diffraction patterns of $CaWO_4$ treatment: (**a**) milled for 1 h in vacuum, (**b**) milled for 100 h with carbon, (**c**) powder (**b**) heated for 1 h at 100°C in argon, (**d**) milled for 100 h under nitrogen and (**e**) powder (**d**) heated for 1 h at 1000°C in argon, ♦ W, ♦MgO, ◊ Fe_3W_3C, ■ WC, • W_2C, [Welham 1999]

These peaks are broad and comparatively weak, which is typical of phases that have a small crysatllite size. On heating this powder for 1 h at 1000°C in argon, the MgO peaks had narrowed considerably. The tungsten crystallite size increased to 21 nm. A simple acid leach of the powder removed all of the MgO, leaving behind either W_2C and Fe_3W_3C or W. The elemental tungsten formed was ∼99% W [Welham 1999].

4.3.5 Titanates

The mineral ilmenite $FeTiO_3$ is a huge resource of rutile TiO_2 for the production of titanium dioxide, which can be used directly as a pigment or in the manufacture of titanium. A number of methods have been proposed for extraction rutile from ilmenite [Chen et al. 1996]. However, most of them involve either a hot chemical process or a high temperature reduction process (Becher process). These processes are both expensive and complicated, and have many associated environmental implications. Therefore, the search for a simple and economic extraction method remains of high interest to industry [Henn and Barclay 1995].

Mechanochemical reduction of ilmenite $FeTiO_3$ by silicon Si has been described [Welham 1998d]. XRD traces for 2:1 $FeTiO_3$:Si powder milled under low and high intensity for up to 200 h are shown in Fig. 4.27.

After 100 h milling under low intensity trace (a) was obtained. The only peaks present are for $FeTiO_3$ and Si, although the peaks are somewhat wider and less intense than the feed powder due to crystallite size refinement. After a further 100 h of milling the trace (b) showed major changes, both $FeTiO_3$ and Si were absent with the only peak present due to elemental iron. The apparent absence of phases for both Ti and Si is probably due to an extremely small crystallite size and the presence of many phases each of low abundance.

Annealing of the 200 h milled powder for 1 h at 800°C under argon gave trace (c), clearly, peaks for TiO_2 have emerged along with a small peak for Si. There may be a small peak for SiO_2 at 305°C but it is overlapped by that of rutile and, in the absence of other SiO_2 peaks, no positive confirmation of quartz can be made. Thus, it would seem that the reaction had occurred after 200 h with the annealing stage causing recrystallization of nanocrystalline phases.

Under high intensity milling there was clearly Fe present after 50 h, trace (d), implying that the reduction of $FeTiO_3$ had started within 50 h. However, the presence of peaks for $FeTiO_3$ and Si show that the reaction was of relatively low extent. After 100 h of milling (e) shows that the reaction was apparently complete, neither $FeTiO_3$nor Si peaks were present, only those for Fe and TiO_2. Further milling to 200 h showed no significant changes in the XRD trace (f), other than a widening and weakening of the TiO_2 peaks as the crystallite size decreased [Welham 1998c].

The chemistry of the whole process of ilmenite reduction by silicon can be described [Chen et al. 1996].

$$2FeTiO_3 + Si \rightarrow 2Fe + 2TiO_2 + SiO_2 \tag{4.35}$$

$$Fe + xSi \rightarrow FeSi_x, x = 1 \text{ or } 2 \tag{4.36}$$

$$2nTiO_2 + Si \rightarrow 2Ti_nO_{2n-1} + SiO_2 \tag{4.37}$$

The final stage is the reduction of the remaining Ti-O compounds to a titanium silicide

$$Ti_nO_{2n-1} + (16n-5)Si \rightarrow 2nTi_5Si_3 + (10n-5)SiO_2 \tag{4.38}$$

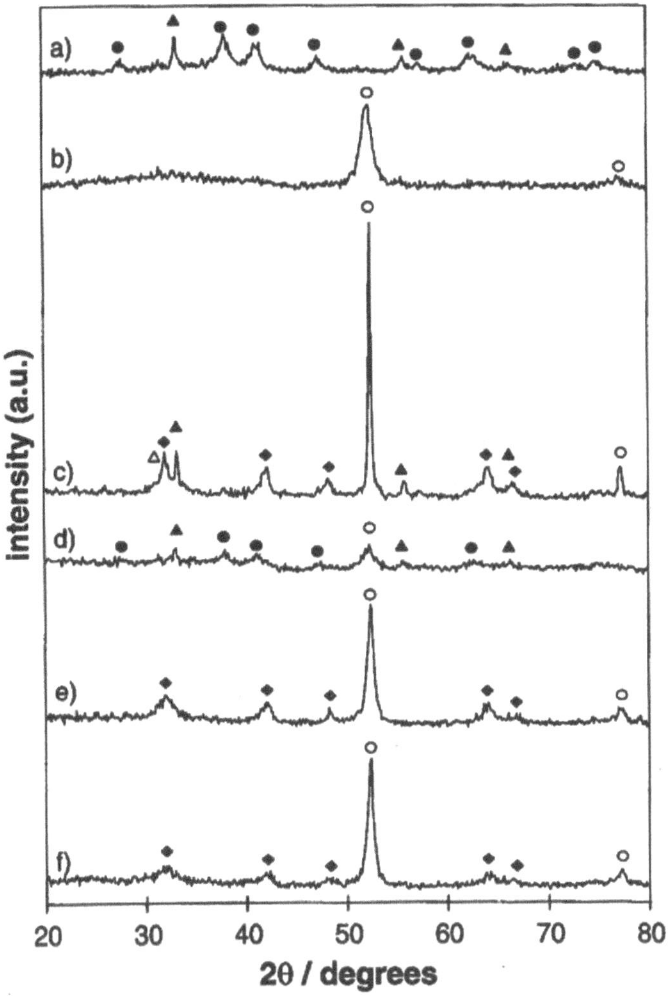

Fig. 4.27 X-ray diffraction patterns of 2:1 molar ratio $FeTiO_3$+Si milled for (**a**) 100 h, and (**b**) 200 h under low intensity, (**c**) sample b annealed for 1 h at 800°C under argon, (**d**) 50 h, (**e**) 100 h and (**f**) 200 h under high intensity; ♦ Si, ♦ TiO_2, ● $FeTiO_3$, ○ Fe, Δ SiO_2 [Welham 1998c]

Mechanochemical reduction of ilmenite $FeTiO_3$ by sulphur has been studied by two groups in Australia [Henn and Barclay 1995; Welham 1998d]. The reaction has been suggested as a route to the formation of rutile and pyrrhotite in geological systems [Frost 1991]. Iron is known to have a very high affinity for sulphur, and a range of iron sulfides has been formed by milling elemental iron and sulphur together at room temperature [Baláž et al. 1995].

Thermodynamical calculation performed in work of Frost (1991) showed two ilmenite-to-sulphur molar ratios were worth investigating. A 3:2 ratio shows

$$3FeTiO_3 + 2S \rightarrow F_2O_3 + FeS_2 + 3TiO_2 \tag{4.39}$$
$$(\Delta G_{298} = -96.5\,\mathrm{kJ\,mol^{-1}})$$

The 4:7 ratio show

$$4FeTiO_3 + 7S \rightarrow FeSO_4 + 3FeS_2 + 4TiO_2 \tag{4.40}$$
$$(\Delta G_{298} = -231\,\mathrm{kJ\,mol^{-1}})$$

For 2:4 ratio the following reaction was suggested

$$2FeTiO_3 + 4S \rightarrow 2FeS_2 + O_2 + 2TiO_2 \tag{4.41}$$

The X-ray diffraction traces of the as-milled products (Fig. 4.28) show $FeTiO_3$ was still a major phase, present in all samples immediately after milling, although the peaks were somewhat less intense and broader then those of the original material. Elemental sulphur was present only in the products of the 4:7 ratio run and its absence in the other products may be due to either amorphization of reaction. TiO_2 was found to be present in all of the samples, but the level was similar to that previously observed as an impurity in the ilmenite starting powder and could not be confirmed as a reaction product [Welham 1998d].

The reaction between S and $FeTiO_3$ during ambient temperature milling was only evident in samples which were milled dry in an inert atmosphere, preferably under pressure. The reaction found to be incomplete, with the formation of FeS_2 and rutile TiO_2 and from industrial viewpoint is not viable. However, ilmenite milled

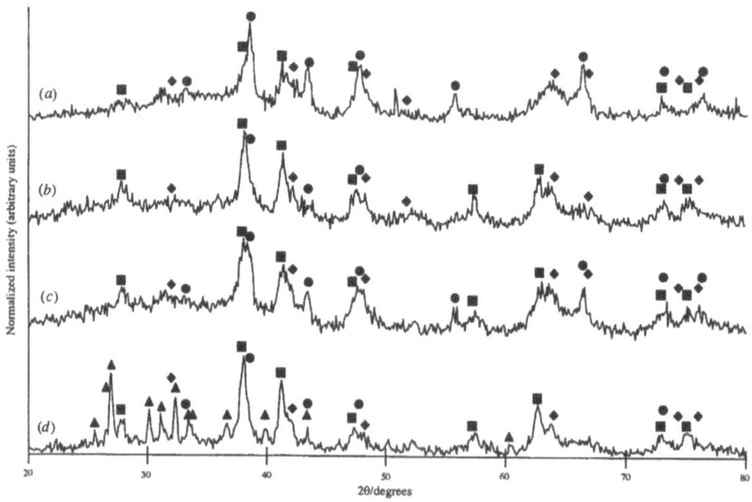

Fig. 4.28 X-ray diffraction patterns of as-milled powders, milling time 100 h. (**a**) 3:2 $FeTiO_3$+S ratio, 100 kPa He; (**b**) 3:2 $FeTiO_3$+S ratio, 5.0 ml H_2O, open to air; (**c**) 3:2 $FeTiO_3$+S ratio, dry vacuum; (**d**) 4:7 $FeTiO_3$+S ratio, dry, vacuum. ■ $FeTiO_3$, ◆ TiO_2, ♦ α-S, • FeS_2 [Welham 1998d]

alone may prove to be a viable feedstock for a thermal reactor in which sulphur is maintained at a high partial pressure.

The observations show that the milling can be alternative to traditional processes. Obviously, further investigation is needed [Welham 1998d].

An ilmenite mineral and Al powder have been mechanically milled together for 100 h in a laboratory ball mill. The reaction

$$FeTiO_3 + 7Al \rightarrow FeAl_2 + TiAl_3 + Al_2O_3 \quad (4.42)$$

has been considered as a guide. The as-milled powder and an unmilled powder of identical composition were annealed at up to 1200°C and examined by X-ray diffraction and DTA methods. The unmilled sample showed Al melted prior to an exothermic reaction starting at ~ 850°C. In the milled powder the aluminium had been completely consumed below its melting point. The milled powder showed no thermal activity, other than a reversible phase transition at 1067 ±4°C, indicating that reaction occured within the mill. The products of both powders were the same, $TiAl_3$, Fe_3Al_{13} and Al_2O_3, although in the milled powder these phases were nanocrystalline until annealing caused crystallite growth [Welham 1998e].

The mechanochemical reduction of $FeTiO_3$ and TiO_2 by Mg metal has been studied [Welham 1998f] for reactions

$$FeTiO_3 + Mg \rightarrow MgTiO_3 + Fe \quad (4.43)$$

$$MgTiO_3 + Mg \rightarrow 2MgO + TiO \quad (4.44)$$

$$TiO + Mg \rightarrow MgO + Ti \quad (4.45)$$

Rutil was predicted to be reduced to TiO, and then to titanium metal by reaction

$$TiO_2 + Mg \rightarrow TiO + MgO. \quad (4.46)$$

The phases present in the as-milled powder changed with the ratio $FeTiO_2$: Mg (Fig. 4.29a). The 1:1 ratio showed that the ilmenite and $MgTiO_3$ could both be present (the peak positions for these phases overlap considerably), the small peaks at $2\theta = 50$ and 52° indicate MgO and Fe respectively. As the ratio increased (Fig. 4.29b and 4.29c), $FeTiO_3$ peaks weakened and peaks for Fe and MgO become more intensive, this is due to the increasing reduction of $FeTiO_3$ to Fe and an unidentified titanium product. After milling, neither mixture of TiO_2 and Mg showed many XRD peaks (Fig. 4.29d and 4.29e). There was only the main TiO_2 peak and the two main peaks for MgO indicating that reduction had occured. The MgO peaks were much more intense than any other peak, implying the main phase present was MgO [Welham 1998f].

In a high temperature process ilmenite is purified by reduction to TiO_2 and elemental iron which is then leached to form an inpure TiO_2. The 88–92% TiO_2 is then chlorinated. To produce metallic titanium the chloride is reacted with molten Mg under an inert atmosphere. The further processes are needed to obtain metallic Ti [Welham 1998f].

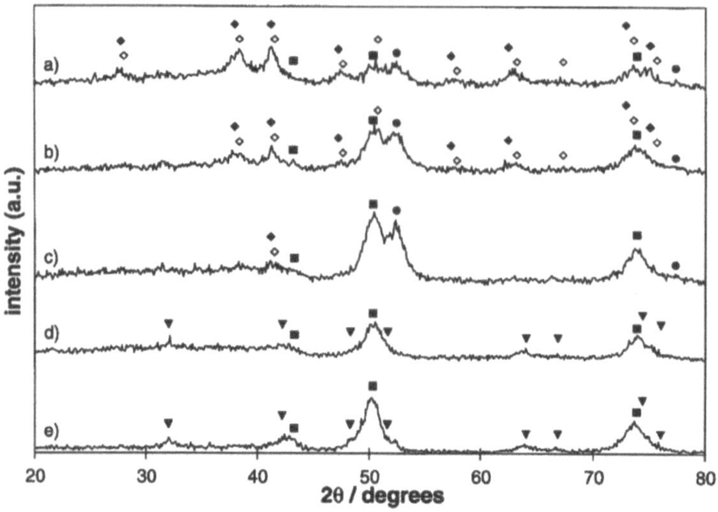

Fig. 4.29 X-ray diffraction patterns of as-milled powders, milling time 100 h. (a) 1:1, (b) 1:2, (c) 1:3 $FeTiO_3$+Mg ratio, (d) 1:1 and (e) 1:2 TiO_2+Mg ratio. ◆ $FeTiO_3$, ◇ $MgTiO_3$, • Fe, ■ MgO, ▼ TiO_2 [Welham 1998f]

4.3.6 Sulphides

Sulphides exhibit a great variety of chemical and physical properties. They display similar structural defects as oxides with cation vacancies, intersticial cations or anionic defects all possible. However, the concentration, structure and mobility of these defects are much more varied in the case of sulphides [Mrowec 1988].

The cationic vacancies change their properties according to conditions. For instance, their influence on electric properties and character of the chemical bond of sulphides is dependent on temperature [Vanjukov et al. 1978]. The formation of interesticial defects, in which metal ions are displaced from a normal lattice position to an intermediate position, depends on strength of the metal-sulphur bond. The Me-S and S-S bond in sulphides are similar to each other and the difference between them decreases with increasing temperature. However, if sulphur is released, the metal present in the normal crystallografic position remains without a partner and the defectiveness of sulphide increases. The influence of defects on solid state reactions is frequently more significant than the influence of the sulphide structure [Tkáčová 1989].

Sulphides play an important role in traditional technological applications as well as advanced materials (Table 4.8).

Solid state reactions of sulphides to prepare the elemental nanometals or their oxides proceed via two routes

$$Me_1S + Me_2 \rightarrow Me_1 + Me_2S \qquad (4.47)$$

$$Me_1S + CaO \rightarrow Me_1O + CaS \qquad (4.48)$$

Table 4.8 Sulphides in traditional technology and as advanced materials

Traditional applications		Advanced materials	
Chemical engineering		High-energy batteries	TiS_2
Hydrodenitrogenation	Co_9S_8	Solar cells	$CuInS_2$, FeS_2, CuS
Hydrodesulfurisation	MoS_2, WS_2	Diagnostic materials	Ag_2S
Dehydratation	MnS, CuS, ZnS	Luminiscence materials	ZnS, CdS: Mn, Cu
		Superconductors	La_2S_3, CuS
Mechanical engineering		Intercalates	TiS_2, TaS_2, NbS_2
Lubricants	MoS_2	Ion-selective sensors	PbS, CdS, CuS
		Optical coatings	ZnS
Extractive metallurgy		Photoconductors	ZnS, Ag_2S
Pyrometallurgy	Cu_3AsS_4, PbS	Lasers	CdS
Hydrometallurgy	$CuFeS_2$, ZnS, $Cu_{12}Sb_4S_{13}$	Photolithographic materials	As_2S_3
Biohydrometallurgy	FeAsS, FeS_2, ZnS	Holographic recordings	SnS
		Optical fibres	As_2S_3
Mineral processing		Infra-red detectors	PbS
Sulfidizers	Na_2S	High-temperature termistors	Cu_2S
		Quantum dots	$Cd_{1-y}Zn_yS$

In reaction (4.47) the reduction of the metal sulphide Me_1S is performed with a reducing element Me_2 (usually Fe, Mg, Al, Si) and is called redox reaction (see Chapter 4.3.1). The reaction (4.48) represents a displacement reaction. Sometimes combined effect of redox and displacement reaction is applied.

4.3.6.1 Thermal Treatment

It has to be mentioned that in modern technology reducing of sulphides by metals is practicized to a limited extent. The most frequently applied reducing agent is iron [Szczygiel et al. 1998]. The process is inefficient because the metal sulphide dissolves in the FeS formed. Nowadays it is used for the reduction of rich antimonite Sb_2S_3 ores or concentrates and to some extent for reduction galena PbS [Habashi 1986].

During the Industrial Revolution in Germany (1750–1850), antimony was sometimes obtained by heating the sulphide with scrap iron, which combined with sulphur and an alkali carbonate. The Sb was purified by fusion with a little antimony sulphide Sb_2S_3 and sodium carbonate Na_2CO_3 [Habashi 1993].

In principle non-metal reducing element can be applied. Reduction of PbS and $CuFeS_2$ by carbon in the presence of calcium oxide has been studied [Kharitidi et al. 1981, 1983].

Reduction of galena by solid carbon in the presence of CaO occured through stages of sulphide sublimation and reaction between the gaseous sulphide and CaO. Carbon did not participate in direct reduction but intensify PbS sublimation and reducing sulphur dioxide to S_2, CaS and CS_2

$$3PbS\,(g) + CaO \rightarrow 3Pb + 2CaS + SO_2 \tag{4.49}$$

Heating in helium to 1200°C at 10°/min were used for reaction (4.49).

Reduction of chalcopyrite $CuFeS_2$ with carbon in the presence of CaO was studied at 1100°C. The overall reaction is governed by the equation

$$CuFeS_2 + 2CaO + 2C \rightarrow Cu + Fe + 2CaS + 2CO \tag{4.50}$$

4.3.6.2 Mechanochemical Treatment

The concept of direct reduction of sulphides to metals by mechanochemical treatment was introduced by Soviet Scientists [Molčanov and Jusupov 1981]. The authors named the process *mechanometallurgy*. By dry milling of cinnabar HgS in a planetary mill equipped with copper vials and balls it was possible to obtain elementary mercury according to the following reaction

$$HgS + 2Cu \rightarrow Cu_2S + Hg \tag{4.51}$$

By the authors the reduction also proceeds in the course of milling of HgS in water using iron vials and balls

$$2HgS + 7H_2O \rightarrow 2Hg + H_2SO_3 + H_2SO_4 + 5H_2 \tag{4.52}$$

The transformations of selected sulphides by milling with the different metals have been studied. The solid state reduction of the sulphides have been performed by room temperature ball milling of powder mixtures [Matteazzi and LeCaër 1992b].

4.3.6.3 Reduction with Aluminium

The most results were obtained with aluminium as reducing element. The reaction of iron sulphide with Al

$$3FeS + 2Al \rightarrow 3Fe + Al_2S_3 \tag{4.53}$$

shows the predominant presence of patterns belonging to bcc α-Fe (Fig. 4.30) with some Al in it, since the lattice parameter $a = 0.2876$ nm is greater than that for pure

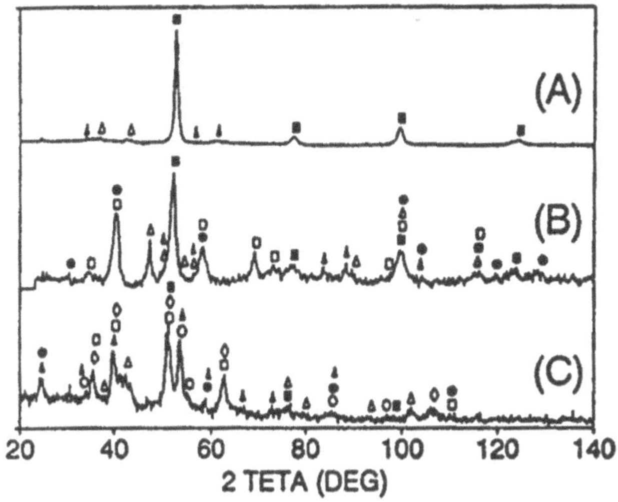

Fig. 4.30 X-ray diffraction patterns of as-milled powders, milling time 24 h. (**A**) 3:2 molar ratio FeS+Al; (**B**) 1:1 molar ratio FeS+Mn; (**C**) 2:1 molar ratio FeS+Si. Symbols: (**A**) ■α-Fe, Δ $FeAl_2O_4$, ▲ Al_2O_3; (**B**) ■α-Fe, Δβ-Mn, ▲γ-Mn,□ MnS, ● MnS_2; (**C**) ■α-Fe, ◊ FeS, □ $Fe_{1-x}S$, ▲ SiS_2, Δ WC, ○ FeSi, ● α-$FeSi_2$ [Matteazzi and LeCaër 1992b]

α-Fe (a = 0.2866 according to JCPDS 6-694) to an Fe-Al alloy with about 7 at % Al. Further weak components are present in the pattern, corresponding to tetragonal Al_2S_3 (JCPDS 24-14) and cubic hercynite $FeAl_2O_4$.

The reaction of copper sulphide

$$3Cu_2S + 2Al \rightarrow 6Cu + Al_2S_3 \quad (4.54)$$

shows clearly the presence of the following products: fcc Cu with a lattice parameter a = 0.3617 nm very close to that of pure Cu (a = 0.3615 nm according to JCPDS 34-348) and crystallite size d = 23 ± 3 nm, γ-Al_2S_3 (JCPDS 34-348).

The reactions of cobalt, lead and zinc sulfides

$$3CoS + 2Al \rightarrow 3Co + Al_2S_3 \quad (4.55)$$

$$3PbS + 2Al \rightarrow 3Pb + Al_2S_3 \quad (4.56)$$

$$3ZnS + 2Al \rightarrow 3Zn + Al_2S_3 \quad (4.57)$$

shows common X-ray diffraction patterns (Fig. 4.31). The patterns reveal, that cobalt sulphides and Al are no longer present and cobalt metasl is present in hexagonal as well as cubic form. Among products Al_2S_3 and cobalt aluminide Co_2Al_5 were identified. For reaction (4.56) Al lines are no longer present and cubic Pb, mixture of hexagonal Al_2S_3 and cubic Al_2S_3 have been formed. For reaction (4.57) the starting ZnS powder was constituted mainly by cubic β-ZnS (sphalerite) and a small amount of hexagonal ZnS (wurtzite). There remains some β-ZnS, while both

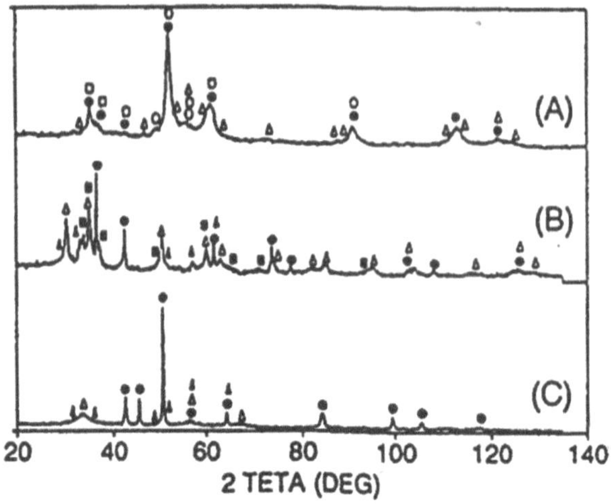

Fig. 4.31 X-ray diffraction patterns of as-milled powders, milling time 24 h. (**A**) 3:2 molar ratio CoS+Al; (**B**) 3:2 molar ratio PbS+Al; (**C**) 3:2 molar ratio ZnS+Al. Symbols: (**A**) • Co (cub); (**B**) • Pb; (**C**) – Zn; ○ Co (hex), ▲ Al_2S_3 (hex), □ Al_2S_3 (tetr), ■ Al_2S_3 (cub); (**A**) –Δ Co_2Al_5, (**B**) –Δ PbS, (**C**) –Δβ-ZnS [Matteazzi and LeCaër 1992b]

the reaction products according to reaction (4.57) are present: hexagonal Zn (most intense pattern) of $d = 32 \pm 5$ nm crystallite size and hexagonal Al_2S_3. As for the two previous cases, a Zn-Al_2S_3 nanocomposite has been obtained.

It was concluded by the authors the reduction of metal sulphides by room temperature ball milling with a suitable reducing agent has been shown to be feasible. The feasibility of a particular reaction by mechanical activation appears not to be limited to negative free energy or enthalpy as shown for FeS reduction by silicium. Metal-(alloys)-(intermetallic compounds)-sulphide nanocomposites are obtained in this way with crystallite sizes in the range 10–30 nm. The ball milling-induced reduction of molybdenite MoS_2 by Al has been investigated in paper [Takacs et al. 2006]. The reduction takes place according to the equation

$$3MoS_2 + 4Al \rightarrow 3Mo + 2Al_2S_3 \tag{4.58}$$

The kinetics of reaction (4.58) is given in Fig. 4.32.

The graph clearly shows that most of the reaction takes place between 20 min and 60 min and it is practically complete after 120 min of milling. Although the reduction of MoS_2 with Al is a highly exothermic reaction that was expected to progress in the form of a self-sustaining process, ignition could not be achieved by ball milling. It is suggested that the reason is in the lubricant property of MoS_2. It prevents the formation of large composite agglomerates where the conditions for the initiation and propagation of the reaction would support MSR. The existing agglomerates are porous and they start breaking up before the reaction could accelerate.

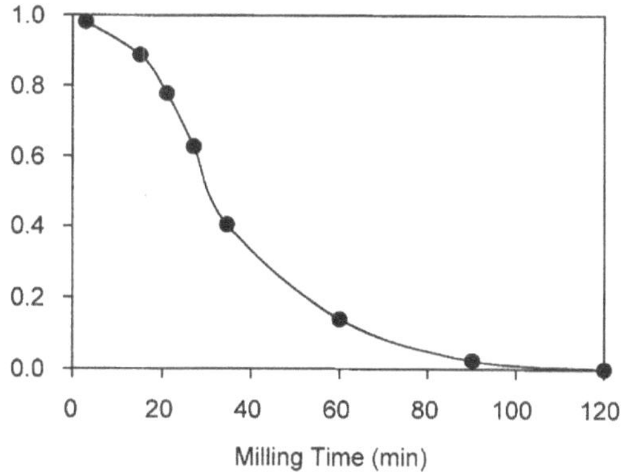

Fig. 4.32 The kinetics of MoS_2 consumption as a function of milling time [Takacs et al. 2006]

4.3.6.4 Reduction with Iron

Several papers have been published using iron metal as reducing element for mechanically induced solid state reduction by sulfides [Baláž et al. 2002a, b, 2004a, b, c, 2005a, 2007; Godočíková et al. 2004a, b].

The mechanochemical reduction of iron disulphide FeS_2 (pyrite) by elemental iron can be described by the simplified equation

$$FeS_2 + Fe \rightarrow 2FeS \tag{4.59}$$

The reaction is thermodynamically possible at ambient temperature, as the enthalpy change is negative. The value $\Delta H^\circ_{298} = -13.3\,\mathrm{kJ\,mol^{-1}}$ was calculated from the thermodynamic data.

Its progress with milling time is illustrated by selected XRD patterns in Fig. 4.33.

The process of product formation is clearly seen for samples milled for 25 and 45 min. Hexagonal troilite FeS (JCPDS 75-0602) has been indicated as the only reaction product. Figure 4.34 shows the room temperature magnetization data for investigated samples as a function of external magnetic field. It is evident that the resulting magnetization curves are well saturated after application of magnetic field higher than 2 T. The difference in the saturation magnetization of the samples milled for different times, as displayed in the inset of this Figure is mainly caused by a various amount of the ferromagnetic component in these samples. The inset shows that the amount of metallic iron decreases continuously for milling time up to 45 min. For longer milling time the saturation magnetization is negligible indicating that the constituent phases are almost entirely paramagnetic i.e. the reaction (4.59) tends to be finished.

The increasing paramagnetism can well be seen in Mössbauer spectra on Fig. 4.35a. Spectrum for sample milled for 5 min is a superposition of a doublet corresponding to FeS_2 and a sextet corresponding to Fe. The spectra during milling are changed and their analysis is given in Fig. 4.35b in the form of Mössbauer parameters dependence on the milling time. There are two characteristics sextets and

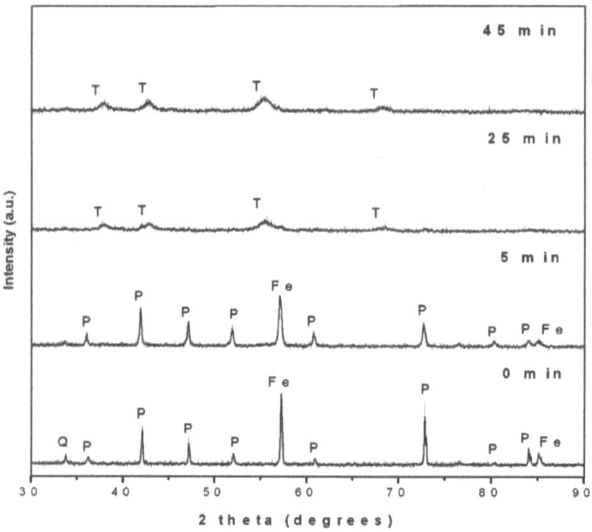

Fig. 4.33 X-ray diffraction patterns at different stages of reaction (4.59) as a function of milling time. T – troilite FeS, P – pyrite FeS_2, Q – quartz SiO_2 [Baláž et al. 2004a]

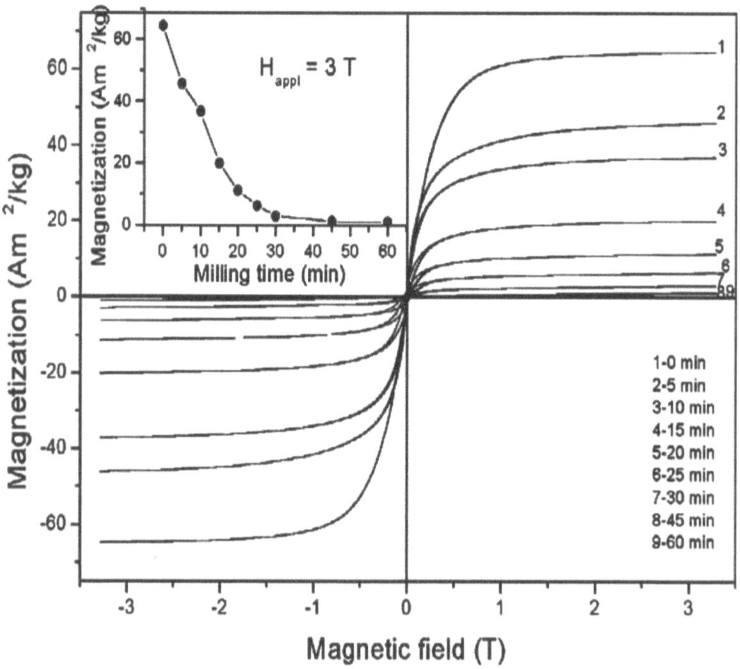

Fig. 4.34 Magnetization as a function of external magnetic field and/or of milling time for reaction (4.59) [Baláž et al. 2004a]

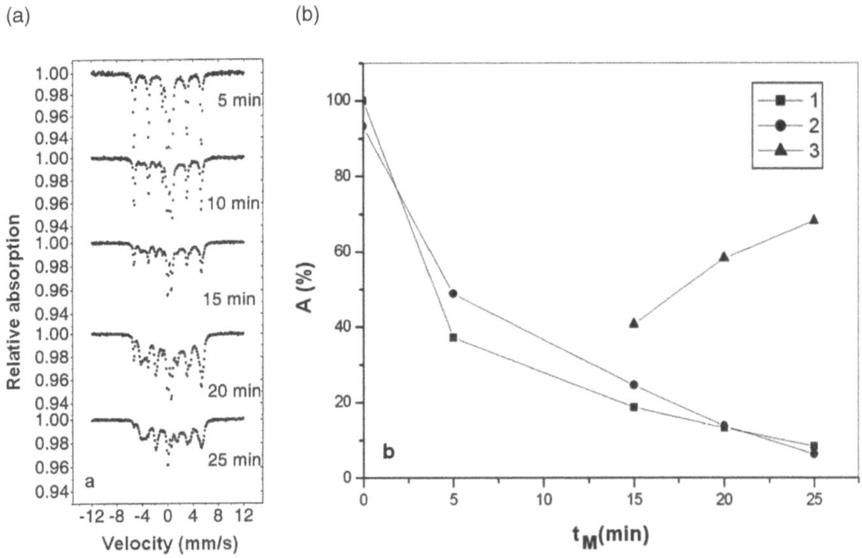

Fig. 4.35 (**a**) Mössbauer spectra for reaction (4.59); (**b**) Mössbauer parameters for reaction (4.59), A – total area of component, t_M – milling time; 1 – pyrite FeS_2 (doublet: IS = 0.31 mm s^{-1}, QS = 0.63 mm s^{-1}), 2 – Fe (sextet A: MF = 32.90 T), 3 – troilite FeS (sextet B: MF = 28.40 T), [Baláž et al. 2004a]

one doublet. The doublet with IS = 0.31 mm s^{-1} and QS = 0.63 mm s^{-1} corresponds to FeS_2 which is consumed during milling.

Elemental Fe which is also consumed, as can be seen from kinetics for sextet A with MF = 32.90 T. Beginning with t_M = 15 min, the new sextet B (MF = 28.40 T) has been identified with an increase in the total area component A. In accordance with XRD patterns in Fig. 4.33 and literature data [Matteazzi and LeCaër 1992b; Baláž et al. 2002b] this sextet can be described as troilite FeS. Stoichiometric FeS is antiferromagnetic with alternate layers of Fe atoms magnetized in opposite direction [Shuey 1975].

In the Mössbauer spectrum taken at 25 min, the traces of elemental Fe are still visible. However it is clear that antiferromagnetic troilite FeS is the dominant phase in the milled powders. From Fig. 4.34, it can be seen that this phase contributes only very little to the overall magnetization of our system.

The conventional high-energy temperature lead recovery process from lead sulphide PbS is based on reduction with iron scrap [Habashi 1986]. Recently, 92–94% lead recovery has been obtained by the direct reduction of lead sulphide with elemental iron in the temperature interval 1000–1150°C for 180 min [Szczygiel et al. 1998].

The mechanochemical reduction of lead sulphide with elemental iron under ambient temperature has been studied [Baláž et al. 2004b; Godočíková et al. 2004a,b].

The reaction can be described by the equation

$$PbS + Fe \rightarrow Pb + FeS \qquad (4.60)$$

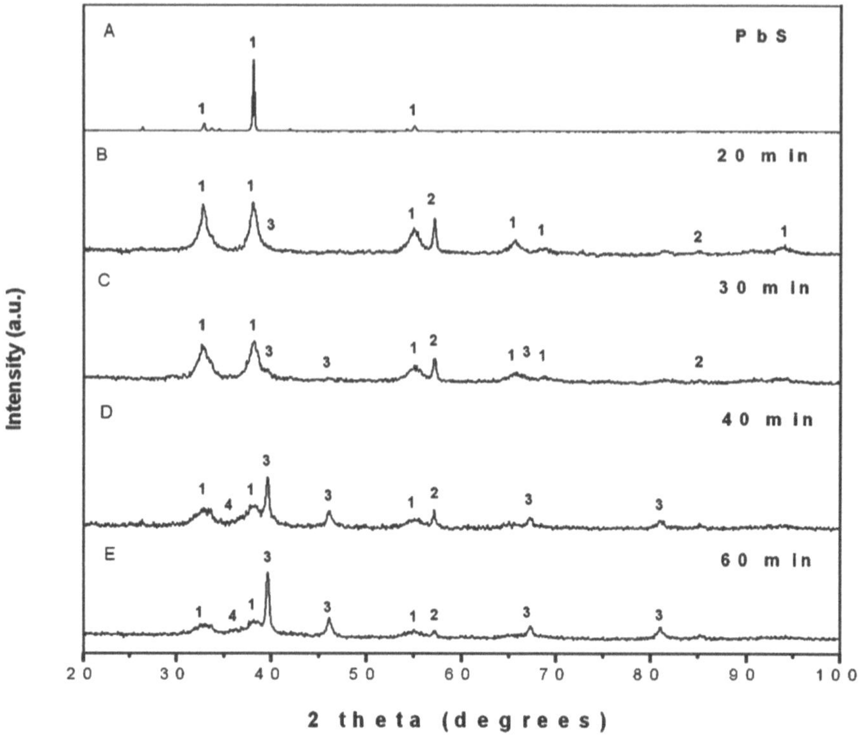

Fig. 4.36 X-ray diffraction patterns at different stages of reaction (4.60) as a function of milling time, A: PbS, B–E: PbS + Fe milled for 20–60 min (1 – lead sulfide, 2 – iron, 3 – lead, 4 – FeS (cubic)) [Godočíková et al. 2004b]

The reaction is thermodynamically possible at ambient temperature, as the enthalphy change is negative ($\Delta H^\circ_{298} = -1.3\,\text{kJ mol}^{-1}$). The process is illustrated by the XRD patterns in Fig. 4.36.

The primary process – the reduction of lead sulfide by iron while lead metal and iron sulfide are formed – is clearly seen. After 20 min of milling, the relative amount of lead sulfide decreased and the lead is formed as indicated by the diffraction peak at 39.9° (JCPDS 04-0686). The process kinetics as described by the conversion degree, β is shown in Fig. 4.37.

Only 19% conversion to lead metal was achieved during the initial 20 min of milling. The intensity of the iron (JCPDS 06-0696) and lead sulphide (JCPDS 78-1897) reflections decreases upon continued milling, but these phases are detectable even after 60 min of milling. After 60 min, the reflections from lead metal and cubic FeS (JCPDS 23-1123) can be observed and 75% conversion to lead was achieved.

The reaction products are present in the form of nanocrystalline particles. Particle size of 13–21 nm for lead have been obtained. Figure 4.38 shows a typical TEM image of the polydisperse Pb/FeS nanoparticles formed by mechanochemical reduction of lead sulphide with iron during 60 min.

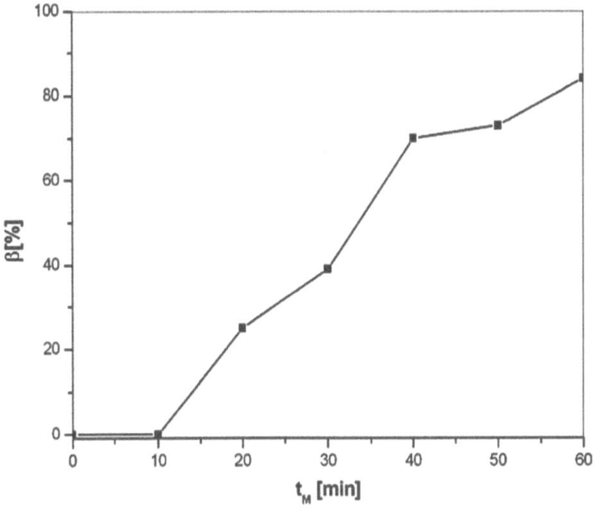

Fig. 4.37 Conversion degree, β for reaction (4.60) as a function of milling time, t_M [Godočíková et al. 2004b]

Examination of TEM shows that the as-milled powder is composed of spherical and rod-like particles. Dimensions of spherical particles are close to nanolead particle size determined by XRD.

It is evident from magnetic measurements (Fig. 4.39) that the magnetization curves are well saturated after the application of a magnetic field higher than 2 T.

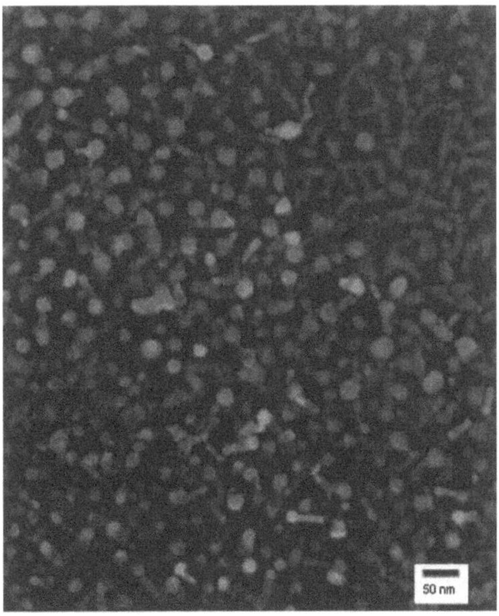

Fig. 4.38 TEM image of Pb/FeS nanoparticles formed by reaction (4.60) after 60 min milling [Godočíková et al. 2004b]

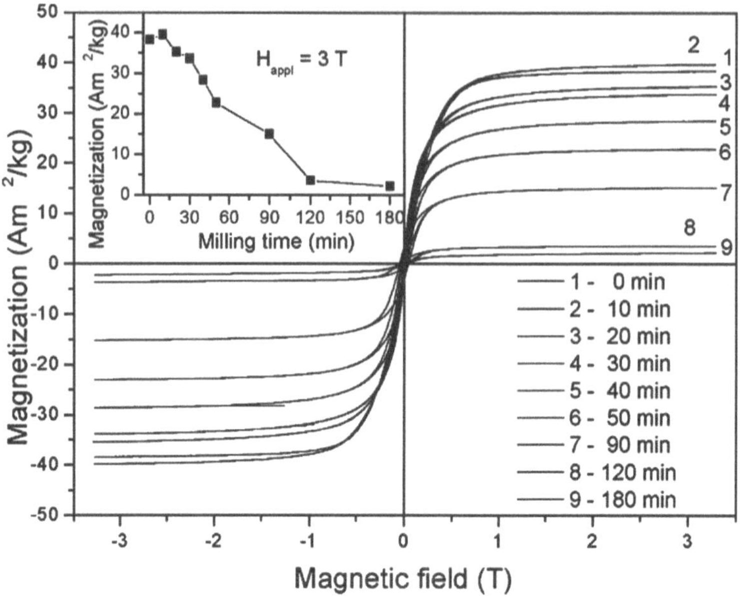

Fig. 4.39 Magnetization as a function of external magnetic field and/or of milling for reaction (4.60) [Godočíková et al. 2004b]

The difference in the saturation magnetization of the samples, as displayed in the inset of this Figure, is mainly caused by the different amount of the ferromagnetic component (bcc-Fe) in the milled samples. For longer milling times, the saturation magnetization is negligible indicating that the constituent phases are almost entirely paramagnetic and the mechanochemical reduction is completed.

The changes in the magnetic nature of the samples can be well seen in Mössbauer spectra in Fig. 4.40a. There is a small doublet in pattern of PbS which belongs to admixtured pyrite FeS_2. With the increasing milling time, the original sextet of iron hyperfine splitting is changing as well. The changes of Mössbauer parameters are given in Fig. 4.40b. Sextet A, with hyperfine magnetic field MF = 32.84 T originating from elemental Fe, is diminishing with milling time. For $t_M \leq 20$ min, iron is preferentially consumed by reaction with the present pyrite FeS_2 in sample and this reaction is thermodynamically more viable in comparison with the reaction (4.60).

Doublet with IS = 0.31 mm s^{-1} and quadrupole splitting QS = 0.59 mm s^{-1} correspond in accordance with literature, to FeS_2 [Marfunin and Mkrtčjan 1967]. At $t_M > 20$ min the formation of pyrrhotite $Fe_{1-x}S$ characterized by a sextet C (MF = 25.37 T) is evident and supported by the XRD patterns in Fig. 4.36. Pyrrhotite $Fe_{1-x}S$ sample rarely consist of only one structural type because at temperatures below 350°C the iron vacancies in the Fe-S system start to order producing a series of superstructures, which could exhibit antiferromagnetic, ferrimagnetic or paramagnetic behaviour. As an example, in our system at $t_M > 30$ min the formation of a further doublet have been observed. In accordance with literature this belongs to cubic

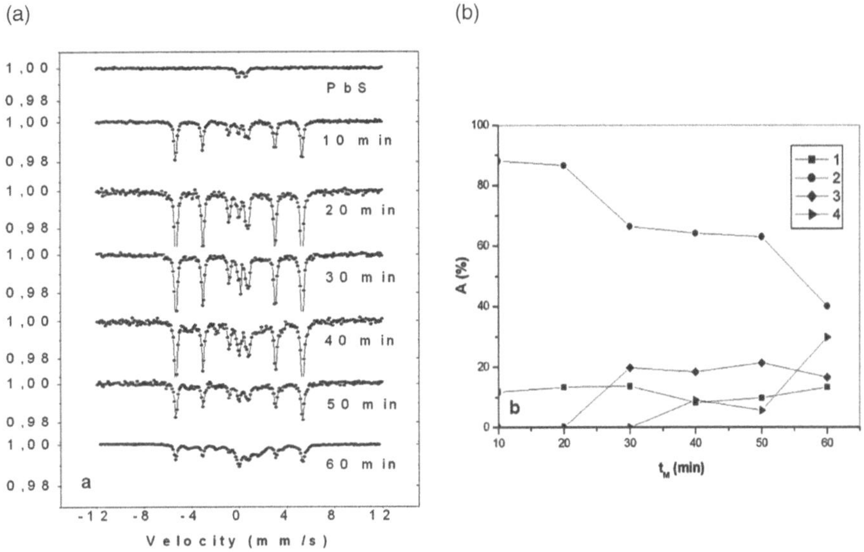

Fig. 4.40 a) Mössbauer spectra for reaction (4.60); b) Mössbauer parameters for reaction (4.60), A – total area of component, t_M – milling time; 1 – pyrite FeS_2 (doublet: IS = $0.31\,mm\,s^{-1}$, QS = $0.63\,mm\,s^{-1}$), 2 – Fe (sextet A: MF = 32.845 T), 3 – pyrrhotite $Fe_{1-x}S$ (sextet C: MF = 25.37 T), 4 – cubic $Fe_{1+x}S$ (doublet: IS = $0.85\,mm\,s^{-1}$, QS = $1.69\,mm\,s^{-1}$), [Baláž et al. 2004a]

$Fe_{1+x}S$ which is known only from synthetic studies. The magnetic characterization of the pyrrhotites thus presents a complex problem [Vaughan and Lennie 1991].

Conventionally, antimony sulphide Sb_2S_3 is reduced via high-energy temperature annealing in the presence of reducing agents such as hydrogen or iron. The reduction of antimony sulphide with hydrogen at atmospheric pressure and at temperature 473–873 K has been studied [Torma and Inal 1979]. Only 60% of Sb_2S_3 was converted to metallic antimony. The reduction with elemental iron is possible at temperatures 773–823 K but the process is inefficient because the metal sulphide dissolves in the already formed iron sulphide [Habashi 1986]. The yield of both processes is low, they need high temperatures and the products have large grain size.

The direct reduction of antimony sulphide has been studied [Baláž et al. 2007]. The reaction with elemental iron during milling proceeds at ambient temperature, according to the equation

$$Sb_2S_3 + 3Fe \rightarrow 2Sb + 3FeS \qquad (4.61)$$

The reaction is thermodynamically feasible, as the enthalpy change for reaction under study is negative ($\Delta H^{\circ}_{298} = -116.7\,kJ\,mol^{-1}$).

The progress of the mechanochemical reaction (4.61) is illustrated by the selected XRD patterns in Fig. 4.41. The process is rather straightforward with elemental antimony Sb (JCPDS 05-0562) and FeS (pyrrhotite-4H, JCPDS 22-1120) being the

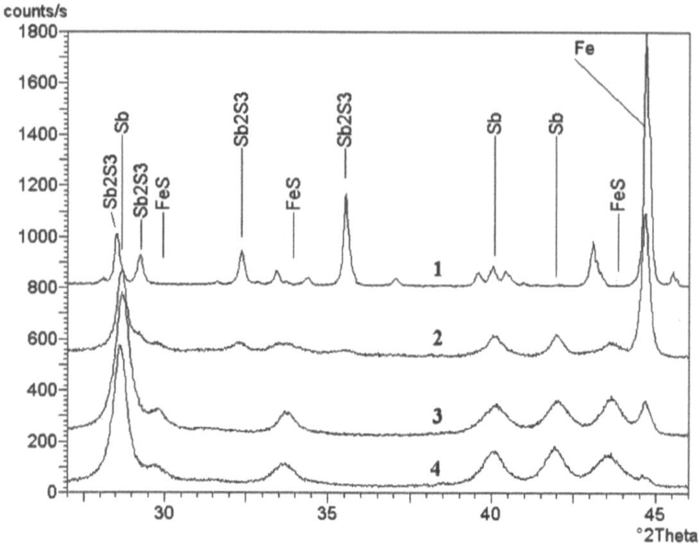

Fig. 4.41 X-ray diffraction patterns at different stage of reaction (4.61) as function of milling time. 1–0 min, 2–20 min, 3–60 min, 4–120 min [Baláž et al. 2007]

only solid state products. The intensity of the metal iron reflexion (110) decreases upon continued milling and only a very small amount of iron is detectable at 60 min (pattern 4). In the starting material (pattern 1) more small peaks are seen which belong to stibnite Sb_2S_3 (JCPDS 42-1393).

The conversion degree for precursors (Sb_2S_3, Fe) and product (Sb) of reaction (4.61) as obtained by normalization of the XRD intensities is shown in Fig. 4.42. It is clear that most of the reduction is complete after 60 min with the intensity of Sb_2S_3 decreasing faster than Fe. There is no reason to expect different phase ratios here as the content of Fe and Sb_2S_3 should change parallel according to Eq. (4.61). A possible explanation is the partial amorphization of stibnite which is indeed manifested in an increased background of the XRD patterns.

Figure 4.43 shows the room temperature magnetization data for investigated samples as a function of external magnetic field. It is evident that the resulting magnetization curves are well saturated after an application of magnetic field higher than 2 T. The variation of the saturation magnetization as a function of milling time, as displayed in the inset of Fig. 4.43, reflects the amount of metallic iron, the only ferromagnetic component in the samples. In good agreement with the XRD results shown in Fig. 4.42, the Fe fraction decreases continuously for milling times up to 180 min, with more than 80% being transformed during the first 60 min. After 180 min of milling the saturation magnetization is negligible indicating that the constituent phases are almost entirely paramagnetic and/or antiferromagnetic, i.e. reaction (4.61) is practically complete.

The XRD line width of the sample milled for 180 min have been analyzed in order to confirm the presence of nanosize particles. The grain size of Sb was found

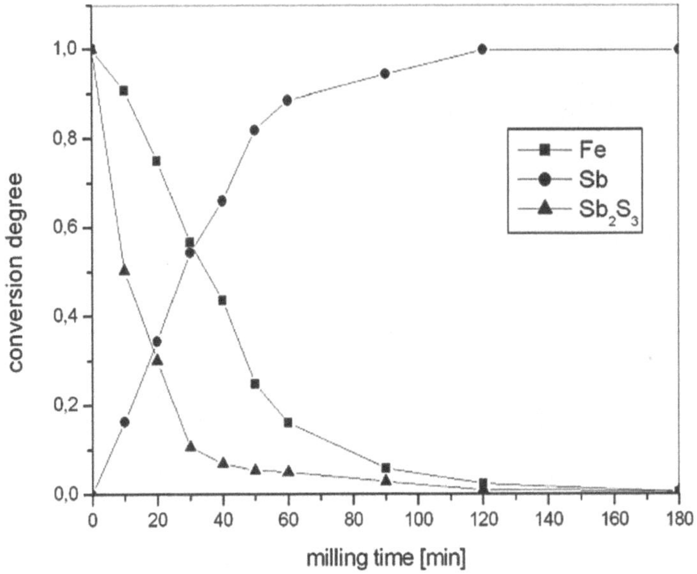

Fig. 4.42 Conversion degree for reaction (4.61) as a function of milling time [Baláž et al. 2007]

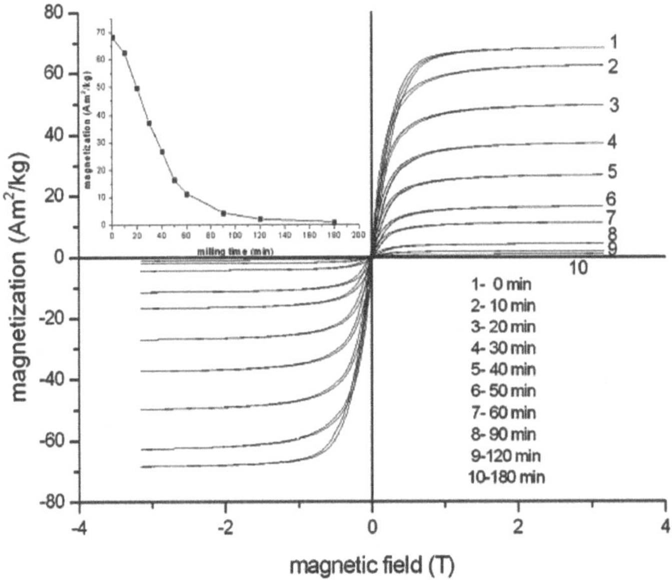

Fig. 4.43 Magnetization as a function of external magnetic field for several milling times. The inset shows the saturation magnetization as a function of milling time [Baláž et al. 2007]

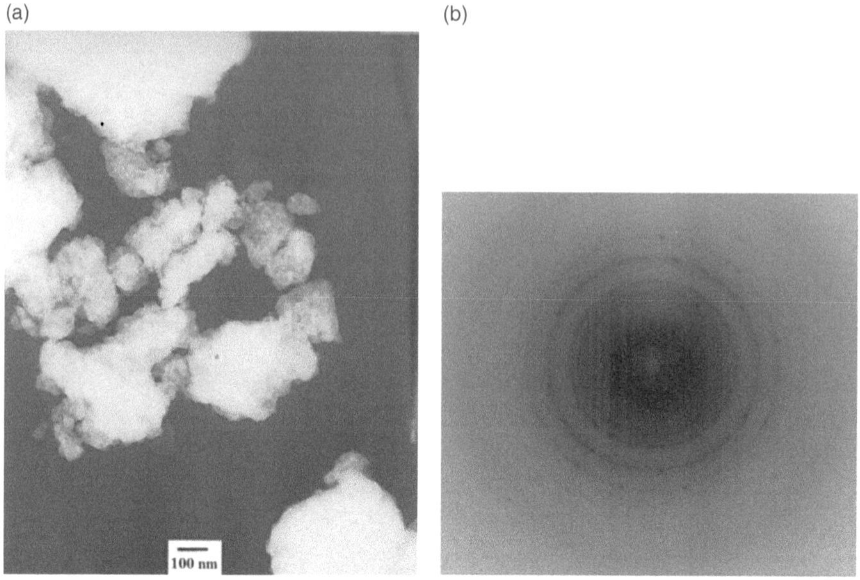

Fig. 4.44 (**A**) Dark field TEM image and (**B**) SAED pattern of nanoparticles formed by reaction (4.61), milling time 180 min [Baláž et al. 2007]

to be 19 nm with 0.35% residual strain and the particle size of FeS is about 10 nm. Figure 4.44A shows a typical dark field TEM image obtained using beams from the diffraction rings of the irregular shaped nanoparticles formed during 180 min of milling. Figure 4.44B shows the SAED pattern of an area containing some nanoparticles. The SAED pattern shows a set of rings instead of spots due to the random orientation of the nanoparticles. The TEM image is consistent with the XRD grain size but also shows the strongly agglomerated nature of the powder. The agglomeration of nanoparticles is a general phenomenon originating from the tendency to reduce the surface area.

The reaction (4.61) was also studied with mineral sample from Romania (deposit Baiut) [Baláž et al. 2005a]. XRD revealed that next to main phase stibnite Sb_2S_3 also quartz SiO_2 was present (Fig. 4.45A). The starting material was milled with elemental Fe in a ratio corresponding to Eq. (4.61) in a planetary mill and the XRD patterns were taken to identify the main components (Fig. 4.45B). Both product phases predicted by Eq. (4.61) were identified: elemental Sb (JCPDS 35-0732) and pyrrhotite–1T (JCPDS 29-0726). In addition, a substantial amount of non-reacted Fe is present. This is explained by the approximately 10% SiO_2 contained in the starting material.

The mechanochemical reaction between arsenic sulphide As_2S_3 and elemental iron can be described by the equation

$$As_2S_3 + 3Fe \rightarrow 2As + 3FeS \tag{4.62}$$

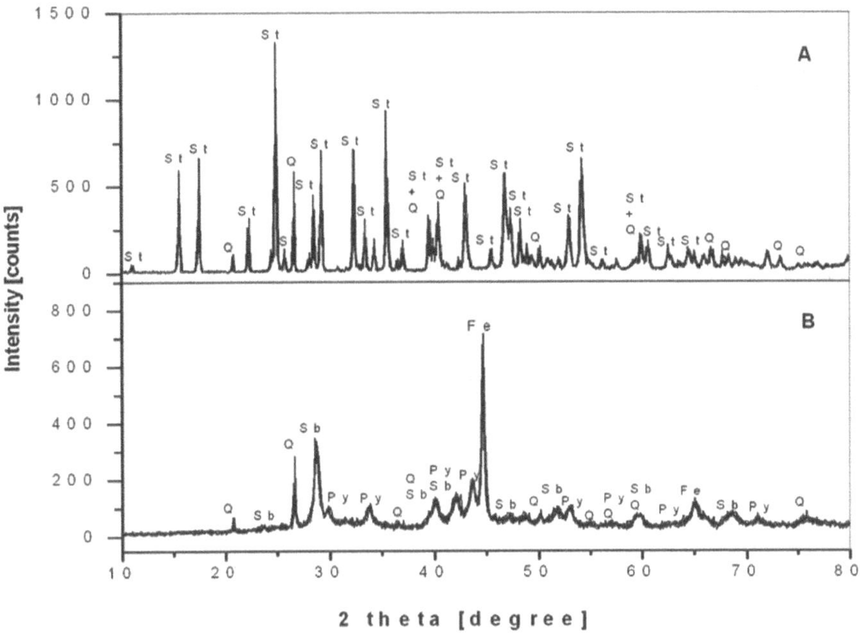

Fig. 4.45 X-ray diffraction patterns for reaction (4.61), **A**: starting stibnite Sb_2S_3 (Baiut, Romania), **B**: Sb_2S_3 with Fe milled for 10 min. St – stibnite, Sb_2S_3, Q – quartz, SiO_2, Sb – antimony, Py – pyrrhotite 1T, Fe – iron [Baláž et al. 2005a]

The reaction is thermodynamically feasible, as the enthalpy change is negative, $\Delta H^{\circ}_{298} = -160.7\,\mathrm{kJ\,mol^{-1}}$.

The starting material was mineral obtained from Manhattan, Nevada (USA). Its XRD analysis (Fig. 4.46) revealed that the main phase was auripigment As_2S_3 with a small amount of realgar As_4S_4. The mineral was milled with elemental Fe in stoichiometric molar ratio in a planetary mill and XRD patterns were measured to identify the main components of the reaction (4.62). Surprisingly, the only reaction product was iron sulphide, identified as pyrrhotite-4M Fe_7S_8 (JCPDS card 24-0079.) Only partial conversion to reaction products was achieved as can be seen from the presence of non-reacted Fe in this figure. Neither elemental arsenic nor any arsenic sulphide is detected among the products of mechanochemical reaction (4.62). This fact can be interpreted by the low thermal stability of arsenic sulphides [Vaughan and Craig 1978] and the possible volatization of the arsenic species during milling.

The mechanochemical reaction between copper sulphide and elemental iron has been studied very carefully [Matteazzi and LeCaër 1992b; Baláž et al. 2002b, 2004a, c, 2005a]. The reaction can be described by equation

$$Cu_2S + Fe \rightarrow 2Cu + FeS \qquad (4.63)$$

The reaction (4.63) is thermodynamically possible, as the enthalpy change is negative, ($\Delta H^{\circ}_{298} = -21.0\,\mathrm{kJ\,mol^{-1}}$).

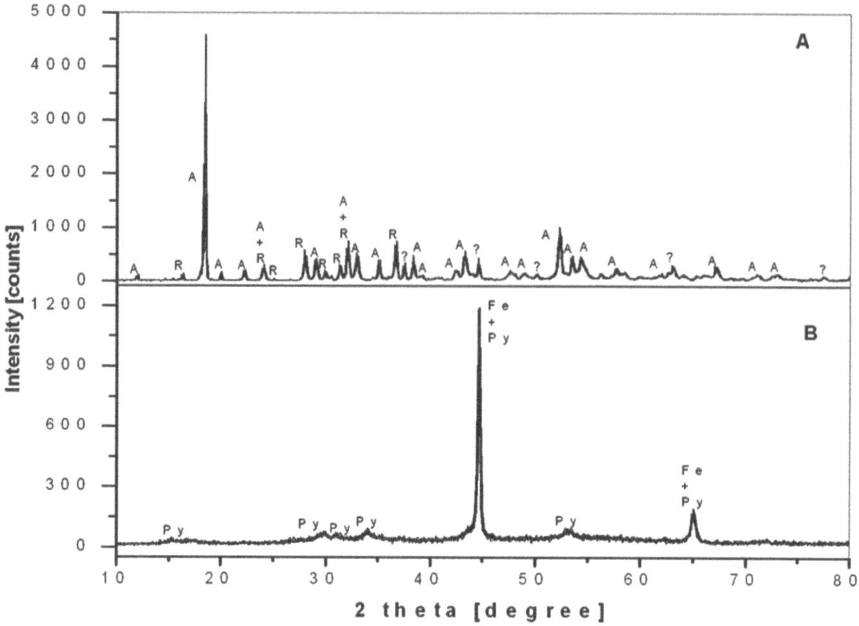

Fig. 4.46 X-ray diffraction patterns for reaction (4.62), **A**: starting mineral auripigment As_2S_3 (**A**) with small amount of realgar As_4S_4 (R), **B**: As_2S_3 with Fe milled for 10 min. Py – pyrrhotite 4M, Fe_7S_8, Fe – iron [Baláž et al. 2005a]

The progress of the mechanochemical reaction is illustrated by the selected XRD patterns in Fig. 4.47.

The primary process – the reduction of copper sulphide by iron while copper metal and iron sulphides are formed – is clearly seen, particularly by inspecting the relative intensities of the diffraction lines of Cu and Fe metal.

The conversion degree, β, can be defined as $\beta = I_{Cu}/(I_{Cu} + I_{Fe})$, where I_{Cu} and I_{Fe} are the intensities of the Cu(111) and Fe(110) diffraction lines. The process kinetics as described by this parameter is shown in Fig. 4.48.

The process is significantly complicated by the existence of several copper sulphide and iron sulphide phases. Any external influence, such as high temperature treatment and ball milling [Baláž et al. 2003], can cause phase transformations in sulfides. Already 1 min of milling decreased the relative amount of djurleite $Cu_{1.94}S$ and increased the fraction of chalcocite Cu_2S. In addition, a new tetragonal phase $Cu_{1.81}S$ (JCPDS 41-0959) formed as indicated by the diffraction peaks at 32.5° and 39°. Another tetragonal phase Cu_2S (JCPDS 72-1071) may also be present.

Most of the copper sulfide is consumed during the first minutes of milling. After 10 min, only the reflections from Cu, cubic FeS (JCPDS 23-1123) and Fe can be observed. The presence of iron without residual copper sulphides requires either significant off-stoichiometry of the iron sulphide or the presence of free sulphur. Unfortunately, free sulphur may remain undetected by XRD due to the low scattering amplitude of sulphur and its tendency toward amorphization. The intensity of

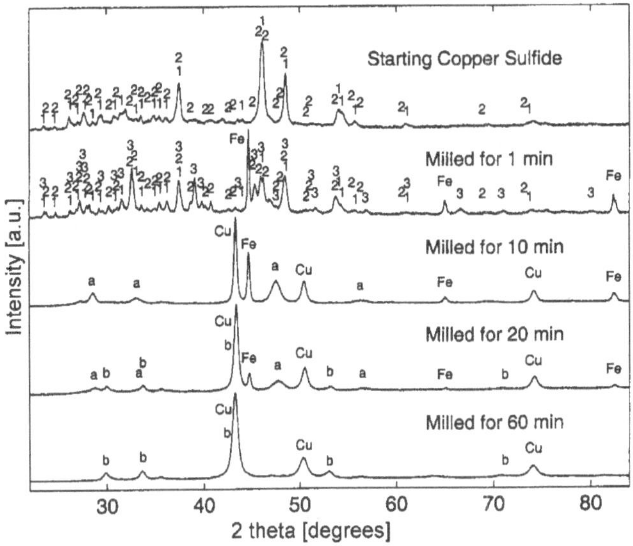

Fig. 4.47 X-ray diffraction patterns for reaction (4.63) as a function of milling time, 1 – djurleite $Cu_{1.94}S$, 2 – chalcocite Cu_2S, 3 – tetragonal $Cu_{1.81}S$, a – cubic FeS, b – hexagonal FeS [Baláž et al. 2002b]

the iron reflections decreases upon continued milling and no iron is detectable after 60 min. Simultaneously, FeS changes to its hexagonal modification troilite (JCPDS 75-0602). This phase is considered the stable modification of stoichiometric FeS [Shuey 1975].

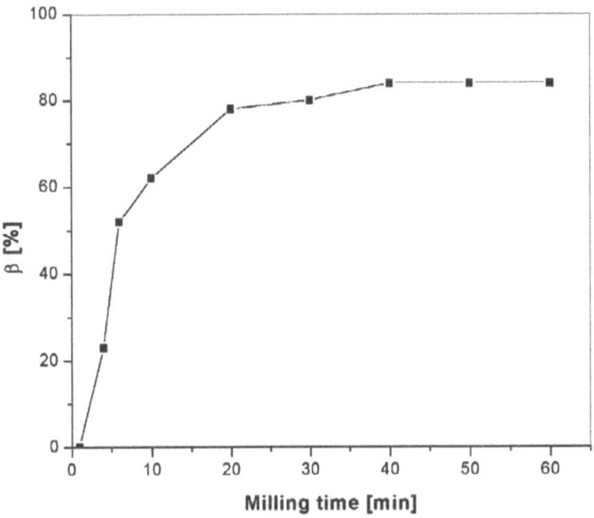

Fig. 4.48 Conversion degree for reaction (4.63) as a function of milling time [Baláž et al. 2002b]

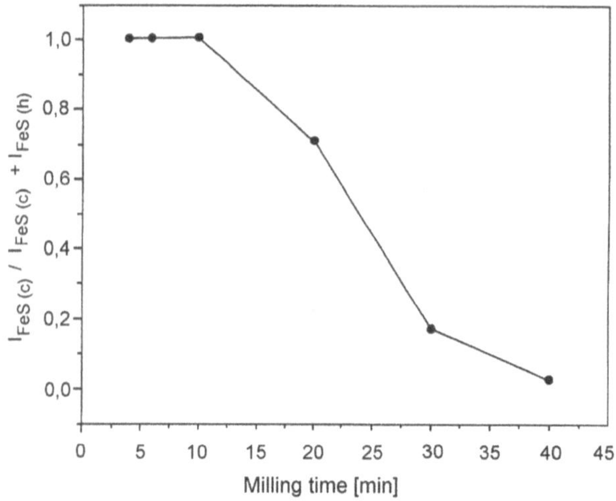

Fig. 4.49 Conversion of cubic FeS (c) to hexagonal FeS (h) as a function of milling time [Baláž et al. 2005a]

The kinetics of the cubic iron sulphide transformation to hexagonal iron sulphide is shown in Fig. 4.49 based on the intensity ration of XRD peaks. Iron sulphides (pyrrhotites) belong to a group of complicated compounds containing ordered iron vacancies. The distribution of vacancies distinguishes the individual pyrrhotite types. At temperatures below 350°C, the vacancies start to order producing a series of superstructures based on the Ni-As type structure [Shuey 1975; Vaughan and Craig 1978; Vaughan and Lennie 1991]. Cubic FeS is a metastable phase that can be synthesized in the laboratory, but it is not found naturally.

The Mössbauer spectra presented in Fig. 4.50 support the interpretation given for the XRD results above. The spectra are dominated by two overlapping sextets with a paramagnetic doublet appearing in the center, especially at milling times between 4 and 10 min. The relative intensities of both sextets as well as the observed doublet are given in Fig. 4.51 as a function of milling time.

The parameters of the high-field sextet (hyperfine field 32.9 ± 0.3 T, isomer shift 0.01 ± 0.01 mm/s and quadrupole shift 0.01 ± 0.01 mm/s) indicate that it originates from elemental Fe. Its relative fraction is diminishing with the milling time (Fig. 4.51) consistent with XRD results. The sextet, which dominates the patterns after longer milling times, is characterized by the hyperfine field between 30.0 T and 30.9 T, isomer shift of 0.75 ± 0.01 mm/s and quadrupole splitting of 0.07 ± 0.01 mm/s.

These parameters are typical of iron monosulphides. The small magnetization observed after prolonged milling is probably due to the ferrimagnetic contribution of Fe-deficient $Fe_{1-x}S$ [Vaughan and Lennie 1991; Jiang et al. 1998]. The isomer shift of the doublet varies from 0.44 mm s^{-1} to 0.50 mm s^{-1} and the quadrupole splitting increases from 0.96 mm s^{-1} to 1.11 mm s^{-1} with increasing milling time. This component probably originates from an increasingly disordered intermediate state. Avvakumov observed a transition from sextet to doublet as the consequence of

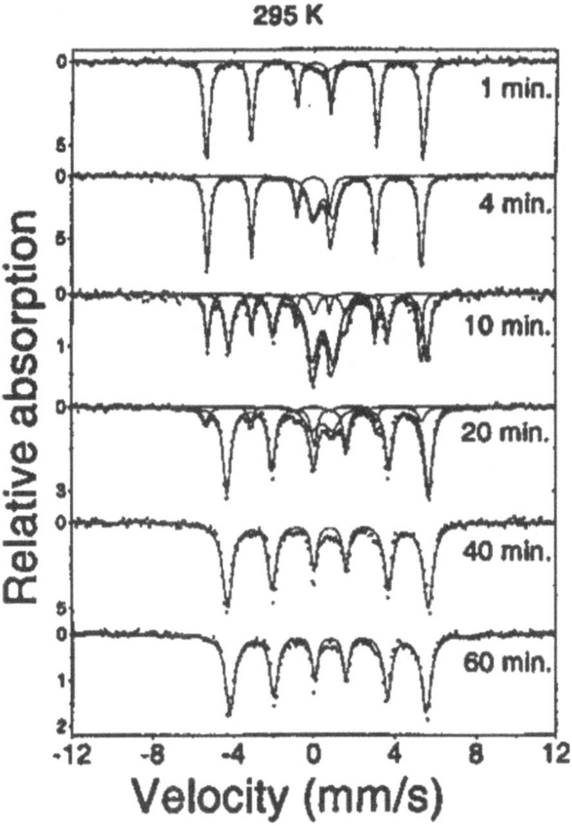

Fig. 4.50 Mössbauer spectra for reaction (4.63) mixtures as a function of milling time [Baláž et al. 2005b]

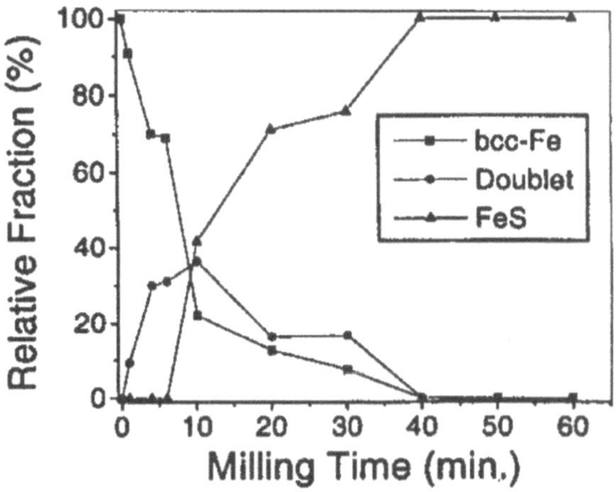

Fig. 4.51 The relative intensities of the Mössbauer sub-spectra as a function of milling time [Baláž et al. 2005b]

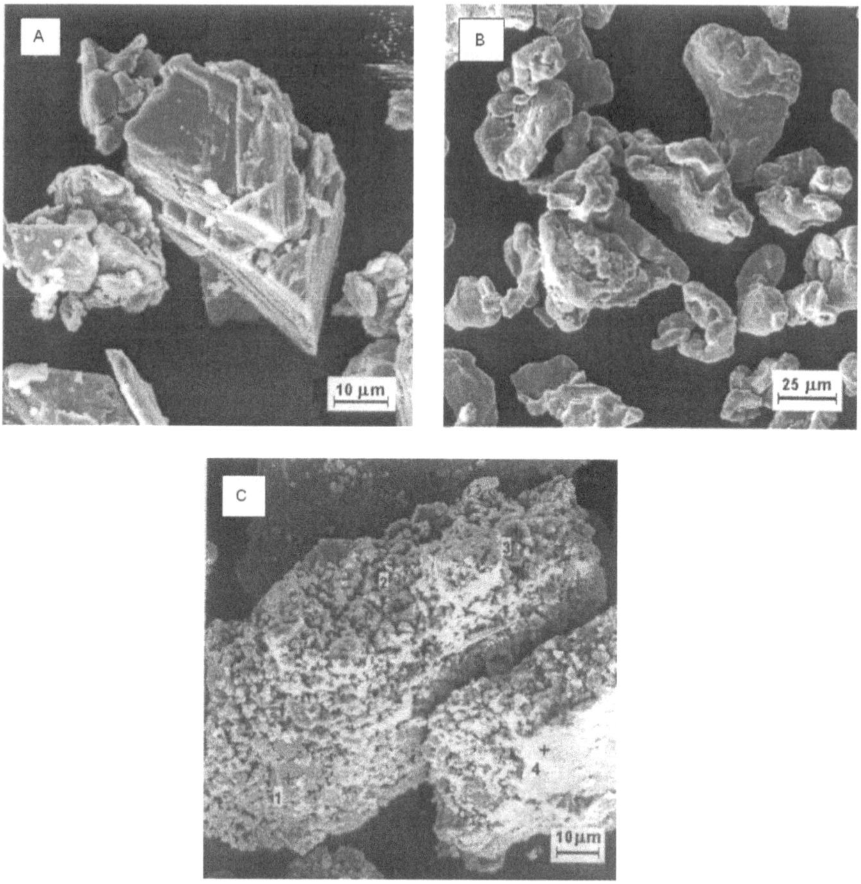

Fig. 4.52 SEM study of the mechanochemical reduction of copper sulphide by iron: (**A**) elemental Fe, (**B**) copper sulphide, (**C**) Cu/FeS nanocomposite [Baláž et al. 2004d]

mechanical disordering of the iron sublattice when milling a single-phase pyrrhotite sample [Avvakumov 1986].

Micrometer aggregates of nanoparticles are formed during milling of copper sulphide with iron (Fig. 4.52). Usually each micrometer-sized powder particle is comprised of a mixture of nanosized grains of the product phases as can be observed in our case [Gaffet et al. 1999; Koch 1991; McCormick et al. 2001]. The dimension of the grains is changing with milling time (Fig. 4.53).

The very similar results have been obtained with chalcocite Cu_2S mineral from deposit Khan (South-western Africa) [Baláž et al. 2004c].

4.3.6.5 Scaling-Up the Process

Several processes have been designed for production of pure metals from compounds including minerals with iron as reducing metal (Fig. 4.54) [Habashi 1986].

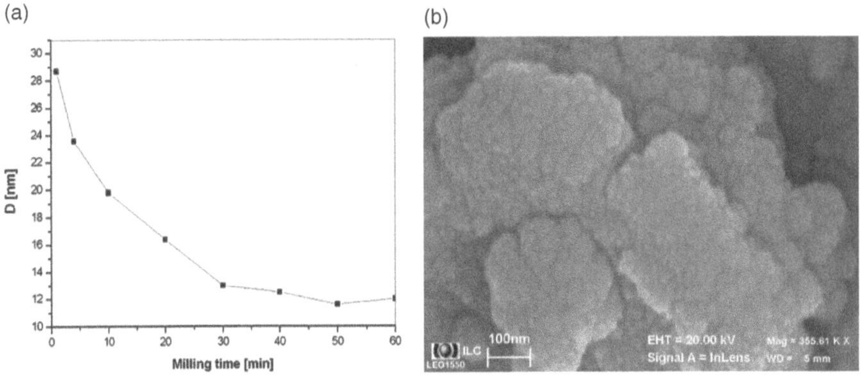

Fig. 4.53 Crystallite size, D, of product copper in reaction (4.63) as a function of milling time [Baláž et al. 2005a]

The selection of a solid–solid reduction process depends on numerous factors and usually a compromise is made between the advantages and the disadvantages. Suitable additives are used to improve the process [Habashi 1986]. However, the conventional metallothermic reduction is not capable of making nano-size metals and or Me/FeS nanocomposites due to the high temperature.

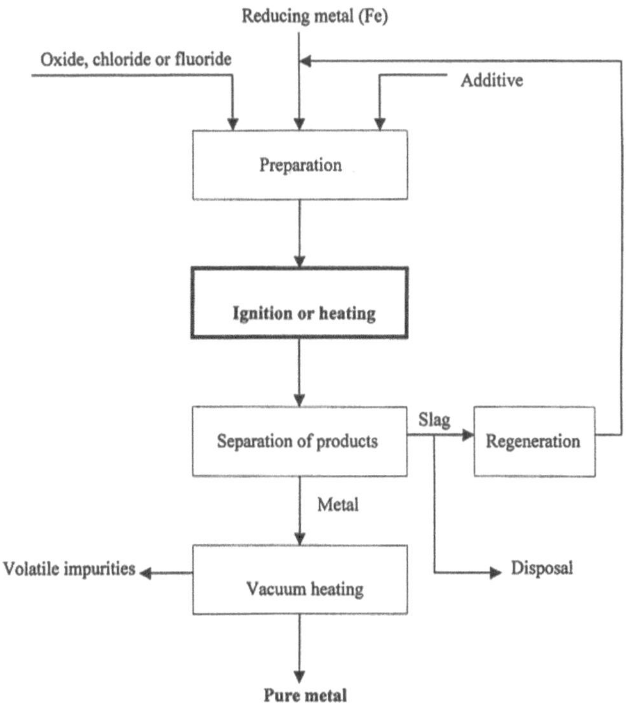

Fig. 4.54 General scheme of metallothermic reduction [Habashi 1986]

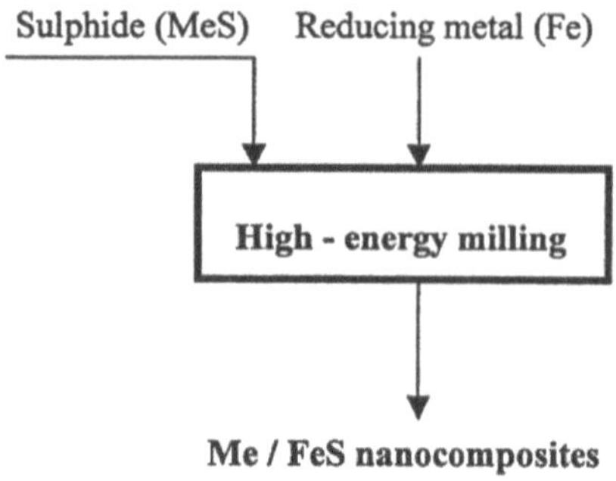

Fig. 4.55 General scheme of mechanochemical reduction [Baláž et al. 2005a]

Mechanochemical reduction as outlined in Fig. 4.55 is a very straightforward, one-step, ambient temperature process without many operations. The key step is the high-energy milling which requires a suitable industrial mill for scaling up the process. In mechanochemical processing of sulfides with elemental iron as reducing agent, the product is in the form of Me/FeS (Me = Pb, Sb, Cu) nanocomposite particles. They can be applied in technology as produced, e.g. as supported metal catalysts in chemical engineering [Weisser and Landa 1972] or have to be separated into components in order to prepare nanometals with unique properties [Baláž et al. 2004d]. The reduction of sulphides in a ball mill is an example of a so-called "dry" technology, as reactions in gas and liquid phases are excluded.

In case of copper sulphide mechanochemical reduction the simplified flowchart can be outlined as in Fig. 4.56. The very straightforward one-step solid state process is characterized by the formation of Cu/FeS nanocomposite. If we need nanocopper,

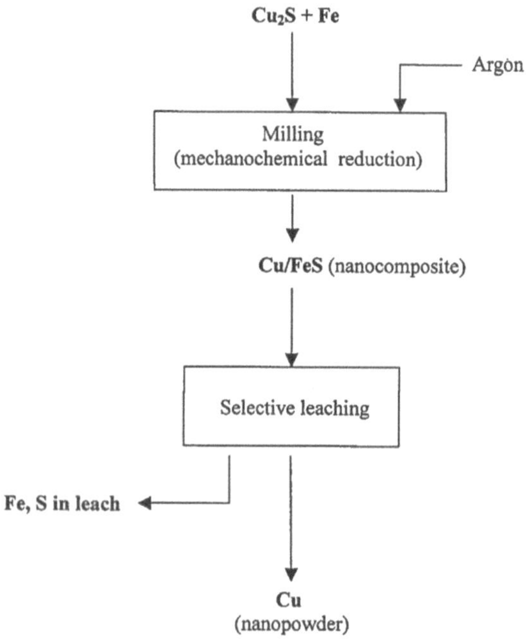

Fig. 4.56 Scheme of copper nanopowder production [Baláž et al. 2004c]

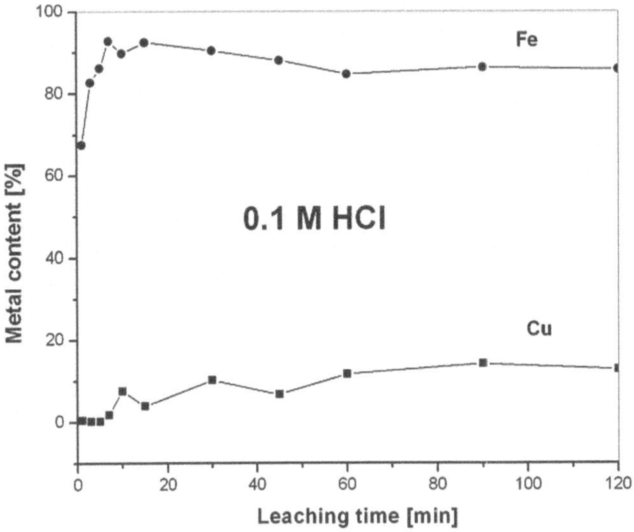

Fig. 4.57 Metal recovery vs. leaching time for Cu/FeS nanocomposite

the product of reaction (4.63) has to be separated into its components. Selective leaching of the iron sulphide can be an appropriate method of producing copper nanopowder. Several acids (H_2SO_4, HNO_3, HCl) have been tested for the selective leaching of iron sulphide from the nanocomposite with the aim to preserve nanocopper in solid form. The best results have been obtained with HCl: more than 86% Fe was dissolved together with the marginal amount of copper (0.2%), Fig. 4.57.

4.3.6.6 Reduction with Magnesium

Ultrafine Mo particles can be produced as a result of the slow burning transition into detonation in reaction of molybdenite MoS_2 with Mg reducing metal according to the reaction

$$MoS_2 + Mg \rightarrow Mo + 2MgS \tag{4.64}$$

The mechanical treatment of the reactant mixture results in a considerable decrease of the ignition temperature and in a strong increase of the burning speed [Gaffet and LeCaër 2004].

The same reducing element has been applied for decomposition of stibnite Sb_2S_3 [Godočiková et al. 2007] following the reaction

$$Sb_2S_3 + 3Mg \rightarrow 2Sb + 3MgS \tag{4.65}$$

The reaction is thermodynamically possible, as the enthalpy change for reaction (4.65) is negative, ($\Delta H^{\circ}_{298} = -872.4\,\mathrm{kJ\,mol^{-1}}$).

The progress of the mechanochemical reduction is illustrated by XRD patterns (Fig. 4.58) of the starting powder (A) and taken after 20 (B), 30 (C), and 120 (D)

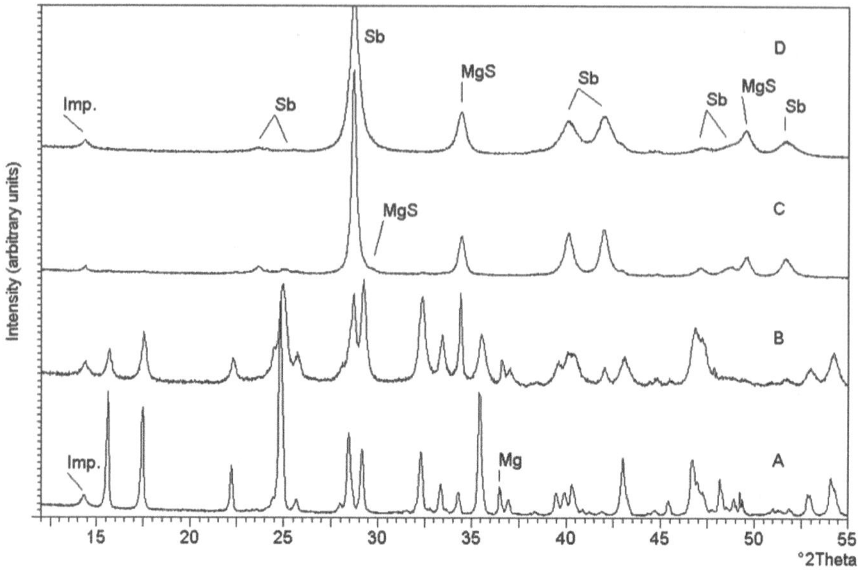

Fig. 4.58 X-ray diffraction patterns as a function of the milling time. A-stoichiometric mixture of Sb_2S_3 and Mg before milling. The milling times for patterns B-D are 20 min, 30 min and 120 min, respectively [Godočiková et al. 2008]

min of milling. Pattern A is dominated by stibnite, Sb_2S_3 (JCPDS 75-1310); all the unmarked lines correspond to this phase. The (101) peak of magnesium (JCPDS 35-0821) and an unidentified peak at about 14.3° are also visible. This latter line probably originates from an inert impurity that does not participate in the reaction. There is very little chemical change during the first 20 min of milling, although the presence of some rombohedral Sb metal (JCPDS 35-0732) is indicated by the (012) peak at 28.7° and the (110) peak at about 42°. The most obvious change is the broadening of the diffraction peaks.

There is a qualitative difference between patterns B and C, indicating a fast chemical change between 20 and 30 min. It seems that most of the reaction takes place as a self-sustaining reaction some time between 20 and 30 min; the transformation is practically complete at 30 min. This is not surprising. The ratio of the reaction heat to the room temperature heat capacity is about 4660 K and values above 2000 K indicate the possibility of a self-sustaining reaction [Takacs 2002]. Elemental antimony and niningerite MgS (JCPDS 35-0730) are the only solid state reaction products. The lines are relatively narrow after 30 min of millling, suggesting that a high temperature process resulted in larger particles. Only the line broadening changes between 30 min and 120 min.

The reaction proceeds according to Eq. (4.65) without intermediate phases and it is complete by 30 min. Just as there is little transformation before 20 min, little Sb_2S_3 remains after at least 30 min of milling. A small shoulder at 43.2° is the only sign that a minor amount of Sb_2S_3 is present after 30 min. The reason can be that a small fraction of the Mg was oxidized and therefore not available as reducing agent. The X-ray patterns were measured to 120°, but we opted to show a narrower range of angles with better resolution.

The detailed analysis of XRD patterns has been performed in order to determine grain sizes and strains using the Williamson-Hall method [Williamson and Hall 1953]. During the first 20 min the particle size of stibnite decreases: it is 65 nm after 10 min and 35 nm after 20 min. The random strain is less than 0.1% as expected for a brittle substance. Right after the reaction at 30 min, the particle size is 130 nm with 0.3% strain. The strain remains the same until 90 min, the particle size drops to 25 nm and remains about the same from 40 min to 90 min. There is a little extra broadening after 120 min of milling, providing 18 nm particle size and 0.35% strain.

The dependence of the specific surface area on milling time is shown in Fig. 4.59. Two phases of the formation of new specific surface area can be identified. In the first period up to 30 min, the specific surface area increases; the effect of the reaction is not obvious. After 40 min the surface area is approximately constant, suggesting a dynamic equilibrium between particle fracturing and agglomeration [Baláž 2000]. In a polydisperse systems like the current one, the presence of very fine particles along with relatively coarse particles greatly promotes the formation of aggregates, due to the enhanced role of the van der Waals forces. Particles in the nanometer size range have a strong tendency to agglomerate due to their relatively large specific surface area.

The surface morphology of the mechanochemically synthesized Sb/MgS nanoparticles is depicted in Fig. 4.60 where the agglomeration of nanoparticles can be clearly seen. The agglomeration increases with the increasing milling time. The

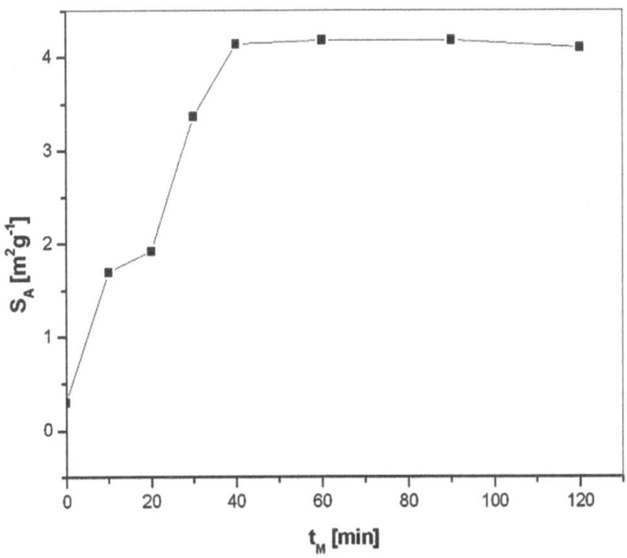

Fig. 4.59 Specific surface area as a function of the milling time [Godočiková et al. 2007]

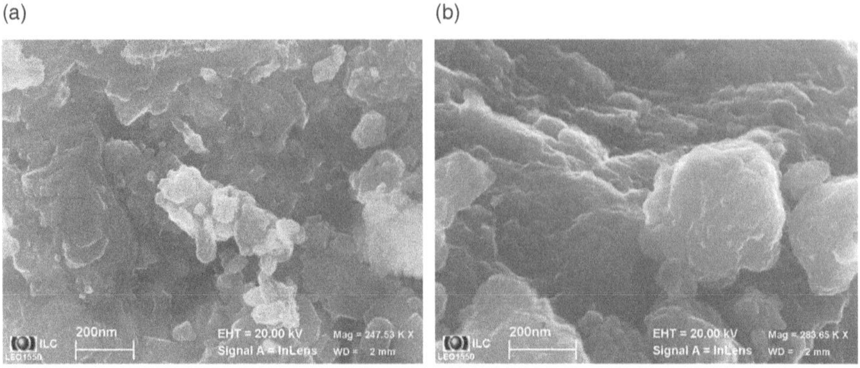

Fig. 4.60 SEM image of Sb/MgS nanocomposite; milling time 30 min (**A**) and 120 min (**B**) [Godočiková et al. 2008]

solid state combination of nanoparticles into agglomerates is a general phenomenon which is one of the ways a nanocrystal system is able to compensate for its unsaturated surface forces via surface reconstruction.

4.3.6.7 Reduction with Silicon

The mechanochemical reduction of copper sulphides with silicon as reducing element has been studied in [Baláž et al. 2005b]. XRD analysis of the starting materials (Fig. 4.61A) revealed that the main present phases were chalcocite Cu_2S, djurleite $Cu_{1.94}S$ and elemental silicon. The behaviour of sulphides and silicon at separate milling is different. There are no new phases at silicon milling on XRD pattern except of tungsten carbide (5) which is a consequence of wear from milling vial and balls. The shape of diffraction peaks and their positions illustrate the effect of particle size diminution as a consequence of milling. However, the phase transformations are unambiguosly observed: anilite Cu_7S_4 and digenites $Cu_{9-x}S_5$ have been identified among the products of milling. It has been described recently [Baláž et al. 2003] that digenite, which is in fact a group of copper sulphides [Gaines et al. 1997] is one of the main phase during copper sulphide milling. Anilite has been observed as an intermediate product and this phase is transformed into low digenite modification due to milling [Morimoto et al. 1969]. The presence of anilite (Fig. 4.61B) can not be excluded also in our case.

The set of starting copper sulphide and silicon mixtures has been milled in different time and the products have been identified by X-ray diffractometry (Fig. 4.62). Djurleite $Cu_{1.94}S$ and chalcocite Cu_2S together with unreacted silicon have been identified in powders milled till 3 min (Fig. 4.62A and B). $Cu_{1.96}S$ phase formed at milling time 6 min (Fig. 4.62C) seemed to be the most stable phase even in case when ternary sulphide is formed during prolonged milling. In higher time of milling (Fig. 4.62D–F) elemental copper (1) and the ternary copper silicon sulphide Cu_8SiS_6 are the main reaction products together with the described $Cu_{1.96}S$ phase.

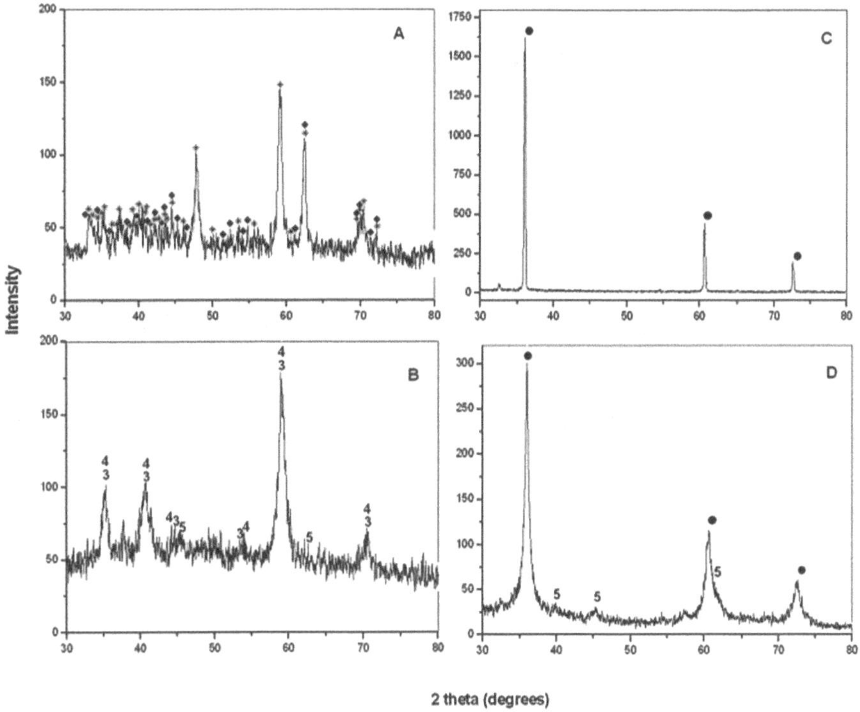

Fig. 4.61 XRD patterns of starting copper sulphides (**A**) and silicon (**C**) and products of their separate milling for 20 minutes (**B** and **D**): ♦ – Cu_2S (chalcocite) (JCPDS 23-962); ∗ – $Cu_{1.9375}S$ (djurleite) (JCPDS 23-0960); • – Si (silicon, syn) (JCPDS 27-1402); 3 – $Cu_{9-x}S_5$; $Cu_{1.765}S$ (digenite) (JCPDS 47-1748); 4 – Cu_7S_4 (anilite) (JCPDS 34-0489); 5 – WC (tungsten carbide) (JCPDS 25-1047) [Baláž et al. 2005b]

The dependance of specific surface area on milling time for starting copper sulphides (A), starting silicon (B) and reaction mixture of copper sulphides with silicon (C) is given in Fig. 4.63. There are small changes in values of S_A for copper sulphide milling in contrast to silicon which seemed to be strong agglomerated after 10 min of milling. Anomalous behaviour of surface area increase for starting mixture milled for 6–20 min ($S_A = 0.4$–$4.9\,m^2g^{-1}$) can be related to the formation of ternary copper silicon sulphide Cu_8SiS_6 because the specific surface area of any milled copper sulphide is not higher than $0.2\,m^2\,g^{-1}$.

The overall process of mechanochemical transformations in copper sulphide-silicon system proceeds via complicated mechanism with several overlapping steps. Phase transformations in copper sulphide phases, the formation of elemetal nanocopper with X-ray determined particle size of 23 nm and creation of ternary sulphide Cu_8SiS_6 with anomalous surface area values are the main products of the hypothetical reaction

$$6Cu_2S + Si \rightarrow 4Cu + Cu_8SiS_6 \qquad (4.66)$$

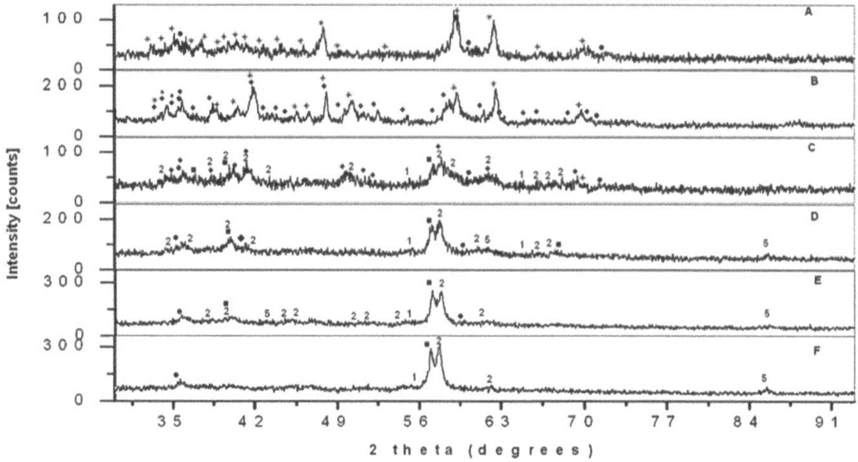

Fig. 4.62 XRD patterns of mixture of starting copper sulphides with silicon milled for different times: A – 1 min; B – 3 min; C – 6 min; D – 10 min; E – 15 min; F – 20 min (silicon has been pre-milled for 10 min for all samples): ♦ – Cu_2S (chalcocite) (JCPDS 46-1195, JCPDS 84-1770, JCPDS 72-1071); ∗ – $Cu_{1.9375}S$ (djurleite) (JCPDS 34-0660); ■ – $Cu_{1.96}S$ – (JCPDS 12-0174); ● – Si (silicon, syn) (JCPDS 27-1402); 1 – Cu (copper, syn) (JCPDS 04-0836); 2 – Cu_8SiS_6 (copper silicon sulphide) (JCPDS 21-0305); 5 – WC (tungsten carbide) (JCPDS 25-1047) [Baláž et al. 2005b]

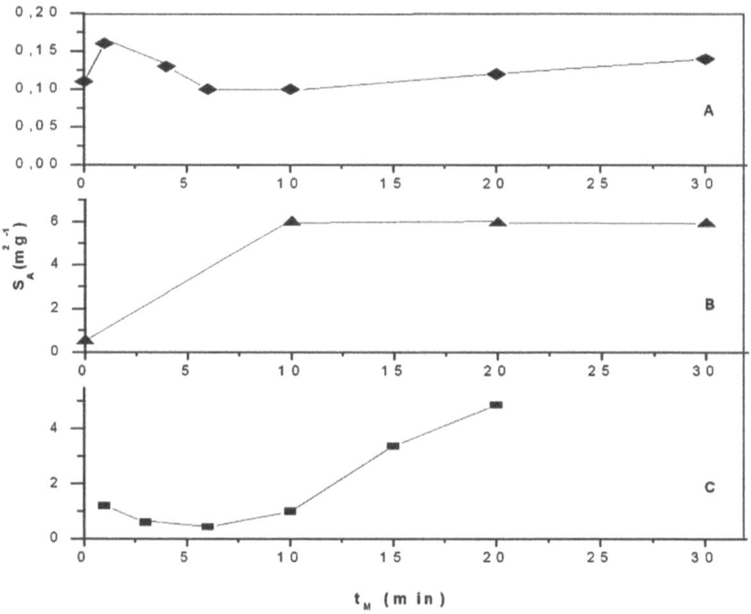

Fig. 4.63 Specific surface area, S_A vs. milling time, t_M: A – copper sulphides; B – silicon; C – mixture of copper sulphides with silicon [Baláž et al. 2005b]

4.3.6.8 Thermodynamics of Sulphide Reduction

The heats of formation of sulphides are much smaller than the corresponding values for oxides (Table 4.9). Consequently, the redox reactions between a sulphide and a more reactive metal is less exothermic than the analogous reaction between the corresponding oxide and the same metal. Nevertheless, in several cases the reaction heats are high enough to permit mechanically induced self-sustaining reactions [Takacs 2002].

Table 4.9 Mechanically induced redox reactions of sulphides

Reaction	ΔH° (kJ mol^{-1})	Reference
$3Cu_2S + 2Al \rightarrow 6Cu + Al_2S_3$	−485	Matteazzi and LeCaër (1992b)
$Cu_2S + Fe \rightarrow 2Cu + FeS$	−21	Matteazzi and LeCaër (1992b)
$3PbS + 2Al \rightarrow 3Pb + Al_2S_3$	−267	Matteazzi and LeCaër (1992b)
$PbS + Fe \rightarrow Pb + FeS$	−1	Baláž et al. (2004b); Godočíková et al. (2004a, b)
$Sb_2S_3 + 3Fe \rightarrow 2Sb + 3FeS$	−117	Baláž et al. (2007)
$Sb_2S_3 + 3Mg \rightarrow 2Sb + 3MgS$	−872	Godočiková et al. (2007)
$As_2S_3 + 3Fe \rightarrow 2As + 3FeS$	−161	Baláž et al. (2005a)
$MoS_2 + 2Mg \rightarrow Mo + 2MgS$	−70	Danielian et al. (1991)
$3ZnS + 2Al \rightarrow 3Zn + Al_2S_3$	−107	Danielian et al. (1991)
$FeS_2 + Fe \rightarrow 2FeS$	−13	Baláž et al. (2004a)
$3FeS + 2Al \rightarrow 3Fe(Al) + Al_2S_3$	−271	Baláž et al. (2004a)
$2FeS + Si \rightarrow 2Fe(Si) + SiS_2$	+88	Baláž et al. (2004a)
$FeS + Mn \rightarrow Fe + MnS$	−117	Baláž et al. (2004a)
$3CoS + 2Al \rightarrow 3Co + Al_2S_3$	−405	Baláž et al. (2004a)
$HgS + Cu \rightarrow Hg + CuS$		Molčanov and Jusupov (1981)

4.3.7 Other Minerals

Thermodynamically, chrysocolla $CuSiO_3{\cdot}2H_2O$ could react with sodium sulfide to form nanocrystalline covellite CuS and water soluble sodium silicate according to the exchange reaction

$$CuSiO_3 \cdot 2H_2O + Na_2S \rightarrow CuS + Na_2SiO_3 + 2H_2O \qquad (4.67)$$

It was shown that the solid state pretreatment using a vibration mill increases the conversion rate of chrysocolla [Banza and Gock 2003]. This procedure can be efficiently used for the recovery of copper in the form of nanocrystalline copper sulphide as an alternative to pre-heating or direct leaching.

Mineral malachite $CuCO_3{\cdot}Cu(OH)_2$ has been also treated with the same sulphur precursor

$$CuCO_3 \cdot Cu(OH)_2 + 2Na_2S \rightarrow 2CuS + Na_2CO_3 + 2NaOH \qquad (4.68)$$

and the overall transformation of malachite into the nanocrystalline covellite CuS was achieved.

Copper sulphate mineral chalcantite $CuSO_4{\cdot}5H_2O$ transformation have been studied by Indian authors using Fe, Mg, Sn, Mn, Co and Pb reducing elements to obtain nanocrystalline copper [Varghese et al. 2001; Varghese et al. 2004].

In ($CuSO_4{\cdot}5H_2O$+Fe) *system* the overall chemical reaction during milling can be written as follows

$$\begin{aligned} CuSO_4 \cdot 5H_2O + Fe &\rightarrow Cu + FeSO_4 \cdot 5H_2O \\ &\rightarrow Cu + FeSO_4 \cdot 5H_2O \end{aligned} \quad (4.69)$$

As milling proceeds (Fig. 4.64), the intensity of $CuSO_4{\cdot}5H_2O$ and Fe peaks gradually diminishes while that of $FeSO_4{\cdot}5H_2O$ and Cu peaks increases. On latter milling times $FeSO_4{\cdot}5H_2O$ is completely converted into $FeSO_4{\cdot}4H_2O$ [Varghese et al. 2001].

In ($CuSO_4{\cdot}5H_2O$+Mg) *system* $MgSO_4{\cdot}4H_2O$ and Cu_2O are formed as final products. An intermediate phase and hydrogen were observed before completion of mechanochemical reduction.

In $CuSO_4{\cdot}5H_2O$ + Sn *system* the reaction can be represented

$$\begin{aligned} CuSO_4 \cdot 5H_2O + Sn &\rightarrow Cu + \beta - brass + SnSO_4 + 5H_2O \\ &\rightarrow Cu + SnSO_4 + 5H_2O \end{aligned} \quad (4.70)$$

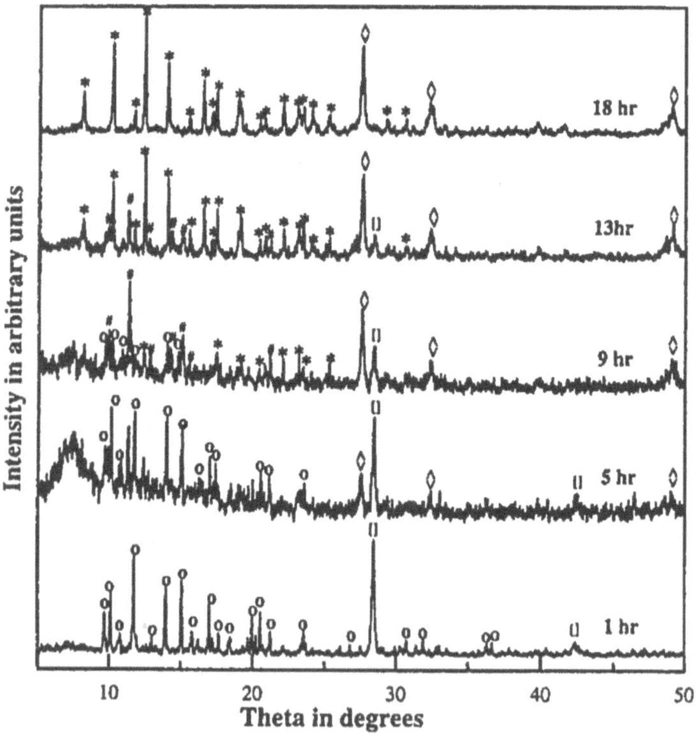

Fig. 4.64 X-ray diffraction patterns of mixture of starting $CuSO_4{\cdot}5H_2O$ with iron milled for different times: $CuSO_4{\cdot}5H_2O$, [ ] Fe, ◊ Cu, # $FeSO_4{\cdot}5H_2O$, ∗ $FeSO_4{\cdot}4H_2O$ [Varghese et al. 2001]

Table 4.10 Grain size, D for various substances in systems (metal sulphate+metal) [Varghese et al. 2001]

Reaction system	D (nm)
$CuSO_4 \cdot 5H_2O + Fe$	27 (Cu)
$CuSO_4 \cdot 5H_2O + Mg$	19 (Cu_2O)
$CuSO_4 \cdot 5H_2O + Sn$	54 (Sn)

There is no water of hydration associated with the product sulfate $SnSO_4$.

The grain size analysis of the products formed during reduction are given in Table 4.10.

In ($CuSO_4{\cdot}5H_2O$+Pb) *system* the reaction (4.71) is extremely rapid

$$CuSO_4 \cdot 5H_2O + Pb \rightarrow Cu + PbSO_4 + 5H_2O \tag{4.71}$$

and practically complete withing 20 min of planetary milling (Fig. 4.65). The reduction of bivalent copper by lead as reducing metal is extremely slow in solution

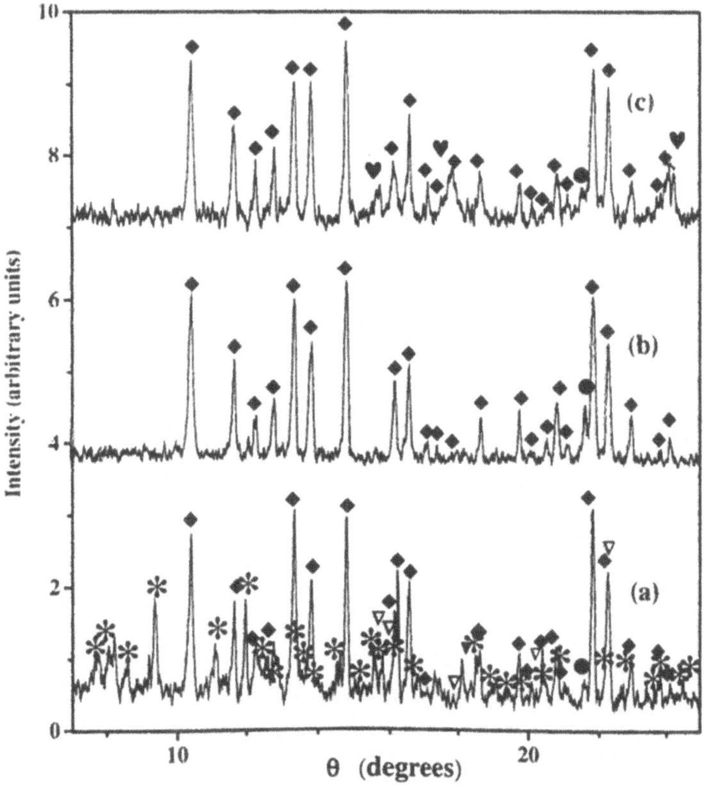

Fig. 4.65 X-ray diffraction patterns of mixture of starting $CuSO_4{\cdot}5H_2O$ with lead milled for different times: (**a**) 5 min; (**b**) 30 min; (**c**) 50 h: ∗ $CuSO_4{\cdot}5H_2O$, ▽ Pb, •Cu, ♦ $PbSO_4$, ⊙ WC [Varghese et al. 2004]

due to the presence of electric double layer at the interface between lead metal and solution where the redox system has to occur. Since a similar electric double layer is absent in solid state, the resistance due to double layer does not play a part in the kinetics while milling ($CuSO_4{\cdot}5H_2O$+Pb) system. Thus, while the reaction (4.71) is slow in solution, it is much faster in solid [Varghese et al. 2004]. For the given process conditions, the kinetics and mechanism of the mechanochemical reduction can vary with the variation in the chemical nature of the reactants as well as products. A careful selection of systems becomes important in the process control [Varghese et al. 2001, 2004].

Bastnaesite $ReCO_3F$ (Re-rare earths) is one of the most important mineral resources containing rare earth elements especially La and Ce. A novel process for extracting rare earths contained in this mineral has been developed in Japan [Zhang and Saito 1998]. This process is composed of three steps combining milling, washing and leaching operations. Milling of bastnaesite and NaOH powders in the first step plays a significant role in achieving a high yield of rare earths elements in the leaching of the washed sample (Fig. 4.66). The leaching is governed by the reaction

$$LaCO_3F + 3NaOH \rightarrow La(OH)_3 + Na_2CO_3 + NaF \tag{4.72}$$

and extraction of the rare earth elements such as La, Ce, Pr, Nd, Sm included in bastnaesite was attempted with NaOH powder using planetary mill.

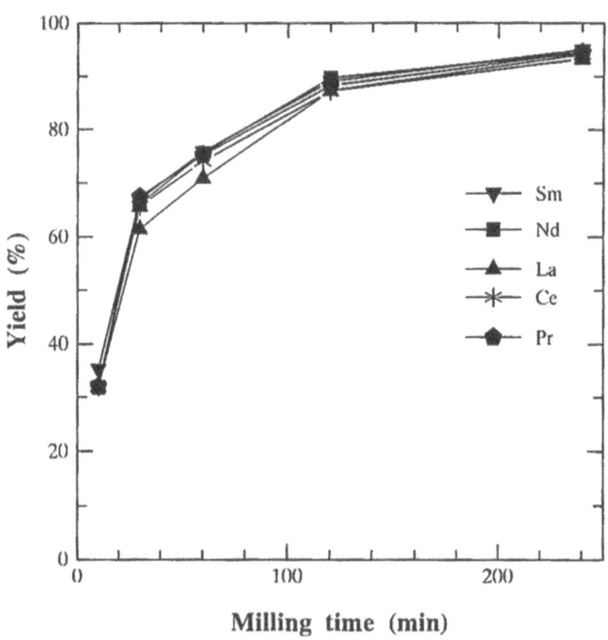

Fig. 4.66 Yield of the rare earth elements extracted from the washed samples by 10 N H_2SO_4 solution as a function of milling time [Zhang and Saito 1998]

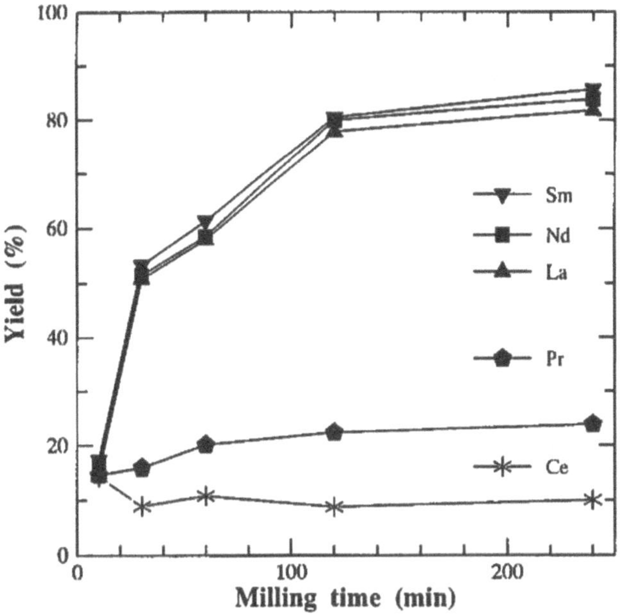

Fig. 4.67 Yield of the rare earth elements extracted from the washed samples by 0.05 N H_2SO_4 solution as a function of milling time [Zhang and Saito 1998]

The change in selectivity of the rare earth elements extraction can be achieved by the change of leaching conditions (Figs. 4.66 and 4.67).

References

Avvakumov, E.G., Djakova, V.E. and Strugova, L.I. (1974) Mechanical activation of solid-state reactions. 4. Solid state reduction of cassiterite. Izvestija SO AN SSSR, seria chimiceskich nauk 2, 26–29 (in Russian).

Avvakumov, E.G. and Strugova, L.I. (1974) Mechanical activation of solid-state reactions. 6. Application of diffusionsless kinetics for mechanochemical reactions in solid mixtures. Izvestija SO AN SSSR, seria chimiceskich nauk 2, 34–38 (in Russian).

Avvakumov, E.G., Matycin, L.I. and Staver, A.M. (1975) Solid state reduction of SnO_2 by compression and mechanical activation. Fizika gorenija i vzryva 6, 922–927 (in Russian).

Avvakumov, E.G. (1986) *Mechanical Methods of the Activation of Chemical Processes*. Nauka, Novosibirsk (in Russian).

Baláž, P., Post, E. and Bastl, Z. (1992) Thermoanalytical study of mechanically activated cinnabar. Thermochimica Acta 196, 371–377.

Baláž, P., Havlík, T., Bastl, Z. and Briancin, J. (1995) Mechanosynthesis of iron sulphides. Journal of Material Science Letters 14, 344–346.

Baláž, P. (2000) *Extractive Metallurgy of Activated Minerals*. Elsevier, Amsterdam.

Baláž, P. and Briančin, J. (2001) Direct reduction of mechanically activated galena and sphalerite with hydrogen. Journal of Thermal Analysis and Calorimetry 65, 769–776.

Baláž, P. and Godočíková, E. (2001) Thermal reduction of mechanically activated cinnabar (HgS) and stibnite (Sb_2S_3). Journal of Thermal Analysis and Calorimetry 65, 51–57.

Baláž, P., Boldižárová, E. and Bajger, Z. (2001c) Non-traditional way of magnetite recovery from metallurgical waste. Acta Metallurgica Slovaca 7, 97–100.

Baláž, P., Godočíková, E., Boldižárová, E., Luxová, M., Bastl, Z. and Jiang, J.Z. (2002a) Characterization of nanocrystalline products prepared by mechanochemical reduction of copper sulphide. Czechoslovak Journal of Physics, Supplement A 52, A65–A68.

Baláž, P., Takacs, L., Jiang, J.Z., Soika, V. and Luxová, M. (2002b) Mechanochemical reduction of copper sulphide. Materials Science Forum 386–388, 257–262.

Baláž, P., Takacs, L., Boldižárová, E. and Godočíková, E. (2003) Mechanochemical transformations and reactivity in copper sulphides. Journal of Physics and Chemistry of Solids 64, 1413–1417.

Baláž, P., Aláčová, A., Godočíková, E., Kováč, J., Škorvánek, I. and Jiang, J.Z. (2004a) Study of magnetic properties of nanopowders prepared by pyrite-troilite transformation via high energy milling. Czechoslovak Journal of Physics, Supplement D, 54, D197–D200.

Baláž, P., Godočíková, E., Aláčová, A., Škorvánek, I., Kováč, J. and Jiang, J.Z. (2004b) Magnetic properties of nanocrystalline pyrrhotite prepared by high-energy milling. Czechoslovak Journal of Physics, Supplement D 54, D121–D124.

Baláž, P., Takacs, L., Luxová, M., Godočíková, E. and Ficeriová, J. (2004c) Mechanochemical processing of sulphidic minerals. International Journal of Mineral Processing 74S, 365–371.

Baláž, P., Godočíková, E., Krilová, L., Lobotka, P. and Gock, E. (2004d) Preparation of nanocrystalline materials by high energy milling. Materials Science and Engineering A 386A, 442–446.

Baláž, P., Godočíková, E., Takacs, L. and Gock, E. (2005a) Mechanochemical preparation of metal/sulphide nanocomposite particles. International Journal of Materials and Product Technology 23, 26–41.

Baláž, P., Boldižárová, E. and Godočíková, E. (2005b) Preparation of nanocrystalline copper and copper silicon sulphide by mechanochemical route. Materials Science Forum 480–481, 453–456.

Baláž, P., Takacs, L., Godočíková, E., Škorvánek, I., Kovác, J. and Choi, W.S. (2007) Preparation of nanosized antimony by mechanochemical reduction of antimony sulphide Sb_2S_3. Journal of Alloys and Compounds 434–435, 773–775.

Banza, A.N. and Gock, E. (2003) Mechanochemical processing of chrysocolla with sodium sulphide. Minerals Engineering 16, 1349–1354.

Basset, D., Matteazzi, P. and Miani, F. (1994) Kinetic effects in mechanically activated solid-state reduction of hematite. Hyperfine Interactions 94, 2235–2238.

Beecroft, L.L. and Ober, C.K. (1997) Advanced nanocomposite materials for optical applications. Chemistry of Materials 9, 1302–1317.

Benjamin, J.S. (1970) Dispersion strengthened superalloys by mechanical alloying. Metallurgical Transactions 1, 2946–2951.

Boldyrev, V.V. (1993) Mechanochemistry and mechanical activation of solids. Solid State Ionics 63–65, 537–543.

Boldyrev, V.V. (1996a) Mechanochemistry and mechanical activation. Materials Science Forum 225–227, 511–520.

Boldyrev, V.V. (1996b) Reactivity of solids and new technologies. In: V.V. Boldyrev (Ed.) *Reactivity of Solids: Past, Present and Future.* Blackwell Science, Oxford, pp. 267–285.

Boldyrev, V.V. and Avvakumov, E.G. (1971) Mechanochemistry of inorganic solids. Uspechi chimiji 40, 1835–1856 (in Russian).

Boldyrev, V.V. (1979) Control of reactivity of solids. Annual Revne of Materials Science 9, 455–469.

Boldyrev, V.V. (1998) Mechanical activation and its application in technology. Materials Science Forum 269–272, 227–234.

Boldyrev, V.V. and Tkáčová, K. (2000) Mechanochemistry of solids: past, present and prospects. Journal of Materials Synthesis 8, 121–132.

Boldyrev, V.V. (2006) Mechanochemistry and mechanical activation of solids. Russian Chemical Reviews 75, 177–189.

Botta, P.M., Aglietti, E.F. and Porto Lopez, J.M. (2000) Thermal and phase evolution of mechanochemical reactions in the Al-Fe_3O_4 system. Thermochimica Acta 363, 143–147

Budnikov, P.P. and Ginstling, A.M. (1971) *Reactions in Solid Mixtures*. Strojizdat, Moscow (in Russian).

Burkin, A.R. (1966) *The Chemistry of Hydrometallurgical Processes*. Spou, London.

Butyagin, P.J. (1971) Kinetics and nature of mechanochemical reactions. Uspechi chimiji 40, 1935–1959.

Butyagin, P.J., Avvakumov, E.G. and Kolbanov, I.V. (1974) Žurnal fiziceskoj chimii 12, 3009.

Butyagin, P.J. (1984) Disordering of structure and mechanochemical reactions in solids. Uspechi chimiji 53, 1769–1789 (in Russian).

Butyagin, P.J. and Yuščenko, V.S. (1987) Kinetika i kataliz 27, 1035–1039.

Campbell, S.J., Kaczmarek, W.A. and Wang, G.M. (1995) Mechanochemical transformation of haematite to magnetite. Nanostructured Materials 6, 735–738.

Cao, G., Garcia, M.E., Alcala, M., Burgess, L.F. and Mallouk, T.E. (1992) Chiral molecular recognition in intercalated zirconium phosphate. Journal of American Chemical Society 114, 7574–7575.

Carter, R.E. (1991) Journal of Chemical Physics 34, 2010.

Cech, R.E. and Tiemann, T.S. (1969) The hydrogen reduction of copper, nickel, cobalt and iron sulfides and the formation of filamentary metals. Transactions of Metalurgical Society of AIME 245, 1727–1733.

Cech, R.E. (1974) Journal of Metals 26, 32–38.

Chen, Y., Marsh, M., Williams, J.S. and Ninham, B. (1996) Production of rutile from ilmenite by room temperature ball-milling-induced sulphurization reaction. Journal of Alloys and Compounds 245, 54–58.

Chunpeng, L., Zhonghua, L. and Zuze, Z. (1988) Reduction kinetics of stibnite with hydrogen and recovery of metallic antimony/lead by evaporation. In: F. Chongyne, L. Jianehun, L. Songren (Eds.) *Proceedings of the Ist International Conference on the Metalurgy and Materials Sciences of Tungsten, Titanium, Rare Earth and Antimony "W-Ti-Re-Sb 88"*, Vol. 1. Pergamon Press, Oxford, pp. 539–544.

Courtney, I.A. and Dahn, J.R. (1997) Electrochemical and in-situ X-ray diffraction studies of the reaction of lithium with tin oxide composites. Journal of Electrochemical Society 144, 2045–2052.

Čižikov, D.M. (1976) *Metallurgy of Non-ferrous Metals*. Nauka, Moscow (in Russian).

Danielian, N.G., Janazian, S.K. and Melnichenko, V.V. (1991) X-ray investigation of metal reduction by mechanical alloying and solid state reaction. Modern Physics Letters 5, 1355–1359.

Dutta, J. and Hofman, H., http:/www.mxsg3.epfl.ch/ltp/Cours/Nanomat/Nanomat/pdf

El-Eskandarany, M.S., El-Bahnasawy, H.N., Ahmed, H.A. and Eissa, N.A. (2001) Mechanical solid-state reduction of haematite with magnesium. Journal of Alloys and Compounds 314, 286–295.

Frost, B.R. (1991) Stability of oxide minerals in metamorphic rocks. Reviews in Mineralogy and Geochemistry 25, 469–488.

Gaffet, E. and Harmelin, M. (1990) Crystal-amorphous phase transition induced by ball-milling in silicon. Journal of Less-Common Metals 157, 201–222.

Gaffet, E., Bernard, F., Niepce, J.C., Charlot, F., Gras, C., LeCaër, G., Guichard, J.L., Delcroix, P., Mocellin, A. and Tillement, O. (1999) Some recent developments in mechanical activation and mechanosynthesis. Journal of Materials Chemistry 9, 305–314.

Gaffet, E. and Bernard, F. (2002) Mechanically activated powder metallurgy processing: A versatile way towards nanomaterials synthesis. Annales de Chimie Sciences des Materiaux 27, 47–59.

Gaffet, E. and LeCaër, G. (2004) Mechanical processing for nanomaterials. In: H.S. Nalwa (Ed.) *Encyclopedia of Nanoscience and Nanotechnology*, Vol. 5. American Scientific Publishers, pp. 91–129.

Gaines, R.V., Skinner, H.C.W., Ford, E.E., Mason, B., Rosenzweig, A., King, W.T. and Gowty, E. (1997) *Dana's New Mineralogy.* Willey, New York.

Gillan, E.G. and Kaner, R.B. (1996) Synthesis of refractory ceramics via rapid metal thesis between solid-state precursors. Chemistry of Materials 8, 333–343.

Godočíková, E., Bastl, Z., Spirovová, I. and Baláž, P. (2004a) A study of mechanochemical reduction of lead sulphide by elemental iron on the surface studied by XPS. Journal of Materials Science 39, 3025–3029.

Godočíková, E., Baláž, P., Boldižárová, E., Škorvánek, I., Kováč, J. and Choi, W.S. (2004b) Mechanochemical reduction of lead sulphide by elemental iron. Journal of Materials Science 39, 5353–5355.

Godočíková, E., Baláž, P., Takacs, L., Šepelák, V., Škorvánek, I. and Gock, E. (2007) Sb/FeS nanocomposite prepared by mechanochemical reduction. Kovové materiály 45, 99–104.

Godočíková, E., Takacs, L., Baláž, P., Kováč, J., Šatka, A. and Briančin, J. (2008) Mechanochemical reduction of antimony sulphide Sb_2S_3 with magnesium in a planetary mill. Reviews in Advanced Materials Science 18, 212–215.

Goya, G.F., Rechenberg, H.R. and Jiang, J.Z. (1998) Structural and magnetic properties of ball milled copper ferrite. Journal of Applied Physics 84, 1101–1108.

Habashi, F. (1969) *Extractive Metallurgy*, Vol. 1, General Principles. Gordon and Breach, New York.

Habashi, F. (1986) *Extractive Metallurgy*, Vol. 3, Pyrometallurgy. Gordon and Breach, New York.

Habashi, F. (1993) *A History of Metallurgy.* Métallurgie Extractive Québec.

Hedvall, J.A. (1938) *Reaktionsfähigkeit fester Stoffe.* Verlag Barth, Leipzig.

Henn, J.J. and Barclay, J.A. (1995) Journal of Applied Chemistry and Biotechnology 25, 561–568.

Jermakov, A.E., Barinov, V.A. and Jurcikov, E.E. (1982) Fizika Metallov i Metallovedenije 54, 90–96 (in Russian).

Jiang, J.Z., Larsen, R.K., Lin, R., Morup, S., Chorkendorff, I., Nielsen, K. and West, K. (1998) Mechanochemical synthesis of Fe-S materials. Journal of Solid State Chemistry 138, 114–125.

Jovanovic, S., Sinadinovic, D. and Durkovic, B. (1986a) Reduction of non-ferrous sulfides by hydrogen (I). Rud Geolog I Metal 37, 594–597 (in Serbian).

Jovanovic, S., Durkovic, B. and Sinadinovic, D. (1986b) Reduction of non-ferrous sulfides by hydrogen (II). Rud Geolog I Metal 37, 1247–1251 (in Serbian).

Kaczmarek, W.A. and Ninham, B.W. (1994) Preaparation of Fe_3O_4 and γ-Fe_2O_3 powders by magnetomechanical activation of hematite. IEEE Transactions on Magnetics 30, 732–733.

Karch, J., Birringer, R. and Gleiter, H. (1987) Ceramics ductile at low temperature. Nature 330, 556–558.

Kharitidi, G.P., Skopov, G.V., Chudjakov, I.F. and Veksler, S.F. (1981) Reduction of chalcopyrite by solid carbon in the presence of calcium oxide. Izvestija AN SSSR, Metalurgija 6, 21–27 (in Russian).

Kharitidi, G.P., Skopov, G.V., Lisina, N.N. and Ovchinnikova, L.A. (1983) Reduction and sublimation of lead sulfide in the lead-sulfide-calcium oxide-carbon system. Žurnal prikladnoj chimiji 56, 729–734 (in Russian).

Koch, C.C. (1991) Mechanical milling and alloying. Materials Sciences and Technology 15, 93–198.

Kosmac, T. and Courtney, T.H. (1992) Milling and mechanical alloying of inorganic nonmetallics. Journal of Materials Research 7, 1519–1525.

Lin, J.J., Nadiv, S. and Grodzian, D.J.M. (1975) Changes in the state of solids and mechanochemical reactions in prolonged comminution process. Minerals Science and Engineering 7, 313–336.

Ma, E., Pagan, J., Cranford, G. and Atzmon, M. (1993) Evidence of self-sustained $MoSi_2$ formation during high-energy ball milling of elemental powders. Journal of Materials Research 25, 1836–1844.

Mankhand, T.R., Singh, G. and Prasad, P.M. (1978) Transactions of Indian Institute of Metals 31, 194–199.

Marfunin, A.S. and Mkrtčjan, A.R. (1967) Mössbauer spectra of Fe^{57} in sulphide minerals. Geochimija 10, 1094–1103 (in Russian).
Matteazzi, P. and LeCaër, G. (1991) Reduction of haematite with carbon by room temperature ball milling. Materials Science and Engineering A 149, 135–142.
Matteazzi, P. and LeCaër, G. (1992a) Synthesis of nanocrystalline alumina-metal composites by room-temperature ball-milling of metal oxides and aluminium. Journal of American Ceramic Society 75, 2749–2755.
Matteazzi, P. and LeCaër, G. (1992b) Mechanically activated room temperature reduction of sulphides. Materials Science and Engineering A 156A, 229–237.
Matteazzi, P., Basset, D., Miani, F. and LeCäer, G. (1993) Mechanosynthesis of nanophase materials. Nanostructured Materials 2, 217–229.
McCormick, P.G. (1995) Application of mechanical alloying to chemical refining. Materials Transactions JIM 36, 161–169.
McCormick, P.G., Tsuzuki, T., Robinson, J.S. and Ding, J. (2001) Nanopowders synthesized by mechanochemical processing. Advanced Materials 13, 1008–1010.
Menzel, M., Šepelák, V. and Becker, K.D. (2001) Mechanochemical reduction of nickel ferrite. Solid State Ionics 141–142, 663–669.
Mills, K.C. (1974) *Thermodynamic Data for Inorganic Sufides, Selenides and Tellurides*. Butterworth, London.
Molčanov, V.I. and Jusupov, T.S. (1981) *Physical and Chemical Properties of Fine Ground Minerals*. Nedra, Moscow.
Morimoto, M., Koto, K. and Shimazaki, Y. (1969) Anilite, Cu_7S_4, a new mineral. American Mineralogist 54, 1256–1268.
Mrowec, S. (1988) On the defect structure and diffusion kinetics in transition metal sulfides and oxides. Reactivity of Solids 5, 241–268.
Mukopadhyay, D.K., Prisbrey, K.A., Suryanarayana, C. and Froyes, F.H. (1996) Ball-milling, a novel extraction process for production of W from WO_3 using magnesium as reductant. In: A. Bose, R.J. Dowding (Eds.) *Tungsten and Refractory Metals 3*. Metal Powder Industries Federation, New York, pp. 239–346.
Mulas, G., Monagheddu, M., Doppiu, S., Cocco, G., Maglia. F. and Anselmi Tamburini, U. (2001) Metal-metal oxides prepared by MSR and SHS techniques. Solid State Ionics 141–142, 649–656.
Nakatani, Y., Sakai, M., Nakatani, S. and Matsuoka, M. (1983) Mechanochemical effect of dry-grinding on the transformation phenomenon from γ-Fe_2O_3 to α-Fe_2O_3. Journal of Materials Science Letters 2, 129–131.
Nasu, T., Tokumitsu, K., Miazawa, K., Greer, A.L. and Suzuki, K. (1999) Solid state reduction of iron oxide by ball milling. Materials Science Forum 312–314, 185–190.
Nazar, L.F., Zhang, Z. and Zinkweg, D.J. (1992) Insertion of PPV in layered MoO_3. Journal of American Chemical Society 114, 6239–6240.
Niihara, K. (1991) New design concept of structural ceramic: ceramic nanocomposites. Journal of Ceramic Society of Japan 99, 974–982.
Onajev, I.V. and Spitčenko, V.S. (1988) *Reduction of Sulfides*. Nauka, Alma-Ata (in Russian).
O'Neill, H.S.C. and Navrotsky, A. (1983) Simple spinels: crystallographic parameters, cation radii, lattice energies, and cation distribution. American Mineralogist 68, 181–194.
O'Neill, H.S.C. and Navrotsky, A. (1984) Cation-distributions and thermodynamic properties of binary spinel solid solutions. American Mineralogist 69, 733–753.
Pardavi-Horvath, M. and Takacs, L. (1992) Iron-alumina composites prepared by ball milling. IEEE Transactions on Magnetics 28, 3186–3188.
Parson, A., Petrashov, V.T. and Sosnin, I.A. (2000) Anomalies in quantum and classical magnetoresistance of semi-metallic nanowires. Physica B: Condensed Matter 284–288, 1744–1745.
Patel, P., Roy, S., Kim, Il-Seok and Kumta, P.M. (2004) Synthesis and characterization of tin and antimony based composites derived by mechanochemical in situ reduction of oxides. Materials Science and Engineering B 111, 237–241.

Pavlyuchin, J.T., Medikov, J.J., Avvakumov, E.G. and Boldyrev, V.V. (1981) Defect formation by mechanical activation studied by Mössbauer spectroscopy. ISO AN ZSSSR, seria chimiceskich nauk 4, 11–16 (in Russian).

Pavlyuchin, I.T., Medikov, I.I. and Boldyrev, V.V. (1982) Cation re-distribution in spinel ferrites as a consequence of mechanical activation. Doklady akademii nauk SSSR 266, 1420–1422 (in Russian).

Pavlyuchin, J.T., Medikov, J.J. and Boldyrev, V.V. (1983) Magnetic and chemical properties of mechanically activated zinc and nickel ferrites. Materials Research Bulletin 18, 1317–1327.

Pavlyuchin, J.T., Medikov, J.J. and Boldyrev, V.V. (1984) On the consequences of mechanical activation of zinc and nickel ferrites. Journal of Solid State Chemistry 53, 155–160.

Pavlyuchin, J.T., Medikov, J.J and Boldyrev, V.V. (1988) Mechanical activation of close-packed inorganic crystals. Review of Solid State Sciences 2, 603–621.

Prasad, P.M. and Mankhand, T.R. (1983) Lime enhanced reduction of metal sulphides. In: H.Y. Sohn, D.B. George, A.D. Zunkel (Eds.) *Proceedings of International Sulphide Smelting Symposium*. The Metallurgical Society of AIME, San Francisco, pp. 371–392.

Saito, F., Zhang, Q. and Kano, J. (2004) Mechanochemical approach for preparing nanostructural materials. Journal of Materials Science 39 (2004) 5051.

Schaffer, G.B. and McCormick, P.G. (1989a) Reduction of metal oxides by mechanical alloying. Applied Physics Letters 55, 45–46.

Schaffer, G.B. and McCormick, P.G. (1989b) Combustion synthesis by mechanical alloying. Scripta Metallurgica 23, 835–838.

Schaffer, G.B. and McCormick, P.G. (1990) Displacement reactions during mechanical alloying. Metallurgical Transactions A 21A, 2789–2794.

Schaffer, G.B. and McCormick, P.G. (1991) Anomalous combustion effects during mechanical alloying. Metallurgical Transactions A 22A, 3019–3023.

Schaffer, G.B. and McCormick, P.G. (1992a) The direct synthesis of metals and alloys by mechanical alloying. Materials Science Forum 88–90, 779–786.

Schaffer, G.B. and McCormick, P.G. (1992b) Mechanical alloying. Materials Forum 16, 91–97.

Schrader, R. and Hoffman, B. (1973) Änderung der Reaktionsfähigkeit von Festkörpern durch vorhergehende mechanische Bearbeitung. In: V.V. Boldyrev, K. Meyer (Eds.) *Festkörperchemie*. Verlag Grundstoffindustrie, Leipzig, pp. 522–543.

Senna, M. and Kuno, H. (1973) Effect of preliminary pressing on isothermal transformation of maghemite to hematite. Journal of American Ceramic Society 56, 492–493.

Shi, Y. and Ding, J. (2001) Journal of Applied Physics 90, 4078.

Shuey, R.T. (1975) *Semiconducting Ore Minerals*. Elsevier, Amsterdam.

Sorescu, M. (1998) Phase transformations induced in magnetite by high energy milling. Journal of Materials Science Letters 17, 1059–1061.

Streleckij, A.N., Butyagin, P.J. and Leonov, A.V. (1996) Koloidnyj žurnal 58–62, 248–253 (in Russian).

Suryanarayana, C. (2001) Mechanical alloying and mixing. Progress in Materials Science 46, 1–184.

Szczygiel, Z., Lara, C., Escobedo, S. and Mendoza, O. (1998) The direct reduction of sulfide minerals for the recovery of precious metals. Journal of Metals 50, 55–59.

Šepelák, V., Steinicke, U., Uecker, D.C., Trettin, R., Wissmann, S. and Becker, K.D. (1997) High-temperature reactivity of mechanosynthesized zinc-ferrite. Solid State Ionics 101–103, 1343–1349.

Šepelák, V. and Becker, K.D. (2000) Mösbauer studies in the mechanochemistry of spinel ferrites. Journal of Materials Synthesis and Processing 8, 155–166.

Šepelák, V., Menzel, M., Becker, K.D. and Krumeich, F. (2002) Mechanochemical reduction of magnesium ferrite. Journal of Physical Chemistry B 106, 6672–6678.

Šepelák, V., Bergmann, I., Kipp, S. and Becker, K.D. (2005) Chemie mit der Hammer-Mechanochemie. Zeitschrift für Anorganische und Allgemeine Chemie 631, 993–1003.

Takacs, L. (1993) Metal-metal oxide systems for nanocomposite formation by reaction milling. Nanostructured Materials 2, 241–249.

Takacs, L. and Pardavi-Horvath, M. (1994) Magnetic properties of nanocomposites prepared by mechanical alloying. In: R.D. Shull, J.M. Sanchez (Eds.) *Nanophases and Nanocrystalline Structures*. TMS, Warendale, pp. 135–144.

Takacs, L. (1996a) Ball milling-induced combustion in powder mixtures containing titanium, zirconium or hafnium. Journal of Solid State Chemistry 125, 75–84.

Takacs, L. (1996b) Combustive mechanochemical reactions with titanium, zirconium and hafnium. Materials Science Forum 225–227, 553–558.

Takacs, L. (1996c) Nanocrystalline materials by mechanical alloying and their magnetic properties. In: C. Suryanarayana, J. Singh, F.H. Froes (Eds.) *Processing and Properties of Nanocrystalline Materials*. TMS, Warrendale, PA, pp. 453–464.

Takacs, L. (2002) Self-sustaining reactions induced by ball milling. Progress in Materials Science 47, 355–461.

Takacs, L., Torosyan, A.R. and Baláž, P. (2006) Ball milling induced reduction of MoS_2 with Al. Journal of Materials Science 7033–7039.

Tamman, G. (1932) *Lehrbuch der Metallkunde*. Verlag Barth, Leipzig.

Terry, B.S., Azubike, D.C. and Chrysanthou, A. (1994) Carbothermic reduction as a potential means for the direct reduction of Fe-WC and FeTaC, NbC metal-matrix composites. Journal of Materials Science 29, 4300–4305.

Thiessen, P., Heinicke, G. and Schober, E. (1970) Zur tribochemischen Umsetzung von Gold und CO_2 mit Hilfe radioaktiver Markierung. Zeitschrift für Anorganische und Allgemeine Chemie 377, 20–28.

Tkáčová, K. (1989) *Mechanical Activation of Minerals*. Elsevier, Amsterdam.

Torma, A.E. and Inal, O.T. (1979) Reduction of stibnite by hydrogen. Journal of Less-Common Metals 64, 107–114.

Torosyan, A.R., Tuck, J.R., Korsunskij, A.M. and Bagdasaryan, S.A. (2002) Metastable, mechanically alloyed and nanocrystalline materials. Materials Science Forum 386–388, 251–256.

Torosyan, A.R. and Takacs, L. (2004) Quantitative comparison of the efficiency of mechanochemical reactors. Journal of Materials Science 39, 5491–5498.

Treece, R.E., Gillan, E.G. and Kanner, R.B. (1995) Materials synthesis via solid-state metathesis reaction. Comments on Inorganic Chemistry 16, 313–337.

Tschakarov, C.G., Gospodinov, G.G. and Bontschev, Z. (1982) Über den Mechanismus der mechanochemischen Synthese. Journal of Solid State Chemistry 41, 244–252.

Urakajev, F.Ch., Takacs, L., Soika, V., Schevchenko, V.S. and Boldyrev, V.V. (2001) Mechanism of formation of "hot spots" in mechanochemical reactions of metal and sulfur. Russian Journal of Physical Chemistry 75, 1997–2001.

Urakajev, F.Ch., Ketegenov, T.A., Petrshin, E.J., Savintsev, Y.P., Tjumentseva, O.A., Chupakhin, A.P., Schevchenko, V.S., Jusupov, T.S. and Boldyrev, V.V. (2003) Complex investigation into the abrasion-reaction modification of quartz particle surface by amorphous iron compounds. Journal of Mining Science 39, 304–314.

Urakajev, F.Ch., Schevchenko, V.S. and Ketegenov, T.A. (2004) Synthesis of chalcogenide nanocomposites. Journal of Physical Chemistry 78, 551–554 (in Russian).

Varghese, V., Sharma, A. and Chattopadhyay, K. (2001) Reaction ball milling of systems involving ionic bonds. Materials Science and Engineering A304–306, 434–437.

Varghese, V., Chattopadhyay, K. and Narayanasamy, A. (2004) Factors influencing the kinetics of electrochemical reactions in milling. Journal of Materials Science 39, 5161–5167.

Vanjukov, A.V., Isakova, R.A. and Bystrov, V.P. (1978) *Thermal Decomposition of Metal Sulfides*. Nauka, Alma-Ata (in Russian).

Varnek, V.A., Strugova, L.I. and Avvakumov, E.G. (1974) Mechanical activation of solid-state reactions. 5. Study of solid-state reduction of SnO_2 Mössbauer spectroscopy. Izvestija SO AN SSSR, seria chimiceskich nauk 2, 26–29 (in Russian).

Vassilion, J.K., Ziebarth, R.P. and Disalvo, F. (1990) Chemistry of Materials 2, 738–746.

Vaughan, D.J. and Craig, J.R. (1978) *Minerals Chemistry of Metal Sulfides*. Cambridge University Press, Cambridge.

Vaughan, D.J. and Lennie, A.R. (1991) The iron sulphide minerals: their chemistry and role in nature. Scientific Progress Edinburgh 75, 371–388.

Weeber, A.W. and Bakker, H. (1998) Amorphization by ball milling. A review. Physica B: Condensed Matter 153, 93–135.

Weisser, O. and Landa, S. (1972) *Sulphide Catalysts. Their Properties and Applications*. Academia, Prague.

Welham, N.J. (1996) A parametric study of the mechanically activated carbothermic reduction of ilmenite. Minerals Engineering 9, 1189–1200.

Welham, N.J. (1997) Enhancement of the Becher process by ball milling. Proceedings of the Australasian Institute of Mining and Metallurgy 302, 61–61.

Welham, N.J. (1998a) Mechanical activation of the solid-state reaction between Al and TiO_2. Materials Science Engineering A 255A, 81–89.

Welham, N.J. (1998b) Mechanical activation of the formation of an alumina-titanium aluminide composite. Intermetallics 6, 363–368.

Welham, N.J. (1998c) Mechanochemical reduction of $FeTiO_3$ by Si. Journal of Alloys and Compounds 274, 303–307.

Welham, N.J. (1998d) Mechanochemical reaction between sulfur and ilmenite. Australian Journal of Chemistry 51, 947–953.

Welham, N.J. (1998e) Mechanochemical reaction between ilmenite (FeTiO3) and aluminium. Journal of Alloys and Compounds 270, 228–236.

Welham, N.J. (1998f) Mechanically induced reduction of ilmenite ($FeTiO_3$) and rutile (TiO_2) by magnesium. Journal of Alloys and Compounds 274, 260–265.

Welham, N.J. (1999) Room temperature reduction of scheelite ($CaWO_4$). Journal of Materials Research 14, 619–627.

Welham, N.J. (2000) Mechanical enhancement of the carbothermic formation of TiB_2. Metallurgical Transactions A 31A, 283–289.

Welham, N.J. (2002) Activation of the carbothermic reduction of manganese ore. International Journal of Mineral Processing 67, 187–198.

Williamson, G.K. and Hall, W.H. (1953) X-ray broadening from filled aluminium and volfram. Acta Metallurgica 1, 22–31.

Yang, H. and McCormick, P.G. (1993) Combustion reaction of zinc oxide with magnesium during milling. Journal of Solid State Chemistry 107, 258–263.

Yang, H. and McCormick, P.G. (1994) Mechanochemical reduction of V_2O_5. Journal of Solid State Chemistry 110, 136–141.

Ye, L.L., Liu, Z.G., Huang, J.Y. and Quan, M.X. (1995) Combustion reaction of powder mixtures of composition $Ni_{20}Ti_{50}C_{30}$ during mechanical alloying. Materials Letters 25, 117–121.

Zhang, Q. and Saito, F. (1998) Non-thermal process for extracting rare earths from bastnaesite by means of mechanochemical treatment. Hydrometallurgy 47, 231–241.

Zhang, Q., Wang, J., Saito, F., Okura, T. and Nakamura, I. (2001) Chemical Letters 160, 700.

Zhang, D.L. (2004) Processing of advanced materials using high-energy mechanical milling. Progress in Materials Science 49, 537–560.

Zviadadze, G.I., Turgenev, I.S., Kabisov, I.Ch. and Vasiljeva, O.J. (1985) Izvestija VUZ, Cvetnaja Metalurgija 1, 60–63 (in Russian).

Zviadadze, G.I., Turgenev, I.S., Kabisov, I.Ch. and Vasiljeva, O.J. (1986) Kinetics of lead sulfide reduction by hydrogen. Izvestija VUZ, Cvetnaja Metalurgija 2, 42–45 (in Russian).

Chapter 5
Mechanochemistry in Minerals Engineering

5.1 Introduction

The traditional scheme of metals extraction from minerals involves some processes of mechanical character ameliorating the accesibility of the valuable component by the leaching agent. Leaching represents the key stage in the extraction scheme and its course may be affected by selection and choice of the method leaching and/or by convenient pretreatment of the solid phase. Thermal and mechanical activation belongs among the most important pretreatment methods which influence solid phase leachability [Baláž 2000].

The hydrometallurgical methods for liberating valuable metals from refractory minerals involve processing of the minerals at high temperatures and/or pressures to facilitate metal recovery. These methods, such as roasting and pressure leaching, are capital intensive and, in the case of roasting, produce environmentally damaging effluent streams.

The thermal activation of sulphidic ores aims at transforming the poorly soluble minerals into more soluble forms. That enables better selectivity in transfer of usable metal into solution, nevertheless it appears that some new problems concerning exploitation of the sulphur emissions arise. In the past three decades enhanced public awareness and governmental pressure have focussed on the problem of sulphur oxide pollution. Sulphidic minerals account for a large fraction of the sulphur oxides. The special problem of the minerals is the presence of small amounts of As, Hg, Te, Se which may be emitted together with sulphur in form of oxides by the thermal activation.

The mechanical activation of minerals makes it possible to reduce their decomposition temperature or causes such a degree of disordering that the thermal activation may be omitted entirely. In this process, the complex influence of surface and bulk properties occurs. The mineral activation leads to a positive influence on the leaching reaction kinetics, to an increase in the measured surface area and to further phenomena, especially the potential mitigation of environmental pollutants which is becoming increasingly important with time [Baláž 2000].

At present, it is not known whether the kinetics of heterogeneous reactions are determined by the contact area, the structure of the mineral, or both. The required

© Springer-Verlag Berlin Heidelberg 2008

modification of the structure can be achieved by mechanical activation of the mineral, typically by high-energy milling. The breaking of bonds in the crystalline lattice of the mineral brings about a decrease (ΔE^*) in activation energy and an increase in the rate of leaching [Zelikman et al. 1975].

$$\Delta E^* = E - E^* \tag{5.1}$$

where E is the apparent activation energy of the non-disordered mineral and E^* is the apparent activation energy of the disordered mineral.

The relationship between the rate of leaching and temperature is usually described by the Arrhenius equation

$$k = Z\exp(-E/RT) \tag{5.2}$$

where k, Z, R and T stand for the rate constant of leaching for the non-disordered mineral, pre-exponential factor, gas constant and reaction temperature, respectively.

For the disordered mineral we can write

$$k = Z\exp(-E^*/RT) \tag{5.3}$$

and after substituting for E^* from (6.1) we obtain

$$k^* = k\exp(\Delta E^*/RT) \tag{5.4}$$

From (5.1) it is clear that exp ($\Delta E^*/RT$) > 1 and thus it follows from Eq. (5.4) that $k^* > k$, i.e., the rate of leaching of a disordered mineral is greater than that of an ordered mineral.

It was Senna who analysed the effect of surface area and the structural disordering on the leachability of mechanically activated minerals [Senna 1989]. In order to solve the problem – whether surface area or structural parameters are predominant for the reactivity, the rate constant is divided by the proper surface area and plot against the applied energy by activation (Fig. 5.1).

If the rate constant of leaching divided by the surface area remains constant with respect to the applied energy, as shown in Fig. 5.1a, then the measured surface area

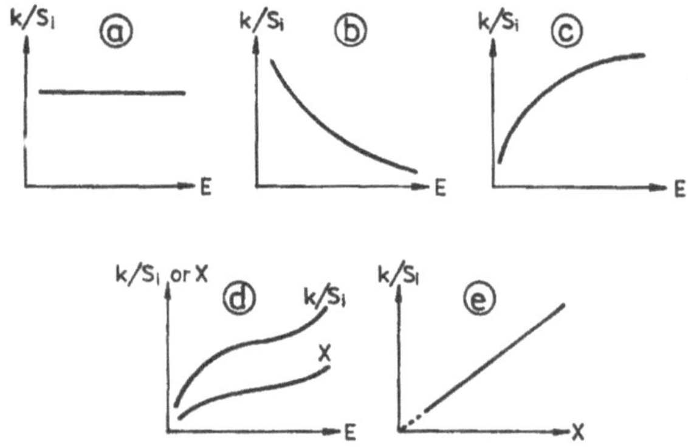

Fig. 5.1 The schematic diagrams representing the mutual dependence of physico-chemical characteristics and reactivity of mechanically activated solids: k – the rate constant of leaching, S_i – surface area, X – structural imperfections, E – applied energy [Senna 1989]

may be the effective surface area and at the same time, the reaction rate is insensitive to structural changes.

If, on the other hand, the value k/S_i decreases with applied energy, as shown in Fig. 5.1b, then the surface area is probably not the effective surface area.

In the third case where k/S_i increases with increasing applied energy, as shown in Fig. 5.1c, the surface area S_i, may be again the effective surface area, with an overlapping effect of the structural imperfection, as a result of mechanical activation.

Alternatively, when k/S_i and X vary parallel to each other with E, as shown in Fig. 5.1d, or the value k/S_i is proportional to X, as shown in Fig. 5.1e, it seems more appropriate to accept the chosen S_i as an effective surface area.

5.2 Mechanochemical Solid–Solid Processing

Mechanochemical processing by high-energy milling is an innovative procedure, that improves the efficiency of mineral processing via several factors, most importantly due to the formation of new surfaces and the creation of lattice defects [Tkáčová 1989; Baláž 2000]. The process is also called mechanochemical processing or mechanical activation and its pecularities are described in Chap. 1. Its benefits include lower reaction temperatures, increased rate and amount of solubility and the formation of water soluble compounds. As a consequence, further processing can be performed in simpler and less expensive reactors during shorter reaction times [Welham 1997, 2001; Aylmore and Lincoln 2000a, b, 2001].

The preparation of tungsten from scheelite and the carbothermic reduction of ilmenite which were described recently are promising examples of direct solid state processing aided by mechanochemical activation [Welham 1996, 1999].

5.2.1 Mechanochemical Reduction

5.2.1.1 Chalcopyrite $CuFeS_2$

According to the XRD pattern obtained from the as received $CuFeS_2$ mineral (deposit Slovinky, Slovakia) its main phase is chalcopyrite with small peaks corresponding to quartz SiO_2 (Fig. 5.2A).

The mechanochemical reduction proceeds according to the equation

$$2CuFeS_2 + Fe \rightarrow Cu_2S + 3FeS \qquad (5.5)$$

and practically total conversion is observed (Fig. 5.2B). Only some trace of chalcopyrite may be present in the sample together with geerite Cu_8S_5 and two forms of pyrrhotite FeS. The lines of the sulphide phases are strongly broadened due to the small grain size [Baláž et al. 2004].

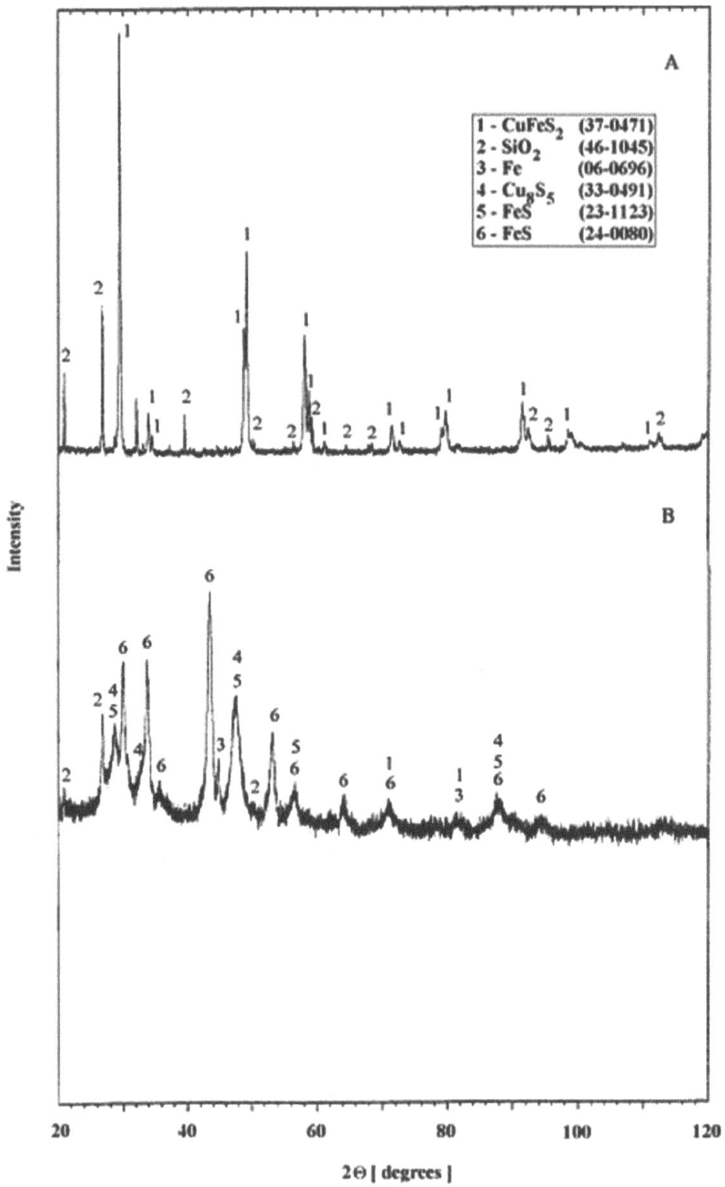

Fig. 5.2 X-ray diffractograms of $CuFeS_2$. **A** – as received, **B** – milled with Fe for 20 min [Baláž et al. 2004]

5.2.1.2 Chalcosine Cu_2S

The mechanochemical reduction of chalcosine Cu_2S (mineral (deposit Khan, South Africa) according to the equation

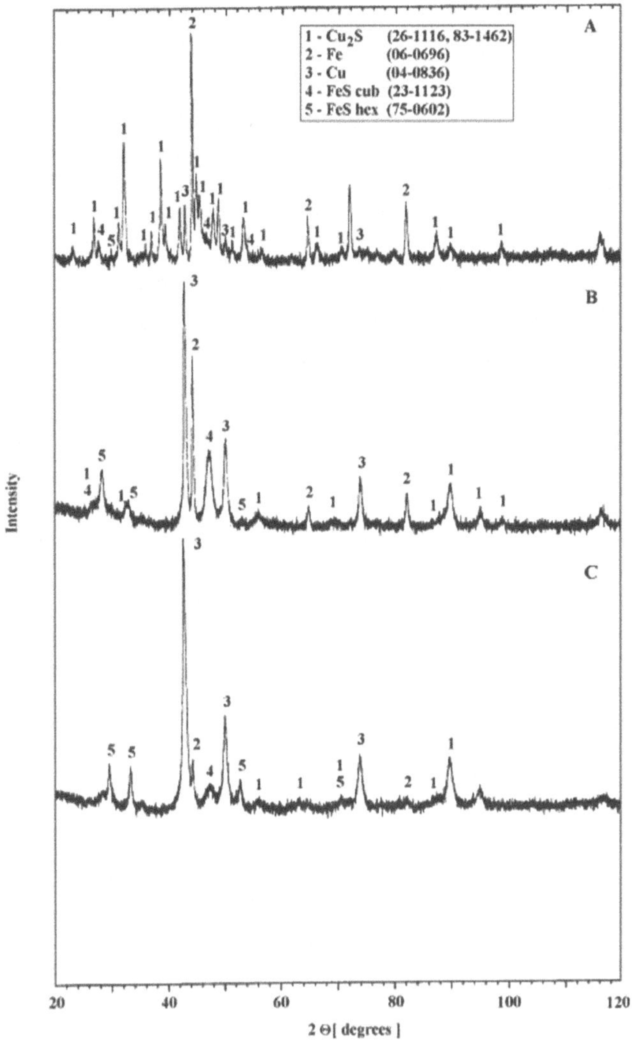

Fig. 5.3 X-ray diffractograms of Cu_2S milled with Fe powder for 10 min at speeds: **A** – 300 min^{-1}, **B** – 400 min^{-1}, **C** – 550 min^{-1} [Baláž et al. 2004]

$$Cu_2S + Fe \rightarrow 2Cu + FeS \tag{5.6}$$

has been attempted at several speeds (and energy inputs) of the mill. The results are illustrated by the selected XRD patterns in Fig. 5.3.

The primary process – the reduction of copper sulphide by iron, while copper metal and iron sulphides are formed – is clearly seen, particularly by inspecting the relative intensities of the diffraction lines of Cu and Fe metal. However, the exact mechanism of the process is quite complex: the formation of bornite

Cu_5FeS_4, the alloy $Fe_{1-x}Cu$ and wüstite $Fe_{1-x}O$ was observed in the preliminary milling tests published earlier [Matteazzi and Le Caër 1992]. In our previous study we observed that the reduction process is accompanied by the transformations of copper and iron sulphide phases as a result of milling [Baláž et al. 2002]. In accord with these results the formation of copper nanoparticles (10–25 nm in size) was accompanied by the cubic modification of FeS (Fig. 5.3A). The cubic FeS transformed to the stable hexagonal form during the later stages of the process (Fig. 5.3B–C).

The chalcosine conversion degree β can be defined as $\beta = I_{Cu}/(I_{Cu}+I_{Fe})$, where I_{Cu} and I_{Fe} are the intensities of the Cu(111) and Fe(110) X-ray diffraction lines. The process kinetics is shown in Fig. 5.4 in terms of this parameter. The dependence of the chalcosine conversion on the milling time is well documented.

The process of chalcosine conversion by the mechanochemical processing can be outlined by the simplified flowchart given in Fig. 5.6. The very straightforward one-step solid state process is characterized by the formation of Cu/FeS nanocomposite (Fig. 5.5). This product has to be separated into its components. Selective leaching [McCormick et al. 2001] of the iron sulphide can be an appropriate method of producing elemental copper. However, the treatment of a liquid phase is necessary in this case. The concept follows a so called "dry" technology strategy in which the steps involving liquids and gases are partially or completely excluded [Boldyrev 1996].

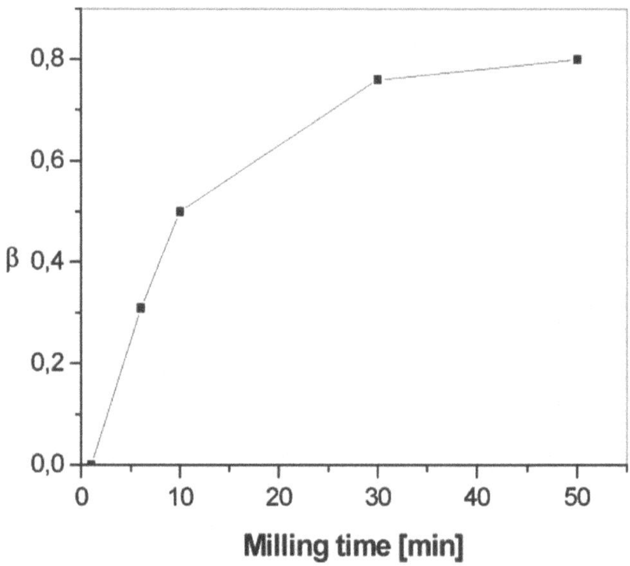

Fig. 5.4 Conversion degree of mechanochemical reduction of Cu_2S, β as a function of milling time [Baláž et al. 2004]

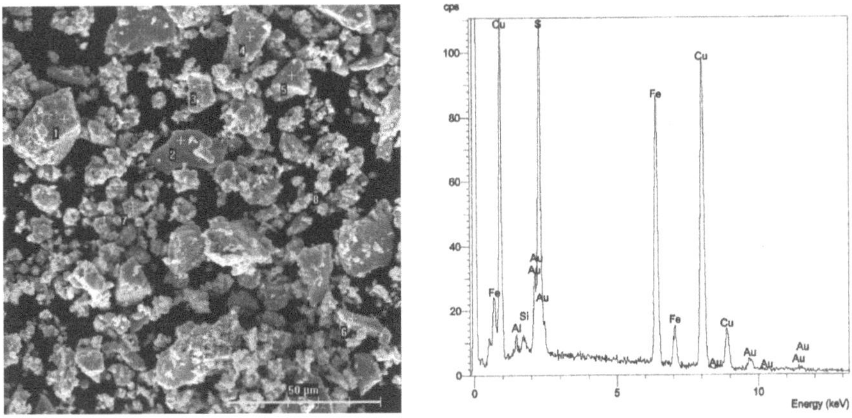

Fig. 5.5 Scanning electron micrographs of products of Cu_2S mechanochemical reduction and corresponding EDAX spectrum imaged on the grain No. 7

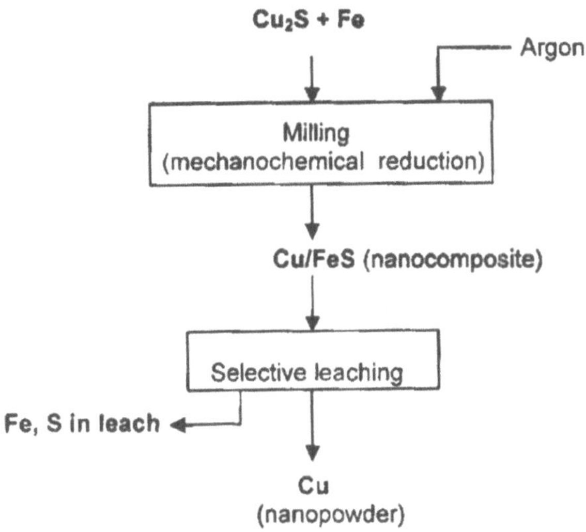

Fig. 5.6 Cu_2S reduction by Fe: mechanochemical solid–solid processing [Baláž et al. 2004]

5.3 Mechanochemical Solid–Liquid Processing

5.3.1 Milling and Leaching as Separated Processes

Leaching belongs to fundamental steps in hydrometallurgy (see also Sect. 6.4.). Leaching should always carried out in conditions that ensure maximum recovery of valuable components at the required selectivity of the process. From the economical point of view, valuable metals should be transferred into the solution within the shortest possible time [Tkáčová 1989].

Mechanical activation seems to be a favourable process for the treatment of minerals prior to leaching with significantly enhanced metal recoveries reported. This increase in leachability may well be due to a combination of mechanically induced structural defects and chemical reactions involving the mineral and the local environment. The differencies in reactivity of the mineral components of ore, as well as the differences in solubility of the reaction products, can be used to selectively leach metals [Welham 1997; Welham and Llewellyn 1998; Hu et al. 2004].

5.3.2 *Oxidative Leaching*

5.3.2.1 Pentlandite $(Fe,Ni)_9S_8$

Pentlandite concentrate composed mainly of pentlandite $(Fe,Ni)_9S_8$ and small amounts of chalcopyrite $CuFeS_2$ and pyrite FeS_2 has been selected to illustrate the effect of mechanical activation on enhancement of leaching [Baláž et al. 2001].

Some special regimes of milling cause not only physico-chemical transformations but also changes in the chemical composition of activated minerals. For sulphides, wet milling can result in so-called mechanochemical oxidation where the formation of water soluble sulphates accompanying the production of insoluble sulphides can contribute to the selectivity of the subsequent leaching process.

Pentlandite and chalcopyrite as economic sources of Cu, Ni and Co are examples of sulphides that are difficult to treat and the considerable effort has been devoted to extraction of these metals [Küzeci et al. 1989; Warner et al. 1992; Mulak 1992]. Owing to the refractory character of the sulphides occurring in pentlandite concentrates, various approaches to influencing the extraction of copper, nickel and cobalt have been attempted by oxidative leaching in combining with mechanical activation [Baláž et al. 1998a, 2000a; Maurice and Hawk 1999a, b].

The oxidative leaching of nickel from pentlandite can be described by the equation [Dutrizac and MacDonald 1974].

$$0.5(Ni,Fe)_9S_8 + 18Fe^{3+} \rightarrow 4.5Ni^{2+} + 22.5Fe^{2+} + 4S \quad (5.7)$$

and the leaching of copper from chalcopyrite by the equation [Tkáčová and Baláž 1988].

$$CuFeS_2 + 4Fe^{3+} \rightarrow Cu^{2+} + 5Fe^{2+} + 2S \quad (5.8)$$

The process of oxidative leaching of the pentlandite concentrate was proceeded by mechanical activation in H_2O (I) in order to dissolve the water-soluble sulphates produced during milling. Then the direct sample was leached out with $Fe_2(SO_4)_3$ solution (II). The obtained Ni, Cu and Co recoveries are given in Table 5.1.

The results indicate the possibility of selective extraction of Ni in principle. Mechanochemical dissolution in water enables us to recover as much as 46% Ni in contrast to 29% Cu and Co under given experimental conditions.

Table 5.1 Metal solubilization from $(Fe,Ni)_9S_8$ after mechanical activation for 60 min [Baláž et al. 2000a]

Metal	Recovery (%)		
	I	II	Total
Ni	45.72	27.28	73
Cu	29.41	70.59	100
Co	29.35	60.65	90

The recovery of Ni, Cu and Co by leaching from the pentlandite concentrate is sensitive to temperature. The kinetics and temperature sensitivity for Co, as an example is given in Fig. 5.7. The recoveries at both applied temperatures reach after 120 min leaching 31% for the non-activated concentrate and 90% for the concentrate subjected to 60 min of mechanical activation.

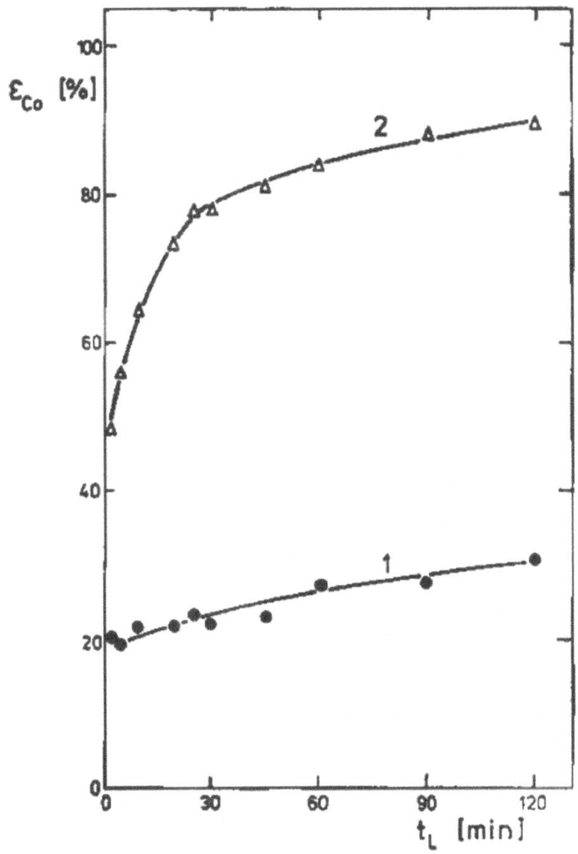

Fig. 5.7 Cobalt recovery, ε_{Co} vs. time of $Fe_2(SO_4)_3$ leaching, t_L for $(Fe,Ni)_9S_8$ mechanically activated for 60 min, 1–45°C, 4–90°C [Baláž et al. 2000a]

5.3.3 Acidic Leaching

5.3.3.1 Arsenopyrite FeAsS

Arsenopyrite is the most common arsenic containing mineral of many ore deposit but it is of limited economic importance and generally is discarded as solid waste during mining operation. When arsenopyrite present in an ore is not associated with gold values, it is viewed as a troublesome impurity. Smelters at present prefer to receive concentrates which are essentially free of arsenic [Nesbitt et al. 1995; Beattie and Poling 1987]. In some sulphide concentrates, arsenopyrite contains "invisible" gold that can be released only by some type of pretreatment, e.g. by chemical oxidation in acidic medium [Robins and Jayaweera 1992; Cruz et al. 1997; Baláž 2000].

The investigation of the influence of mechanical activation of arsenopyrite FeAsS on kinetics and leaching selectivity of As and Fe in H_2SO_4 medium has been performed [Achimovičová and Baláž 2005]. Arsenopyrite of Slovak origin (deposit Pezinok) with small amounts of ankerite $(Fe,Ca)CO_3$, calcite $CaCO_3$ and quartz SiO_2 was used. Non-activated arsenopyrite is very resistant against the influence of H_2SO_4. Figure 5.8A shows that probably only the solubilization of Fe present in ankerite occurred following the equation

$$2(Fe,Ca)CaCO_3 + H_2SO_4 \rightarrow FeSO_4 + CaSO_4 + 2H_2O + 2CO_2 \qquad (5.9)$$

Arsenic and iron in FeAsS are not leached out from non-activated sample.

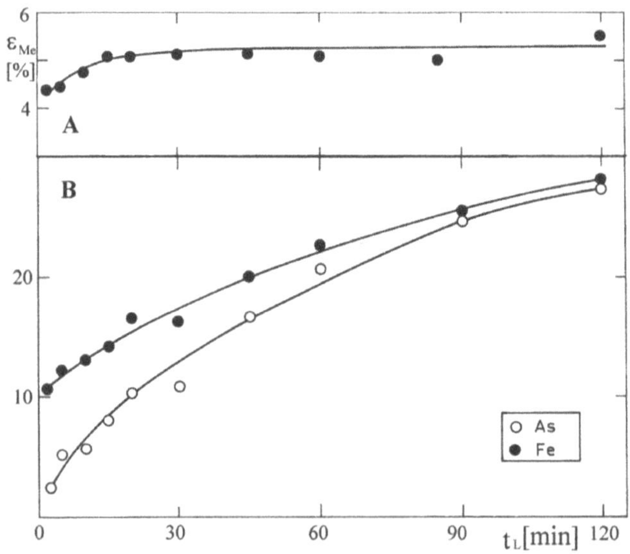

Fig. 5.8 Recovery of metals into the solution ε_{Me}, vs. time of leaching, t_L for FeAsS: **A** – as received, **B** – mechanically activated 30 min [Achimovičová and Baláž 2005]

In case of mechanically activated sample, As and Fe from arsenopyrite are transferred into the leachate. The maximum yield is 29% for As and 31% for Fe, provided the sample was mechanically activated for 30 min (Fig. 5.8B). The chemical processes of FeAsS leaching are very complicated as all the three elements present in the mineral usually form a number of insoluble and soluble compounds. Thus, e.g. Fe enters the leachate as Fe^{2+} or Fe^{3+}, As can be present as As^{3+} or As^{5+}, sulphur in sulphides can be present in the form of S^{2-} or according to the level of oxidation up to S^{6+}.

Mechanical activation as a method of sulphide mineral pretreatment results in the increase of reactive surface and deformation of crystalline structure of minerals. This-in general-leads to accelerating the leaching process [Baláž 2000]. At the end, the increased reactivity of mechanically activated sulphides leads to the decrease of activation energy and the acceleration of leaching of the metals presented in the mineral. This in turn affects also the selectivity of leaching.

Figure 5.9 shows initial reaction rate constant values, k_{Me} of Fe and As leaching from arsenopyrite in relation to the duration of mechanical activation. It can be seen that with the increase of activation time, the reaction rate for both metal increase. The maximum values are achieved after 20 min of activation. It follows that leaching after mechanical activation for 20 min is the most rapid and effective.

The selective leaching of FeAsS as defined by the As/Fe ratio in relation to specific surface is shown in Fig. 5.10. The decreasing trend indicates that with the increase of specific surface, the As/Fe yield rate decreases. It means that in contrast to the As yield, the Fe yield is increasing with the duration of mechanical activation.

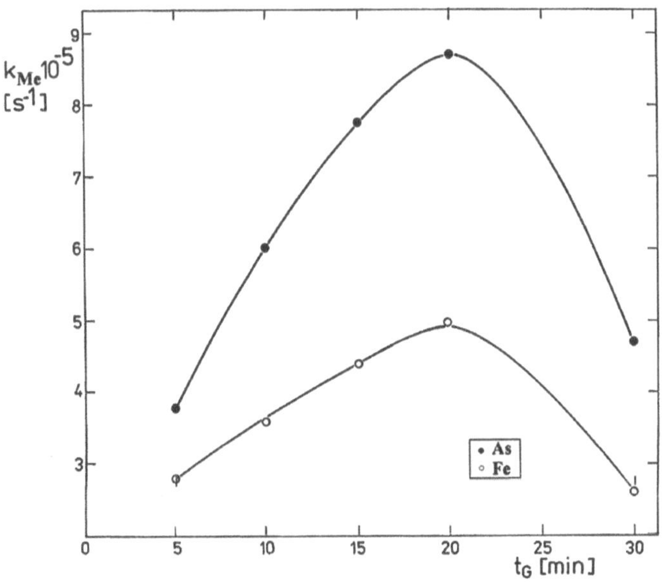

Fig. 5.9 Dependance of leaching rate constants of metals k_{Me} (Me = Fe,As) vs. time, t_G of mechanical activation of FeAsS [Achimovičová and Baláž 2005]

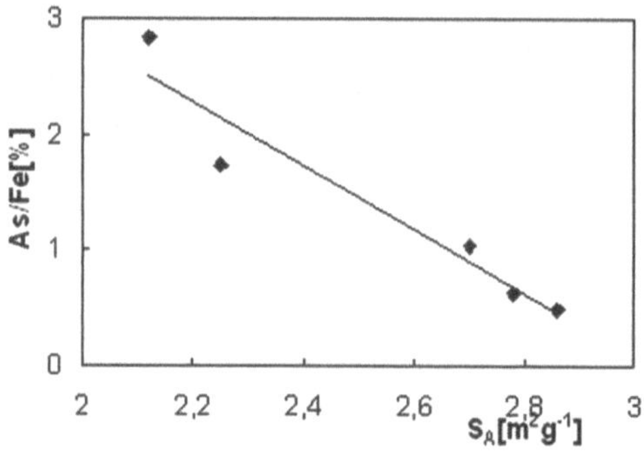

Fig. 5.10 Selectivity of leaching, As/Fe from mechanically activated FeAsS vs. specific surface area, S_A (time of leaching, $t_L = 20$ min) [Achimovičová and Baláž 2005]

Mechanical activation makes the crystalline structure of FeAsS instable and in this process, in all probability the bonds between As and Fe are broken. This in turn makes their transfer to the leachate easier. The mutual ratio of both the discussed elements is also probably linked to the complicated leaching chemistry.

5.3.4 Alkaline Leaching

Complex sulphides tetrahedrite $Cu_{12}Sb_4S_{13}$, jamesonite $FePb_4Sb_6S_{14}$ and enargite Cu_3AsS_4 are categorized as sulfosalts with enargite and tetrahedrite being economically attractive because of the content of copper for extracting. However, the content of antimony and arsenic in these minerals is undesirable as pyrometallurgical smelter feed because they effect the conductivity and ductility of refined copper and also form a scum on electrolytic cells allowing impurities to carry over to the cathodes [Dayton 1982; Habashi 1996, 2002, 2003].

In this regard, alkaline leaching of the minerals in sodium sulphide medium offers a hydrometallurgical approach to solving this problem because it provides selective transfer of antimony and arsenic into leach solutions [Anderson et al. 1994; Javorka and Fröhlich 1996; Robins and Jayaweera 1992]. Chalcocite (Cu_2S), which is formed as a solid residue from these sulphides, is then a more suitable material for pyrometallurgical production of copper. Likewise, jamesonite decomposes during alkaline leaching into galena and pyrrhotite because the antimony is selectively transferred into solution. The alkaline chemistry of these minerals with Na_2S can be described by the following reactions:

$$Cu_{14}Sb_4S_{13} + 2Na_2S \rightarrow 7Cu_2S + 4NaSbS_2 \quad (5.10)$$

$$FePb_4Sb_6S_{14} + 3Na_2S \rightarrow FeS + 4PbS + 6NaSbS_2 \quad (5.11)$$

$$2Cu_3AsS_{13} + 3Na_2S \rightarrow 3Cu_2S + 2Na_3AsS_4 \quad (5.12)$$

There are several methods to enhance leachability of minerals including mechanical activation. The primary effect of mechanical activation is the comminution of solids into ultrafine dimensions. Nanosized particles lead to changes in a great number of physico-chemical properties of particular systems [Baláž 2000]. In our previous work, we found that mechanical activation of sulphides in high-energy mills enhanced the rate of arsenic and antimony extraction from different sulphidic concentrates and the recovery of metals into solution reached maximum values [Baláž et al. 1994, 1995, 1999, 2000a].

Figure 5.11 presents dependencies of specific surface area of the minerals under study as a function of mechanical activation time.

It is evident that the rate of formation of new surface is limited by both time of mechanical activation and type of sulphide. For enargite and tetrahedrite, the specific surface area increases quickly in the first 5–10 min and reaches a constant afterwards. For jamesonite, this phenomenon occurs at a lower time and at lower specific surface area.

The kinetics of new surface area formation is a very complex phenomenon-especially if milling is accompanied by physico-chemical changes which also effect the mechanism of particle interaction. Generally, in mechanically activated systems, the first stage of milling occurs with rapid increase of new surface and the second step commences with aggregation [Juhász and Opoczky 1990].

Here, the new surface area produced is not proportional to the milling time because particle interaction causes aggregation. SEM photographs in Figs. 5.12B

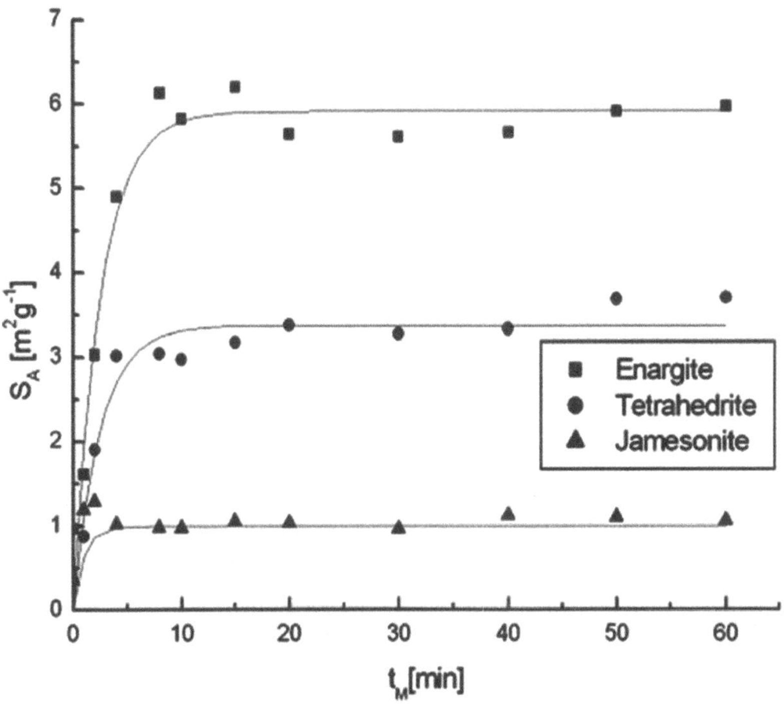

Fig. 5.11 The specific surface area, S_A of enargite Cu_3AsS_4, tetrahedrite $Cu_{12}Sb_4S_{13}$ and jamesonite $FePb_4Sb_6S_{14}$ vs. time of mechanical activation, t_M [Baláž and Achimovičová 2006a]

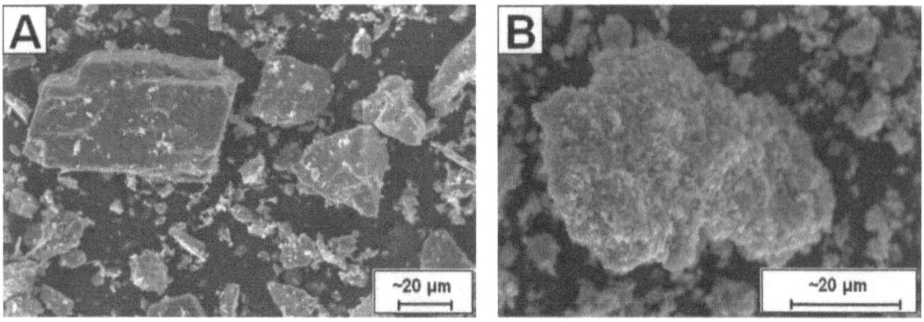

Fig. 5.12 Scanning electron micrographs of $Cu_{12}Sb_4S_{13}$. (**A**) Non-activated ($< 170\,\mu m$) and (**B**) mechanically activated (milling time 5 min) [Baláž and Achimovičová 2006a]

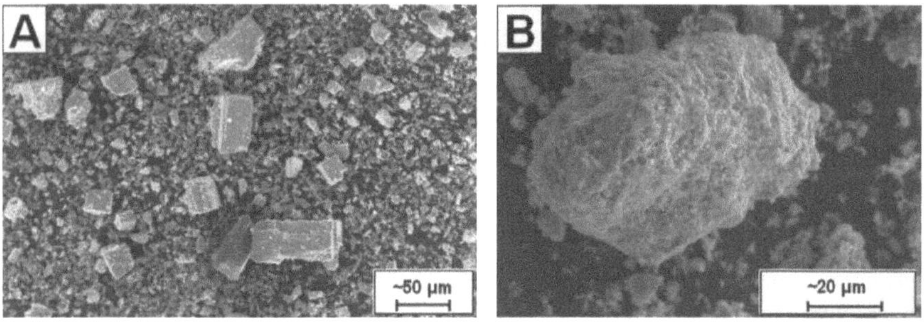

Fig. 5.13 Scanning electron micrographs of Cu_3AsS_4. (**A**) Non-activated ($< 170\,\mu m$) and (**B**) mechanically activated (milling time 5 min) [Baláž and Achimovičová 2006a]

and 5.13B show aggregates formation for tetrahedrite and enargite. The sizes of these aggreagtes even exceed the size of original particles.

The kinetics of tetrahedrite, enargite and jamesonite alkaline leaching in sodium sulphide medium according to Eqs. (5.10–5.12) was followed as a function of milling time.

5.3.4.1 Tetrahedrite $Cu_{12}Sb_4S_{13}$

From Fig. 5.14, it is evident that the influence of mechanical activation of tetrahedrite on the recovery of antimony is unambiguous. While the highest recovery after 10 min of leaching at 313 K was 7% for non-activated sample, it was 43% for sample 40 min of mechanical activation. There is a straightforward increase in the rate and recovery of antimony until approximately 15 min of milling time. This correlates with increase of surface area as shown in Fig. 5.11, however, at higher degrees of activation, this statement cannot be applied. Thus, at higher milling times, mechanical activation of the mineral surface must be taking place, which makes leaching better.

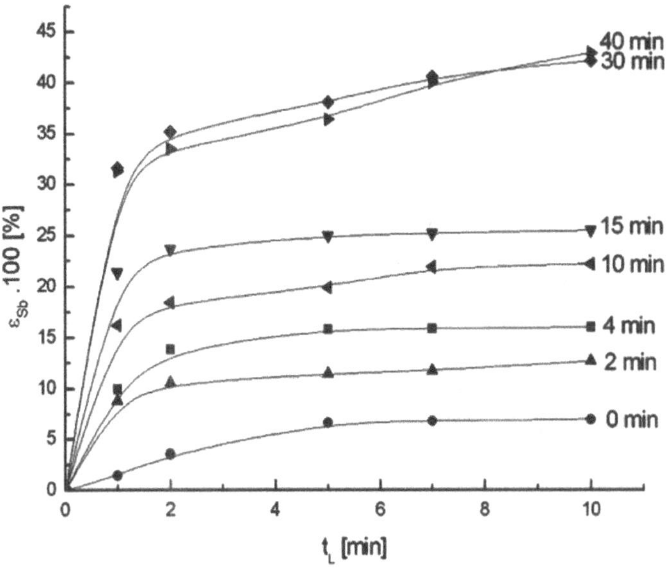

Fig. 5.14 $Cu_{12}Sb_4S_{13}$ leaching: recovery of antimony into leach, ε_{Sb}, vs. time of leaching, t_L. Temperature of leaching-313 K (milling time is given next to the leaching curves) [Baláž and Achimovičová 2006a]

5.3.4.2 Jamesonite $FePb_4S_6S_{14}$

This mineral behaves similar as tetrahedrite. For non-activated sample, the recovery of Sb is only 8%; however, for mechanically activated sample, 40 min yields Sb recovery of 63% after 10 min of leaching (Fig. 5.15). Again, the leaching data on Fig. 5.15 are compatible with dependence of surface area on milling time as shown in Fig. 5.11.

In principle, ores rich in jamesonite can be treated hydrometallurgically. Figure 5.16 shows a simplified flowsheet of this process that consists of the following operations: mechanical activation, alkaline leaching in Na_2S solution at moderate temperature and atmospheric pressure and solid/liquid separation of antimony-free lead sulphide and iron leach residue. After purification of the leach solution, the pressure oxidation and precipitation of $Na[Sb(OH)_6]$ or recovery of antimony cathodically by electrowinning are possible alternative routes. The leach residue after washing is suitable for further pyrometallurgical treatment.

5.3.4.3 Enargite Cu_3AsS_4

Considerable effect of mechanical activation on the recovery of arsenic from enargite is illustrated in Fig. 5.17 but only for shorter times of mechanical activation. The highest recovery of 87% As was achieved for sample mechanically activated 20 min

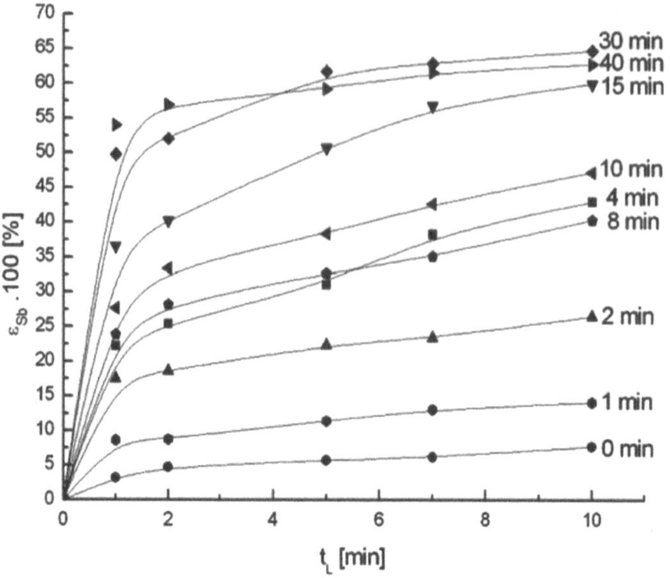

Fig. 5.15 $FePb_4Sb_6S_{14}$ leaching: recovery of antimony into leach, ε_{Sb}, vs. time of leaching, t_L. Temperature of leaching-313 K (milling time is given next to the leaching curves) [Baláž and Achimovičová 2006a]

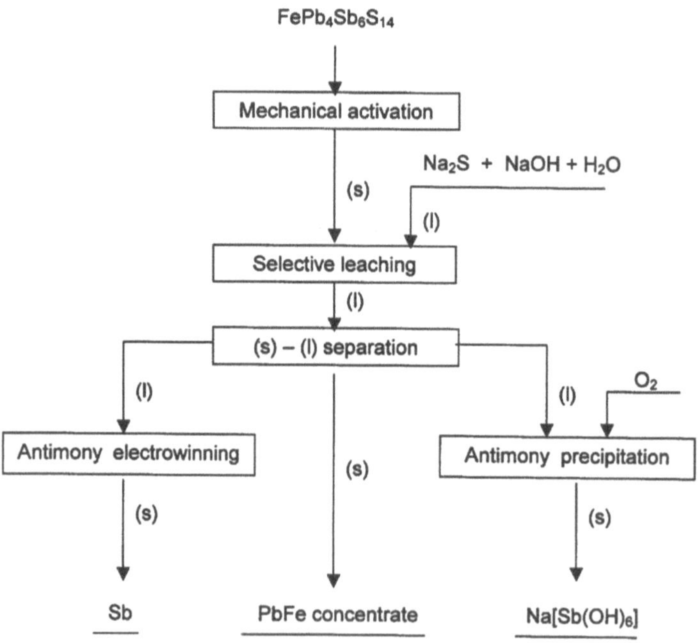

Fig. 5.16 Simplified hydrometallurgical process flowsheet of $FePb_4Sb_6S_{14}$ using alkaline sulphide [Baláž and Achimovičová 2006b]

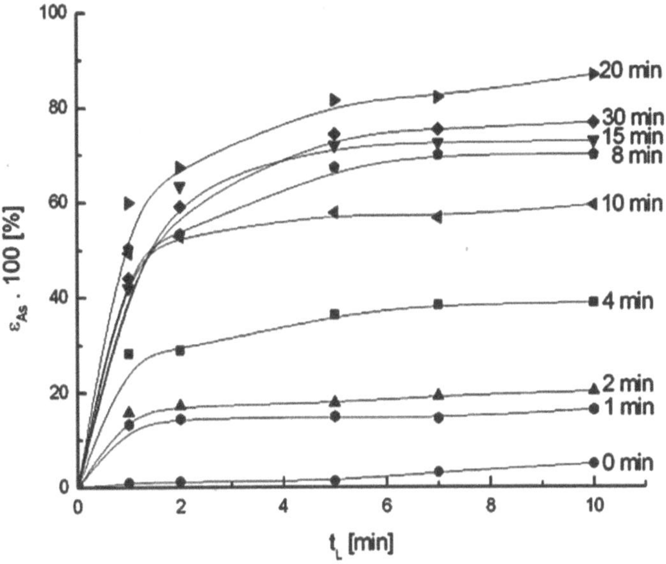

Fig. 5.17 Cu_3AsS_4 leaching: recovery of arsenic into leach, ε_{As}, vs. time of leaching, t_L. Temperature of leaching-328 K (milling time is given next to the leaching curves) [Baláž and Achimovičová 2006b]

in comparison to the non-activated sample which yielded 5% As after 10 min of leaching. However, in this case, no correlation with the dependence of surface area on milling time as documented by Fig. 5.11 has been found. On the other hand, the differences in values of arsenic recovery are not dramatic and all slopes at the beginning of leaching for samples milled for 8 min and longer are practically the same.

On the basis of the laboratory tests, a simplified flowsheet has been proposed (Fig. 5.18) involving: alkaline leaching in Na_2S solutions at moderate temperature and atmospheric pressure; solid–liquid separation of the arsenic-free copper sulphide leach residue; and pressure oxidation and precipitation of calcium or iron arsenates after purification of the leach solution. The leach residue after washing is suitable for further pyrometallurgical treatment [Baláž et al. 2003a].

5.3.5 Selective Leaching of Bismuth

Bismuth, widely used in chemicals, pigments and catalyst, is only recovered as a by-product. World output is about 6500 t/y principally from China (as a by-product in tungsten concentrates). Modest quantities are produced in Peru by pyrometallurgical way [Ellis 2004].

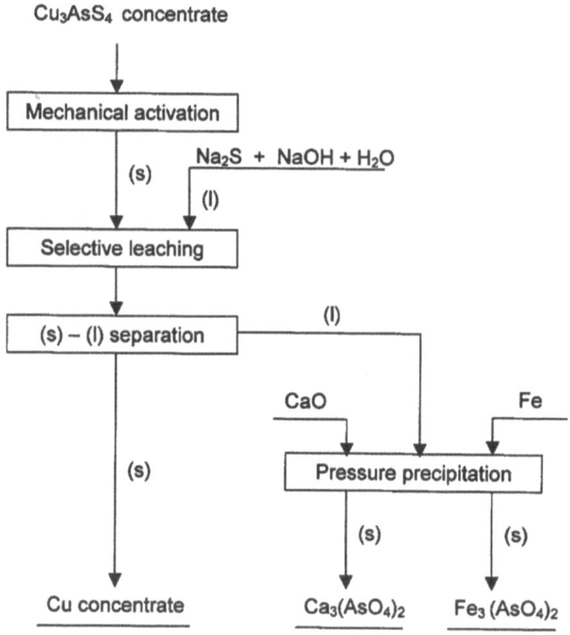

Fig. 5.18 Simplified hydrometallurgical process flowsheet of Cu_3AsS_4 concentrate using alkaline sulphide [Baláž et al. 2003a]

Bismuth presence is very frequently associated with the occurence of lead sulphide. However, when lead is pyrometallurgically recovered from sulphide concentrates, bismuth is a harmful element. That is why selective leaching of bismuth makes sence. The key stage for Bi leaching from the concentrate is the way of its pretreatment and selection of leaching agent. HCl and (H_2SO_4+NaCl) belong among the fundamental leaching agents for Bi dissolution [Jansen and Taylor 1995; Medkov et al. 1999].

The possibility of Bi leaching from lead sulphide concentrate (1.32% Bi, 51.3% Pb, 1.54% Cu, 8.42% Zn, 3.91% Fe, 2210 g^{-1} Ag and 18.75% S) was examined. Different leaching agents and mills have been tested in order to obtain selective extraction of bismuth with its high recovery into leach. The results are summarized in Table 5.2.

The results show that selectivity of Bi extraction was low at applying HCl and (H_2SO_4+NaCl) leaching agents despite of sufficient Bi recovery in several cases. However, with sodium thiosulphate $Na_2S_2O_3$ solution applied at 25°C and mechanochemical pretreatment made in an attritor it was possible to achieve 99% recovery of Bi even in 3 min of leaching time (Fig. 5.19). With the exception of copper where the recovery was 22%, the recoveries for all the other metals were 5% Pb, 1% Zn, 0.5% Fe and 1% Ag. It illustrates very good selectivity of sodium thiosulphate application for Bi recovery.

Table 5.2 Specific surface area, S_A, leaching conditions (leaching time 60 min, solid/liquid ratio 1:400; 0.2 M HCl; H_2SO_4 = 15 gL^{-1}, NaCl = 200 gL^{-1}, $Na_2S_2O_3 \cdot 5H_2O$ = 124 gL^{-1}; T=25°C) and recovery of metals into leach, ε_{Me} for lead sulphide concentrate mechanically activated in different mills (leaching time 60 min) [Baláž et al. 2005a]

Mill/milling time	S_A [m^2g^{-1}]	Leaching reagents	ε_{Me} [%]					
			Bi	Pb	Cu	Zn	Fe	Ag
–	0.7	HCl	1.8	3.7	0.2	1.2	7.6	0.6
		H_2SO_4+NaCl	60	7.5	24	2	17	1.6
		$Na_2S_2O_3.5H_2O$	38	3.5	5	2	1	0.6
Eccentric vibration mill/15 min	1.5	HCl	1.8	11	0.1	3.5	19	0.6
		H_2SO_4+NaCl	92	50	4	10	32	2
		$Na_2S_2O_3.5H_2O$	47	2	6.5	0.5	1.6	2
Planetary mill/15 min	2.9	HCl	1.8	11	0.1	4.1	13	0.6
		H_2SO_4+NaCl	63	42	7.3	9.2	25	2
		$Na_2S_2O_3.5H_2O$	57	3.6	12	1	3	1.6
Attritor mill/30 min	12	HCl	98	33	3	4	71	0.6
		H_2SO_4+NaCl	99	57	4	7	87	1.5
		$Na_2S_2O_3.5H_2O$	99	5	22	1	0.5	1

The plot in Fig. 5.20 describes the effect of new surface area formation on recovery of Bi as a consequence of mechanical activation in different mills. The plot appears to be linear for all activated as well as non-activated samples of the concentrate. It probably suggest that Bi recovery is simply due to the increase of surface

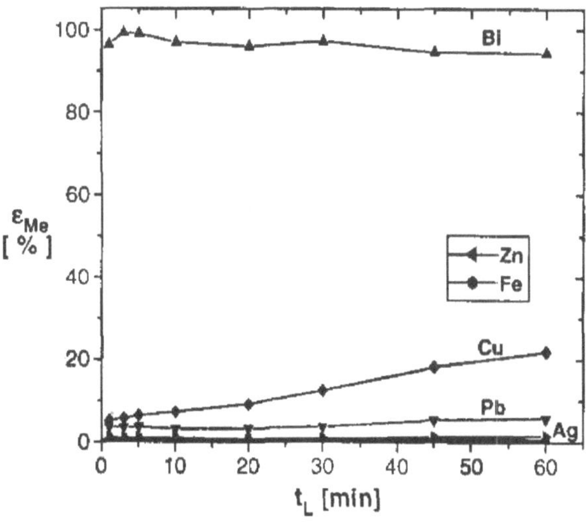

Fig. 5.19 Recovery of Bi, Cu, Pb, Zn, Fe and Ag, ε_{Me} vs. leaching time, t_L for lead sulphide concentrate mechanically activated in an attritor for 30 min, leaching temperature 25°C [Ficeriová et al. 2005a]

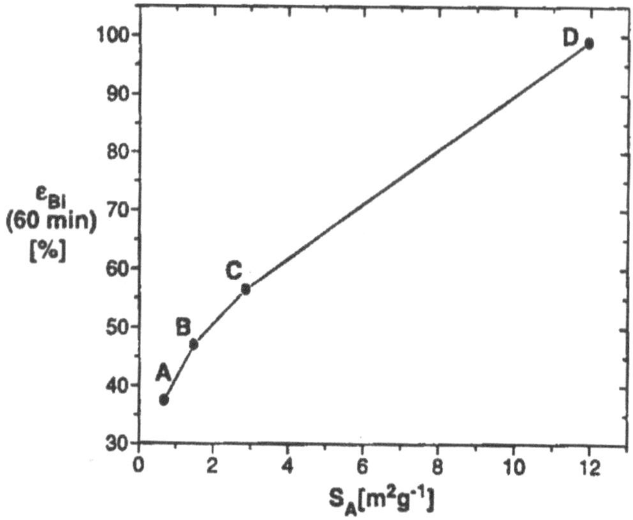

Fig. 5.20 Recovery of bismuth after 60 min leaching, ε_{Bi} in sodium thiosulphate solution at 25°C vs. specific surface area, S_A in various types of mills: "As-eceived" sample (A); Eccentric vibratory mill (B); planetary mill (C) and attritor-stirring ball mill (D) [Baláž et al. 2005a]

area. The highest value of Bi recovery was obtained for the sample with the highest surface area using milling in an attritor. It was shown [Beckstead et al. 1976] by milling of the chalcopyrite concentrate that the limit for ball milling was $4\,m^2g^{-1}$, whereas the specific surface area of attritor-milled products continues to increase to at least $12\,m^2g^{-1}$.

These values are very close to the obtained results. It is known that milling in aqueous environment and/or the use of small mill balls (attritor) is more favourable for new surface formation whereas dry milling and/or the use of larger mill balls (vibration mill and/or planetary mill) favour bulk disorder of minerals [Baláž 2000; Tkáčová 1989; Chodakov 1972].

5.3.6 Non-Cyanide Leaching of Precious Metals

5.3.6.1 Gold

Gold occurs in sulphides in a wide range of forms. It may be physically included in intercrystalline spaces of sulphides that result from defects in their structure, or it can be chemically bonded in solid solutions and compounds. There are a lot of sulphides in which gold is present as an admixture. Leaching, as a key step in the hydrometallurgy of gold, belongs to the heterogeneous reactions of the solid–liquid type, and its course can be influenced through the solid, as well as liquid phase.

Chemical, biological and physical pretreatments are applied as intervention steps directed to the solid phase, the goal of which is to change the composition and the particle size of the gold-bearing sulphides and thus, to facilitate subsequent leaching [Marsden and House 1992; La Brooy et al. 1994]. Simultaneously, with examination of chemical, biological and physical pretreatment, new processes of mechanochemical pretreatment (Fig. 5.21) are being successfully applied to fundamental research, as well as to plant operations. In this process the minerals are subjected to high-energy milling. The milling results in particle disintegration and chemical or physicochemical transformations, which significantly affect the subsequent mineral processing operations [Tkáčová 1989; Baláž et al. 1995, 1996; Welham 1996; Linge and Welham 1997; Baláž et al. 1998a; Welham 1998; Baláž 2000].

Alkaline cyanidation continues to be dominant method in hydrometallurgy for gold dissolution [Ubaldini et al. 1996; Habashi 2008]. Cyanide leaching has dominated gold processing for over 100 yeras and will probably continue to do so in the future, despite the fact that the cyanide is coming under the close scrutiny of environmental legislators [Potter and Salisbury 1974].

The use of thiosulphate as a gold leachant represents an alternative method [Marsden and House 1992; La Brooy et al. 1994; Hiskey and Atluri 1988; Abbruzzese et al. 1995; Breuer and Jeffrey 2000]. Economic and technical evaluation of plant tests of the Patera and Newmont processes showed great promise for thiosulphate leaching [Block-Bolten and Torma 1986; Wan and Brierley 1997]. Leaching of gold in thiosulphate solution results in the formation of a stable complex and is described by the equation

$$Au + 5S_2O_3^{2-} \rightarrow Au(S_2O_3)_2^{3-} + e^- \quad (5.13)$$

The dissolution step in ammoniacal thiosulphate solution is an electrochemical reaction and is promoted by the presence of copper ions [Aylmore and Muir 2001;

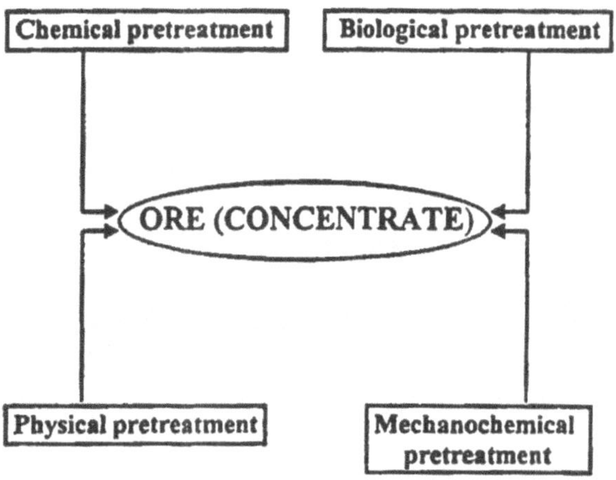

Fig. 5.21 Different ways of pretreating precious metal-bearing ores (concentrates) [Baláž 2000]

Aylmore 2001]. The role of copper (II) ions in the oxidation of metallic gold is shown in the following reaction

$$Au + 5S_2O_3^{2-} + Cu(NH_3)_4^{2+} \rightarrow Au(S_2O_3)_2^{3-} + 4NH_3Cu(S_2O_3)_3^{5-}. \quad (5.14)$$

A gold-bearing CuPbZn complex sulphide concentrate from Banská Hodruša (the Svetozár vein, Slovakia) was selected as a model material for testing the effect of mechanochemical pretreatment on the subsequent thiosulphate leaching of gold. The chemical composition of the concentrate was as follows: 353 gt^{-1} Au, 170 gt^{-1} Ag, 0.93% Cu, 4.08% Pb, 3.57% Zn, 20.06% Fe, 44.15% S, 0.2% Sb, 0.17% Hg, 0.02% Bi, 0.12% As, 0.03% Mn, 0.02% Co, 0.07% Mg and 5.5% SiO_2.

Mineralogical analysis (Fig. 5.22) showed the presence of chalcopyrite $CuFeS_2$, galena PbS, sphalerite ZnS, tetrahedrite $Cu_{12}Sb_4S_{13}$, pyrite FeS_2 and quartz SiO_2 in the concentrate. Gold occurs primarily free in the form of wiry, flat and flaky aggregates filling up the intergrain space in sulphides, carbonates and quartz. Some small gold inclusions are also present in the sulphides, predominantly sphalerite and galena [Jágerský 1999]. A small amount of gold is associated with chalcopyrite, while pyrite is regarded as barren in this respect. The investigation of the presence of invisible gold was beyond the scope of this chapter.

Figure 5.23 shows the effect of leaching time on gold recovery for various energy inputs during mechanical activation experiments. In the as-received concentrate, only 54% of the gold were recovered after 120 min leaching (curve 1). The results for the mechanically activated samples (curve 2–4) indicated that the

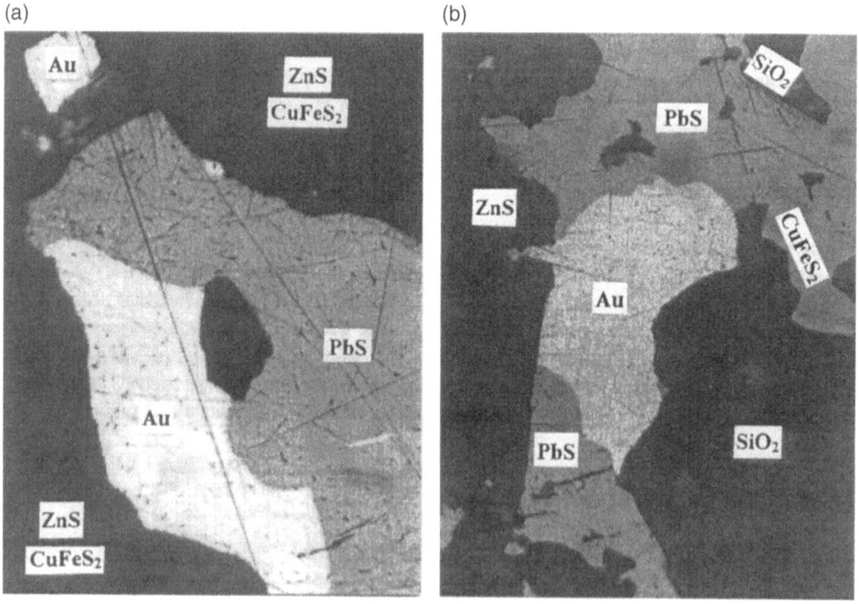

Fig. 5.22 Gold associations with sulphide minerals [Ficeriová et al. 2002]

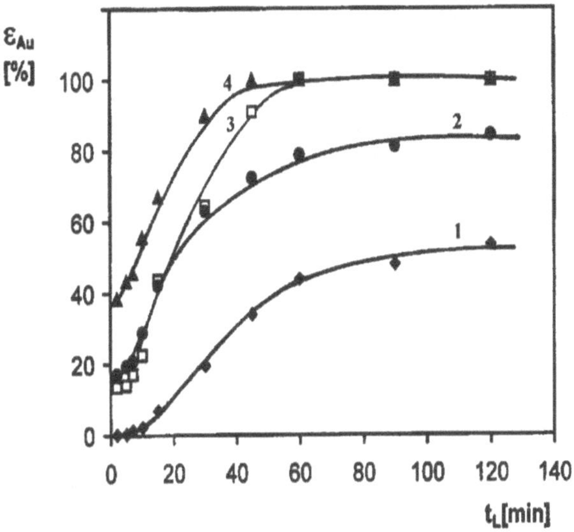

Fig. 5.23 Recovery of gold, ε_{Au} versus leaching time, t_L for mechanically activated concentrates. Energy input, E_M: (1) 0 kWht^{-1} (as-received sample), (2) 202 kWht^{-1}, (3) 335 kWht^{-1} and (4) 403 kWht^{-1} [Ficeriová et al. 2002]

physico-chemical changes of gold bearing minerals brought about an acceleration of the process of thiosulphate leaching. It was possible to achieve a gold recovery of 99% within 1 h for activated samples (curve 3 and 4).

Figure 5.24 shows the relationship between the rate constant of leaching and the energy input of milling of the mechanically activated samples investigated. The

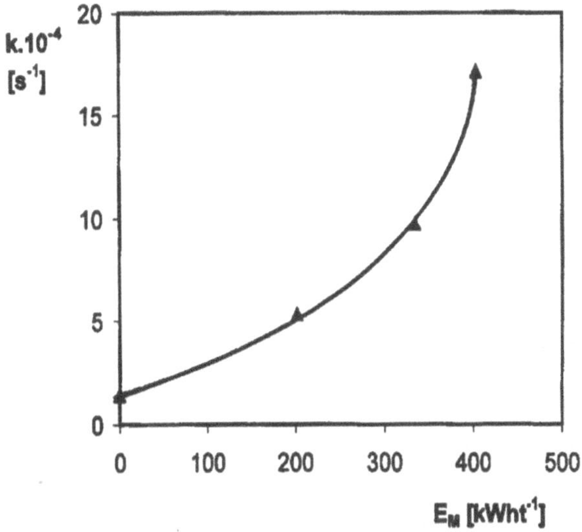

Fig. 5.24 Rate constant of gold leaching (k) versus energy input (E_M) [Ficeriová et al. 2002]

results show that the extraction of gold from CuPbZn concentrate strongly depends on energy consumption by milling. It is important to note that the values of energy input were calculated for the batch attritor whereas the actual energy input for continuous operating attritors are usually 10 times lower [Baláž 2000].

The physicochemical changes of CuPbZn concentrate due to mechanical activation have an influence on both the rate of extraction and the recovery of gold from this gold-bearing concentrate when leached with ammonium thiosulphate. It was possible to obtain 99% gold recovery after 45 min leaching of an activated sample, which compares very favourably with 54% recovery from the as-received concentrate in 120 min.

5.3.6.2 Silver

There are at least 200 minerals bearing silver in major, minor and trace amounts. However, of these, only 10–12 minerals are of practical importance. In order of leachability, these are elemental silver, silver halides and silver sulphides [Marsden and House 1992]. The sulphides occurring in the form of sulfosalts (proustite, pyrargyrite, tennantite, tetrahedrite, etc.) cause considerable problems in the leaching of silver. Classical cyanide leaching of these minerals results in 5–10% Ag extraction (Lodejščikov and Ignateva 1973).

Several silver-bearing complex sulphide concentrates were used to test the effect of mechanical activation and mechanochemical leaching on the subsequent leaching of silver (Table 5.3).

The leaching of silver in thiourea $CS(NH_2)_2$, thiosulphate $Na_2S_2O_3$ and cyanide NaCN solutions can be described by equations [Murthy and Prasad 1996].

$$Ag + Fe^{3+} + 3CS(NH_2)_2 = Ag\left[CS\left(NH_2\right)_2\right]_3^+ + Fe^{2+} \tag{5.15}$$

$$2Ag + 4S_2O_3^{2-} + H_2O + 1/2O_2 = 2Ag(S_2O_3)_2^{3-} + 2OH^- \tag{5.16}$$

$$2Ag + 4CN^- + H_2O + 1/2O_2 = 2Ag(CN)_2^- + 2OH^- \tag{5.17}$$

Figure 5.25 shows the effect of the leaching time on silver recovery for various leaching medium for as-received sample. The leaching for 180 min in an cyanide solution resulted in 3.5% Ag dissolution. In solution of thiourea and ammonium thiosulphate, 5 and 5.7% Ag was dissolved.

Table 5.3 Chemical composition of the concentrates under study [Ficeriová et al. 2005b]

Concentrates	Components (%)										
	Ag (gt^{-1})	Au (gt^{-1})	Bi	Cu	Pb	Zn	Fe	Sb	As	S	SiO_2
Casapalca (Peru)	15500	0.9	–	19.5	16.5	8.8	3.4	8.3	3.2	29	0.7
B.Hodruša (Slovakia)	353	170	0.02	0.9	4.1	3.6	20	0.2	0.12	44	5.5
Atacocha (Peru)	2210	–	1.3	1.5	51.3	8.4	3.9	–	–	19	–

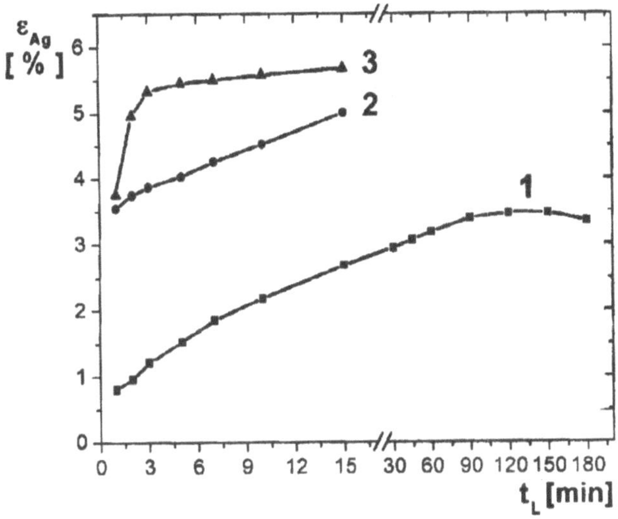

Fig. 5.25 Recovery of silver into leach, ε_{Ag} vs. leaching time, t_L for the as-received sample. Medium: 1 – NaCN, 2 – $CS(NH_2)_2$, 3 – $(NH_4)_2S_2O_3$ [Ficeriová et al. 2004]

Figure 5.26 indicates that the mechanically induced physicochemical changes of silver-bearing minerals brought about an acceleration of the process of the subsequent leaching. By application of thiosulphate and thiourea leaching, 99% recovery of Ag was reached already after 3–5 min of leaching (curves 2, 3). In cyanide medium, recovery of Ag was only 40% during 180 min of leaching (curve 1).

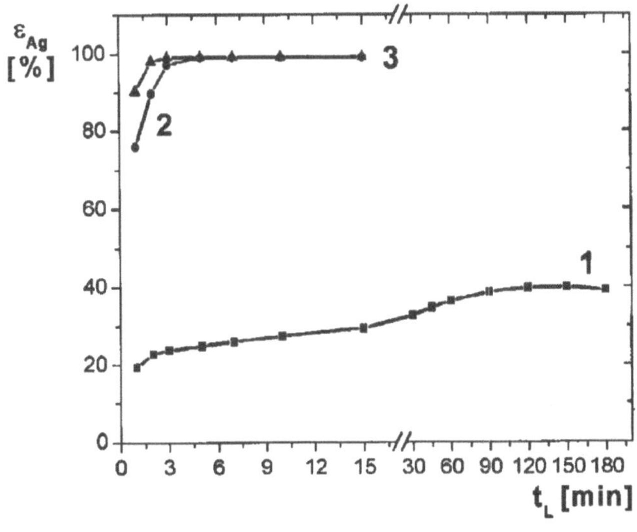

Fig. 5.26 Recovery of silver into leach, ε_{Ag} vs. leaching time, t_L for mechanochemically pretreated (60 min) sample. Medium: 1 – NaCN, 2 – $CS(NH_2)_2$, 3 – $(NH_4)_2S_2O_3$ [Ficeriová et al. 2004]

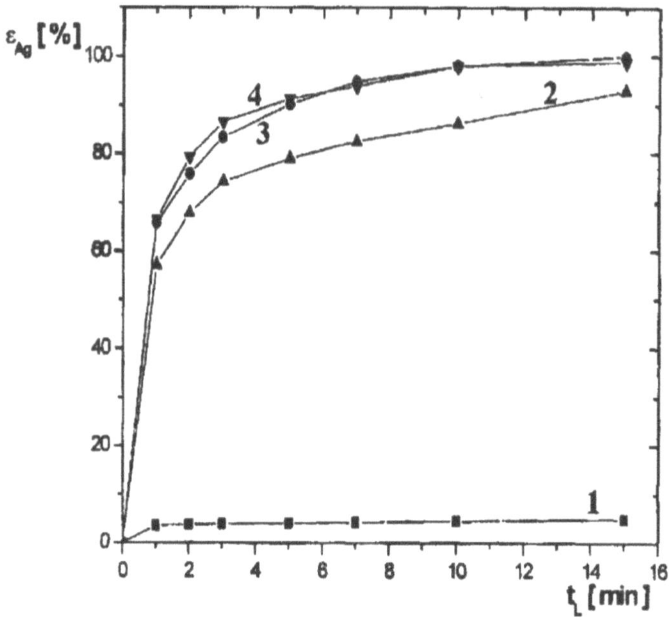

Fig. 5.27 Recovery of silver into leach, ε_{Ag} vs. leaching time, t_L for the mechanochemically activated samples using milling energies of: 0 kWh ton^{-1} as-received sample (1); 562 kWh ton^{-1} (2); 1123 kWh ton^{-1} (3); 2246 kWh ton^{-1} (4) [Baláž et al. 2003b]

The dependence of silver recovery in thiosulphate solutions on milling time is shown in Fig. 5.27 from where it can be seen that only 5% Ag was dissolved from the "as-received"concentrate. However, over 90% of the silver could be solubilised within 10 min from the mechanochemically treated concentrates. A maximum recovery of 99% was achieved with the most energetically treated sample.

The degree of tetrahedrite decomposition D can be defined by the equation

$$D = \left(1 - \frac{Sb_x}{Sb_0}\right).100[\%] \tag{5.18}$$

where Sb_x and Sb_0 are the contents of antimony in mechanochemically treated tetrahedrite and in as-received sample, respectively.

The silver leaching curves were fitted and the rate constants were calculated [Baláž et al. 2003b]. Figure 5.28 shows the dependence of the rate constant of silver leaching (k_{Ag}) on parameter D calculated by Eq. (5.18). It follows from this plot that there is certainly some structural sensitivity for silver leaching from tetrahedrite by thiourea solution. If the rate constant of leaching remains constant with respect to the mineral decomposition as shown in the left part of Fig. 5.28 ($D < 30\%$), the reaction rate is insensitive to structural changes. However, it follows from the right part of Fig. 5.28 ($D > 30\%$) that the tetrahedrite decomposition contributes to the silver leachabality and the reaction is structural sensitive. Similar results were obtained

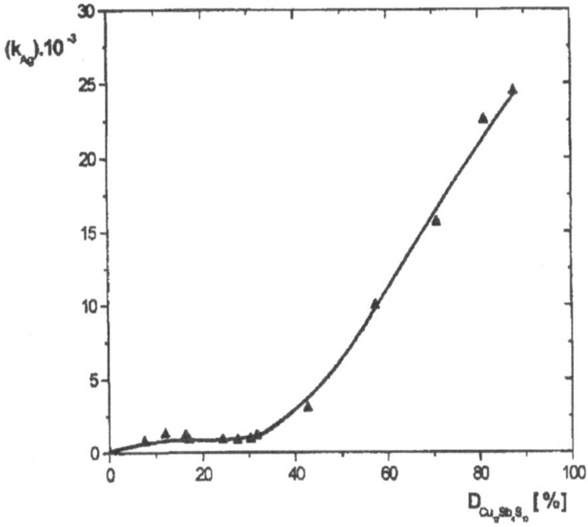

Fig. 5.28 Rate constant, k_{Ag} for silver leaching vs. degree of mechanochemical destruction D of $Cu_{12}Sb_4S_{13}$ [Baláž et al. 2003b]

for thiosulphate leaching of mechanically activated tetrahedrite where the structural sensitivity has been evident for $D > 60\%$ [Ficeriová et al. 1998].

5.3.7 Simultaneous Milling and Leaching (Mechanochemical Leaching)

During mechanical activation the crystal structure of a minerals usually disordered and generation of defects or other metastable forms can be found. These effects are not stable and have different relaxation times (Fig. 5.29). If the process of mechanical activation is separated from leaching in time, then a number of highly excited states may form and decay before leaching. This is in accordance with the concept of slowly changing states after interrupting mechanical activation [Lyakhov 1993]. According to this concept, mechanical activation deals only with the long lived slowly relaxing states (Fig. 5.30). All the short-lived states are not at disposal for subsequent chemical leaching.

On the other hand, if the mechanical activation and chemical leaching are integrated into a common step all the excitation states can be utilized. In addition to the improvement of milling performance the leaching agent serves also as milling additive. Thus there is a possibility that a common mechanical activation and leaching step-called mechanochemical leaching-contributes to operational benefits and economics of the overall process.

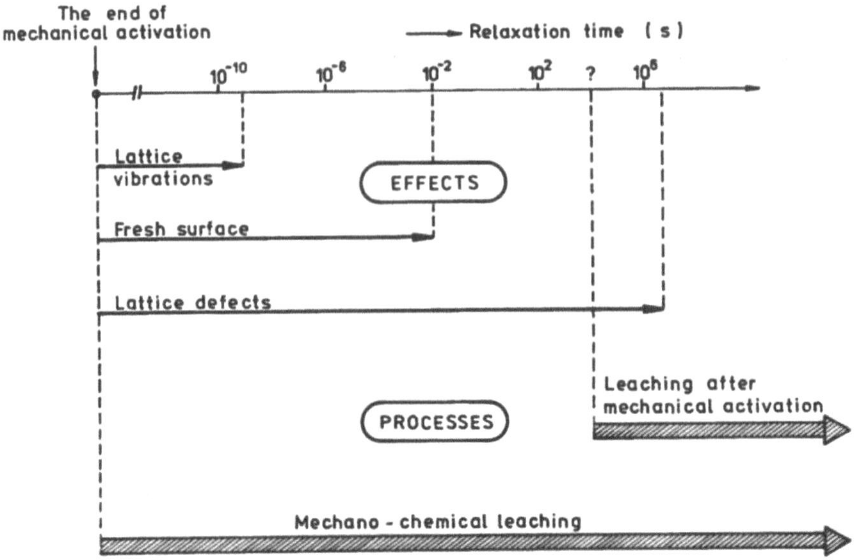

Fig. 5.29 The period and duration of excitation states (effects) after termination of mechanical activation [adapted after Heinicke 1984]

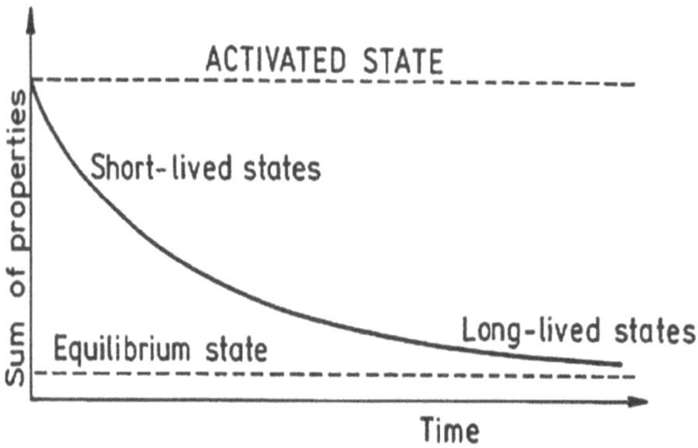

Fig. 5.30 A generalized relaxation curve for mechanically activated solid [Lyakhov 1993]

5.3.7.1 Tetrahedrite $Cu_{12}Sb_3S_{13}$

The principle of mechanochemical leaching of tetrahedrite has been verified in an attritor working in a batch mode [Baláž et al. 1994, 1998b]. The experiments with alkaline leaching of tetrahedrite concentrate under conditions close to hydrometallurgical practice also showed good results. The data presented in Fig. 5.31 illustrate that nearly 80% Sb extraction can be achieved by mechano-chemical leaching for 70 min, compared with 40% Sb extraction for chemical leaching for 120 min.

This high extraction of antimony means that only a small fraction of tetrahedrite remains in the solid residue, which corresponds with the XRD data in Fig. 5.32,

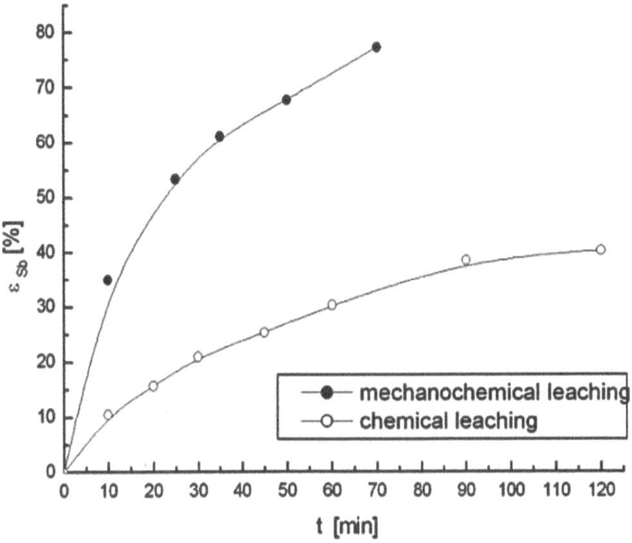

Fig. 5.31 Recovery of Sb from $Cu_{12}Sb_4S_{13}$ concentrate leach, ε_{Sb} vs. time, t [Baláž and Achimovičová 2006b]

where only small peaks of tetrahedrite (T) were detected. Copper sulphide has not been detected by XRD but its occurrence is possible in an amorphous state.

The concept of mechanochemical leaching of tetrahedrite concentrates (named Melt process) was developed and verified in a laboratory attritor [Baláž et al. 1994] and in semi-industrial attritor [Baláž et al. 1995, 1997; Baláž 2000]. It was further tested in a pilot plant hydrometallurgical unit in Rudňany (Slovakia). For more details in this technology see Sect. 6.3.3.

5.3.7.2 Celestite $SrSO_4$

Celestite ($SrSO_4$) and strontianite ($SrCO_3$) are the two naturally occurring strontium minerals but only the former is commercially mined and is mainly used for the production of strontium carbonate, from which other strontium chemicals (strontium nitrate, chloride, hydroxide, etc.) are obtained (Hong 1993).

Strontium carbonate can be synthesized according to the equation

$$SrSO_4 + Na_2CO_3 \rightarrow SrCO_3 + Na_2SO_4 \qquad (5.19)$$

Mechanochemical leaching has been performed according to the flowsheet in Fig. 5.33.

Figure 5.34 shows the XRD patterns of the original celestite concentrate and the residues obtained after 10, 20 and 40 min of wet milling at constant $SrSO_4$:Na_2CO_3 mole ratio of 1:1.2. The as-received sample immediately starts to be converted into

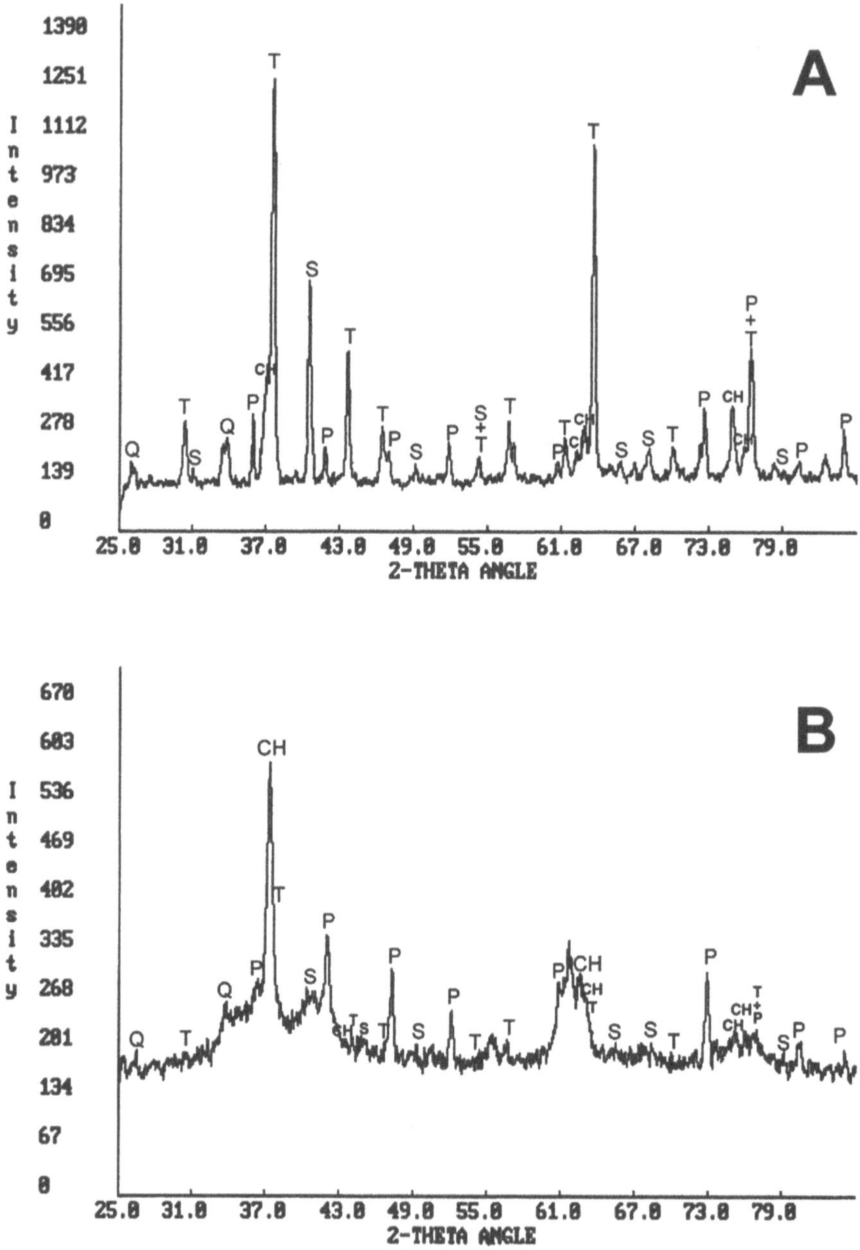

Fig. 5.32 XRD of $Cu_{12}Sb_4S_{13}$ concentrate: **A** – non-activated, **B** – after mechano-chemical leaching: Q – quartz, T – tetrahedrite, CH – chalcopyrite, P – pyrite, S – siderite [Baláž and Achimovičová 2006b]

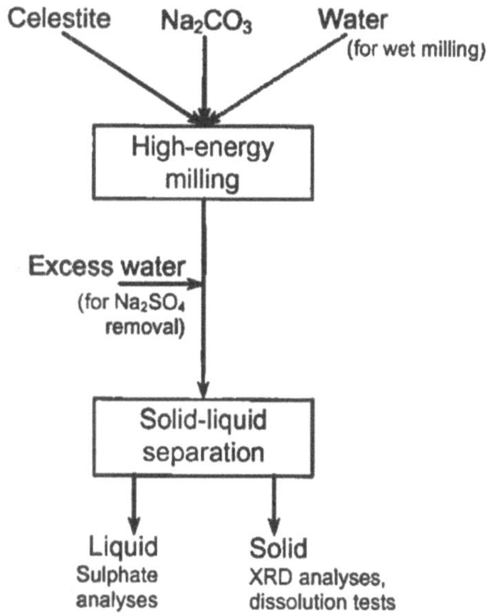

Fig. 5.33 Simplified hydrometallurgical flowsheet of $SrSO_4$ concentrate treatment [Obut et al. 2006]

$SrCO_3$ with the initiation of high-energy milling and after 20 min of wet milling most of the conversion is practically completed.

The arrows in Fig. 5.34 show the main celestite peaks used to follow the conversion reaction. The peak intensities, especially the peak near $2\Theta = 27.15°$, decrease with milling time and after 40 min of milling the peaks of celestite disappear, which is in good agreement with the results of dissolution tests.

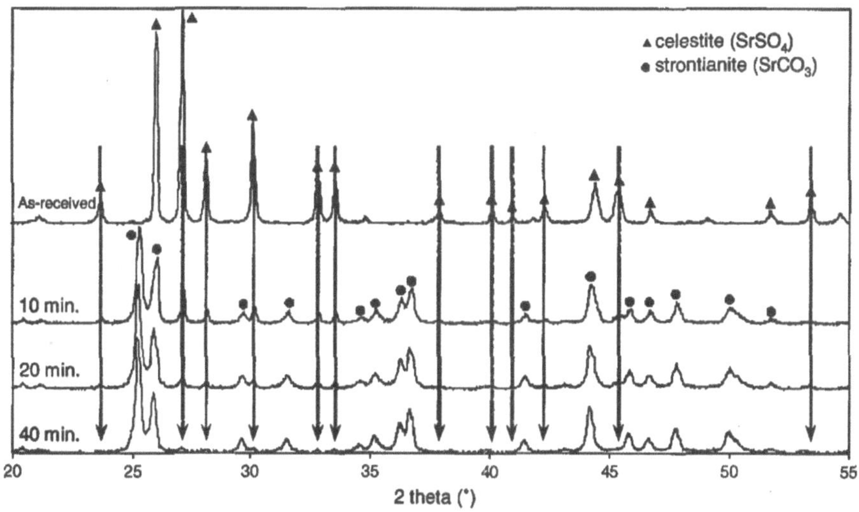

Fig. 5.34 XRD of celestite concentrate [Obut et al. 2006]

5.4 Sorption

The sorption of heavy metals in the environment can cause serious environmental pollution problems. Many methods have been proposed for heavy metal removal, e.g. precipitation, membrane filtration, ion exchange and adsorption [Habashi 1993]. Removal of heavy metal based on sorption on mineral surfaces has been suggested as an alternative approach [Zoubolis and Kydros 1993a, b; Mellah and Chegrouche 1997; Garcia-Sanchez et al. 1999; Vengris et al. 2001; Prieto et al. 2003]. The adsorption process is used especially in the water treatment field and the investigation has been made to determine inexpensive and good adsorbents. Sorption is defined as a surface process irrespective of mechanism, adsorption or precipitation. Both mechanism involve characteristic reactions of some metals with solid surfaces.

The properties of fine particles with a high surface area has led to their utilization as sorbents for heavy metals [Baláž 2000]. While many physical properties of these particles are now well understood, the influence on the chemical properties such as adsorption and chemical reactivity remain poorly understood [Zhang et al. 1999].

5.4.1 Zinc Sorption on Calcite

The modification of calcite solid-state properties by high-energy milling and their influence on the process of zinc adsorption has been studied [Baláž et al. 2005b].

The dependence of the zinc uptake E on milling time for planetary and vibratory mill and corresponding surface and bulk changes of calcite during mechanical activation are given in Table 5.4.

More drastic changes in surface and bulk of calcite samples induced in mineral by planetary milling cause the higher zinc uptake in comparison with less pronounced effect of vibratory milling. Generally, there is a strong evidence of positive influence of the overall disordering of calcite by milling on zinc uptake during sorption experiments as manifested in Fig. 5.35.

Table 5.4 Zinc uptake, E specific surface area, S_A crystallinity, X and S_A/X ratio for calcite milled in a planetary and vibratory mill, respectively, t_M-milling time [Baláž et al. 2005b]

Mill	t_M (min)	E (%)	S_A (m^2g^{-1})	X (%)	S_A/X (m^2g^{-1})
Planetary	0	35	0.9	100	0.9
	0.6	59	4.3	71	6.1
	3	65	6.0	62	9.7
	15	69	5.7		
	30	85	4.9	40	12.4
Vibratory	0	30	0.1	100	0.1
	10	40	3.3	59	5.6
	15	49	4.0	55	7.3
	30	52	4.9	46	10.7

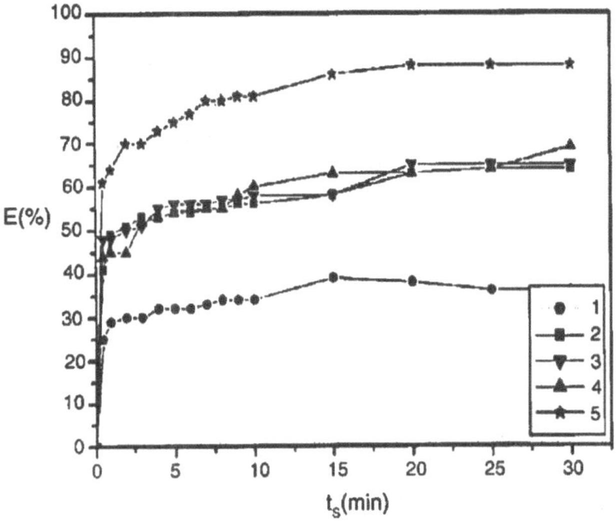

Fig. 5.35 Zinc uptake, E vs. sorption time, t_S for calcite milled in a planetary mill, temperature 25°C, milling time: (1) 0 min, (2) 0.6 min, (3) 3 min, (4) 15 min and (5) 60 min [Baláž et al. 2005b]

The Freundlich model is frequently used to describe the adsorption of gas molecules onto metal surfaces [Freundlich 1906]. However, the model has found successful application in many other sorption processes, including sorption from liquid media. The Freundlich isotherm robustly fits most experimental adsorption-desorption isotherm curves, and is especially good for fitting data from highly heterogeneous sorbent systems, including minerals. In a "batch" aqueous adsorption experiments, the data may be fitted to the Freundlich adsoprtion isotherm.

The shape of this isotherm strongly suggest that a description of the adsorption is in a form

$$a = kc^{1/n} \tag{5.20}$$

where a is the amount of metal ion adsorbed per unit weight, k and n the Freundlich constants and c is the equilibrium concentration of Zn^{2+}. Equation (5.20) can be linearized to the form

$$\log a = \log k + \frac{1}{n} \log c \tag{5.21}$$

The slope of line expressed by Eq. (5.21) gives the value 1/n and intercept of the line gives the value log k. The Freundlich constants for zinc ions sorption and correlation coefficient were calculated.

The Freundlich isotherm is an empiric formula with two parameters k and n, respectively. The main drawback for the isotherm is that the parameters have not exact meaning and have to be determined experimentally. However, recently a fundamental analysis of the isotherm for the adsorption of phenolic compounds on thermally activated carbon and correlation of its parameters with molecular properties of adsorbate and adsorbent have been performed [Otake et al. 2001]. The modified

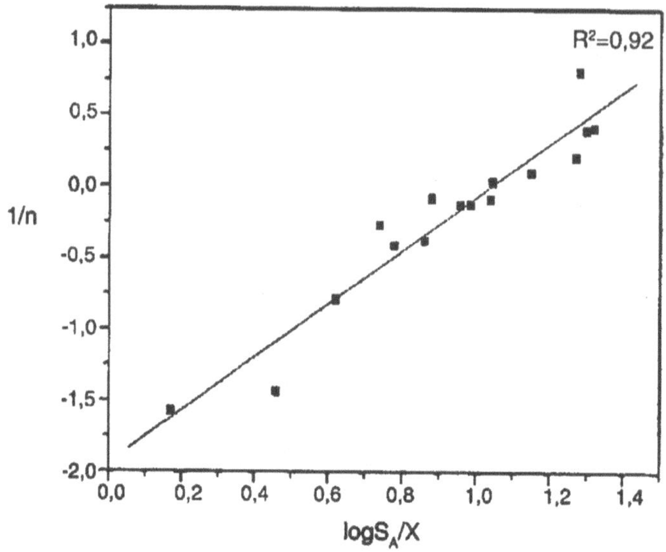

Fig. 5.36 Linear correlation between constant 1/*n* of Freundlich equation and logarithm of specific surface area and crystallinity ratio, S_A/X, calcite milled in a planetary mill [Baláž et al. 2005b]

Freundlich parameter (1/*n*) was found to have an inverse linear relationships with the electron density of phenolics calculated from molecular orbital theory [Furuya et al. 1997].

In this chapter, the different approach has been applied. We have modified the solid-state properties of calcite by mechanochemical activation and characterized it by the structure sensitivity parameters S_A/X (see Sect. 3.1.3). One solute (Zn^{2+} ions) was then applied for a series of mechanically activated calcites of different origin. The Freundlich parameter, 1/*n* was calculated from the linearization of adsorption isotherms and plotted against structure sensitivity parameter, $\log(S_A/X)$. The good linear correlation between 1/*n* versus $\log(S_A/X)$ values has been found (Fig. 5.36).

In the paper by Fyruya et al. Freundlich parameter (1/*n*) was related to the affinity between the adsorbate and the adsorbent [Furuya et al. 1997]. In our work, there is an evidence of relation of this parameter with the solid-state properties of adsorbent.

5.4.2 Sulphides

The special properties of fine particles with a high surface area and their significant occurence as the by-product waste from mixed sulphides processing plants has led to the suggestion that they may be utilized as sorbents for toxic metals [Zoubolis et al. 1993a]. The pyrite fines which are usually stockpiled in the mine area for copper ion separation and arsenic removal from solutions were studied [Zoubolis et al. 1992, 1993b].

Exchange Reactions of the Type

$$Me_IS + Me_{II}SO_4 \rightarrow Me_ISO_4 + Me_{II}S \tag{5.22}$$

(Me = Fe, Ni, Cu, Co, Ni, platinum metals, etc.) usually take place with sulphides Me_IS in the presence of excess free acid at increased temperature and pressure. Pentlandite, pyrrhotite, chalcopyrite, pyrite, sphalerite and galena were tested as active sorbents of type Me_IS [Kulebakin 1988].

The estimation of sorption properties has shown that mechanically activated pyrrhotite is an exceedingly efficient sorbent and for the sorption of platinum metals a high activity and selectivity can be achieved. According to [Kulebakin 1988] the capacity of mechanically activated pyrrhotine for platinum metals is as follows (mgg^{-1}): Os – 3000; Pd – 2300; Pt – 23.2; Rh – 14.0; Ru – 5.0; Ir – 2.2. A pyrrhotite concentrate was effectively used for copper sorption from solutions using the unique sorption properties of activated pyrrhotite in the two stage counter-current flowsheet shown in Fig. 5.37.

When nickel powder is in contact with the copper ions the cementation reaction proceeds by the equation

$$Cu^{2+} + Ni \rightarrow Cu + Ni^{2+} \tag{5.23}$$

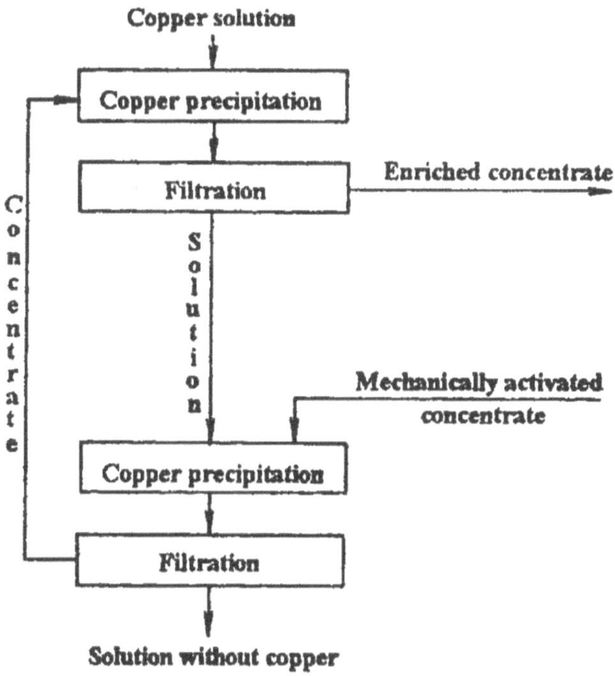

Fig. 5.37 The counter-current flowsheet of copper sorption by mechanically activated pyrrhotite concentrate [Kulebakin 1988]

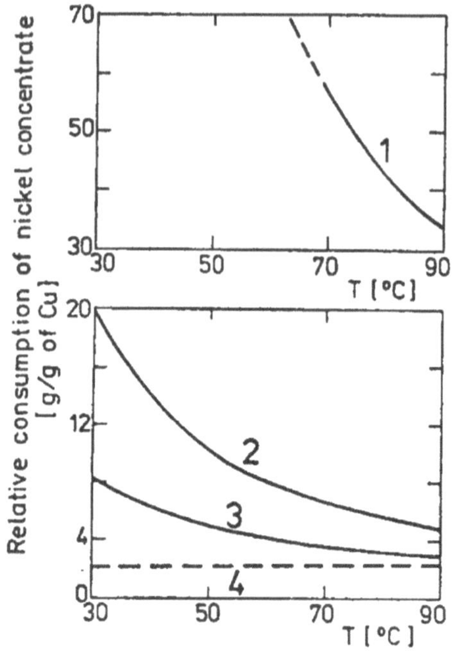

Fig. 5.38 The influence of mechanical activation and temperature on relative consumption of nickel concentrate: 1 – as received concentrate; 2,3 – concentrate mechanically activated in a planetary mill for 2 and 7 min, respectively; 4 – theoretical calculation based on the equation: $Cu^{2+} + NiS \Leftrightarrow Ni^{2+} + CuS$ [Kulebakin 1988]

The application of nickel concentrate instead of nickel powder has been developed. The as received concentrate is composed mainly of Ni_3S_2 and NiO. After mechanical activation the new phase NiS is formed. The observed changes in surface area and amorphization depend on regime of the mechanical activation. The activated concentrate has a positive effect on copper sorption (Fig. 5.38).

Mechanical activation for 2 min has reduced the consumption of nickel concentrate from 34 to 4.7 gg^{-1} Cu at 90°C and from 57 to 7.1 gg^{-1} Cu at 70°C. The activity of as received concentrate at low temperatures is low, whereas the activated concentrate has considerable activity even at 30°C. The mechanical activation of nickel concentrate brings about the changes of surface microstructure. XRD analysis of copper sorption products has shown the occurence of new sulphidic phases $Cu_{7.2}S_4$, Cu_9S_5 and NiS [Kulebakin 1988].

References

Abbruzzese, C., Fornari, P., Massida, R., Veglio, F. and Ubaldini, S. (1995) Thiosulfate leaching for gold hydrometallurgy. Hydrometallurgy 39, 265–276.

Achimovičová, M. and Baláž, P. (2005) Influence of mechanical activation on selectivity of acid leaching of arsenopyrite. Hydrometallurgy 77, 3–7.

Anderson, C.G., Harrison, K.D. and Krys, L.E. (1994) Process integration of sodium nitrite oxidation and fine grinding in refractory precious metal concentrate pressure leaching. SME Annual Meeting Albuquerque, New Mexico, Preprint No. 94–145.

Aylmore, M.G. and Lincoln, F.I. (2000a) Mechanochemical milling-induced reactions between gases and sulfide minerals: I. Reaction of SO_2 with arsenopyrite, pyrrhotite and pyrite. Journal of Alloys and Compounds 309, 61–74.

Aylmore, M.G. and Lincoln, F.I. (2000b) Mechanochemical milling-induced reactions between gases and sulfide minerals: II. Reaction of CO_2 with arsenopyrite, pyrrhotite and pyrite. Journal of Alloys and Compounds 314, 103–113.

Aylmore, M.G. (2001) Treatment of a refractory-gold-copper sulfide concentrate by copper ammoniacal thiosulfate leaching. Minerals Engineering 14, 615–637.

Aylmore, M.G. and Muir, D.M. (2001) Thiosulphate leaching of gold – a review. Minerals Engineering 14, 135–174.

Baláž, P., Kammel, R. and Achimovičová, M. (1994) Mechano-chemical treatment of tetrahedrite as a new ecologically non-pollution way of metals recovery. Metall 48, 217–220.

Baláž, P., Sekula, F., Jakabský, Š. and Kammel, R. (1995) Application of attrition grinding in alkaline leaching of tetrahedrite. Minerals Engineering 8, 1299–1308.

Baláž, P., Ficeriová, J., Šepelák, V. and Kammel, R. (1996) Thiourea leaching of silver from mechanically activated tetrahedrite. Hydrometallurgy 43, 367–377.

Baláž, P., Kammel, R., Sekula, F. and Jakabský, Š. (1997) Mechanochemical leaching: the possibility to influence the rate of metals extraction from refractory ores. In: H. Hoberg, von H. Blottnitz (Eds.) Proceedings of the XX International Mineral Processing Congress, Aachem, pp. 149–159.

Baláž, P., Števulová, N., Kammel, R. and Malmström, R. (1998a) Extraction of Ni, Cu and Co from mechanically activated pentlandite. Metall 52, 620–623.

Baláž, P., Achimovičová, M., Ficeriová, J., Kammel, R. and Šepelák, V. (1998b) Leaching of antimony and mercury from tetrahedrite $Cu_{12}Sb_4S_{13}$. Hydrometallurgy 47, 297–307.

Baláž, P., Achimovičová, M., Sanchez, M. and Kammel, R. (1999) Attrition grinding and leaching of enargite concentrate. Hydrometallurgy 53, 53–56.

Baláž, P. (2000) Extractive Metallurgy of Activated Minerals, Elsevier, Amsterdam.

Baláž, P., Boldižárová, E., Achimovičová, M. and Kammel, R. (2000a) Leaching and dissolution of a pentlandite concentrate pretreated by mechanical activation. Hydrometallurgy 57, 85–96.

Baláž, P., Achimovičová, M., Bastl, Z., Ohtani, T. and Sanchez, M. (2000b) Influence of mechanical activation on the alkaline leaching of enargite concentrate. Hydrometallurgy 54, 205–216.

Baláž, P., Takacs, L., Jiang, J.Z., Soika, V. and Luxová, M. (2002) Mechanochemical reduction of copper sulphide. Materials Science Forum 386–388, 257–262.

Baláž, P., Ficeriová, J. and Villachica, C. (2003b) Silver leaching from a mechanochemically pretreated complex sulphide concentrate. Hydrometallurgy 70, 113–119.

Baláž, P., Takacs, L., Luxová, M., Godočíková, E. and Ficeriová, J. (2004) Mechanochemical processing of sulphidic minerals. International Journal of Mineral Processing 74S, S365–S371.

Baláž, P., Achimovičová, M., Ficeriová, J. and Godočíková, E. (2003a) Mechanochemistry of copper minerals: new challenges in extractive metallurgy. In P.A. Riveros, D. Dixon, D.B. Dreisinger, J. Menacho (Eds.) Proc. Int. Conference *Copper 2003-Cobre 2003* Santiago, Chile, vol. VI, pp. 353–365.

Baláž, P., Aláčová, A., Achimovičová, M., Ficeriová, J. and Godočíková, E. (2005a) Mechanochemistry in hydrometallurgy of sulphide minerals. Hydrometallurgy 77, 9–17.

Baláž, P., Aláčová, A. and Briančin, J. (2005b) Sensitivity of Freundlich equation constant 1/n for zinc sorption on changes induced in calcite by mechanical activation. Chemical Engineering Journal 114, 115–121.

Baláž, P. and Achimovičová, M. (2006a) Selective leaching of antimony and arsenic from mechanically activated tetrahedrite, jamesonite and enargite. International Journal of Mineral Processing 81, 44–50.

Baláž, P. and Achimovičová, M. (2006b) Mechano-chemical leaching in hydrometallurgy of complex sulphides. Hydrometallurgy 84, 60–68.

Beattie, M.J. and Poling, G.W. (1987) A study of the surface oxidation of arsenopyrite using cyclic voltametry. International Journal of Mineral Processing 20, 87–108.

Beckstead, L.W., Munoz, P.B., Sepulveda, J.L., Herbst, J.A., Miller, J.D., Olson, F..A. and Wadsworth, M.E. (1976) Acid ferric sulfate leaching of attritor-ground chalcopyrite concentrates. In: J.C. Yannopoulos, J.C. Agarwall (Eds.) Proceedings of the International Symposium *Extractive Metallurgy of Copper*, vol. 2, The Metallurgical Society of AIME, New York, pp. 611–632.
Block-Bolten, A. and Torma, A.E. (1986) Thiosulfate leaching of gold from sulfide wastes. Metall 40, 687–689.
Boldyrev, V.V. (1996) Mechanochemistry and mechanical activation. Materials Science Forum 225–227, 511–520.
Breuer, P.L. and Jeffrey, M.I. (2000) Thiosulfate leaching kinetics of gold in the presence of copper and ammonia. Minerals Engineering 13, 1071–1081.
Chodakov, G.S. (1972) *Physics of Grinding*. Nauka, Moscow (in Russian).
Cruz, R., Lázaro, I., Rodrigues, J.M., Monroy, M. and Gonzálles, I. (1997) Surface characterization of arsenopyrite in acidic medium by triangular scan voltammetry on carbon paste electrodes. Hydrometallurgy 46, 303–319.
Dayton, S. (1982) How Sb and As are purged from a high-silver copper concentrate before smelting. Engineering and Mining Journal, January, 78–83.
Dutrizac, J.E. and McDonald, R.J.C. (1974) Percolation leaching of pentlandite ore. CIM Bulletin, March, 169–175.
Ellis, R. (2004) Latin America speciality. Mining Magazine, March, 49–51.
Ficeriová, J., Baláž, P. and Boldižarová, E. (1998) Structural sensitivity of silver leaching from tetrahedrite by ammonium thiosulphate. Acta Metallurgica Slovaca 4, 65–69.
Ficeriová, J., Baláž, P., Boldižárová, E. and Jeleň, S. (2002) Thiosulphate leaching of gold from a mechanically activated CuPbZn concentrate. Hydrometallurgy 67, 37–43.
Ficeriová, J., Baláž, P. and Boldižárová, E. (2004) Influence of mechanochemical pretreatment on leaching of silver in cyanide and non-cyanide medium. Journal of Materials Science 39, 5339–5341.
Ficeriová, J., Baláž, P. and Villachica, C. (2005a) Thiosulphate leaching of silver, gold and bismuth from a complex sulfide concentrates. Hydrometallurgy 77, 35–39.
Ficeriová, J., Baláž, P. and Boldižárová, E. (2005b) Combined mechanochemical and thiosulphate leaching of silver from a complex sulphide concentrate. International Journal of Mineral Processing 76, 260–265.
Freundlich, H. (1906) Über die Adsorption in Lösungen, Zeitschrift für Physikalische Chemie 57, 385–470.
Furuya, E.G., Chang, H.T., Miyra, Y. and Noll, K.E. (1997) A fundamental analysis of the isotherm for the adsorption of phenolic compounds on activated carbon. Separation and Purification Technology 11, 69–78.
Garcia-Sanchez, A., Ayuso, E. and Blas, O. (1999) Sorption of heavy metals from industrial waste water by low-cost mineral silicates. Clay Minerals 34, 469–477.
Habashi, F. (1993) *A Textbook on Hydrometallurgy*. Métallurgie Extractive Québec, Enr., Québec, Canada.
Habashi, F. (1996) *Pollution Problems in the Mineral and Metallurgical Industries*. Métallurgie Extractive Québec, Sainte-Foy, Québec.
Habashi, F. (2002) *Textbook of Pyrometallurgy*. Métallurgie Extractive Québec, Sainte-Foy, Québec.
Habashi, F. (2003) *Metals from Ores. An Introduction to Extractive Metallurgy*. Métallurgie Extractive Québec, Sainte-Foy, Québec.
Habashi, F. (2008) A short history of hydrometallurgy. 79, 15–22.
Heinicke, G. (1984) *Tribochemistry*. Akademie-Verlag, Berlin.
Hiskey, J.B. and Atluri, V.P. (1988) Dissolution chemistry of gold and silver in different lixiviants. Mineral Processing and Extractive Metallurgy Review 4, 95–134.
Hong, W. (1993) Celestite and strontianite. Industrial Minerals, November, 55–59.
Hu, H., Chen, Q., Yin, Z., Zhang, P. and Wang, G. (2004) Effect of grinding atmosphere on the leaching of mechanically activated pyrite and sphalerite. Hydrometallurgy 72, 79–86.

Jágerský, I. (1999) Technology of the gold concentrate treatment in gravitational and flotation mill of Slovenská Banská Co. Hodruša-Hámre. Mineralia Slovaca 31, 353–356.

Jansen, M. and Taylor, A. (1995) Key elements in the selection of sulphide concentrate leach processes for copper concentrate leaching. Minerals Engineering 16, 361–368.

Javorka, J. and Fröhlich, L. (1996) Pressure leaching of tetrahedrite raw materials. Mineralia Slovaca 28, 447–449.

Juhász, A.Z. and Opoczky, L. (1990) *Mechanical Activation of Minerals by Grinding: Pulverizing and Morphology of Particles*. Ellis Horwood Limited, Chichester.

Kulebakin, V.G. (1988) *Application of Mechanochemistry in Hydrometallurgical Processes*, Nauka, Novosibirsk (in Russian).

Küzeci, E., Li, X.M. and Kammel, R. (1989) Acid leaching of nickel sulphide concentrate. Metall 43, 434–439 (in German).

La Brooy, S.R., Linge, H.G. and Walker, G.S. (1994) Review of gold extraction from ores. Minerals Engineering 7, 1213–1241.

Linge, H.G. and Welham, N.J. (1997) Gold recovery from arsenopyrite ore by in-situ slurry oxidation. Minerals Engineering 10, 557–566.

Lodejščikov, V.V. and Ignateva, I.D. (1973) *Processing of Silver Bearing Ores*. Nedra, Moscow, 224 p. (in Russian).

Lyakhov, N.Z. (1993) Mechanical activation from the viewpoint of kinetic reaction mechanism. In: K. Tkáčová (Ed.) Proc. Ist Int. Conference on Mechanochemistry *INCOME'93*, vol. I, Košice, Slovakia, pp. 59–65.

Marsden, J. and House, I. (1992) *The Chemistry of Gold Extraction*. Ellis Horwood, New York.

Matteazzi, P. and Le Caër, G. (1992) Mechanically activated room temperature reduction of sulphides. Materials Science and Engineering A 156A, 229–237.

Maurice, M. and Hawk, J.A. (1999a) Ferric chloride leaching of mechanically activated chalcopyrite. Hydrometallurgy 52, 289–312.

Maurice, M. and Hawk, J.A. (1999b) Simultaneous autogenous milling and ferric chloride leaching of chalcopyrite. Hydrometallurgy 52, 371–378.

McCormick, P.G., Tsuzuki, T., Robinson, J.S. and Ding, J. (2001) Nanopowders synthesized by mechanochemical processing. Advanced Materials 13, 1008–1010.

Medkov, M.A., Steblevskaya, N.I., Smolkov, A.A., Grishchenko, D.N. and Belobetskaya, M. (1999) Extraction of bismuth, silver and europium from technological solutions. CIM Bulletin, 89, 76–79.

Mellah, A. and Chegrouche, S. (1997) The removal of zinc from aqueous solutions by natural bentonite. Water Research 31, 621–629.

Mulak, W. (1992) Kinetics of dissolution of Ni_3S_2 in acidic potasium dichromate solutions. Hydrometallurgy 28, 309–322.

Murthy, D.S.R. and Prasad, P.M. (1996) Leaching of gold and silver from Miller Process dross through non-cyanide leachants. Hydrometallurgy 42, 27–33.

Nesbitt, H.W., Muir, I.J. and Pratt, A.R. (1995) Oxidation of arsenopyrite by air and air-saturated, distilled water, and implications for mechanism of oxidation. Geochimica et Cosmochimica Acta 59, 1773–1786.

Obut, A., Baláž, P. and Girgin, I. (2006) Direct mechanochemical conversion of celestite to $SrCO_3$. Mineral Engineering 19, 1185–1190.

Otake, Y., Kalili, N., Chang, T.H. and Furuya, E.G. (2001) Relationship between Freundlich type equation constants and molecular orbital properties. In: *Proc. of the Annual Meeting AICHE*. Reno, Nevada, Poster 712.

Potter, G.M. and Salisburg, H.B. (1974) Innovations in gold metallurgy. Mining Congress Journal, p. 54–55.

Prieto, M., Cubillas, P. and Fernandez-Gouzales, A. (2003) Uptake of dissolved Cd by biogenic and abiogenic aragonite: a comparison with sorption onto calcite, Geochimica et Cosmochimica Acta 67, 3859–3869.

Robins, R.G. and Jayaweera, L.D. (1992) Arsenic in gold processing. Mineral Processing and Extractive Metallurgy Review 9, 255–271.

Senna, M. (1989) Determination of effective surface area for the chemical reaction of fine particulate materials. Particles & Particle Systems Characterization 6, 163–167.

Tkáčová, K. and Baláž, P. (1988) Structural and temperature sensitivity of leaching of chalcopyrite with Fe(III) sulphate. Hydrometallurgy 21, 103–112.

Tkáčová, K. (1989) *Mechanical Activation of Minerals*. Elsevier, Amsterdam.

Ubaldini, S. Massida, R. Abbruzzese, C., Veglio, F.. Toro, L. (1996) Gold-recovery from finely disseminated ore by use cyanidation and thioureation. In: M. Kemal, V. Arslan, A. Akar, N. Canbazoglu (Eds.) *Changing Scopes in Mineral Processing*, Balkema, Rotterdam, pp. 559–562.

Vengris, T., Brinkiene, R. and Sveikanskaite, A. (2001) Nickel, copper and zinc removal from waste water by a modified clay sorbent. Applied Clay Science 18, 183–190.

Wan, R.Y. and Brierley, J.A. (1997) Thiosulphate leaching following biooxidation pretreatment for gold recovery from refractory carbonaceous-sulfidic ore. Minerals Engineering 15, 76–80.

Warner, T.E., Rice, N.M. and Taylor, N. (1992) An electrochemical study of the oxidative dissolution of synthetic pentlandite in aqueous media. Hydrometallurgy 31, 55–90.

Welham, N.J. (1996) A parametric study of the mechanically activated carbothermic reduction of ilmenite. Minerals Engineering 9, 1189–1200.

Welham, N.J. (1997) The effect of extended milling on minerals. CIM Bulletin 90, 64–68.

Welham, N.J. (1998) Mechanochemically induced reaction between alkaline earth metal oxides and TiO_2. Journal of Materials Research 13, 1607–1613.

Welham, N.J. and Llewellyn, D.J. (1998) Mechanical enhancement of the dissolution of ilmenite. Minerals Engineering 11, 827–841.

Welham, N.J. (1999) Non-thermal production of tungsten from scheelite. Materials Science and Technology 15, 456–458.

Welham, N.J. (2001) Enhanced dissolution of tantalite, columbite following milling. International Journal of Mineral Processing 61, 145–154.

Zelikman, A.N., Voldman, G.M. and Beljajevskaja, L.V. (1975) *Theory of Hydrometallurgical Processes*. Metallurgija, Moscow (in Russian).

Zhang, H., Penn, R.L., Hammers, R.J. and Banfield, J.F. (1999) Enhanced adsorption of molecules on surfaces of nanocrystalline particles. Journal of Physical Chemistry B 103, 4656–4662.

Zoubolis, A.I., Kydros, K.A. and Matis, K.A. (1992) Adsorbing flotation of copper hydroxo precipitates by pyrite fines. Separation Science and Technology 27, 2143–2155.

Zoubolis, A.I., Kydros, K.A. and Matis, K.A. (1993a) Removal of toxic metal ions from solutions using industrial solid by-products. Water Science and Technology 27, 83–91.

Zoubolis, A.I. and Kydros, K.A. (1993b) Arsenic (III) and arsenic (V) removal from solution of pyrite fines. Separation Science and Technology 28, 2449–2463.

Chapter 6
Applied Mechanochemistry

6.1 Introduction

Mechanical activation is an innovative procedure where the improvement in technological processes governed by its application can be attained via a combination of several effects which influence the properties of applied solids.

This chapter is devoted to the application of broad possibilities of mechanochemistry in mineral processing, extractive metallurgy, chemical engineering, building industry, coal industry, materials engineering, agriculture, pharmacy and waste treatment. Nanoparticles are often formed by this technological approach which add value to the processed solids. The environmental aspects of these processes are particularly attractive [Lomovskij 1994; Lomovskij and Boldyrev 2006].

The main advantages in comparison with the traditional technological procedures are

- decrease in the number of technological stages
- excluding the operations that involve the use of solvents and gases and
- the possibility of obtaining a product in the metastabile state which is difficult (or impossible) to obtain using traditional technological methods [Boldyrev 2006].

Simplification of the processes, ecological safety and the product extraordinariness characterize the mechanochemical approach in technology.

6.2 Mineral Processing

The efficiency of mineral processing depends on the separation of individual mineral components and on exposure of their surface. The production of flotation concentrates, with particle sizes of tens of micrometers, is not sufficient for many following processes to operate at their optimum.

It is well known that the degree of dispersion and other physico-chemical properties of minerals are changed by mechanical stress in milling maschines. These effects not only have significance in enhancing dispersion, but also affect particle

© Springer-Verlag Berlin Heidelberg 2008

sizes typically encountered in flotation since the structural defects may be concentrated within this surface layers [Jusupov and Heegn 1997].

6.2.1 Sulphides

The sulphidic concentrates are milled to flotation fineness in industrial plants, as a rule, in balls mills where the milling effect is achieved by rubbing and crushing. As vibratory mills are often used for mechanical activation work using a similar regime, their application to modification of the surface properties of minerals was investigated [Molchanov and Jusupov 1981].

It was stated as early as in 1960s [Ray and Formanek 1960] that galena PbS milled in an iron mill was much less floatable than galena milled in a ceramic mill. This fact was elucidated recently [Peng et al. 2002].

It has been found that the adsorption capacity of galena for the flotation collector potassium butylxanthate decreasing with decreasing size of particles [Plaksin and Šafejev 1962]. This may be due to a change in electron concentration in the surface layer of PbS. The higher electron concentration in the case of finer particles makes the electron transitions between the xanthate anions and the surface of galena energetically unfavourable owing to which the formation of a solid particle-xanthate bond is hindered. However, the relationship between size of the particles and their adsorption properties is not unambiguous [Glembockij 1972]. In addition to the conditions of milling, a number of other factors play a role and many of these factors are antagonistic.

The influence of mechanical activation on flotability of PbS and sfalerite ZnS was investigated [Ocepek and Milosavljkevic 1977). Experiments performed in a flotation cell showed that there is an optimum degree of mechanical activation that increases the flotability of sulfides compared with non-activated sulfides. The content of amorphous phase increases with the time of mechanical activation and results in a decrease of flotability. For PbS, the decrease in flotability can be explained by mechanochemical formation of anglesite $PbSO_4$.

Mechanical activation of galena concentrate involving a closed cycle with classifier has been described [Auerbach et al. 1987] which showed that greater selectivity and increased lead recovery could be achieved. Moreover, it was possible to reduce the consumption of electric energy by 17%, the abrasion of mill lining to 68% and the milling flowsheet to a single stage. The increase in recovery and selectivity may be due to a considerable decrease in iron content in the pulp due to abrasion reduction and the shorter contact of particles with aqueous phase.

The contact phenomena of galena with milling media has been studied [Kostovic and Pavlica 1995]. The galena flotation response after contact with stainless or mild steel depends on contact time and conditions of the system in which contact occurred. In this system, stainless and especially mild steels are effective in galena depression. Oxidative products are sulphate or oxide or hydroxide species coating the surface of galena that had contacted stainless or mild steel. These products provoke a depression of galena and decrease the adsorption of xanthate from aqueous solution on galena surface.

A specialized mill to control the milling environment has been applied [Peng et al. 2002]. The degree of the galena oxidation during milling was modified by different milling media and different purging gases [nitrogen, air or oxygen). A correlation was developed between hydroxide consumption and the extent of galena oxidation on flotation recovery. This correlation may be used to predict oxidation during milling as well as galena recovery over a range of experimental conditions.

Electrochemical aspects play important role in milling applied before flotation [Čanturija and Vigdergauz 1994]. The wet milling is usually applied for concentration of sulphide ores by milling in steel rod and ball mills. The wet milling and application of iron balls bring about not only structural transformation of sulphides due to close contact between sulphide and milling balls but also other effects [Baláž 2000].

The wet milling bring about a loss in weight of the balls as a consequence of corrosion and abrasion [Adam et al. 1986a]. However, it is difficult to estimate the relevance of this effect. Moreover, it is known that the sulphidic minerals are nobler than most steels used for making the milling balls and must therefore accelerate the anodic dissolution of metals [Adam et al. 1986b]. The authors studied the electrochemical aspects of milling media and observed the galvanic coupling of mild steel medium with magnetite Fe_3O_4 or pyrrhotite $Fe_{1-x}S$ resulted in the formation of iron hydroxide $Fe(OH)_3$ species on the mineral surface [Natarajan and Iwasaki 1984]. Oxygen accelerated the corrosion of the milling media, thus more iron hydroxide species were found on the mineral surface [Peng et al. 2002]. The results of investigation of the wet milling of pyrrhotite $Fe_{1-x}S$ were used for designing the model of corrosion of milling balls which is represented in Fig. 6.1. This model assumes the corrosion on the surface itself of milling ball (A) and the corrosion in the course of interaction between milling ball and sulphide [Natarajan and Iwasaki 1984]. These effects are likely still more significant in the case of mechanical activation in wet medium.

It has been indicated the possibility to influence the flotation properties of sulphidic ores by control of the electrochemical potential in the pulp during the milling process [Čanturija and Vigdergauz 1994; Čanturija and Lunin 1983; Elisejev

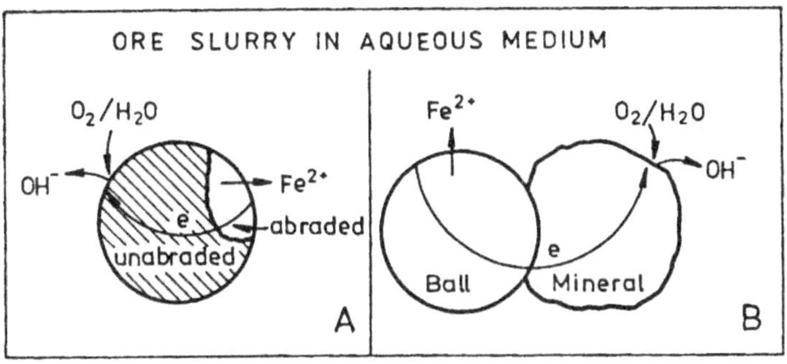

Fig. 6.1 Corrosion model for interaction of milling balls with pyrrhotite $Fe_{1-x}S$, **A** – the differential abrasion cell, **B** – the ball mineral cell [Adam et al. 1986a]

et al. 1980]. The pilot plant experiments demonstrated the possibility of raising the concentration of defects and weakening the bonds at the interface of mineral grains and was shown to work on several sulphidic concentrates.

6.2.2 Other Minerals

The flotability of quartz SiO_2, cassiterite SnO_2 and feldspar $KalSi_3O_8$ in confrontation with mechanical activation of the minerals has been studied [Jusupov and Heegn 1997]. Solubility in HF and NaOH has been selected as a method for characterization of solid surface structure. Table 6.1 shows the data for the rate constant of dissolution k for SiO_2 activated in different mills. The calculated span is determined between 0.22 $\mu m\,h^{-1}$ for dry ball milling and 0.37 $\mu m\,h^{-1}$ for planetary milling.

The data in Table 6.1 represent the solubility of bulk SiO_2.

For characterization of surface layers the initial solubility is more adequate method. Figure 6.2 shows the thickness of surface layer calculated from the solubility

Table 6.1 The solubility data for quartz SiO_2 mechanically activated in different mills [Jusupov and Heegn 1997]

Milling mode	k [$\mu m\,h^{-1}$]
Ball milling	0.22
Vibratory milling	0.33
Jet milling	0.33
Disintegrator	0.26
Planetary milling (small balls)	0.30
Planetary milling (big balls)	0.37

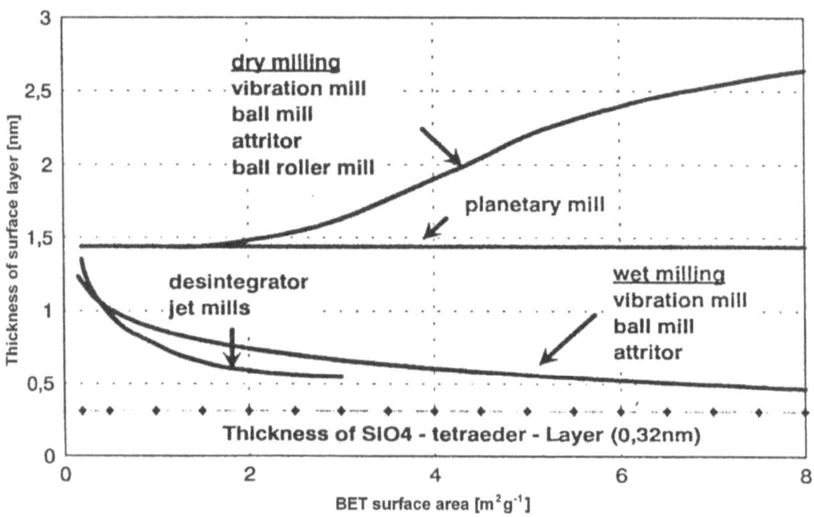

Fig. 6.2 Thickness of surface layer of mechanically activated quartz SiO_2 in relation to its surface area [Jusupov and Heegn 1997]

Table 6.2 Flotation of quartz SiO_2 after mechanical activation [Jusupov and Heegn 1997]

Mechanical activation	Recovery of SiO_2 [%]		
	pH 3.5	pH 6.5	pH 8.5
–	60.3	76.2	50.7
Dry milling	13.3	21.3	0.2
Wet milling	1.4	23.0	37.0

data of SiO_2 in 0.1 N NaOH after 20 min of action. Various modes of mechanical activation have been applied. For comparison the thickness 0.32 nm of a monomolecular layer of SiO_4-tetraheders is given in Fig. 6.2.

Mechanical activation of quartz leads to decrease of flotability. This is shown by data of mineral flotation by cationic flotation agent ANP-14 (Table 6.2). The quantity of flotation agent was not optimal in this case.

The flotation strategy which applies the combination of different pretreatments has led to improvement in flotability [Jusupov et al. 1977, 1987; Cases et al. 1985]. Example is illustrated by Fig. 6.3.

It is accepted that decreasing of quartz flotability is not only the result of crystal structure breakage but also the iron hydroxide layer formation on SiO_2 as a consequence of mechanical activation. Thus mechanical activation plays an important role in manipulation of flotation processes.

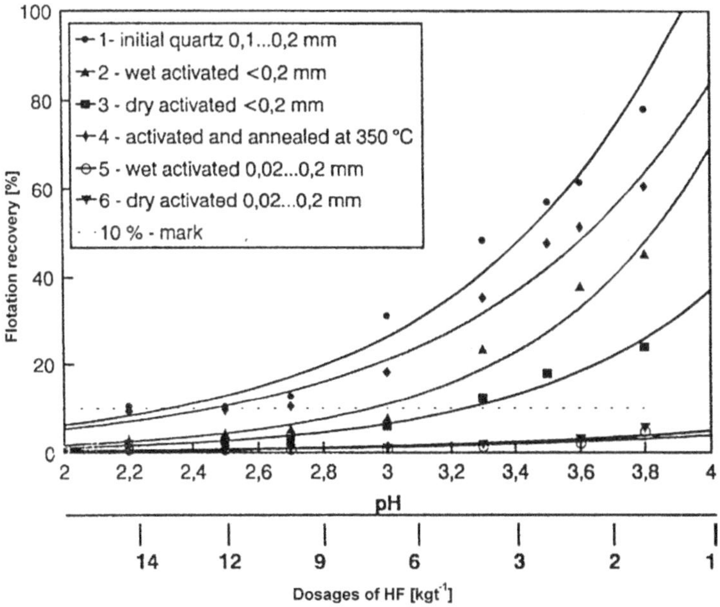

Fig. 6.3 Effect of various models of mechanical activation on the quartz SiO_2 flotability [Jusupov and Heegn 1997]

6.3 Extractive Metallurgy

According to classical view extractive metallurgy is the art and science of extracting metals from their ores by chemical methods [Habashi 1993]. It is actually divided into three sectors: hydrometallurgy, pyrometallurgy and electrometallurgy. Hydrometallurgy is the technology of extracting metals from ores by aqueous methods, pyrometallurgy by dry thermal methods, and electrometallurgy by electrolytic methods.

In general, hydrometallurgy involves two distinct steps (Fig. 6.4)

- selective dissolution of the metal values from an ore – process known as *leaching* and
- selective recovery of the metal values from the solution, an operation that involves a *precipitation* method.

Sometimes a purification/concentration operation is conducted prior to precipitation. These processes are aimed at obtaining a pure and a concentrated solution from which the metals values can be precipitated effectively. The methods used are: adsorption on activated charcoal, sorption on ion exchange resins, extraction by organic solvents as well as other advanced processes of metals extraction [Havlík 2005].

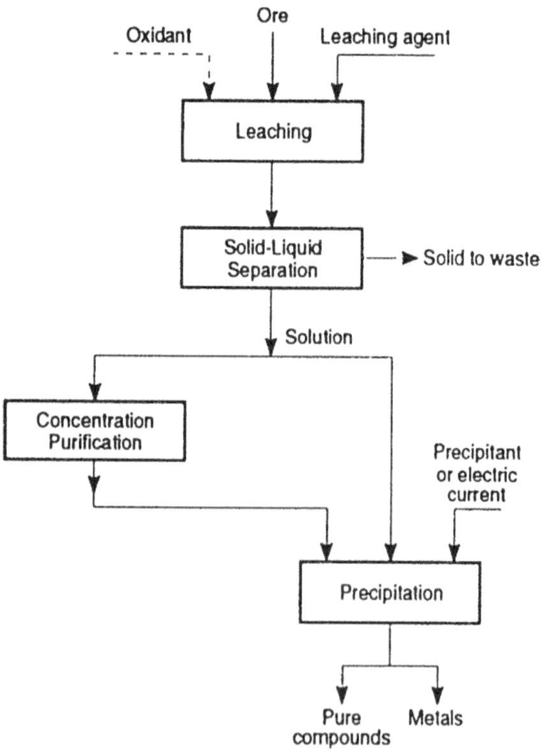

Fig. 6.4 General outline of hydrometallurgical process [Habashi 1993]

6.3.1 Mechanical Activation as Pretreatment Step for Leaching

6.3.1.1 Ore Minerals

The pilot-plant and plant application of mechanical activation as a method of ore concentrates was intensively investigated by German scientists [Gerlach and Gock 1973; Gerlach et al. 1973; Pawlek 1976a, b, 1977, 1978; Gock 1977, 1978; Gerlach 1982; Pawlek et al. 1989; Kammel et al. 1987a; Gärtner et al. 1991].

Wet milling of chalcopyrite concentrate in an attritor reduces the particle size to 0.1–1 μm in a short time with the efficiency of activation increased by addition of NaOH. During subsequent acid leaching in an autoclave (0.1 MPa and 110°C) full extraction of copper into solution was achieved in 30 min. Pawlek published the flowsheet (Fig. 6.5) and considers the wet milling to be convenient for the step of mechanical activation.

The technological parameters of the combined process of mechanical activation in a vibration mill and subsequent oxidative leaching of sulfidic concentrates of chalcopyrite, sphalerite and molybdenite are summarized in Table 6.3.

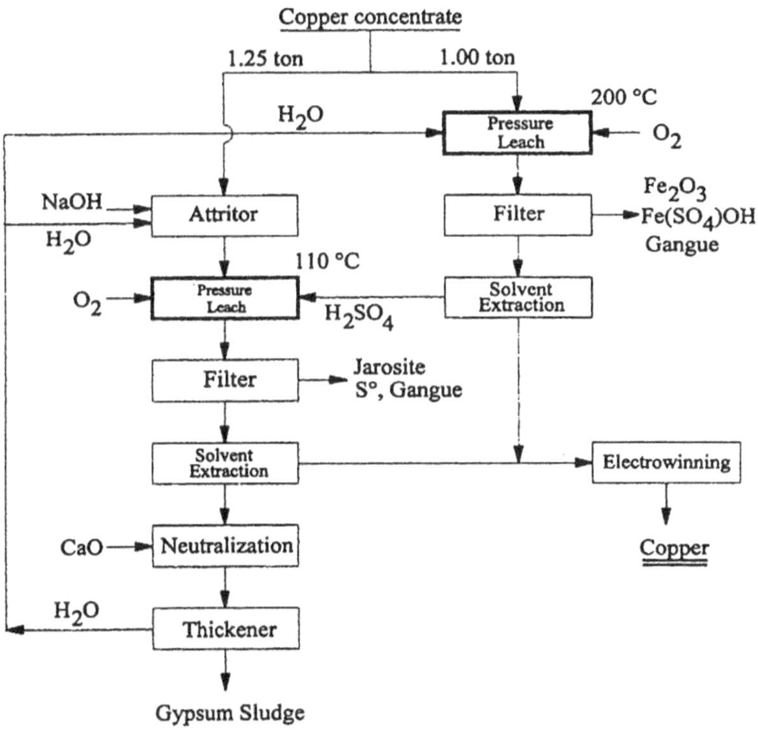

Fig. 6.5 Flowsheet for treatment of chalcopyrite concentrate mechanically activated in attritor followed by high temperature oxidative pressure leaching and low temperature acid pressure leaching [Pawlek 1976a]

Table 6.3 Technological parameters of mechanical activation in a vibration mill followed by oxidative leaching of flotation concentrates of $CuFeS_2$, ZnS and MoS_2 [Gock 1977]

Flotation concentrate	$CuFeS_2$	ZnS	MoS_2
Mechanical activation	Vibration mill		
Amplitude (mm)	6	6	6
Revolutions (s^{-1})	1,000	1,000	1,000
Ball charge (%)	85	85	85
Mass of balls to mass of milled material ratio	47	47	47
Energy input of mill ($kWht^{-1}$)	224	75	224
Relative acceleration	6.8	2.73	–
Oxidative leaching	Cylinder autoclave		
Content of solid phase (gl^{-1})	150	100	50
Initial concentration of H_2SO_4 (gl^{-1})	65	185	–
Temperature (°C)	120	120	180
Partial pressure of oxygen (MPa)	2	1	1
Time of leaching (min)	120	120	120
Results of leaching			
Solution	$CuSO_4Fe_2(SO_4)_3$	$ZnSO_4Fe_2(SO_4)_3$	H_2SO_4
Residue	$Fe(OH)SO_4$,S	S	MoO_3 or $MoO_3.2H_2O$
Recovery (%)	100	100	97.2
Metal output	Reductive electrolysis		Reduction with carbon or hydrogen

Complex Cu-Pb-Zn sulfide ores represent an important source of non-ferrous metals and typically consist of fine grains of chalcopyrite, galena and sphalerite dispersed in a matrix of pyrite, phyllites and quartz [Wills 1984].

Flotation of such ores may cause some complications. Liberation of the individual minerals occurs during energy intensive fine milling, typically to bellow 75 μm. The finely milled ore has large surface area leading to a high reagent consumption during flotation. Selective flotation is multistage and produces low grade concentrates with low metal recoveries making a single sulfide flotation stage a far more convenient way of ore dressing prior to metal extraction.

Due to the complex nature of these ores, hydrometallurgical techniques seem to be the most convenient way of extracting metals [Hiskey 1986]. In comparison with pyrometallurgical techniques the hydrometallurgical way offers greater flexibility and, more importantly, can process low-grade ores, will extract major metals and trace elements and is amenable to automatic control.

Mechanical activation seems to be a favourable process for the treatment of minerals prior to leaching with significantly enhanced metal recoveries reported. This increase in leachability may well be due to a combination of mechanically induced structural defects and chemical reactions involving the mineral and the local environment. The differences in reactivity of the mineral components of ore, as well as the differences in solubility of the reaction products, can be used to selectively leach metals [Welham 1997; Welham and Llewellyn 1998].

The possibility of applying both these principles to selective extraction of zinc from a complex Cu-Pb-Zn concentrate was investigated [Tkáčová et al. 1993; Baláž et al. 1989] using hydrogen peroxide as a model strong oxidative lixiviant. In accordance with data from the literature the overall reactions of sulphide MeS with H_2O_2 can be described by the following two equations

$$MeS \rightarrow Me^{2+} + S^0 + 2e \tag{6.1}$$

$$MeS + 4H_2O \rightarrow Me^{2+} + SO_4^{2-} + 8H^+ + 8e \tag{6.2}$$

Reaction (6.2) manifests itself to a lesser extent. The Cu-Pb-Zn concentrate reacts with hydrogen peroxide to give soluble copper and zinc sulphates, insoluble $PbSO_4$ and partially soluble iron hydroxide.

Figure 6.6 depicts the recoveries of zinc and copper from as-received complex Cu-Pb-Zn concentrate plotted versus duration of leaching in 4 and 30% hydrogen peroxide. After 5 h of leaching 61% Zn and 40% Cu enter the solution. The metal recoveries are relatively low and the reagent consumption high.

Figure 6.7 shows the results of peroxide leaching the same concentrate after mechanical activation in a vibration mill. With 30% H_2O_2 as a leaching agent, 96% Zn and 90% Cu entered the solution.

The tests represented in Fig. 6.7 by full curves seem to be very promising as they suggest that the more dilute leaching agent (4% H_2O_2) can extract metals with high recoveries. Furthermore, the curves depicted in Fig. 6.7 also suggest that the weaker leaching agent has a favourable effect on selectivity of leaching, particularly during the initial stages of leaching. For instance, after 60 min of leaching by 30%

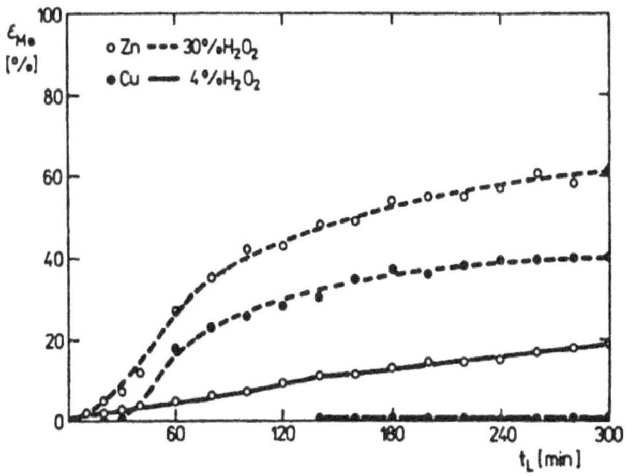

Fig. 6.6 Zinc and copper recovery, ε_{Me} vs. leaching time, t_L for as-received Cu-Pb-Zn concentrate [Baláž et al. 1989]

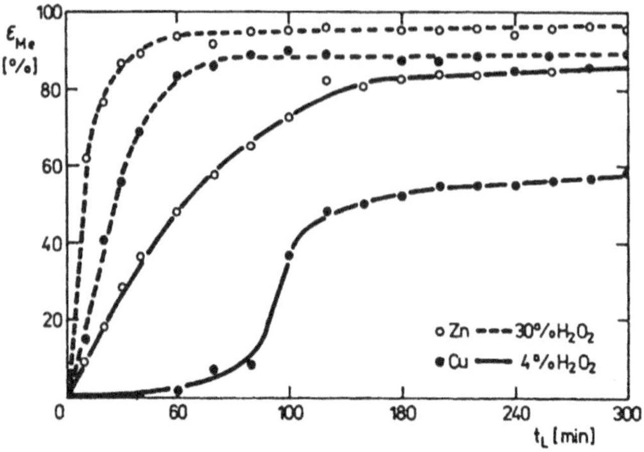

Fig. 6.7 Zinc and copper recovery, ε_{Me} vs. leaching time, t_L for mechanically activated Cu-Pb-Zn concentrate [Baláž et al. 1989]

H_2O_2 the ratio Zn/Cu equals 1.1 whereas when leaching in 4% H_2O_2 it amounts to 24.5. The beneficial effect of mechanical activation of the concentrate at randomly selected conditions of activation is documented in Fig. 6.7.

A systematic study on the effect of the milling conditions on the recoveries of copper and zinc into the solution by leaching concentrate in 4% H_2O_2 was carried out following the 2^2 plan of experiments (Table 6.4).

The effect of mechanical activation was studied at four levels of the energy input to the mill which were adjusted by selecting the amplitude and the speed of rotation of the mill. Table 6.5 summarizes the results for milling times of 0.125–8 h. High metal recoveries suggest that leaching is feasible even in the strong diluted lixiviant (4% H_2O_2).

Figure 6.8 shows the dependence of the dissolution of copper, zinc and iron with time for the most intensively activated sample i.e. Test No. 4 milled for 480 min. The temperature of the solution measured during this tests showed two maxima (at t = 10 and t = 240 min) which corresponded with the highest rates of recovery of zinc and copper respectively.

Leaching of sphalerite or chalcopyrite from the concentrate by H_2O_2 solution is an exothermic process hence the temperature of the solution rises. Of great

Table 6.4 Plan of experiments 2^2 (vibration mill, milling time 60 min)

Test No.	Parameters of the mill	
	Amplitude (mm)	Frequency (min^{-1})
1	3.9	590
2	5.8	590
3	3.9	1100
4	5.8	1100

Table 6.5 Copper and zinc recovery vs. milling time, t_M for Cu-Pb-Zn concentrate mechanically activated in accordance with conditions in Table 6.4 (leaching time 300 min)

t_M	Copper (%)				Zinc (%)			
(min)	No. 1	No. 2	No. 3	No. 4	No. 1	No. 2	No. 3	No. 4
0	0.92				18.61			
7.5	2.30	17.02	4.15	44.10	24.99	63.37	45.16	83.50
15	3.12	19.87	44.27	56.40	40.81	71.28	79.12	85.00
30	13.56	39.82	44.29	68.40	64.46	82.15	86.62	86.70
60	33.70	54.63	54.12	67.00	71.70	81.83	86.05	89.90
120	42.90	61.21	57.39	58.50	79.45	87.50	88.39	89.20
240	66.46	63.00	65.87	57.34	86.31	90.78	97.11	91.79
480	65.44	75.44	70.14	69.00	89.09	98.03	88.80	83.00

importance is the fact that zinc is selectively leached out at $t_L < 180$ min and the solubilization of iron is minimal (2% after 6 h of leaching).

Mechanical activation of the sample brings about a reduction of the particle size of the concentrate. Table 6.6 summarizing the fraction passing 40 μm in the milled samples. For all milling times and conditions the fraction of concentrate below 40 μm was 2–3 times greater than in the as-received sample.

The reaction of Cu-Pb-Zn concentrate with H_2O_2 solutions represents a convenient model system to study the conditions of selective dissolution of a particular metal from mechanically activated complex concentrates. Hydrogen peroxide reacts with galena and iron to give insoluble or poorly soluble products which can be separated from soluble sulfates in a single leaching stage. The selective separation of soluble zinc and copper sulfates can be accomplished by multi-step leaching process,

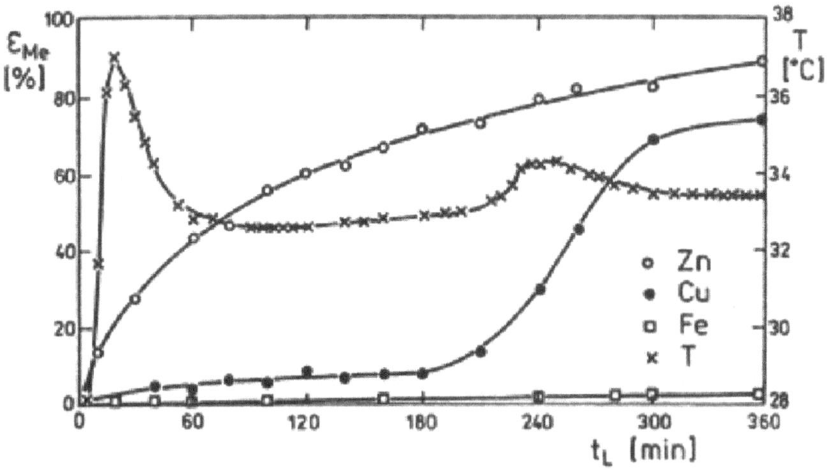

Fig. 6.8 Zinc, copper and iron recoveries, ε_{Me} and temperature of leach solution, T vs. leaching time, t_L for mechanically activated Cu-Pb-Zn concentrate (Test No. 4, $t_M = 480$ min) [Baláž et al. 1989]

Table 6.6 Percent of occurrence of fines passing 40 μm for mechanically activated Cu-Pb-Zn concentrate

Test No.	Milling time (min)						
	7.5	15	30	60	120	240	480
–	31						
1	76	92	90	84	84	81	80
2	77	78	74	69	69	69	65
3	85	79	63	56	67	55	64
4	73	73	62	62	61	61	64

provided the differences in reactivity of sphalerite and chalcopyrite are conveniently regulated by mechanical activation [Tkáčová et al. 1993].

Cu-Ni complex concentrates also serve as a suitable model for selectivity study. The mechanical activation of pentlandite concentrate results in increased specific surface area and amorphization of individual mineral components. The similar behaviour of pentlandite $(Fe, Ni)_9S_8$ and chalcopyrite $CuFeS_2$ after mechanochemical treatment towards ferric sulfate leaching may be used to selectively leach of nickel and copper [Baláž et al. 1998a]. The differences between cobalt and iron during oxidative sulphuric leaching of mechanically activated complex Cu-Ni concentrate from Akarema (Egypt) were described [Amer 1995].

The method for treating molybdenite MoS_2 concentrates containing rhenium has been developed with including the step of mechanical activation [Kähler et al. 1996]. The core of the process lies in the replacement of ore roasting by pressure leaching with oxygen. In this way, it is possible to process rhenium and molybdenum simultaneously. Solvent extraction is used in conjunction with a modified secondary amine to separate Mo and Re. A complete precipitation of ammoniumoctamolybdate is obtained using sulphuric acid. The ammonia can then be recovered, and the process water recycled. In contrast to the ore-roasting process, no hazardous residues are generated by ammoniacal leaching. This results in minimal environmental impact.

A flowsheet of the process is in Fig. 6.9. Molybdenite concentrate with 91.5% MoS_2 and 0.007% ReS_2 has been applied.

6.3.1.2 Gold and Silver Containing Ore Minerals

The technologies hitherto used encounter a problem in processing complex sulphide ores containing gold and silver in an economic way providing sufficient recovery. One of the problems of gold and silver extraction from sulphidic minerals is associated with the form in which the precious metals occur.

Gold and silver are frequently physically-locked within sulphides, may form defects in their structure or can be chemically bonded in the form of solid solutions or compounds [Hiskey and Atluri 1988; Lorentzen and van Deventer 1992; Marsden and House 1992]. Different types of associations of gold with sulphidic minerals are given in Fig. 6.10.

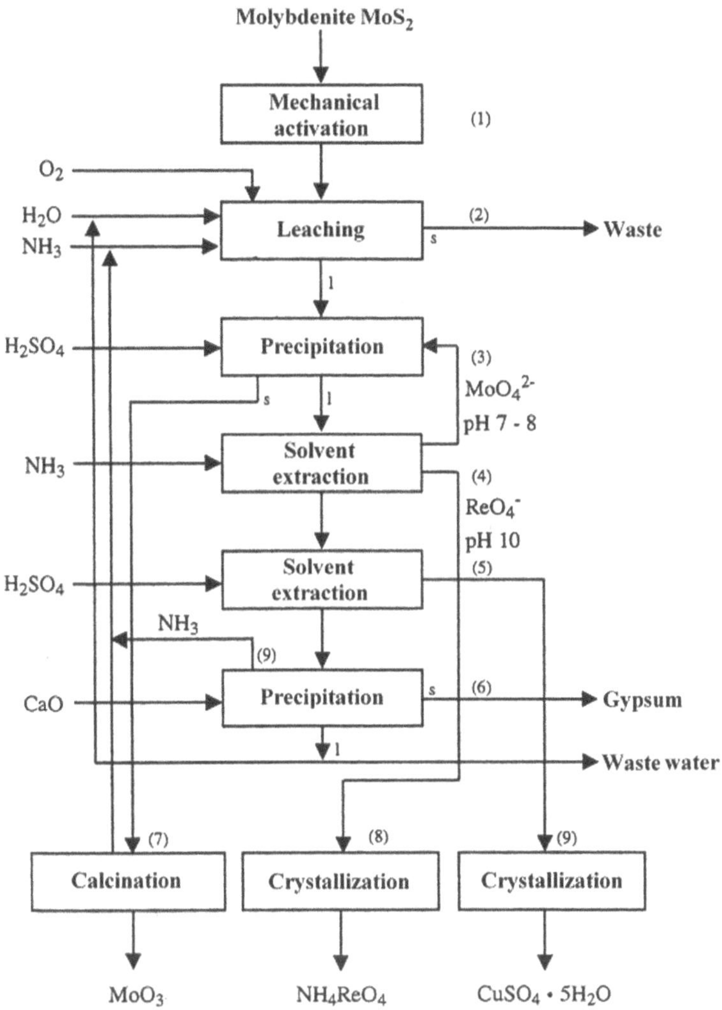

Fig. 6.9 Flowsheet for the treatment of molybdenite concentrate with Re content by vibratory milling [Kähler et al. 1996]

Sulphides are also a considerable natural resource of silver. Around 200 minerals bearing silver in major, minor and variable amounts were detected [Gasparrini 1995]. However, of these, only 10–12 minerals are of practical importance. These are, in order of leachability: elemental silver, silver halides and silver sulphides [Wyslouzil and Salter 1990].

The contact of gold and silver (in form of metals or compounds) with leaching reagent plays a fundamental role in the hydrometallurgical extraction of the precious metals. An improvement of the contact can be achieved by pretreatment. The choice of pretreatment method significantly depends on locality and mineralogy

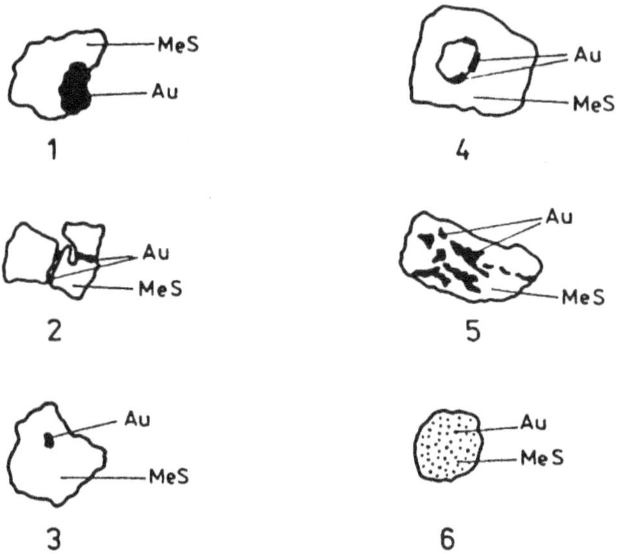

Fig. 6.10 Gold associations with sulphide minerals: 1 – readily liberatable gold, 2 – gold along crystal grain boundaries, 3 – gold grain enclosed in pyrite/sulphide (random position), 4 – gold occurrence at the boundary between sulphide grains, 5 – gold in concretionary pyrite (or other sulphide) along fractures and/or crystal defects, 6 – gold as colloidal particles or in solid solution in sulphide [Marsden and House 1992]

of the ore deposit. The oxidizing pretreatment, which can, in principle, proceed in the pyrometallurgical or hydrometallurgical way, belongs among classical methods [Marsden and House 1992]. Pyrometallurgical pretreatment is the oldest application and consists of an oxidizing roast to convert sulphides to oxides. However, because of environmental demands roasting is becoming more and more suppressed, irrespective of technical innovations [La Brooy et al. 1994].

Chemical and biological pretreatments are applied in hydrometallurgical processes, the goal of these processes is to disintegrate the sulphide and thus to facilitate the subsequent extraction of gold and silver. Pressure oxidation can be used for this but has high capital and operating costs. At present, one of the increasingly used processes is biological oxidation; this is considered acceptable from the viewpoint of environmental considerations and lower economic demands (see Table 6.7).

Table 6.7 Relative costs of pretreatment processes [Wall and Pattison 1997]

Pretreatment	Costs	
	Capital	Operating
Roasting	1.0–1.5	0.75–0.80
Pressure oxidation	1.20–1.25	0.90–0.95
Biological oxidation	1.0	1.0

However, long reaction times and appropriate design of bioreactors can create difficulties.

The process of fine and ultra-fine milling for pretreating gold and silver concentrates has been used very often in recent years. This process requires finer milling than that attained in the ball mills typically used in comminution for flotation. Particle size of 1–20 μm can be produced in intensive milling where size reduction is accompanied by mechanical activation of mineral components [Mullov and Lodejščikov 1979]. The economics of milling is dependent on the metal content of ore, energy and reagent consumptions and amortization of the process plant. Vibration, planetary, jet impact mills and attritors have all been used in ultrafine milling.

The mechanical activation of arsenopyrite concentrate containing 21 gt^{-1} of gold by a jet impact mill has been studied [Minejev et al. 1976]. The subsequent processing with protein hydrolyzate in mixture with 0.15% solution of cyanide made it possible to solubilize 85% of gold.

Jet mill treatment of Au-Ag-As concentrate containing 40% of sulphides gave a size range of 8–30 μm after activation [Minejev 1985]. After arsenopyrite decomposition, cyanidation of the residue resulted in the 96–97% recovery of gold.

The possibilities of intensification of the cyanide leaching of refractory pyrite-arsenopyrite concentrates was studied [Syrtlanova et al. 1979]. In this case, gold was associated in the form of fine dispersion and direct cyanidation was ineffective with only 8–10% gold recovery. A combination of mechanical activation, alkaline leaching and cyanide leaching resulted in 81% decomposition of the arsenopyrite during the alkaline leaching and in significant improvement in gold extraction in the second stage of leaching. The results obtained by different methods of activation are summarized in Table 6.8.

It is clear that the jet impact mill proved to be the most effective. These mills are widely used in industry, they have high capacity (25 th^{-1} and more), simple construction and are not expensive. It seems that planetary milling is less effective because of surface passivation of gold particles as has been shown to occur during mechanical activation of arsenopyrite [Jusupov et al. 1998].

Gold extraction from a pyrite-arsenopyrite concentrate has been studied [Kušnierová et al. 1993]. Mechanical activation in a vibration mill showed relatively little improvement in the recovery of gold, despite considerable amorphization of FeS_2 and FeAsS. Mechanical activation brought about an increase in gold recovery

Table 6.8 Influence of different methods of mechanical activation on FeAsS destruction and subsequent gold extraction [Syrtlanova et al. 1979]

Mechanical activation	FeAsS decomposition (%)	Au cyanidation (%)
–	12.5	8–10
Ball mill	15.0	–
Vibration mill	21.2	–
Planetary mill	81.2	48–50
Jet impact mill	–	62–80

from 4 to 29%. A similar experiment carried out with biologically pretreated concentrate resulted in a recovery increase from about 66 to 93% Au after activation.

The influence of particle size of an arsenopyrite-pyrite concentrate activated in the centrifugal-planetary mill and subsequently classified on gold extraction has been studied [Jusupov et al. 1998]. The results are summarized in Table 6.9.

The optimal range of particle size for gold extraction is clearly different for cyanide and thiourea leaching. For cyanide leaching particles of 5–10 μm appeared to be the most efficient. When employing thiocarbamide as the lixiviant, the decrease in gold extraction can be explained by the presence of a surface layer of elementary sulphur formed during dissolution of arsenopyrite and pyrite in acid medium. The particle size range 40–50 μm appeared to be the most favourable size for leaching of the concentrate.

The effect of combination of fine milling in ball mill, classification and mechanical activation in attritor on gold and silver recovery from a chalcopyrite concentrate has been studied [Pawlek 1976a]. The proposed flowsheet is in Fig. 6.11.

Mechanical activation was used to examine whether intensification of gold and silver leaching from arsenopyrite concentrates with high carbon content could be achieved [Gock and Asiam 1986]. It was found that, after activation, the concentrate can effectively be leached at 100–120°C and an oxygen pressure of 0.5–1.0 MPa. Arsenic and iron quantitatively pass into the leaching solution and the residue is enriched with gold and silver. After decarbonization of the residue by roasting, the precious metals can be completely extracted by cyanidation. The applied activation-leaching process bypasses the drawbacks of traditional pyrometallurgical pretreatment of arsenopyrite concentrates such as emissions of SO_2 and As_2O_3 and gold losses. According to authors the roasting of arsenopyrite at 802°C leads to the loss of 33.7% of gold with fly ashes. The proposed flowsheet is in Fig. 6.12.

The processing of sulphidic concentrate (5–6% $CuFeS_2$, 53–55% FeS_2, 8–10% FeAsS) has been studied by the following strategy: two-step cyanidation-mechanical activation-air conditioning-cyanidation [Kulebakin et al. 1999]. The proposed flowsheet is in Fig. 6.13, condition are given in Table 6.10.

Table 6.9 Gold extraction from mechanically activated and classified arsenopyrite-pyrite concentrate

Grain size (μm)	Leaching agent: Cyanide	Leaching agent: Thiocarbamide
	Au (%)	
Non activated concentrate	72.5	–
−50 + 40	80.1	98.0
−40 + 20	83.6	95.8
−20 + 10	92.3	95.2
−10 + 5	100	89.7
−5	100	58.5
Non-classified activated concentrate	93.6	87.6

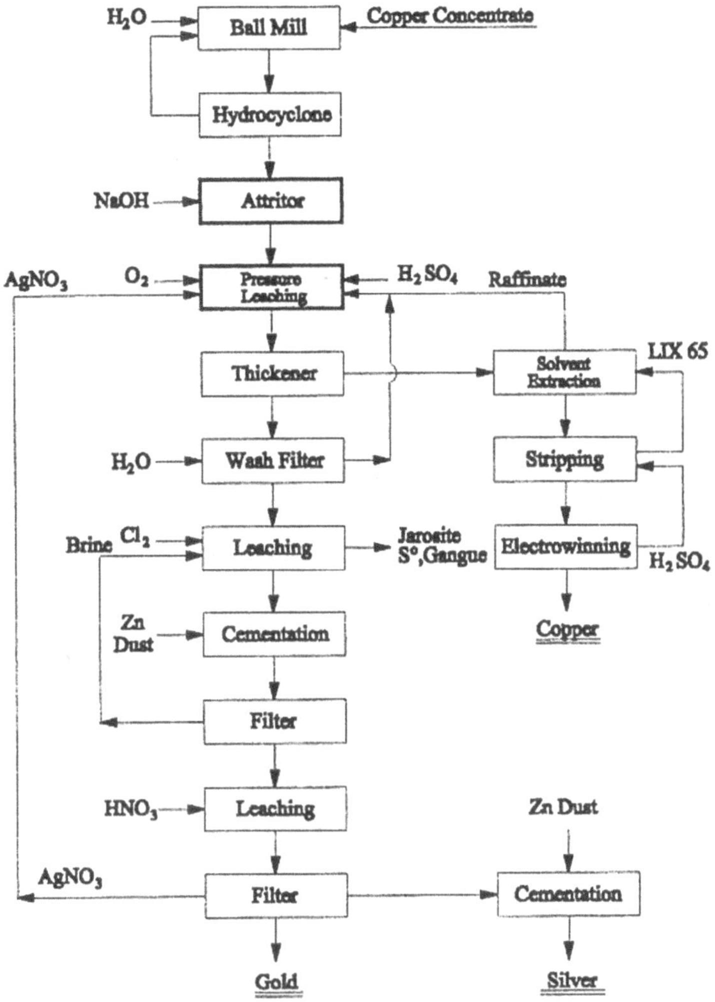

Fig. 6.11 Flowsheet for the treatment of chalcopyrite concentrate by attrition and acid pressure leaching with catalyst ($AgNO_3$) including recovery of Ag and Au [Pawlek 1976a]

The different behaviour of pyrite FeS_2 and arsenopyrite FeAsS during milling was pointed out [Welham 1997; Welham and Llewellyn 1998]. Study of the minerals has shown that selective dissolution was possible by a simple leaching process directly after milling. The separation of both minerals is possible by milling in oxygen, the arsenopyrite decomposing and the pyrite remained essentially unreacted. This has a great potential for processing gold bearing pyrite/arsenopyrite ores whete selective oxidation of the gold bearing arsenopyrite would liberate the gold for cyanide leaching.

Mechanical activation as pretreatment step for leaching is an effective procedure for treatment of various types of concentrates. It will be shown later (Sects. 6.3.3

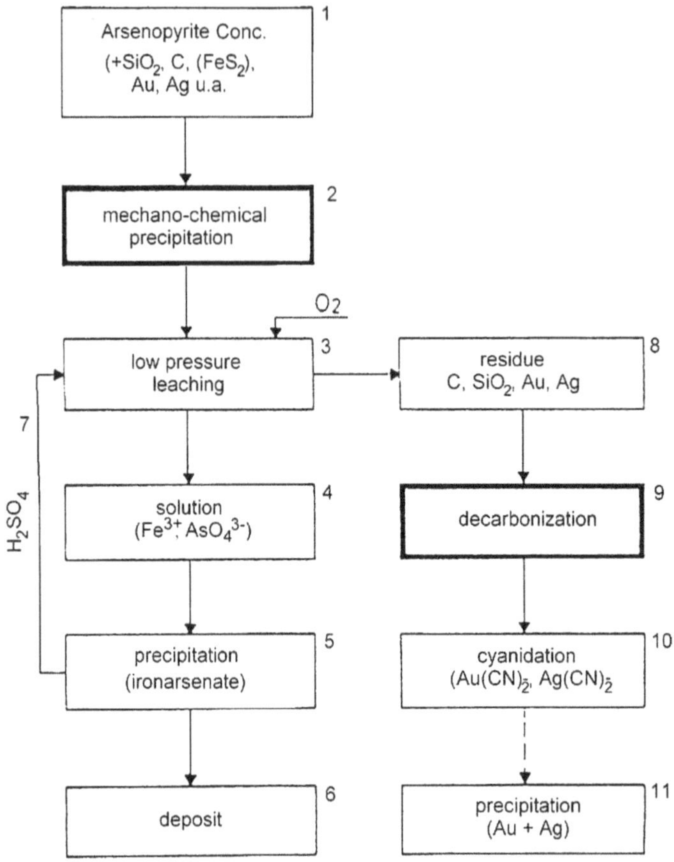

Fig. 6.12 Flowsheet for the treatment of arsenopyrite concentrate by vibration milling for recovery of Au and Ag [Gock and Asiam 1988]

and 6.3.4) that several commercially operating processes utilize a mechanical activation step in their technological flowsheets.

6.3.1.3 Chromites

Chromite $FeCr_2O_4$ is the source mineral for the production of chromium metal, ferrochromium, special alloys and Cr chemicals. The annual worldwide production of chromium chemicals is estimated to be approx. 1 million tons, as sodium dichromate $Na_2Cr_2O_7$ [Münster 1997]. The only relevant process for production of chromates is the oxidative digestion of chromites using Na_2CO_3 at 1000–1100°C. In order to prevent the melt from stick, a leaning agent such as $CaCO_3$ is added to the digestion mixture, to give the melt a loose structure and to facilitate the diffusion of the oxygen to the chromite. The process is governed by reactions

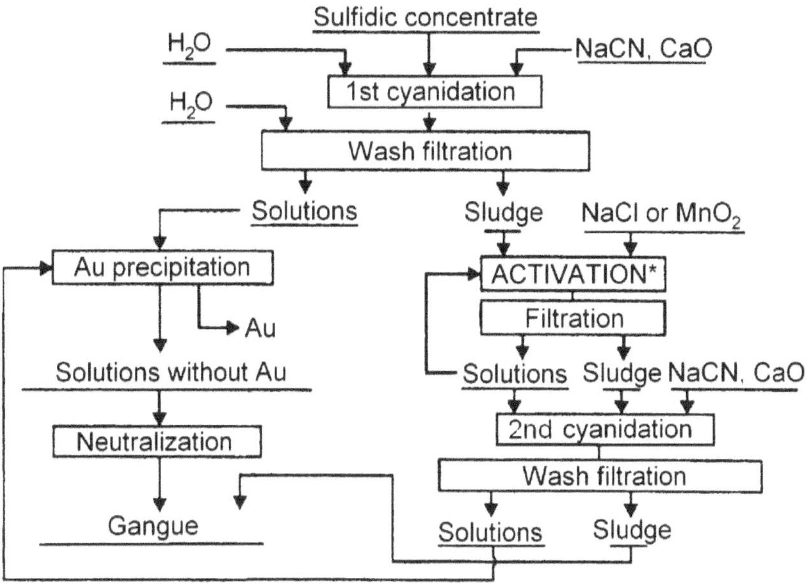

Fig. 6.13 Flowsheet for the treatment of sulphidic concentrate by vibration milling for gold recovery [Kulebakin et al. 1999]

Table 6.10 Results of gold extraction from cyanidation of sulphidic concentrate [Kulebakin et al. 1999]

Parameter	Mechanical activation in NaCl solution (60 gl^{-1})	Mechanical activation with MnO_2 addition (1%)	
Revolutions of the planetary mill (min^{-1})	800	800	600
Au in solid rest after 1st cyanidation (gt^{-1})	9.8	9.4	9.4
Cyanidation after 1st activation and pulp aeration			
NaCN (gl^{-1})	0.7	0.4	0.3
Leaching time (h)	19	18	16
Au in solid rest (gt^{-1})	4.4–6.0	3.6–5.0	4.6–5.0
Cyanidation after 2nd activation and pulp aeration			
NaCN (gl^{-1})	0.3	0.5	0.5
Leaching time (h)	22	18	26
Au in solid rest (gt^{-1})	3.4–3.8	3.0–4.0	3.2–4.0
Au recovery (%)	93.5	93.7	93.5

$$FeCr_2O_4 + 4Na_2CO_3 + 4CaO + 7O_2 \rightarrow 4Na_2CrO_4 + 4CaCrO_4 + 2Fe_2O_3 + 4CO_2 \quad (6.3)$$

$$2Na_2CrO_4 + 4CO_2 + H_2O \Leftrightarrow Na_2Cr_2O_7 + 2NaHCO_3 \quad (6.4)$$

Calcium chromate $CaCr_2O_4$, as opposed to sodium chromate Na_2CrO_4 is poorly soluble, so that the recovery of chromate may be impaired during the leaching of the digestion cake. $Na_2Cr_2O_7$ is won as a crystallizate and is the source material for the production of most chromium compounds.

The disadvantage of the above-described process is that it is impossible to achieve a complete digestion of the chromite ores. Mechanical activation as pre-treatment step to improve the digestion step has been applied [Dincer et al. 2000].

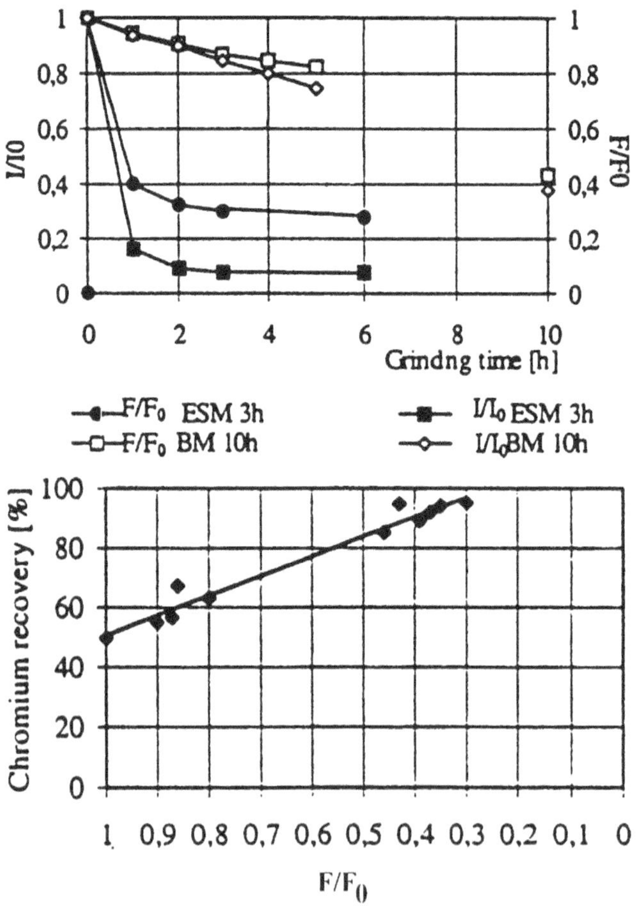

Fig. 6.14 Relative X-ray intensities I/I_0 and F/F_0 for lattice plane (311) of chromite $FeCr_2O_4$ for BM and ESM milling (*left*) and chromium recovery after digestion of chromite vs. F/F_0 ratio (*right*) [Dincer et al. 2000]

Two types of mills were used for comparison, a laboratory scale ball mill (BM) and an industrial eccentric vibratory mill (ESM). Figure 6.14 (left) shows the relative X-ray intensities I/I_0 ratio and the peak area F/F_0 ratio for lattice plane (311) of chromite $FeCr_2O_4$ in dependence on the milling time. The more intensive effect in ESM mill has been achieved. The samples were subsequently applied for digestion tests. Figure 6.14 (right) shows the correlation between the X-ray peak area ratio F/F_0 and chromium recovery for the performed investigations. It follows from the results that the ratio F/F_0 must be at least 0.35 in order to achieve a chromium recovery of more than 95%.

In Fig. 6.15, the modified flowsheet for the production of sodium dichromate is given. The introduction of mechanical activation step enabled the reduction of the mass flows due to complete digestion and the improved landfill-suitability of the tailings.

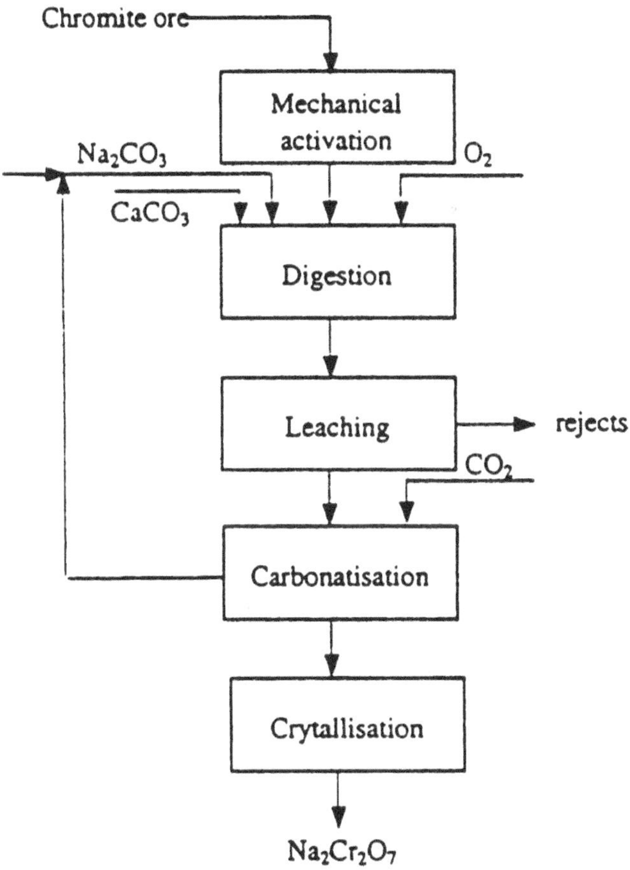

Fig. 6.15 Flowsheet for sodium dichromate $Na_2Cr_2O_7$ production with introduction of mechanical activation step [Dincer et al. 2000]

6.3.1.4 Hydroxides

Investigations were performed in Germany in order to elucidate possibilities for improving processing steps of the hydrometallurgical treatment of bauxite ores by the Bayer process with mechanochemical method [Szanto and Schultes 1969; Pawlek et al. 1992].

Bauxite $Al(OH)_3$ is usually a composition of three minerals: hydrargillit α-$Al(OH)_3$, böhmit γ-$AlO(OH)$ and diaspor α-$AlO(OH)$ with varying leachibility. The use of NaOH to leach bauxite was invented by Bayer in 1892 as a process for obtaining pure aluminium hydroxide which can be calcinated to pure Al_2O_3 suitable for processing to metal [Habashi 1993]. Approximately 2 tons bauxite yield 1 ton Al_2O_3 from which 0.5 ton aluminium is produced.

The reaction in NaOH leaching are following

$$Al(OH)_3 + OH^- \rightarrow [AlO(OH)_2]^- + H_2O \qquad (6.5)$$

$$AlO(OH) + OH^- + H_2O \rightarrow [AlO(OH)_2]^- \qquad (6.6)$$

Tests in an attritor revealed that the pressure leaching extraction of the aluminium hydroxides in the bauxite by sodium hydroxide solutions could already be achieved with high extraction rates at 90°C after relatively short treatment times. Additional favourable results indicate that CaO additions to the bauxite are effective to avoid undesirable sodium silicate formation in the sodium aluminate solution and that the black residual sludge contains up to 70% hematite Fe_2O_3 and nearly negligible aluminium and sodium amounts which might allow to consider further applications which could avoid the disposal problems encountered with the conventional red mud precipitates [Pawlek et al. 1992; Kumar et al. 2004, 2005].

The comparison of the Bayer process and the mechanochemical process is given in Table 6.11.

6.3.2 Attritors in Hydrometallurgy

Attritors were patented in the fifties in the USA and in 1956 the license for their manufacturing was transferred to Netzsch Company in Germany. This mill type was originally used for applications in the chemical and pharmacy industry [Netzsch 1994a] and later for powder metallurgy [Zoz 1995a, b; Zoz et al. 1997, 1998; Kaup et al. 2002b] and mineral processing [Svedala 1996; Ani-Metprotech 1998; Liddell and Dunne 1988].

Attritors use the comminution intensity between the contact surfaces of moving balls, similar to the operation of conventional ball mills, but without the disadvantages of the latter [Püpke 1971; Netzsch 1994a; Klimpel 1997]. An increase in contact points and therefore of contact surfaces is achieved by the use of small milling balls (2–4 mm diameter). Unlike the ball motion in the rotating drum body of conventional ball mills, the balls in the attritor are brought to a higher degree of

Table 6.11 Comparison of the Bayer and mechanochemical processes [Pawlek et al. 1992]

Parameters	Bayer process	Mechanochemical process
Wet milling to (−200 μm)	+	+
SiO_2 removal	+	−
Chemical decomposition	250°C, 100 bar	90°C, 1 bar
Na_2O content in leaching medium (gL^{-1})	140	250
Maximum content of Al_2O_3 in leaching solution (gL^{-1})	185	300
Residence time (min)	3	15
Thickener	+	+
Filtration	+	+
Rest	2 t red mud/t Al_2O_3	0.75 t black mud/t Al_2O_3
Cooling to 50°C	+	+
Colour of leaching solution	Brown	colourless
Addition of inoculation substance	300 gL^{-1}	−
Conditioning (h)	50	4
Filtration	+	+
Filtrate to disposal	+	+

acceleration by a rotating stirring device in a stationary mill container surrounded by a cylindrical cooling chamber Fig. 6.16.

A new series of attritors (Fig. 6.17) were developed by Netzsch Company for continuous mode of operation. The main features of these are: completely enclosed design, newly developed separating elements, horizontally mounted milling chamber, mechanical seal and improved cooling system [Netzsch 1994b].

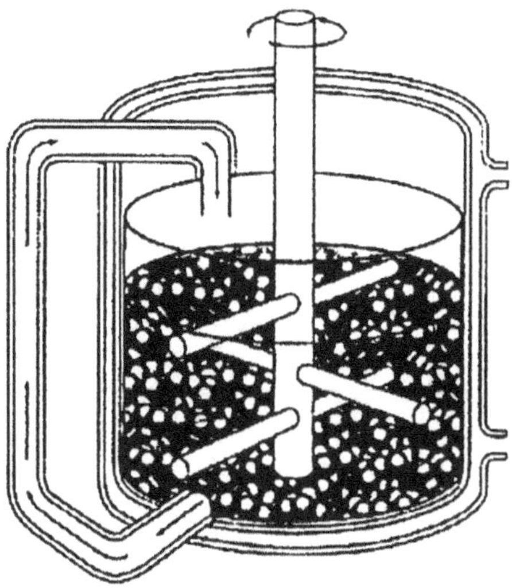

Fig. 6.16 Schematic arrangement of the Union Process attritor-batch mode of operation [Klimpel 1997]

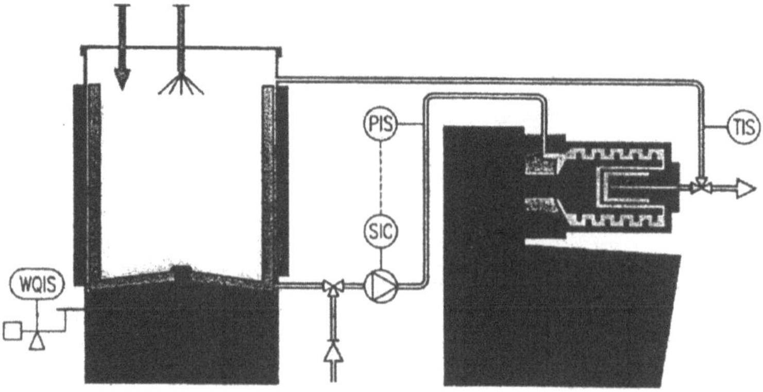

Fig. 6.17 Schematic arrangement of the Netzsch attritor-continuous mode of operation [Netzsch 1994a]

It is possible to vary the kinetic energy of attritor by varying the rotation rate of the stirring mechanism and the media mass by changing the density and diameter of the milling balls. The parameters allow the mill may to be used so that particle size, surface area and size distribution may be optimised, but can also modify the structure of solids during milling. Through special construction of the mill fitted with an eccentric annular discs, it is possible to imparts both centrifugal and a centripetal accelerative forces to the milling elements, thereby resulting in highly intensive mechanical activation of the solids throughout the entire milling chamber.

Attritors are suitable for fine milling in the size range from 45 microns down to 5 microns or less. These types of mills currently working in Australia and New Zealand have installed power up to 3 MW and milling chamber volumes up to 6,000 l.

Improvements in particle liberation and the generation of high surface area provide for enhanced metal recovery by low-pressure, low-temperature autoclave oxidation, or by alternative processes such as bacterial oxidation or cyanidation.

The application of an industrial attritors has been effectively applied by hydrometallurgical processes. The works [Enderle et al. 1997; Johnson et al. 1998] were aimed to obtain concentrates from Australian orebodies suitable for subsequent metallurgical process. Many investigation failed to produce either separate lead and zinc concentrates or a single mixed Zn-Pb concentrate. The principal reason was the extremely small average grain size for galena (1 μm) and for non sulfide gangue (9 μm) in conjuction with the relatively small grain size of sfalerite (26–38 μm) and pyrite (27–45 μm) depending on the orebody deposit.

A new type of a horizontal attritor mills (IsaMill) of 3000 l (1.1 MW) and 10000 l (2.6 MW) have been developed (Fig. 6.18) and tested for the Australian ores [Peace et al. 2005, 2006]. Operating conditions and the efficient energy utilization are listed in Table 6.12.

Fig. 6.18 The horizontal stirred mill-IsaMill [http://www.isamill.com]

Table 6.12 Operating conditions and specific energy consumption for IsaMill [Enderle et al. 1997]

Site	Mill pressure (kPa)	% Solid (by weight)	Temperature of exciting pulp (°C)	Size reduction ratio	Specific energy[3] ($kWht^{-1}$)
Mount Isa	225	40	40–43[2]	1.67	7.6
McArthurRiver					
-open circuit	300[1]	20	40–50[2]	3.75	28
-close circuit	425	20	40–50[2]	4.30	36

[1] Approximate value
[2] Depending on ambient temperature
[3] Based on net power draw (gross power-no load power)

6.3.3 Mechanochemical Processes for Ore Treatment

A growing list of hydrometallurgical processes are lining up to seize the ultimate prize in the treatment of primary sulphide concentrates, out of hands of the smelters.

During the International Conference Copper Cobre 2003 in Chile the current status of copper hydrometallurgy and the most commercially attractive potential applications were reviewed [Peacey et al. 2003]. The main processes applying sulphate leaching are listed in Table 6.13. Among them Activox and Nenatech processes (see in detail later) apply fine milling. The particle size of concentrate into leaching is usually less than 20 microns. The features of these processes include

- leaching temperature in these process is lower than sulphur melting temperature (119°C)
- element sulphur formed in the processes remains in the leaching residue which leads to the difficulty to recover precious metals from leaching residue
- some sulphate is produced that is precipitated as gypsum
- iron precipitation is as goethite and jarosite

Table 6.13 New sulfate processes for chalcopyrite concentrate leaching [Peace et al. 2003]

Process	Leach pressure	Sulphur product	Proposed mechanism	Level of development	Note
Activox	Low	S^0, SO_4^{2-}	Direct	Laboratory	Fine milling
Nenatech	Atmospheric	S^0, SO_4^{2-}	Direct	Laboratory	Fine milling
Dynatec	Pressure (medium)	S^0	Direct	Laboratory	With coal
AAC/UBC	Pressure (medium)	S^0	Direct	Pilot plant (planed)	With surfactant
Placer Dome	Pressure (high)	SO_4^{2-}	Direct	Demo plant (started 2003)	High temperature
Biocop	Atmospheric	SO_4^{2-}	Indirect	Commercial (underway)	Bacterial
Bactech/Mintek	Atmospheric	S^0, SO_4^{2-}	Indirect	Demo plant	Bacterial
Geocoat	Atmospheric	SO_4^{2-}	Indirect	Pilot plant	Bacterial

6.3.3.1 Lurgi-Mitterberg

Chalcopyrite $CuFeS_2$ is among the most refractory minerals with respect to leaching agents and even under high pressures and temperatures the recovery of copper is only to about 20% [Habashi 1978]. The leaching of chalcopyrite flotation concentrate was tested for the first time on an industrial scale in the Lurgi-Mitterberg process (Fig. 6.19) [Biangardi and Pietsch 1975, 1976; Tü and Fischerrke 1978]. According to proposers of the process, the leachability of copper can be improved by mechanical activation of the concentrate in a vibration mill. If the performance of the mill is sufficient, a complete extraction of copper can be achieved in single step at temperatures below the melting point of sulphur.

The flowsheet of the process is represented in Fig. 6.20. The chalcopyrite flotation concentrate from the Mitterberg deposit was dried to ~1% moisture and activated in a vibration mill. The energy necessary for copper extraction was dependent upon the mineralogical composition of the concentrate. For the Mitterberg concentrate, an energy input of about 300 $kWht^{-1}$ was needed for the recovery of 96% of copper. If a predominantly chalcocite Cu_2S concentrate was tested, the energy input necessary for obtaining at minimum 95% of copper decreased to 100 $kWht^{-1}$.

The product obtained after milling the chalcopyrite concentrate is mixed with the reversible electrolyte from the extraction of copper. The electrolysis proceeds at pressures of 1–2 MPa and residence time of 2 h, the efficiency being 0.6 tm^{-3} for 24 h. The partially dissolved iron is precipitated in the autoclave together with arsenic, antimony, bismuth and other contaminants and remains in the solid residue. The solution obtained by pressure leaching is pure enough to be electrolysed following the solid-liquid separation. The cathode copper is produced using a current density of 200 Am^{-2}.

Fig. 6.19 Pilot plant of the Lurgi-Mitterberg process at Mühlbach (Austria): the main building (*up*) the vibration mill (*down*) [Biangardi and Pietsch 1976]

The Lurgi-Mitterberg process was in operation in 1974–76, the capacity was 1 t of cathode copper daily. The plant was closed because of high operational costs with the concentrate transport and high power consumption.

6.3.3.2 Irigetmet

Extensive studies of the influence of mechanical activation on the efficiency of subsequent cyanide leaching of some sulphidic concentrates containing gold were made in Irigetmet in the former Soviet Union [Mullov and Lodejščikov 1979]. It was found that the time of cyanization can be decreased several fold after optimum mechanical activation. Optimising the regime of milling was essential because the consumption of NaCN increased with the extent of activation owing to mechanochemical destruction of sulphidic minerals. The optimum regime of milling

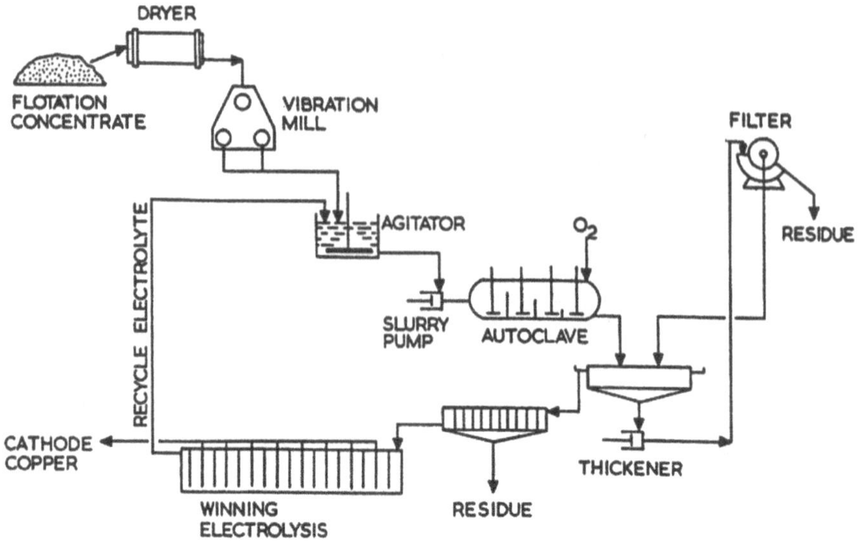

Fig. 6.20 Flowsheet of the Lurgi-Mitterberg process [Biangardi and Pietsch 1976]

consisted of attaining 95% 20–40 μm. A continuous planetary mill working at relative acceleration $b/g = 50–70$ using a revolution rate of 1500 min^{-1} and rate of pulp feed of 2 dm^3h^{-1} proved to be satisfactory. The recovery of gold increased by 11% while the time of cyanidation shortened threefold. It is interesting that a significant increase in NaCN consumption did not appear.

Irigetmet subsequently developed an amalgam-free technology for gold extraction from sulphidic gravity concentrates. The technology comprises thermal decomposition of sulphides followed by milling and gravity separation of free gold. A planetary mill proved to be the best for the selective milling. In the course of testing an arsenopyrite concentrate, a gold recovery of 97.9% was achieved with practically all gold particles larger than 20 μm recovered. By cyanidation of the finely ground residue the total gold recovery was >99.9%. This technology offers the possibility that besides an increase in gold extraction the highly toxic process of amalgamation can be eliminated from the technological cycle.

6.3.3.3 Activox

The Activox process was developed in Australia as an alternative to the pretreatments of sulphidic concentrates by roasting and bacterial oxidation [Corrans and Angove 1991, 1993a; Angove 1993; Corrans et al. 1993b, 1995; Johnson et al. 2000; Palmer and Johnson 2005]. The process has been applied to the recovery of non-ferrous and precious metals from concentrates and calcines. A principle idea of Activox is shown in Fig. 6.21.

Activox is a hydrometallurgical process combining ultra fine milling to a P_{80} of ~10 μm with a low temperature (100°C), low pressure (1000 kPa) oxidative leach

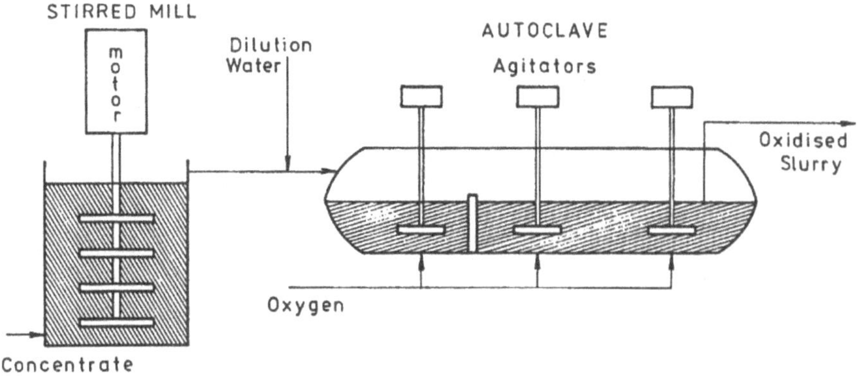

Fig. 6.21 A principle idea of Activox process [Corrans et al. 1995]

to liberate metals from a sulphide matrix. Base metals, (i.e. copper, zinc, nickel and cobalt) are extracted into the leach liquor, while gold and silver remain in the leach residue in a form suitable for further processing. The Activox conditions favour the formation of elemental sulphur over sulphate, thereby using less oxygen (usually less than 1.5 kg of O_2 per kg of S) than required for complete oxidation to sulphate (typically 2.2 kg O_2 per kg S). Other features include rapid oxidation times (1–2 h), clean pregnant liquors and the possibility to treat environmentally hazardous species such as arsenopyrite to produce stable ferric arsenate residues [Evans and Johnson 1999].

As mentioned above it is not necessary totally oxidise all the reduced species to achieve satisfactory results. Elemental sulphur in the leach residues can be recovered as a valuable by-product. The second advantage is the reduction in oxygen demand by the mild conditions of oxidation. Oxygen is one of the major operational costs in oxygen-leaching and typical data for oxygen consumption are given in Table 6.14.

The Activox process offers a number of potential advantages [http://www.wmt.com.au; Palmer and Johnson 2005]. Some of them are

- oxygen consumption is often much less than the requirements for total sulphide oxidation
- most of the iron precipitates selectively as hematite or goethite, which can be readily stabilized
- slurries, post oxidation, settle and filter well
- recoveries of metals such as Ni, Co and Cu into solution exceeded 95%

Table 6.14 Oxygen consumption during pressure oxidation of nickel concentrate [Corrans et al. 1993]

	Conventional pressure oxidation	Activox^R pressure oxidation
Sulphide sulphur (%)	22.5	22.5
O_2/sulphide sulphur (mass ratio)	2.2:1	1.3:1
Reduction ratio of O_2 usage	1	0.6

- Au recoveries from residue by cyanidation after oxidation are usually 90–95% and cyanide consumptions are moderate
- mild conditions of oxidations allows the use of comparatively low cost materials for construction
- studies have indicated substantial reduction in capital and operating costs for Activox, compared to other technologies. Downstream processes can be tailored to suit the Activox products
- Activox can operate with highly saline process water, and water contaminants that are detrimental to bacteria can be tolerated.

One of the first examples of the Activox application was to improve Ni recovery from pentlandite concentrate from Western Australia. The concentrate was milled to various sizes in an attritor and subsequently oxidised in an autoclave for one hour at oxygen pressures of below 1000 kPa and temperatures below 100°C [Corrans et al. 1993]. After milling to 80% – 5 microns and mild oxidation for 40–60 min the recovery was 97% Ni. The recovery of nickel under the same leaching conditions but without milling was no more than 50%Ni.

Later a much wider range of applications which include Cu, Ni, Co, Zn and Au concentrates were tested. As for activation equipments, three different mills including Bradkeen-Metprotech, Netzsch-IsaMill and Metso-Mineral were tested [Adams and Johnson 2001; Palmer and Johnson 2005].

The Activox based flowsheet for the production of Ni, Co and Cu from sulphide concentrates is shown in Fig. 6.22. The flowsheet is based on proven unit operations

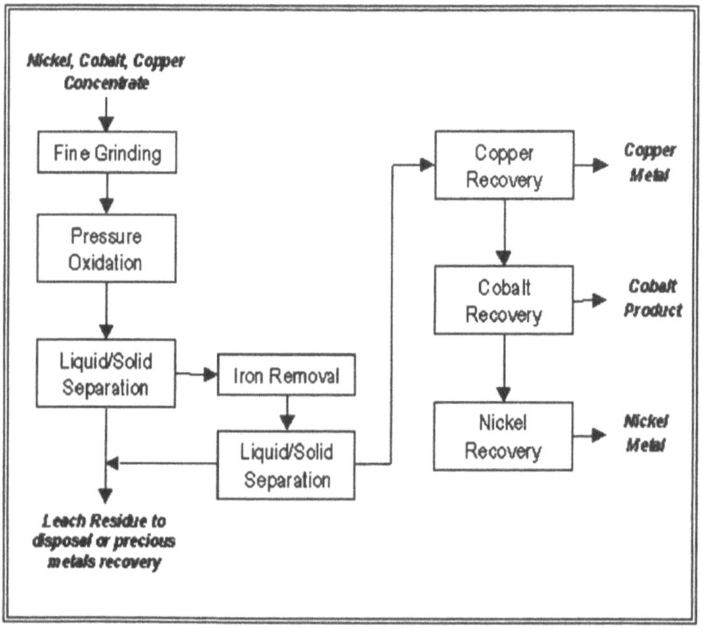

Fig. 6.22 Flowsheet of the Activox process for production of Ni, Co and Cu [http://www.wmt.com.au]

and only uses readily available commercial reagents. The flowsheet can handle a wide range of feed materials and is not sensitive to short or long term variations in feed grades [http://www.wmt.com.au].

The Activox process is also able to achieve the liberation of encapsulated gold from milled sulphide minerals. A typical flowsheet is shown in Fig. 6.23.

An arsenopyrite-pyrite concentrate was fine milled and treated by the cyanidation process (Table 6.15).

It can be seen from the data in Table 6.15 that fine milling on its own does not substantially improve the gold extraction over the as-received concentrate, whereas fine milling and relatively mild conditions of pressure ($P < 1000\,kPa\ O_2$, $T < 100°C$) improved the gold recovery to 91% during the subsequent cyanidation. Generally 60–90 min retention time (in milling or leaching) is required to effect liberation of the gold by oxidation of the sulphides.

Many other Au bearing concentrates have been treated by this method. Cyanide leaching of gold usually exceeds 90% and is often above 95%.

The Activox process has been already successfully demonstrated in several major pilot plant campaigns, the first in 1995 for the Yakabindie Nickel Project, followed

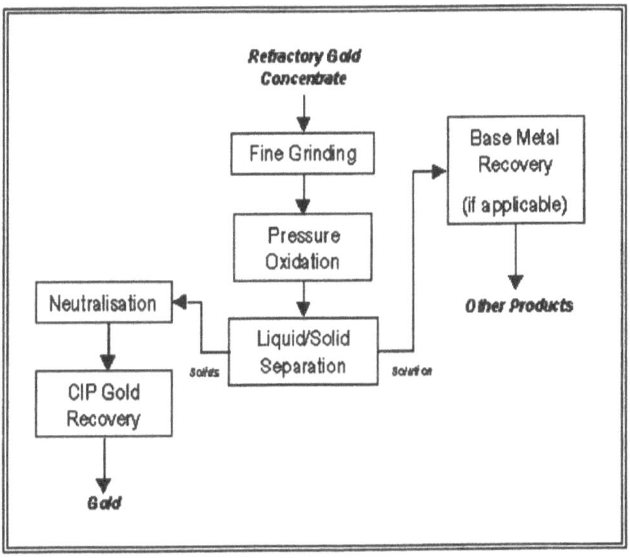

Fig. 6.23 Flowsheet of the Activox process for production of Au [http://www.wmt.com.au]

Table 6.15 Cyanidation of arsenopyrite-pyrite concentrate [Corrans et al. 1993]

Process	Grind size P_{80}(μm)	NaCN (kgt^{-1})	Au recovery (%)
–	75	16.1	60
Fine milling	17.6	19.2	66
Fine milling	3.7	19.2	68
Activox	5	14	91

by campaigns in 1998, 1999 and 2000 on mixed Ni, Co and Cu bearing concentrates from an operating mines in Africa [http://www.wmt.com.au].

The recent extention of the Activox process is the Tati Hydrometallurgical Demonstration Plant (HDP) [http://www.wmt.com.au; Kloiber et al. 2005]. HDP is located 40 km from Francistown in north-eastern Botswana and treats 310 kg/h dry nickel concentrate of composition: 4.22% Ni, 0.127% Co, 2.53% Cu, 33.6% Fe and 21.7% S^{2-}. The leaching results are given in Table 6.15a.

After leaching, the HDP then follows conventional hydrometallurgical steps of solid/liquid separation, iron removal, solvent extraction and electrowinning to recover metal values. The simplified flowsheet is presented in Fig. 6.24.

In the ultra-fine milling step, the technology applies all three types of vertically and horizontally stirred mills mentioned above. Each mill conforms to common design basis: 500 kg/t concentrate, 50% solids, $P_{80} = 10\,\mu m$ and power consumption 45 kWh/t.

An economic evaluation of refractory gold processes based on fine milling pre-treatment has been performed [Johnson et al. 1995]. From the evaluation it follows

Table 6.15a Activox leaching results at HDP [Kloiber et al. 2005]

Base metal recovery	96% Ni, 91% Co, 78% Cu
Operating pressure	1000 kPa
Operating temperature	100–110°C
Residence time	2.6 h
Chloride addition target	4 g/L in autoclave discharge
Acid addition rate	44 kg/t to autoclave feed
Oxygen addition rate	83 kg/h

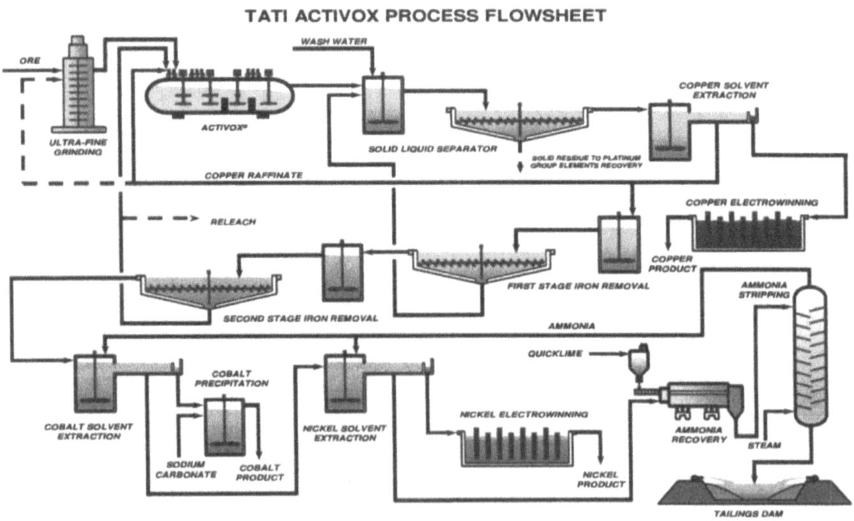

Fig. 6.24 Tati HDP Activox flowsheet [Kloiber et al. 2005; Palmer and Johnson 2005]

- less oxygen is required by the Activox process than by pressure oxidation, biological oxidation or roasting (when coupled with an acid plant for sulfure capture),
- lower reagent consumptions are required for the neutralization step as the extent of oxidation is less,
- the power for fine milling is not excessive, and accounts for only 15% of the total power consumed by the process. Considerably more energy is saved by not having to generate the oxygen required to achieve full sulphur oxidation.

Energy consumption is relatively insensitive to the type of media being used (Fig. 6.25), but is affected by pulp density and hardness of the activated material.

Capital and operating costs for ultrafine milling have been estimated [Corrans et al. 1993]. It must be emphasized that the costs are very specific to both size and mineralogy, so that the data in Table 6.15b should be viewed as a guide only.

In applying stirred mills the three main operational costs are [Corrans et al. 1993]

- energy – minimised by choice of media size, optimization of classification, optimisation of mill design

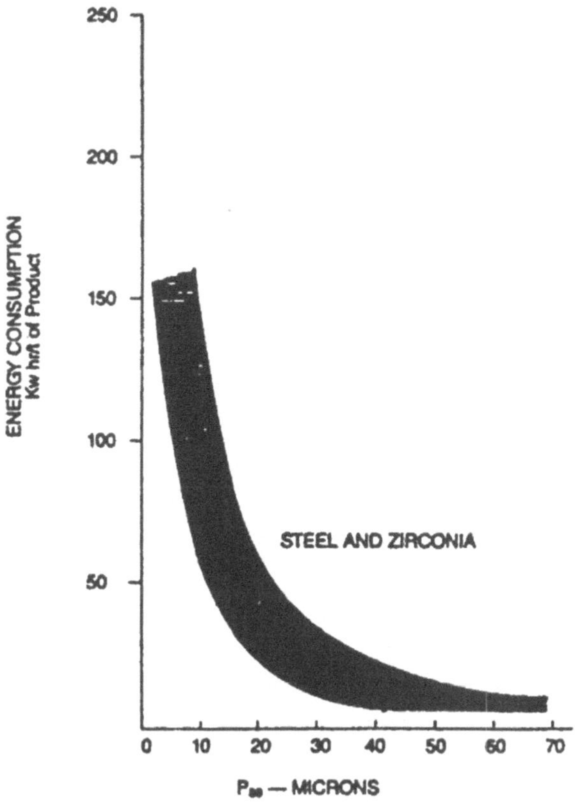

Fig. 6.25 Energy consumption during ultrafine milling [Corrans and Angove 1991]

Table 6.15b Relative economics of different pretreatment processes for FeS_2/FeAsS concentrate [Corrans et al. 1993]

Costs	Ultrafine milling (Activox) (%)	Pressure oxidation (%)	Roasting (%)	Bacterial oxidation (%)
Capital	60–80	100	90	60
Operating	60–80	100	75	90

- media – minimised by selection of media to provide maximum wear resistance at the lowest initial price, and
- wear parts – minimised by selecting the materials that have the best wear characteristics for their particular location in the mill.

A typical operating cost determined for ultrafine milling of 13 th^{-1} copper concentrate from a feed 80% of 60 microns to a product 80% of 6 microns in closed circuit classification is given in Table 6.16.

Table 6.16 Operating cost for ultrafine milling of chalcopyrite/pyrite concentrate to 6 μm [Liddell 1997]

Operation	Operating costs (A$)
Energy consumption	2.00
Media consumption	1.43
Wear parts	1.06
Miscellaneous	0.22
Total operating cost (excluding labour and capital cost)	4.71

To put the operating cost in context with the other unit operations, the ultrafine milling direct operating cost is 1.02 Australian cents per pound of cathode copper produced. This is approximately only 2.5% of the total combined crushing, milling, flotation, ultrafine milling, oxidation, solvent extraction and electrowinning operating cost [Liddell 1997].

Engineering studies have indicated that the hydrometallurgical Activox process is practical and cost effective to build and will be considerably cheaper to operate than the more traditional pyrometallurgical approach thus becoming an attractive alternative for processing sulphide mineral concentrates.

6.3.3.4 Albion (Nenatech)

The process was developed in 1993 in Australia and it is known by two names-Albion and Nenatech (http://www.albionprocess.com; Hourn et al. 1999]. Originally the technology was developed to treat the Nena deposit in Papua New Guinea. Technology owner calls it the Albion process-after the suburb in Brisbane.

The key to the Albion process is the ultrafine milling stage. The process of ultrafine milling results in a high degree of strain being introduced into the mineral lattice. As a result, the number of grain boundary fractures and lattice defects in the minerals increase by several orders of magnitude, relative to unmilled minerals. The increase in the number of defects within the mineral lattice activates the mineral, facilitating leaching. The rate of leaching is also enhanced, due to the dramatic increase in the mineral surface area.

Passivation of the mineral surface by sulphur based leaching products is also minimised by ultrafine milling. Typically, milling precipitates that form on the surface of a leaching mineral will slowly passivate the mineral, by preventing the access of chemicals to the mineral surface. Passivation is normally complete once this precipitated layer is 2–3 microns thick. Ultrafine milling of a mineral to a particle size of 80% pasing 8–12 microns will eliminate passivation, as the leached mineral will disintegrate prior to the precipitate layer becoming thick enough to passivate it.

The oxidative leaching stage is carried out in non-pressurised agitated tanks. Oxygen is introduced to the leach slurry to assist oxidation. Leaching is carried out autothermally, in that the temperature of the leach slurry is set by the amount of heat released in the leaching reaction. Heat is not added to leaching vessel from external sources. Temperature is controlled by the rate of addition of oxygen, and by the leach slurry density.

The Albion based flowsheet for the production of copper from sulphide concentrates is shown in Fig. 6.26.

The Albion process can apply the same activation equipments as in Activox process: Metso-Mineral mill, Bradken-Metprotech mill and Netzsch-Isa mill (Fig. 6.18) to produce an activated, finely milled concentrate at relatively low specific energy inputs. The specific energy input is 20–50 kWh/t to produce a concentrate milled to P_{80} of 15 microns. The exact value depends on type of the concentrate.

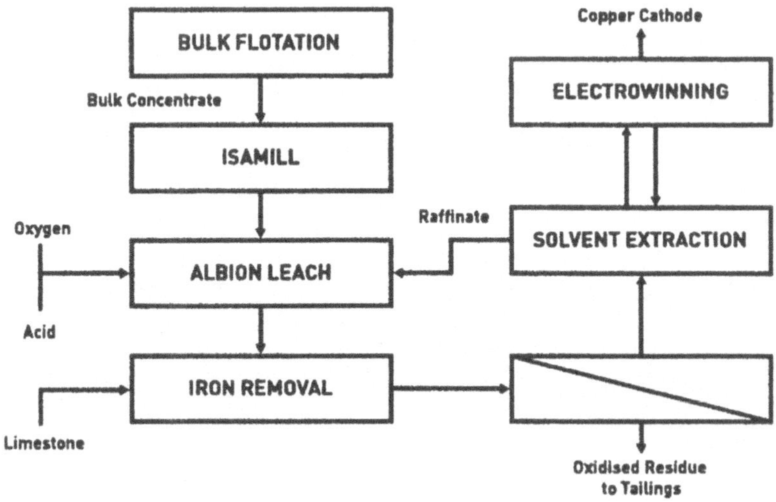

Fig. 6.26 Flowsheet of the Albion process [http://www.albionprocess.com]

Table 6.17 Performance of the Albion process on the sulphide concentrates [Hourn and Halbe 1999]

Sulphide concentrate	Sulfur oxidation (%)	Metal recovery (%)
Chalcopyrite	98	98
Bornite	95	97
Chalcocite	98	98
Enargite	88	92
Pentlandite	90	95
Pyrite	85–90	95
Cobaltite	90	85

Base metal recoveries are typically 95–99%. Gold and silver recoveries are typically 90–95% with reagent consumption typically a half to one quarter of bacterially leached residues. Typical metals recoveries for selected refractory minerals are listed in Table 6.17.

Capital costs for Albion process are specific to each project's location, throughoutput and concentrate grade. As for milling, several installations of 1 MW Isa Mill have been performed. A single mill would process approximatelly 200–250000 tonnes per annum of bulk concentrate, which of grading 15% Cu is equivalent 50–65 US$ of capital pre annual tonne of Cu produced. This illustrates the relatively low capital cost associated with attritors as an enabling technology for the hydrometallurgical treatment of refractory sulphides [Hourn and Halbe 1999].

Capital costs for an Albion process plant are less than 50% of a comparative pressure leach, bacterial or roasting plant. A comparison of capitals costs for a 50000 t/a Cu leach plant between the Albion process and the other oxidative leach technologies is shown in Fig. 6.27.

Some principal components of the operating costs for a 15% Cu concentrate are summarized in Table 6.18.

Typical operating costs for a 15% copper concentrate treated by Albion process in schema milling-leaching-solvent extraction-electrowinning is equivalent 308 US$.

6.3.3.5 Sunshine

In 1984 the Sunshine Mining and Refining Company introduced a new concept to the hydrometallurgical treatment of complex sulphidic concentrates with antimony, copper and silver content. Later the concept was modified for the sulphuric acid oxygen pressure leaching with the application of nitric acid [Anderson et al. 1993]. This treatment allows the recovery of silver and copper from the solid residue after alkaline leaching of tetrahedrite.

It was shown that the milling had an important role by the introduction of a regrind circuit before the pressure leach improving the overall recovery of silver. The milling in tube mill led to size reduction from original P_{80} of 25 microns to P_{80} of 10 microns. The installation of the regrind circuit and application of sodium nitrite

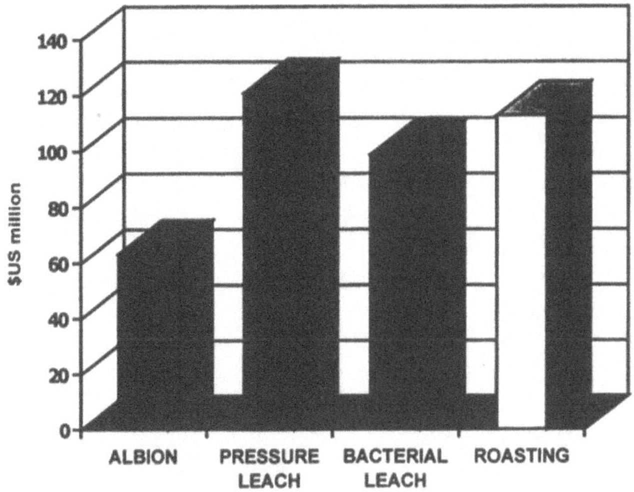

Fig. 6.27 Comparative capital costs for the Albion process and other oxidative leach processes [Hourn and Halbe 1999]

Table 6.18 Typical operating costs for Cu production in Albion process

Operation	US$/tonne of Cu
Milling including pumps and media feeders	11
Milling media	6–8
Electrowinning	125
Leaching and neutralization	15
Limestone	20–40
Oxygen/air	35/15
Labour, maintenance, minor consumables	100

in the plant allowed the silver recovery to increase from the historically recorded value of 87.5–92.1%. At full production, these enhancements allow recovery of an extra 230,000 troy ounces of silver per year [Anderson et al. 1993; Nordwick and Anderson 1993; Allen 1998].

Sunshine Mine in Idaho which was the richest silver mine in USA history was shut down in 2001 because of financial problems and rapid decrease of silver prize on world market. However, in October 2005 Sterling Mining Company announced a three-stage plan to restore activity at the Sunshine Mine. Production, on a preliminary scale, is planned to start late 2007 or early 2008.

6.3.3.6 Melt

Melt process has been developed in Slovakia for hydrometallurgical treatment of tetrahedrite sulfidic concentrates [Baláž et al. 1994c; Baláž 2000; Baláž and

Achimovičová 2006]. The process name Melt is an abbreviation: Mechanochemical Leaching of Tetrahedrites.

Tetrahedrite, in simplified formula $Cu_{12}Sb_4S_{13}$ is one of the most common sulfominerals. The general formula representing the naturally occurring tetrahedrite-tennantite series is $(Cu,Ag)_{10}(Cu,Zn,Fe,Cd,Hg,Cu)_2(Sb,Bi,As)_4S_{13}$ denoted $M^{+}_{10}M^{2+}_{2}M^{3+}_{4}S_{13}$. In most natural tetrahedrites, the two Cu^{2+} atoms are replaced by a divalent element such as Zn^{2+} which usually occupies a tetrahedral site in sulphides [Pattrick and Hall 1983].

Tetrahedrites represent the most important source of copper and antimony and are also of economical interest due to their content of silver and mercury.

The tetrahedrite concentrates from the Rožvnava deposit are produced by non-selective sulphide flotation and have a high content of copper (~27%), antimony (~16%) and silver (~4000 gt^{-1}) [Fábry 1992].

With the aim to developing a processing route which allows extraction of most of the metal values from these tetrahedrite concentrates various hydrometallurgical methods have been considered too. The variable presence and proportions of metals in the tetrahedrite and the accompanying sulphides complicate the leaching and subsequent metal recovery steps. The devise a method for solving this problem requires a multistep hydrometallurgical process which exhibits high selectivity in individual stages.

Hydrometallurgical treatment of tetrahedrite is possible in acid oxidative [Dutrizac and Morrison 1984; Havlík et al. 1991, 1994, 1998] or in alkaline solution [Baláž et al. 1994a; Holmes 1944; Melnikov 1977; Pawlek 1983]. By acid oxidative leaching, e.g. in acidified ferric chloride solutions, copper and iron enter into solution, while antimony is partially precipitated as a compound with the composition similar to the mineral tripuhyite $FeSbO_4$. The overall leaching reaction proceeds very slowly and the kinetics is complicated [Havlík et al. 1998; Havlík and Škrobian 2000].

Alkaline leaching in sodium sulfide medium dissolves selectively antimony leaving copper and iron in the solid residues. Arsenic and mercury are also solubilized as complex anions. This process has a high selectivity with copper and precious metals remaining in the solid residue which is suitable for smelter treatment.

The chemistry of the reaction between tetrahedrite (on simplified formula Cu_3SbS_3) and Na_2S can be described by equations [Anderson and Nordwick 1996]

$$2Cu_3SbS_3 + Na_2S \rightarrow 3Cu_2S + 2NaSbS_2 \tag{6.7}$$

$$NaSbS_2 + Na_2S \rightarrow Na_3SbS_3 \tag{6.8}$$

The soluble Na_3SbS_3 containing trivalent antimony is oxidized to a product containing pentavalent antimony by the polysulfide ions present in the leaching liquor

$$(x-1)Na_3\,SbS_3 + Na_2\,S_x \rightarrow (x-1)Na_3\,SbS_4 + Na_2S \tag{6.9}$$

The behaviour of arsenic leaching from tennantite $Cu_3AsS_3$by Na_2S may be described by equation

$$2Cu_3AsS_3 + 3S^{2-} \rightarrow 3\,Cu_2S + 2AsS_3^{3-} \tag{6.10}$$

or in the presence of polysulphide anions by equation [Hiskey and Atluri 1988]

$$Cu_3\,AsS_3 + 5/(2n-2)\,S_n^{2-} + (3n-8)/(2n-2)S^{2-} \rightarrow 3CuS + AsS_4^{3-} \tag{6.11}$$

However in addition to tennantite, arsenic derived from tetrahedrite, orpiment and realgar is also solubilised in the alkaline solution – arsenopyrite is resistant to leaching in this medium.

The leaching of mercury sulfide with sodium sulfide gives a soluble complex too, according to the equation [Melnikov 1977]

$$HgS + Na_2S \rightarrow Na_2HgS_2 \tag{6.12}$$

This salt is prone to hydrolysis and its solution stability necessitates the presence of a base, usually NaOH.

The refractoriness of tetrahedrite requires the application of concentrated leaching agents, high temperatures and long leaching times for efficient dissolution of the valuable antimony content. In preliminary experiments, the recovery of antimony into alkaline leach was not higher than 40% after two hours of leaching (Fig. 6.28).

Alkaline leaching of tetrahedrite with a solution of Na_2S has been applied by the Sunshine Mining and Refining Company [Ackerman et al. 1993; Nordwick and Anderson 1993]. Leaching was carried out using a 280–300 gl^{-1} sodium sulphide solution at boiling point (104°C) and atmospheric pressure. The process is run in batch mode with a 12 h residence time solubilising 90–95% Sb and 60% As [Gould 1955].

In 1992, the Institute of Metallurgy of Technical University in Berlin and the Institute of Geotechnics, Slovak Academy of Sciences in Košice started to test a new method which combined milling and leaching in a batch process within a stirred ball mill (attritor). The results, summarized in Table 6.19, reveal 52–99% recoveries of

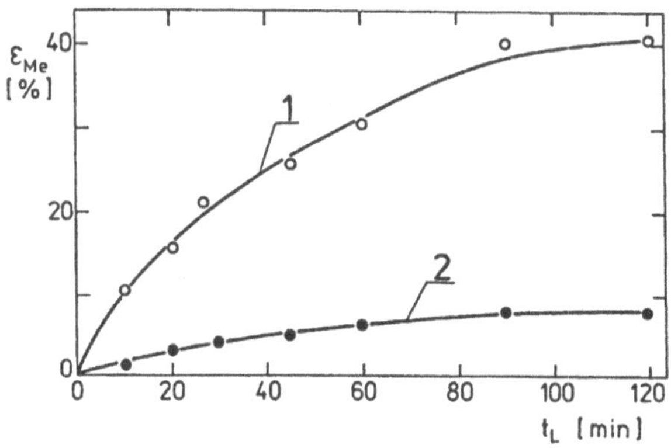

Fig. 6.28 Recovery of antimony (1) and mercury (2) into leach, ε_{Me} vs. time of leaching, t_L for as received tetrahedrite concentrate, temperature 90°C [Baláž 2000]

Table 6.19 Experimental conditions (temperature, T, Na_2S concentration, c) and recoveries of metals by mechanochemical leaching of tetrahedrite concentrate [Baláž et al. 1994a]

T	C	Metal in leach (%)				
(°C)	(gl^{-1})	Sb	Cu	Fe	Hg	Ag
60	100	52.09	<0.1	<0.1	<0.1	0.60
60	150	83.90	<0.1	<0.1	0.16	0.46
60	200	82.64	<0.1	<0.1	0.33	0.45
80	100	99.64	<0.1	<0.1	0.62	0.40
80	150	99.78	<0.1	<0.1	0.11	0.47
80	200	92.76	<0.1	<0.1	0.11	0.38
95	100	84.31	<0.1	<0.1	0.12	0.45
95	150	95.02	<0.1	<0.1	<0.1	0.49
95	200	92.96	<0.1	<0.1	0.14	0.74

antimony after 60 min of mechanochemical action [Baláž et al. 1994a, b]. Antimony is selectively leached while the other valuable metals remain almost entirely in the solid residue.

This new process was further developed and tested in semi-industrial and industrial scale attritors using a continuous mode of operation [Baláž et al. 1994a, b, c, 1995, 1997a, b, 2003; Baláž and Sekula 1996; Sekula et al. 1998; Baláž 2000; Baláž and Achimovičová 2006]. The principal flowsheet of operation, in which mechanochemical (I) leaching is followed by chemical (II) leaching, is shown in Fig. 6.29.

The mechanochemical leaching of the concentrate is characterized by the typical leaching plots for antimony, arsenic and mercury in Fig. 6.30, as well as by the data in Table 6.19. The process was performed in a combined regime of milling and leaching. The initial stage of mechanochemical leaching (I) for 18 min was succeeded by the combined mechanochemical and chemical leaching (I+II). The presented results show that almost total extraction of antimony can be achieved by leaching for 40 min, leaving 0.25% Sb in the solid residue (Table 6.20).

The behaviour of mercury is not consistent with its expected solubility in the alkaline leaching solution. The effect of a hindered mercury extraction was thought to be due to electrochemical effects, e.g. mercury cementation with iron balls or iron

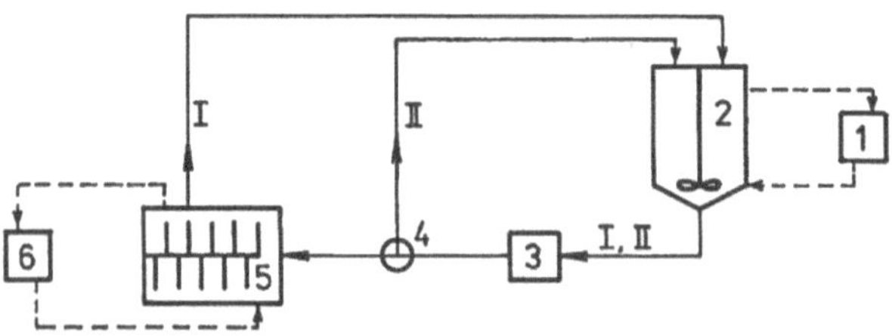

Fig. 6.29 Flowsheet of the leaching unit: 1 – heating, 2 – chemical reactor, 3 – pump, 4 – valve, 5 – attritor, 6 – cooling. The working regimes: I-mechanochemical leaching, II chemical leaching [Baláž et al. 1995]

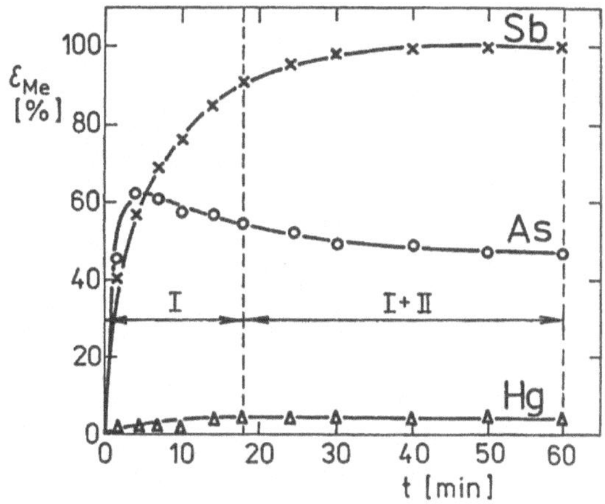

Fig. 6.30 Recovery of Sb, As and Hg, ε_{Me} vs. time of leaching, t. I – mechanochemical leaching, II – chemical leaching. Leaching conditions: 300gl^{-1} Na_2S+50gl^{-1} NaOH, T= 88°C, liquid/solid ratio 3:3, power input 245 kWht^{-1} [Baláž et al. 1997b]

Table 6.20 Chemical composition of tetrahedrite concentrate [Baláž et al. 1997b]

Element	Me (%)	
	As received	Mechanochemically treated
Cu	27.36	26.00
Sb	15.93	0.25
Fe	14.58	16.46
Bi	0.33	0.33
As	1.02	0.27
Hg	0.74	0.11

wear [Baláž et al. 1994a]. This effect was confirmed by experiments with separate mechanochemical and chemical leaching stages. After interruption of the attritor operation at 70 min (see Fig. 6.31), mercury starts to dissolve with higher rates compared to its rate in the mechanochemical regime. At t>200 min nearly total mercury extraction is attained and its amount in solid residue after leaching is below 0.01%.

The behaviour of arsenic in the alkaline leaching process depends on arsenic origin. The recovery of this metal is always lower than 70% as indicated in Fig. 6.32 and it can be assumed that probably only the arsenic present in tetrahedrite or tennantite is extracted, the remaining arsenic (~30%) occurs in other minerals which are insoluble in alkaline sulfide solution (e.g. arsenopyrite) [Anderson and Nordwick 1996; Dayton 1982].

The concept of mechanochemical leaching of tetrahedrite concentrates was developed and verified in a laboratory attritor [Baláž et al. 1994a) and in semi-industrial attritor [Baláž et al. 1995, 1997b; Baláž 2000]. It was further tested in a pilot plant hydrometallurgical unit in Rudvnany [Slovakia) (Fig. 6.33). Figure 6.34 summarizes the flowsheet of this process. Tests have shown that alkaline Na_2S leaching for 30 min at 86°C and atmospheric pressure is required for achieving

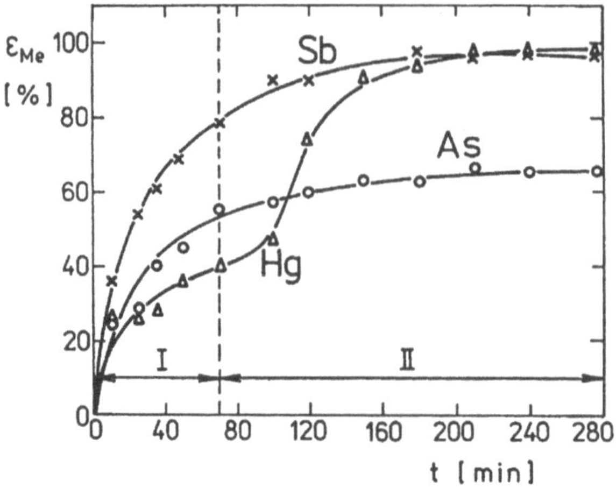

Fig. 6.31 Recovery of Sb, As and Hg, ε_{Me} vs. time of leaching, t. I – mechanochemical leaching, II – chemical leaching. Leaching conditions: 300 gl^{-1} Na_2S+50 gl^{-1} NaOH, T= 95°C, liquid/solid ratio 3:8, power input 120 $kWht^{-1}$ [Baláž et al. 1997b]

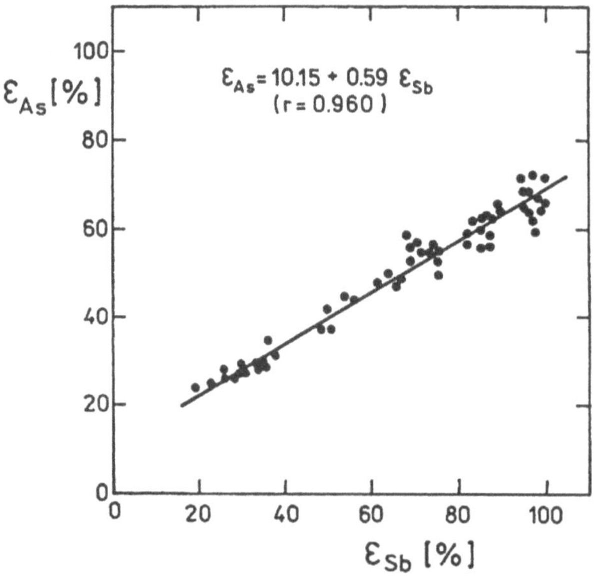

Fig. 6.32 Recovery of arsenic, ε_{As} vs. recovery of antimony, ε_{Sb} in experiments of chemical leaching. Leaching conditions: 300 gl^{-1} Na_2S+50 gl^{-1} NaOH, T= 95°C, liquid/solid ratio 3:3–6:0 [Baláž et al. 1997b]

total extraction of antimony. The other valuable metals (Cu, Ag, Au) form the main economic components of solid residue. The liquid/solid ratio (2:5–4:8) and power inputs (150–250 kWh/t concentrate) that was applied in optimised mechanochemical experiments are acceptable for plant operation [Baláž 2000].

Fig. 6.33 Pilot plant unit at Rudňany (Slovakia): vision of unit and LME Netzsch attritor in combination with chemical reactor [Baláž et al. 1997b)

A similar leaching process Sunshine has been applied by Sunshine Mining and Refining Company in the USA [Anderson et al. 1993; Anderson and Nordwick 1996]. This process applied a slightly modified alkaline Na_2S leaching system where Na_2CO_3 was added to improve the performance, but without intervention of mechanochemical treatment. Table 6.21 compares the leaching conditions for the Melt and Sunshine processes.

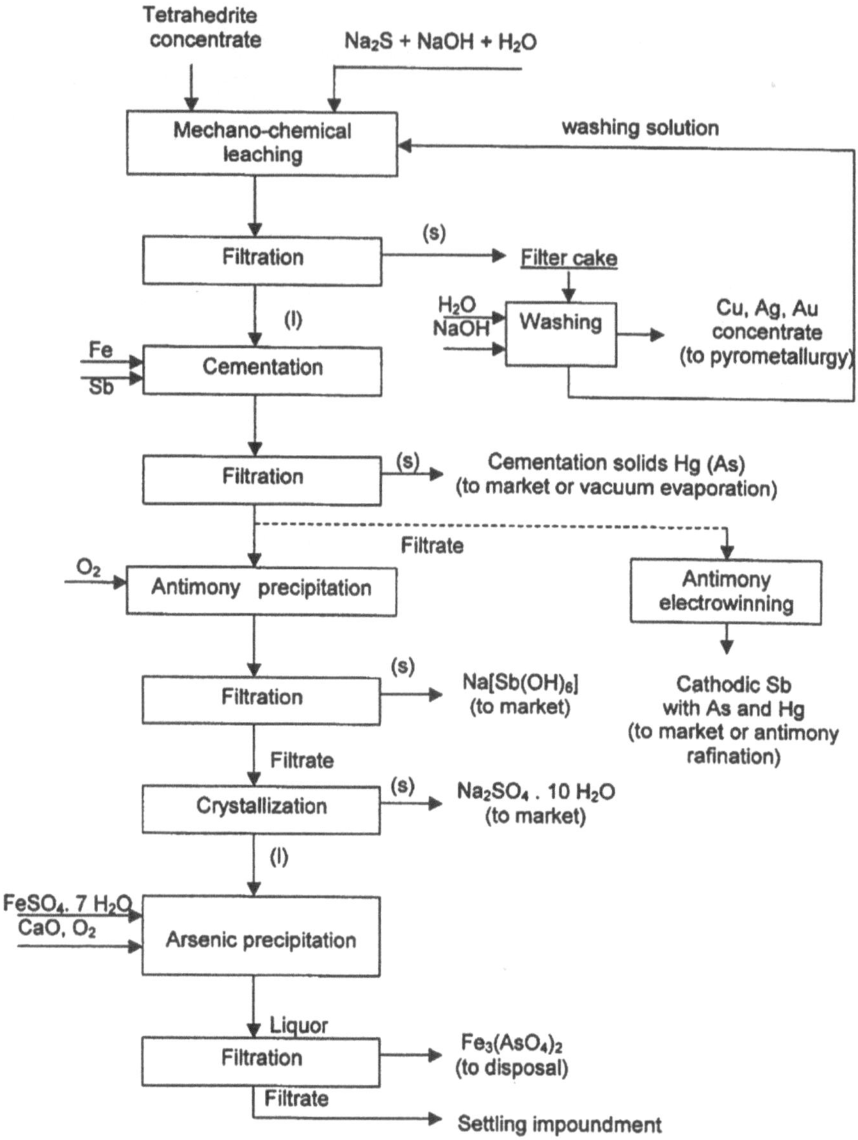

Fig. 6.34 Flowsheet of the Melt process [Sekula et al. 1998; Baláž and Achimovičová 2006]

After inspection of Table 6.21 it is possible to conclude that a much shorter leaching time is needed for the Melt process in comparison with the Sunshine process with relatively small differences in leach conditions. This can bring important economical benefits for the whole technology.

Table 6.21 Comparison of leaching conditions for Melt and Sunshine processes [Baláž and Achimovičová 2006]

Parameter	Melt	Sunshine
Mineralogical composition	Tetrahedrite, pyrite, chalcopyrite, siderite, quartz	Tetrahedrite, pyrite, galena, bournonite
Leach conditions:		
Initial particle size	70%–200 mesh	60%–200 mesh
Leach time [min]	60	720
Temperature [°C]	88–105	105
Na_2S [g/L]	300	100
NaOH [g/L]	50	15
Na_2CO_3 [g/L]	–	25
Solid/liquid ratio [g/L]	300	200

6.4 Chemical Engineering

6.4.1 Crystal Engineering

Mechanochemical solid-state reaction of molecular crystals is a rather new but promising direction. The potential importance of this field relates not solely to the traditional fields of applications of molecular crystals, such as pharmaceutics (see Sect. 6.9) but also to the newly discovered possibilities of their utilization as functional materials for molecular electronics, molecular magnets, etc. [Boldyrev 2006].

In comparison with inorganic solid-state reactions, the interactions in organic molecular crystals are mainly of the non-covalent type, e.g. van der Waals, hydrogen bonds and π-stackings [Braga and Grepioni 2004, 2005; Braga et al. 2006]. No significant activation energies have to be introduced in the exoenergetic reactions of molecular crystals [Kaupp et al. 2002a). Importantly, products formed by co-milling of crystals in the absence of liquid can be different from those obtained from solutions or melts [Braga et al. 2006].

Among the earliest pioneering investigations of solid-state organic reactions were reactions between naphthalene, phenanthrene, anthracene, naphtols, and other substituted hydrocarbons and picric acid [Rastogi et al. 1963; Rastogi and Singh 1966, 1968; Paul and Curtin 1975]. The kinetics studies showed that these reactions were diffusion controlled and that diffusion takes place by surface migration which favors small molecules of regular shape [Braga et al. 1999].

Useful contributions into recent literature of organic solid-state reactions, included those activated mechanochemically have been published [Toda 2002; Tanaka 2003; Kaupp 2005]. Various mills were tested for activation, especially those providing the high energy input such as vibratory mills, planetary mills and attritors. It was established that many types of organic reactions can be realized with these devices. Furthermore, in some cases, the mechanochemical method has proven to be not only simpler and faster than the traditional synthesis, but more selective as well [Boldyrev 2006].

More than 1000 stoichiometric organic solid-state reactions proceeded with 100% yield and do not require purifying workup, i.e. they are solvent-free and waste-free [Kaupp 2005]. Some of these reactions have been successfully scaled-up with application of industrial mills.

Figure 6.35 shows a pilot set-up for performance of solid-state reactions. The collection of the milling product occurs in a cyclone with an internal gas cycle if continuous or semi-continuous operation is desired for large scale production. This equipment allows for complete recovery of the product without charging it with impurities as will be important for synthesis and industrial production. The large mill can presently be upscaled to 400 l volume of the milling chamber [Kaupp et al. 2002b; Kaupp 2006].

Several examples of organic solid-state reactions performed by milling are given bellow [Kaupp 2003a, b, 2005].

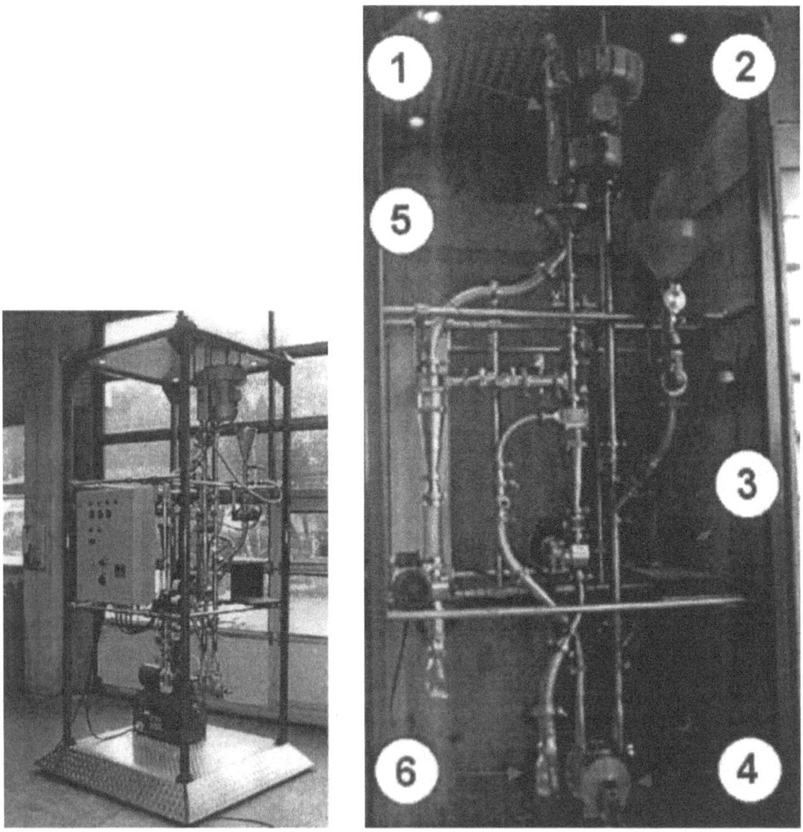

Fig. 6.35 Pilot set-up of a high-energy mill (Simoloyer^R, ZOZ, Germany) with carrier gas cycle and separation/classification system. 1 – side channel turbine, 2 – powder charging, 3 – feeder, 4 – high-energy mill, 5 – cyclon, 6 – powder discharging [Kaupp et al. 2002b; Kaupp 2006]

Glucose and urea formed the stoichiometric crystalline 1:1 complex (reaction 6.13) within minutes. Without milling, only small amounts of products have been obtained by tedious crystallization that took six months.

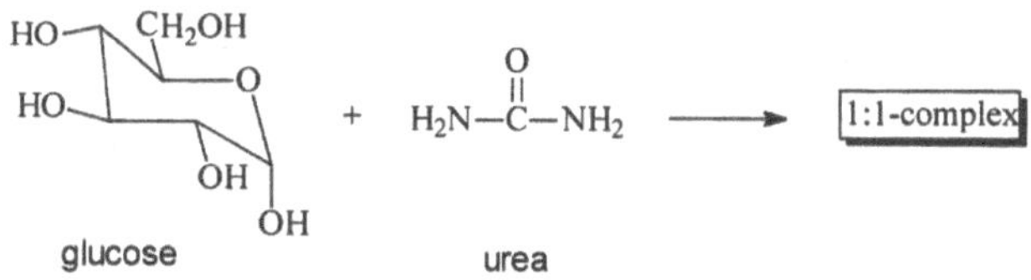

(6.13)

Sodium tartrate dihydrate can be obtained in the form of powder by milling of stoichiometric mixtures of L-(+)-tartaric acid with Na_2CO_3.

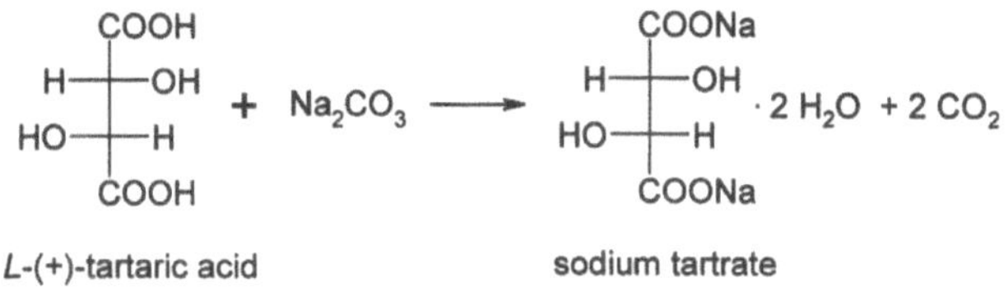

(6.14)

Sodium tartrate is a well known food additive (E 335). The crystals are in use for the preparation of powders or tablets for refreshing drinks. Energy consuming evaporations are avoided that burden the common techniques in water. No excessive heating was observed in reaction (6.14). The liberated CO_2 was released by a safety-valve in a mill.

Imines (Schiff's bases) are versatile building blocks that have been quantitatively synthesized by milling. The scaling-up of p-hydroxybenzaldehyde condensation with p-aminobenzoic acid yielded substituted solid imine hydrate that could be dried in a vacuum at 80°C temperature if desired [Kaupp et al. 2002a]. The water of reaction from the condensation reaction (6.15) does not disturb the procedure by a liquid phase both in the small or large scale runs, as it is incorporated in the product lattice. The reaction proceeds smoothly and efficiently at near ambient condictions without an intervening liquid phase and further up-scaling in larger mills appears possible. This favourable waste-free technical procedure is remarkable, as the previous synthesis of the product of reaction (6.15) required 12 h boiling in ethanol and no yield was reported [Cevasco and Thea 1999].

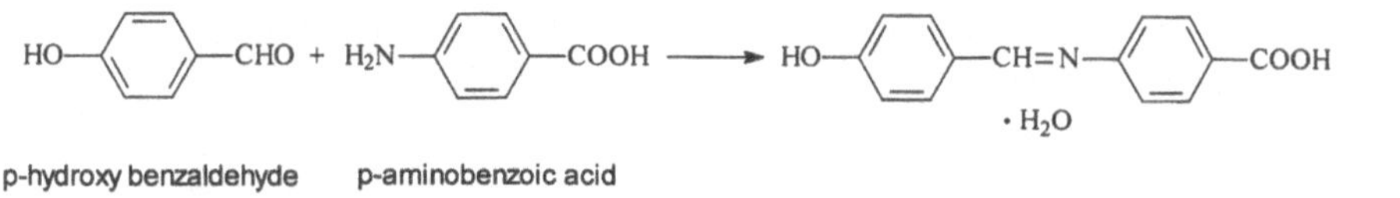

(6.15)

Reaction (6.16) represents an example of solvent-free technique for the heterocyclic ring formation. Phenylacylbromide and thiourea were milled at room temperature. The aminothiazole hydrobromide salt was quantitatively obtained in a complicated cascade reaction sequence (substitution, cyclization, dehydration, isomerization) and represents a versatile building block for organic synthesis with the acid catalyst already attached to it.

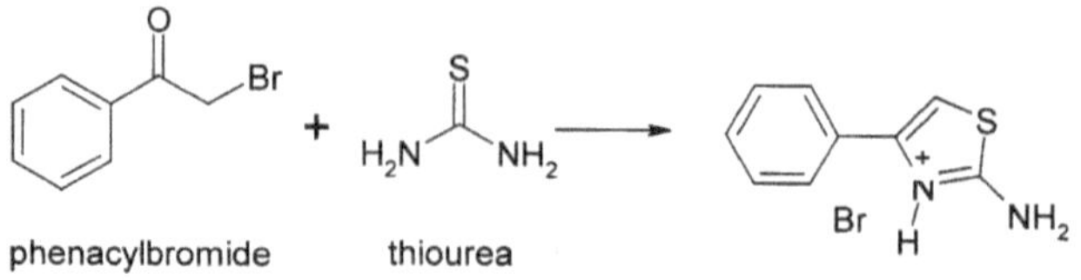

(6.16)

Reaction (6.17) is an example of mechanochemical synthesis of organometallic compound zinc bismethionate in an industrial mill (Fig. 6.36) [Kaufmann et al. 2006]. This compound is prepared by solid-state synthesis at ambient conditions without intervening liquid phase and connected operations like heating, washing, destilation, etc.

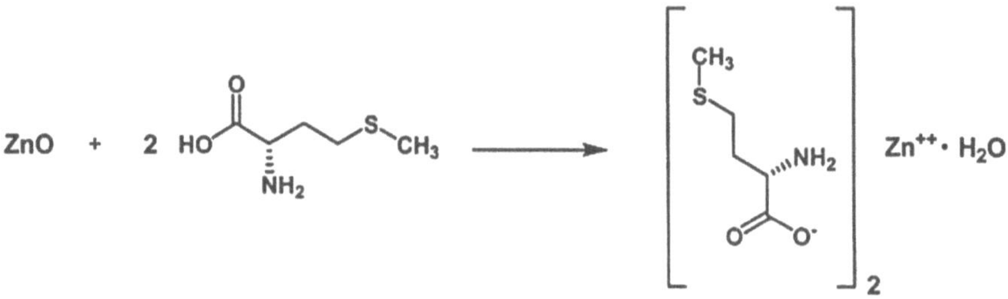

(6.17)

High energy milling represents in organic solid-state synthesis an environmentally benign versatile technique that can be performed at the kg scale. The size of mills with carrier-gas operations may be increased to more than 400 l. The solid-state techniques supercede all other techniques in terms of sustainability, avoiding wastes, energy, time and work savings, easiness of conduction, safety and unmatched performance in the synthesis of highly sensitive products. Furthermore,

Fig. 6.36 Mechanochemical synthesis of zinc bismethionate hydrate in an industrial eccentric vibration mill ESM (Siebtechnik, Germany)

solid-state synthetic procedures are often viable alternative routes to the bottom-up construction of molecular materials from molecular or ionic-building blocks, which is the core of molecular crystal engineering [http://www.zoz.de; Kaupp et al. 2002a, b; Braga and Grepioni 2004; Kaupp 2005, 2006].

6.4.2 Heterogeneous Catalysis

Heterogeneous catalysis comprises the phenomena of the enhancement of the rates of chemical conversions in fluids by the presence of solid surfaces, which are called catalysts [Madey et al. 1976].

The research in heterogeneous catalysis became the basis of modern çatalytical large-scale production techniques like, for example, the synthesis of ammonia, sulphuric acid and of a number of other products of the organo-chemical and fuel industries. In history, hydrogenation of carbon monoxide over Ru, Ni, Co, Fe and Th oxide catalysts became one of the most important sources of gasoline, methane and other chemicals. The development of efficient catalysts for the most important catalytical processes, however, is related to the problems of deactivation and catalysts to poisoning (contamination) their regeneration and reactivation [Weisser and Landa 1972]. Contamination of metal catalyst by small amounts of carbon can completely change the nature of the reaction intermediates and the product distribution during the decomposition of organic compounds [Bond 1962].

The categories of elementary processes customarily used to govern heterogeneous catalysis have been listed as follows [Boudart and Burwell 1974]

- diffusion of reactants to the surface
- chemisorption of at least one of the reactants on the surface
- surface diffusion of chemisorbed reactants and subsequent surface reaction
- desorption of products from the surface
- diffusion of products from the surface

The kinetic data (preexponential factors and activation energies) permit to identify the rate-determining process.

In principle all steps mentioned above can be influenced by mechanical activation of solid catalysts. It has been established already in earlier studies [Heinicke 1984; Heinicke et al. 1967], that mechanical activation affects both the sorption of gases (step 2) upon their interaction with various catalysts and subsequent chemical reactions (step 3).

In catalytic chemical reactions the increase of reaction rate together with decrease of activation energy has been frequently observed [Schrader and Hoffmann 1973]. Strong reduction of the activation energy and, accordingly a strong increase in activity for SO_2 oxidation on mechanically activated V_2O_5 catalyst is documented in Table 6.22.

It is important to stress that not only the reaction rate but also the mechanism of the catalytic reaction changes due to the mechanical activation. This fact makes it possible to increase the selectivity of catalysts with the mechanical activation.

Table 6.22 Activation energy, E_a of the SO_2 oxidation at a mechanically activated V_2O_5 catalyst [Paudert 1965]

Mechanical activation (min)	Treatment	E_a (kJ mol^{-1})
–	–	144
5	After interruption of milling	139
10	After interruption of milling	135
30	After interruption of milling	132
5	During milling	21

The enhancement of adsorption of gases on solid surfaces upon mechanical activation is accompanied is some cases by penetration of gas molecules into the mechanically deformed lattice. The sorption time vs. duration of mechanical activation is characterized for many gases by the existence of an optimum value due to the competition of the mechanically activated adsorption and desorption processes as the active surface centers are gradually filled with gas molecules [Boldyrev 2006; Lomovskij and Boldyrev 2006].

6.4.2.1 Catalysts by Means of Mechanochemical Reactions

The influence of mechanochemical treatment on the catalytic activity of a solid may consist in the fact that new chemical compounds or changed phases are formed due to the treatment having an increased catalytical activity as compared with the initial substances. The formation of metallic carbonyls belongs to this group of mechanochemical reactions.

The mechanism of iron carbonyl $Fe(CO)_5$ formation by milling may be represented by the following sequence of reactions [Heinicke et al. 1970]

$$Fe + S \rightarrow FeS \tag{6.18}$$

$$FeS + CO \rightarrow Fe_3S_2(CO)_9 \tag{6.19}$$

$$Fe_3S_2(CO)_9 \rightarrow FeS + Fe(CO)_5 \tag{6.20}$$

As Fig. 6.37 shows, the sulfphr induced reaction velocity being by the factor of about 100. Since 6–8% of sulphur in the input material is sufficient for the accelerated reaction of the used Fe, the sulphur has catalytic effect on the formation of carbonyl. FeS represents an intermediate product which is regenerated during the disproportionation reaction (6.20).

Carbonyls of various metals (Ni, Fe, W and Mo) have been synthesized by this route. In case of nickel carbonyl $Ni(CO)_4$ a mechanochemical technology (Fig. 6.38) was proposed and tested [Heinicke 1984].

The heart of the plant is a vibration mill designed as a reactor. Here nickel is transformed into $Ni(CO)_4$ by continuous milling of Ni powder in CO atmosphere.

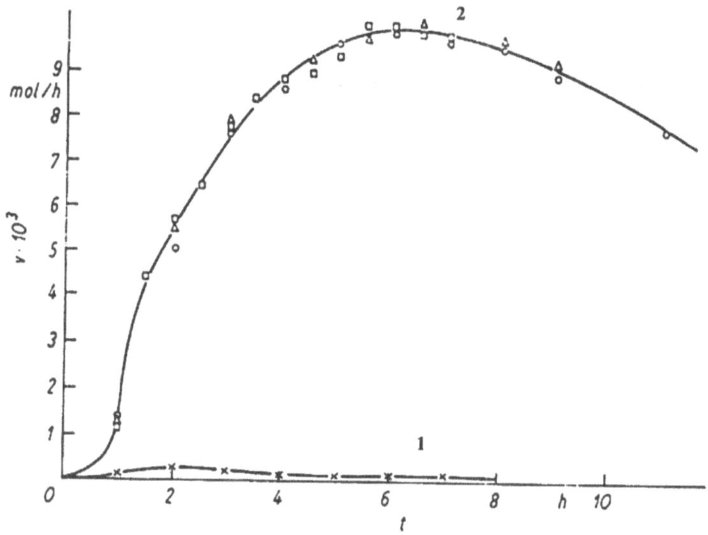

Fig. 6.37 The velocity, v of the $Fe(CO)_5$ formation during the mechanochemical treatment (t – milling time, 1 – pure Fe, 2 – FeS formed by reaction (6.18) [Heinicke et al. 1970]

6.4.2.2 Surface Size Changes of Catalysts by Mechanical Activation

The value of the specific surface area, and consequently the catalytic activity may be increased up during mechanical activation of catalysts. If there is a relatively large specific surface area at the beginning of the mechanical stress, as is the case for the majority of the applied supported catalysts having values of some $100\,m^2g^{-1}$, than a reduction of the specific surface area and consequently of the catalytic activity will appear under the influence of mechanical activation. In many cases this reduction will not be compensated by the activity-increasing effect of the lattice defects (see later). For instance, the initial surface area of an α-Al_2O_3 catalyst which is in the range of about $200\,m^2g^{-1}$ decreases to a value 20–30 m^2g^{-1} after milling 10 h.

However, in some cases this phenomena does not influence the catalytical activity. The hydrodesulphurization (HDS) and hydrogenation (HYD) activities of Mo-containing catalysts are higher for mechanochemically treated sample in spite of lowering its surface area (Fig. 6.39).

A short-time milling causes an increase in the surface area in spite of the relatively large area of about $100\,m^2g^{-1}$ for V_2O_5 catalyst in reaction

$$2SO_2 + O_2 \xrightarrow{V_2O_5} 2SO_3 \tag{6.21}$$

As is shown in Fig. 6.40, the regeneration of the catalyst used up or lowered in their activity is possible in this way.

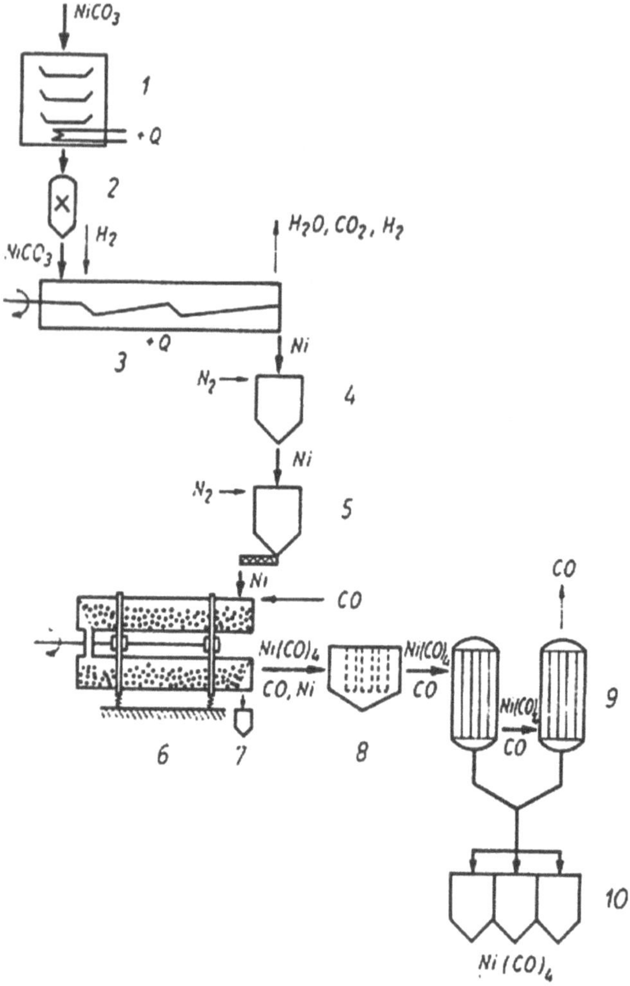

Fig. 6.38 Mechanochemical technology of nickel carbonyl $Ni(CO)_4$ synthesis, 1 – drying, 2 – comminution, 3 – reducing, 4 – bunkering, 5 – dosing, 6 – mechanochemical reactions, 7 – residue, 8 – dust removal, 9 – cooling, 10 – storing. [Heinicke et al. 1975]

6.4.2.3 Perovskite and Zeolite Catalysts

Highly active and stable in hostile environments, monolith perovskite catalysts were elaborated. Pilot tests and bench-scale experiments have demonstrated these catalysts can efficiently operate in the temperature range up to 900–1300°C and at space velocities in the range of 10000/h in the processes of flameless fuels combustion, environmental control as well as in the industrial chemical processes. Applications include fuel combustion, toxic waste incineration, hydrocarbons reforming and sulphur dioxide reduction [Pauli et al. 1992].

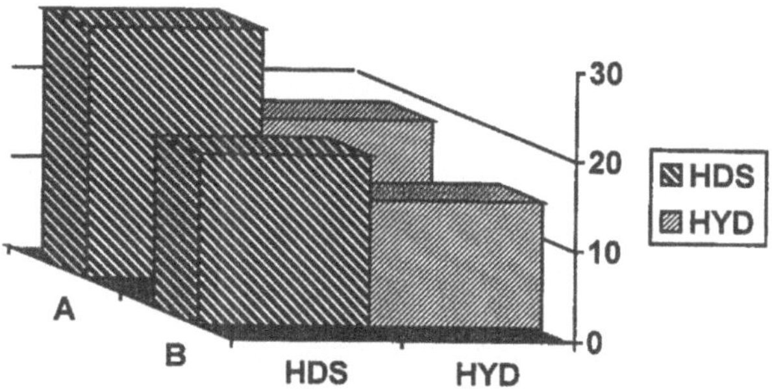

Fig. 6.39 HDS and HYD activities of Mo-containing catalysts synthesized by mechanochemical approach ($S_A = 254\,m^2g^{-1}$) and by impregnation ($S_A = 270\,m^2g^{-1}$) [Kostova et al. 2007]

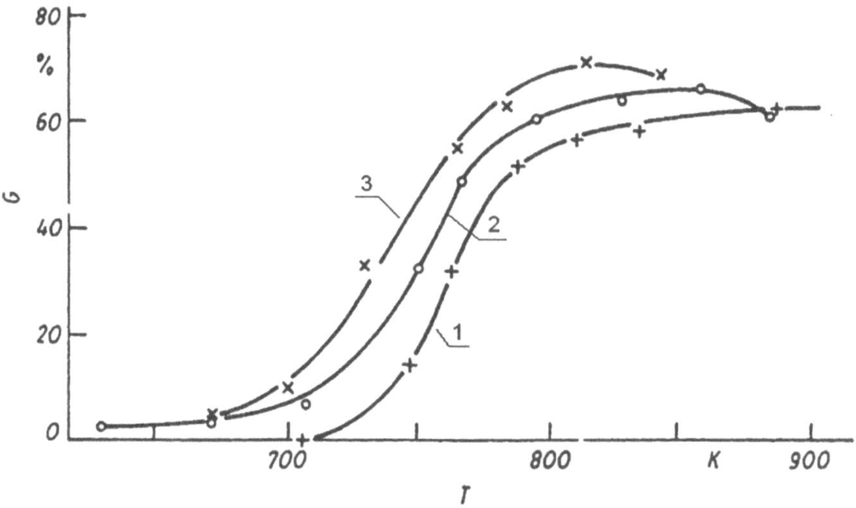

Fig. 6.40 SO_2 oxidation at V_2O_5 catalyst milled in a jet mill, G – degree of conversion, 1 – unmilled, 2 – milled for 2-times, 3 – milled for 5-times [Heinicke 1984]

To elaborate the technology of perovskite monolith honeycomb catalysts production, the methods of preparation of highly dispersed, chemically active powders with uniform phase composition and narrow particle size distribution should be invented. Technique of mechanical activation of solid starting compounds in high-intensity ball mills with a subsequent thermal treatment matches these requirements. Monolith perovskite catalysts were found to have high activity and good stability in a number of industrial catalytic processes such as methane reforming and sulphur dioxide reduction (Fig. 6.41).

Zeolites with their well developed internal surface are good candidates for catalytic applications. The typical question that arises when the activity of zeolite catalysts is studied is whether the reaction proceeds on the relatively small but easily accessible external surface of crystals, or on the very developed "internal" surface formed by a system of intercrystalline micropores, whose size is close to molecular dimensions [Parton et al. 1989].

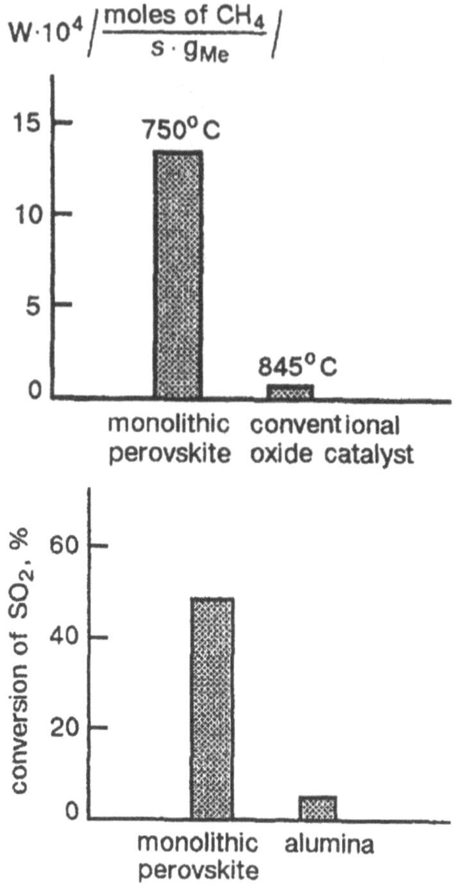

Fig. 6.41 Catalytic activity of the catalysts in methane reforming (*left*) and reduction of methane (*right*) [Isupova et al. 1996]

Dispersion of the zeolite by milling could either increase its external surface area, or decrease its crystallinity (and therefore, intercrystalline microporous space) due to amorphization. The comparison of catalytic activity with the evolution of these factors was supported to help in the location of active sites.

The catalytic properties of Fe ZSM-5 zeolite in benzene oxidation to phenol by N_2O has been studied. The milling of catalyst resulted in a practically complete amorphization and disappearance of its catalytical activity. While catalytical activity for non-activated catalyst was 25% for benzene conversion, this value went down to 1.5% for milled sample. In spite of such a large changes of reaction rate, the selectivity toward phenol remains nearly constant [Kharitonov et al. 1995].

6.4.2.4 Metal Catalysts

Several papers have been devoted to the elucidation of the influence of mechanical activation on catalytic activity of metals by German scientists [Schrader 1966;

Schrader and Tetzner 1961; Schrader et al. 1966; Heinicke 1984). The model reaction of benzene hydrogenation was tested for catalytical activity of mechanically activated Ni and Co powders

$$C_6H_6 + 3H_2 \xrightarrow{\text{Ni,Co}} C_6H_{12} \tag{6.22}$$

The dependence of benzene conversion on reaction temperature is shown in Fig. 6.42.

There is no conversion to hexan for non-activated Ni-catalyst (O′) below 160°C. At 220°C the conversion is only 10%. The activity of the catalyst above this temperature goes down. However, all the activated Ni-catalysts (milling time 10–140 min) show much higher conversion for reaction (6.22) which proceeds by much lower temperatures in this case. Moreover, the working interval is broader in comparison with non-activated Ni.

However, the decrease in catalytical activity of mechanically activated metal catalysts have been also observed. The question of activity is connected not only with total surface area of catalysts but is closely related to their active centers. Ni catalyst treated in the presence of air is no longer effective, so that no catalytic effect will be found without mechanochemical treatment in a subsequent steps. As Fig. 6.43 shows, the steady-state level measured before the catalyst was poisoned will be reached again after the revival of the treatment process [Heinicke 1984].

It was shown that mechanical activation is efficient in creating waste-free energy-saving methods for the preparation of heterogeneous catalysts. Not only the reaction rate but also the mechanism of the catalytic reaction changes due to the mechanical activation. This fact makes it possible to increase the selectivity of catalysts with the mechanical activation [Molchanov and Buyanov 2000, 2001; Boldyrev 2006].

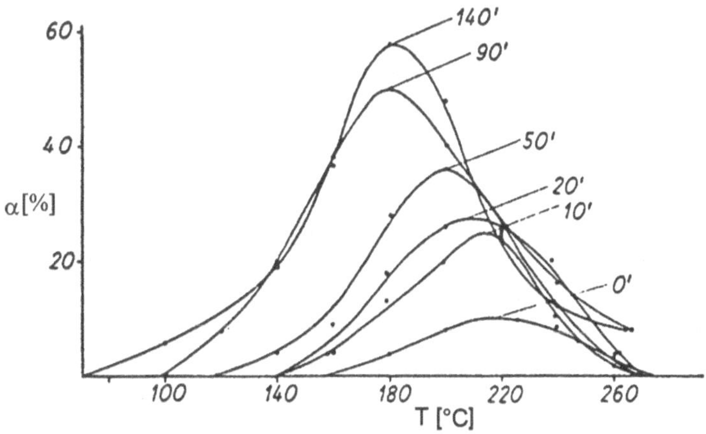

Fig. 6.42 Benzene C_6H_6 conversion, α as a function of reaction temperature T, parameter: milling time (min) [Schrader 1966]

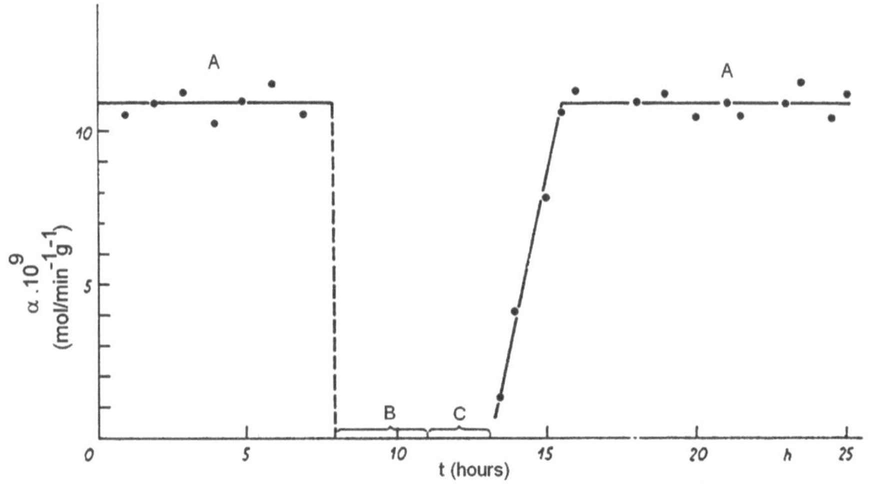

Fig. 6.43 The conversion of C_6H_6 at a mechanically activated Ni catalyst during an intermediate treatment in air: α – catalytical activity, t – reaction time, A – reaction at mechanical treatment, B – treatment in air, C – leaving in H_2/C_6H_6 mixture [Heinicke 1984]

6.5 Materials Engineering

Non-equilibrium processing of materials has attracted the attention of a number of scientists and engineers due to the possibility of producing better and improved materials than is possible by conventional methods [Suryanarayana 1999]. One of the processing methods cited in literature under the term mechanical alloying (MA) is very frequently used for the preparation of advanced materials in materials engineering (see also Sect. 4.3.3). MA started as an industrial necessity in 1966 to produce oxide dispersion strengthened (ODS) nickel- and iron-based superalloys for applications in the aerospace industry. The technique of MA was used for industrial applications from the beginning and the basic understanding and mechanism of the process is only now beginning to be understood. One of the greatest advances of MA is in the synthesis of novel alloys, e.g. alloying of normally inmiscible elements, which is not possible by any other technique. The MA products find applications in various industries and these are summarized in Fig. 6.44.

Mechanical alloying for commercial production is carried out in the mills of up to more than 1000 kg capacity. The ball mills to produce oxide dispersion strengthened nickel- and iron-based alloys, is carried at facilities in the USA and Great Britain (Fig. 6.45).

The major advantages of mechanically alloyed powders include achievement of a homogeneous and fine distribution of metallic and ceramic components to an economical and cost-effective manner. Several application are given below.

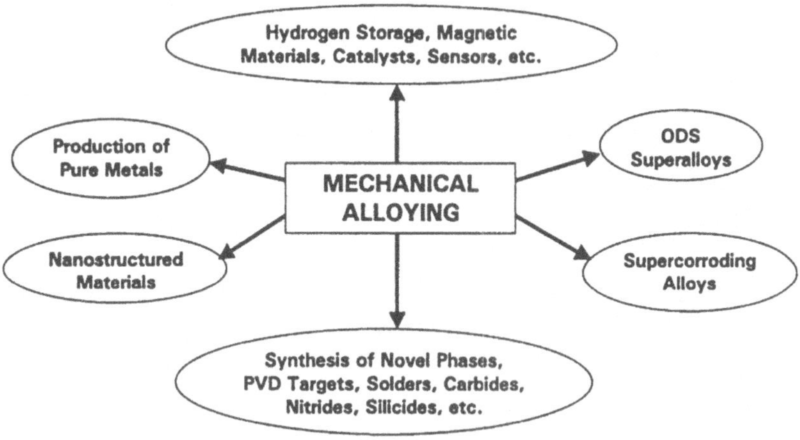

Fig. 6.44 Typical current and potential applications of MA products [Suryanarayana et al. 2001]

Fig. 6.45 Commercial production-size ball mills used for mechanical alloys [Suryanarayana et al. 2001]

6.5.1 Thermoelectric Power Generator Materials

A thermoelectric power generator is a reliable source of electric power. Several materials have been investigated for use in the thermoelectric power generation. β-$FeSi_2$ is one of the most promising materials for energy conversion in the temperature range of 200–900°C. Its biggest disadvantage is the way of its preparation by melting method. The obtained product is porous and brittle and not homogeneous because of segregation problems. Mechanically alloyed powder will not have the segregation problem because no liquid phase is involved during processing. In addition, the number of processing steps is also minimized. For example, Fe, Si and the dopant powder (to make the material either p-type or n-type semiconductor) can be mechanically alloyed and then subjected to sintering followed by β-form annealing to produce the desired material. Figure 6.46 illustrates two different ways of making β-$FeSi_2$ thermoelectric generator material [Suryanarayana 2004].

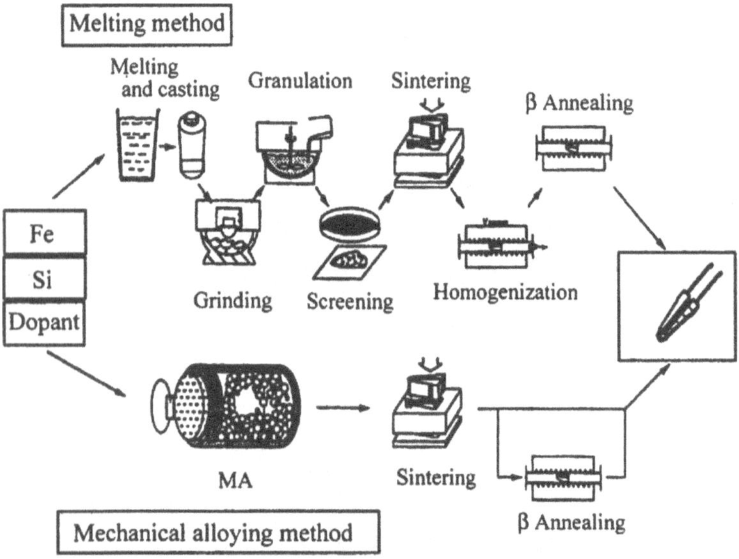

Fig. 6.46 The traditional (melting) and advanced (mechanical alloying) methods to prepare β-$FeSi_2$ [Suryanarayana 2004]

6.5.2 Room Temperature Solders

Reactions between Ga metal or liquid Ga-based eutectics with metal powder such as Cu, Ag, Au, Ni and Sn have been investigated for possible use as low-temperature solders or as dental filling materials. However, it was reported that neither pure metal powders nor equilibrium commercial Cu-Sn alloyes mixed with Ga-based liquids are suitable for practical use due to the long solidification times and poor mechanical properties [Ivanov et al. 1996]. It has been shown that the application of mechanical alloying to such systems significantly improved processing and solding properties of such materials [Suryanarayana 2004].

6.5.3 *Bearings*

In the USA the slide bearings are used based on Stellite (Stellite is a trademark registered in the USA and owned by Deloro Stellite Company). Stellite is 50–60 wt.% Co-matrix alloy containing Cr and W carbides and other composites. Stellite is used as the bearing material although its properties are not satisfactory. MA was applied to improve the properties and the properties were improved [Suryanarayana 2004].

6.6 Coal Industry

Mechanical activation of coal induces the various changes in its structure which is possible to utilize in application of coal as advanced material.

Initially, it considered that increase of chemical activity of mechanically activated coal is result of its milling, formation of the new and accessing of inaccessible pores and increase of external and internal surface. Howver, it turned out that reaches to other physico-chemical changes: disruption of chemical bonds of organic compounds present in the coal, change in its distribution, formation of soluble products and volatile substances, creation of intensive electric fields, emision of electron and formation of free radicals, etc. [Lomovskij and Boldyrev 2006; Chrenkova 1993].

High-energy milling also causes liberation of gases such as O_2, CO_2, CO, H_2 and formation of hydrocarbons (CH_4, C_2H_6, C_3H_8, C_4H_{10}, C_nH_{n+2}, C_nH_{2n}). Their distribution depends on coal composition and condition of mechanical activation.

In the case, that after mechanical activation of coal in technological cycle follows thermal processing, the velocity of thermal reaction increases and its temperature decreases. Change of coal reactivity is conditioned by amorphization and by decreasing of molecular weight in process of destruction and formation of new function groups.

Technologically interesting possibility for utilization of mechanical activation of coal are advanced applications in preparation of compounds with potential therapeutic effect (humic acids and their derivates).

6.6.1 *Humic Acids*

The process of formation of humic acids in coal in alkaline media was studied intensively in the USA. Its basic principle resides in the oxidation of coal mass, while the efficiency of the process depends on the type of coal, conditions of oxidation (even in the presence of catalysts), and technological equipment (e.g. fluid system) (Bailey et al. 1954]. In the USA, alkaline coal oxidation in water solutions, in the presence of O_2 and increased temperatures 200–300°C, has been developed into a pilot operating test [Franke et al. 1952]. It is currently known that the effective industrial production of sodium salt of humic acid or humic acids from coal is limited by the content of humic acids in coal batch at the level of 70% [Friedman and Kinney 1950].

Mechanical activation has significantly stimulated advances in coal processing for special processes. However, a cleaning process is important mainly for sulfur removal.

The desulphurization of coal by chemical methods is accomplished by various acid or alkaline reagents. The study of the processes applying hot alkaline solutions for the purification of coal is not a recent development. As early as in the 1940s, this process was used in Germany to prepare a low-ash product from coal suited for the production of electrode carbon. It has been stated [Crawford 1951; Reggel et al. 1992] that the content of ash in coal can be reduced from the original 98 to 0.7% by using 10% NaOH at increased pressure and temperature of 225°C. The application of NaOH solutions at increased temperatures (150–250°C) and pressures (0.6–2.5 MPa) was investigated in several laboratories.

Another possibility of using alkaline medium for cleaning of coal involves atmospheric pressure. The higher temperatures are applied in this case. Molten NaOH (or its mixture with KOH) is used and the effective temperature is about 400°C. The ability of molten NaOH to desulphurize coal has been known for more than three decades [Masciantonio 1965]. The liberation of the pyritic sulphur starts at temperatures above 150°C while the organic sulphur is released from the coal matrix at temperatures above 200°C. The process known as molten caustic leaching (MCL) was tested in the USA [Chiotti and Markuszewski 1985; Shan et al. 1989; Chriswell et al. 1994]. The objective was to remove mineral components, pyritic and organic sulphur by the reaction of coal with a mixture of molten sodium and potassium hydroxides. For leaching time of 60–180 min and temperatures of 370–400°C, the effectiveness of MCL process for removal of individual components was as follows: 90–95% of pyritic sulphur, 70–90% of organic sulphur and 90–99% of ash.

Due to the presence of sulphidic admixtures in coal and the fact that organic sulphur is firmly built in its matrix, the coal itself represents the refractory material from the viewpoint of chemical cleaning.

The concept of utilization of the synergistic effect of milling and leaching operations was published [Baláž 2000]. This concept utilizes all the excitation states, which are formed in a solid during intensive milling. In addition to the improvement of milling performance (the leaching agent works also as milling additive), there is the possibility that a common milling and leaching step contributes to operational benefits and to the economy of the overall process.

The principle of simultaneous grinding alkaline chemical leaching (GACL) process was successufully applied for coal cleaning: the reduction 41% of total sulphur, 95% of arsenic and the ash content by 43% in the brown coal from Nováky (Slovakia) was achieved. The GACL process is carried out at atmospheric pressure and at a temperature of 90°C. The consumption of NaOH used as an alkaline

Table 6.23 Chemical analysis of coal samples after GACL treatment [Baláž et al. 2001]

Coal	GACL treatment		Chemical analysis						
	Milling time (min)	Milling medium	Ash^d (%)	S^d (%)	As^d (ppm)	Fe^d (%)	Na^d (%)	Humic acids (%) Free	Humic acids (%) Total
Nováky (Slovakia)	Untreated	–	28.2	3.0	264	1.4	0.1	14.3	29.3
	30	Glass	27.4	2.3	46	1.2	0.3	20.5	26.9
	60	Glass	21.5	1.8	45	0.9	0.1	23.9	32.4
	90	Glass	40.6	1.8	13	1.6	0.5	25.7	40.3
	120	Glass	35.3	1.5	11	2.5	0.2		
Pittsburgh No. 8 (USA)	Untreated	–	7.0	2.0	4	1.0	0		
	60	Glass	23.0	0.9	2	1.2	0.1		

leaching medium was six times lower compared with the MCL process [Baláž et al. 1998; Turčániová et al. 2000].

The GACL process was later tested also for American coal (Pittsburgh No.8). The results are summarized in Table 6.23.

Different milling times were chosen. The process of dearsenification of minerals arsenopyrite FeAsS, realgar As_4S_4 and auripigment As_2S_3 contained in Nováky coal seemed be more sensitive to the GACL treatment than sulphur removal [Verbich 1998].

The GACL treatment led to an increase of humic acids content, which is the consequence of the destruction of coal matrix and new bond formation in the treated material. The total humic acids content increased from 29.3% in an untreated sample to 40.3% for Nováky coal treated by GACL process for 120 min [Turčániová and Baláž 2000].

Two samples of Pittsburgh coal No.8 were also analyzed: sample after milling and caustic leaching treatment and an untreated sample for comparison. A reduction of total sulphur content from 2.0 to 0.9% after GACL treatment was determined (Table 6.23). However, an increase in the ash content was also detected, which can be related to the alkaline content in the sample as well as wearing of the glass milling balls.

The GACL process proceeds under technically and economically friendly conditions: atmospheric pressure, temperature of 90°C and dilute solution (5% NaOH) are needed. In comparison with the MCL process, the consumption of NaOH is six times lower. However, the drawback of the up-to-date application of the GACL treatment is the wear of milling media, which contributes to ash-forming compounds in treated coal. Thorough washing of the treated coal is also a critical point.

An attempt has been made to improve economical feasibility of the GACL as for NaOH consumption [Turčániová et al. 2001]. An industrially scalable attritor mill (see Chap. 2) has been applied. The results are summarized in Table 6.24.

The reduction of leaching agent concentration (application of extremely diluted NaOH solutions) will favourably effect the material costs of leaching process. From the data published about GRAVIMELT process (which is a high-temperature coal

Table 6.24 Mechanical activation of brown coal Novaky (Slovakia) using GACL process

Milling conditions			Increase in humic acid content compared to non-treated coal
Time (min)	Temperature (°C)	NaOH (%)	
60	75	1	4.4x
60	75	0.2	3.9x
60	75	0.5	4.3x
60	60	0.2	4.8x
60	60	0.05	3.5x

treatment process applying coal+solid NaOH) it is known, that the concentration of NaOH limits the economy, the price of treated coal and practical implementation of the process [Nowak et al. 1994; Chriswell et al. 1994].

As shown recently, the mechanically activated brown coal can be used for preparation of special diterpene compounds (kaurane derivates) with cytostatic activity for pharmaceutical effectors [Turčániová et al. 2004].

The mechanochemical treatment of coal is a perspective non-traditional method applied in order to utilize coal substances in chemistry, pharmaceutical industry, etc. The combined effect of milling and alkaline treatment favourably influence the maximum destruction of carbonaceous and sulphur organic compounds in coal matrix. As for brown coal in addition to environmental effect of desulphurization, dearsenification and reduction of content of ash matters, an increase in humic acid content and special pharmaceutical compounds has been obtained. Coal mechanochemistry gives invokes to development in the field of clean coal technologies, nanotechnology, materials science, etc.

6.7 Building Industry

Development in the field of building materials is focused on a more effective consumption of energy, on utilization of secondary raw materials and, last but not least, on improvement of the technical and service parameters of the product [Tkáčová 1989].

The first applications of mechanochemistry in building industry are related to the benefit of silicate concrete (Silicalcite) production [Hint 1962]. The main advantage is in the mechanical activation of quartz which has positive influence on calcium hydrosilicates (CHS phase) formation [Juhász and Opoczky 1982, 1990].

Application of concrete with content of waste components is interesting from environmental point of view. Partial or total substitution of cement leads to suppression or elimination of problematic thermal processes. Application of slags and power station ashes as addition to cement clinker is a typical example. The quality of binders prepared from these materials depends on the correct selection of mixture composition (it should correspond to the composition of the cement) and on proper mixing providing the components to give hydrated compounds when coming in contact with water [Avvakumov et al. 2001]. It was shown that mechanical activation of mixtures of various industrial wastes with composition close to that of cement, allows substantially increase in their binding properties [Sulimenko et al. 1994].

Mechanochemical procedures significantly accelerate the solid state synthesis and high temperature compacting processes of building materials. They reduce the temperature of these processes simultaneously with enhancement of product quality [Tkáčová 1989].

6.7.1 Mechanical Activation of Quartz SiO_2

Lime and sand are among the building materials which have been used by man for thousands of years. As archeological investigations have shown, a lime-sand mixture having the ratio 1:3 to 1:5 was used already in ancient times as mortar for the body of wall stones. Noteworthy is the great strength which these binders have maintained in the course the thousands of years. However, the compressive strength of this material is not enough for the manufacture of large size or self-supporting structural units. The compressive strength of $3–4.10^{-7}$ Pa required for this could so far only be reached for cement concrete. The main reason for this is that the chemical reactions between lime, SiO_2 and water leading to solidification run too slowly or in too small extent if quartz sand is used [Heinicke 1984].

A significant step forward in the production of high-strength lime-sand building materials was made when the SiO_2 component began to be mechanically activated. This leads to the increased formation of CHS phases and, consequently, to an increased solidification of the building material structure. As an example, the compressive strength of lime-sand building materials depending upon on the treatment time using a vibration mill, and on the hardening time at a temperature of 373 K is represented in Fig. 6.47.

Whereas the compressive strength of non-activated samples does not exceed the value of 2.10^7 Pa even after a longer hardening time (10 h), a short-term mechanical treatment of a few minutes is already sufficient to obtain a concrete strength of 3.10^7 Pa at hardening times of 0.5 days [Heinicke 1984].

It has been described in mechanochemical literature that the reactivity of quartz, is, in the first instance, not determined by its dispersity and surface size, but decisevely by the degree of amorphization. Comparative investigations of the surface sizes occuring in the building material industry have demonstrated that a linear correlation between the specific surface area and the degree of amorphization of quartz similarly exists. Figure 6.48 shows that the compressive strength increases almost proportionally with the amorphous part of mechanically activated SiO_2.

The results clearly show the addition of fine milled quartz has a profound effect on concrete properties. The fine quartz fillers influence the hydration. The wet milled quartz in a slurry form acts differently from dry milled or dried formely wet milled quartz. Presumably, the fresh wet milled quartz is more reactive than the dry quartz fillers. The higher heat development at least during the first 25 h means that the cement is used better or that the energy built into the mechanically activated quartz

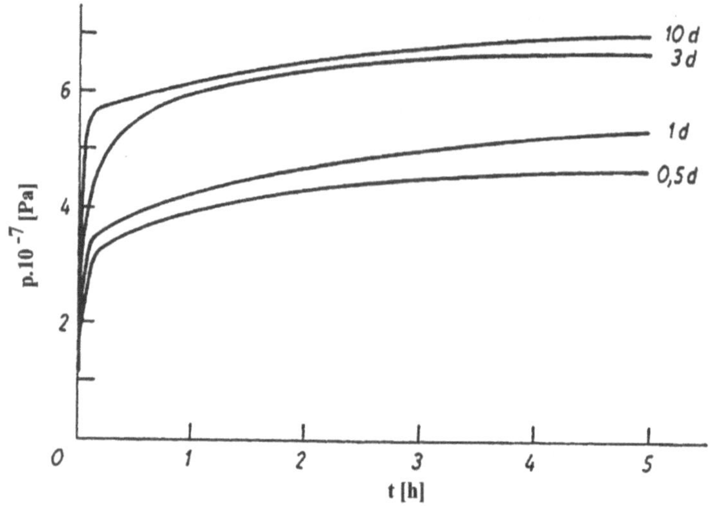

Fig. 6.47 Compressive strength, P at silicate concrete hardened for differently long times, d (in days) in dependence on SiO_2 component milling, t in a vibration mill [Schlegel 1976]

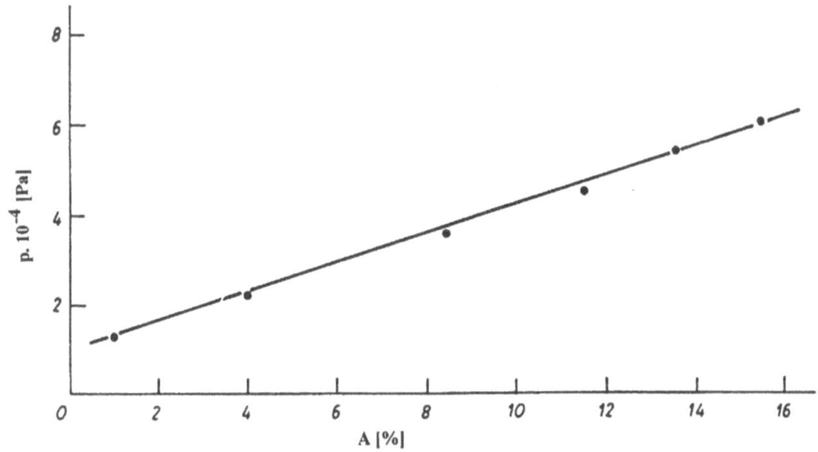

Fig. 6.48 Compressive strength, P in dependence on the amorphous part A of mechanically activated SiO_2 [Koch and Schrader 1970]

is released. In any case, the physical properties of the cement paste are improved [Lagerblad and Vogt 2004; Lagerblad et al. 2004].

6.7.2 *Mechanical Activation of Cement*

Experiments of mechanical activation were performed in several mills using Portland cement from regular production. Generally, mechanical properties of milled cement were improved. The hughest compressive and bending strengths were achieved for the cement mechanically activated in a vibration mill (55 MPa and 9 MPa in comparison with the initial values 43 MPa and 7.5 MPa) as registered after 28 day solidification period [Sekulic et al. 1988].

The effect of short-term milling of Portland cement in a vibration mill on the solidification is shown in Fig. 6.49. The initial compressive strength determined after one day of solidification increased from 3.9 to 16.7 MPa, the late strength determined after 28 days solidification period increased from 39.2 to 58.8 MPa. These results have found industrial application [Greenberg 1961]. The reason for this are mechanically released changes in surface and structure which for their part, accelerate of the hydration velocity of the cement during solidification.

The decisive factors in obtaining a high compressive strength are, however, the chemical processes running during the solidification.

It is interesting to compare the results with up-to day research in this field. High energy milling applied for the milling of cement leads to substantial refinement (<2 μm) and mechanical activation of the powder particles (Fig. 6.50).

The non-refined cement appears as a very heterogeneous material with a huge number of large sized particles with the prevailing particle size 30–60 μm. On the contrary the milled cement shows a much more homogeneous microstructure at very small particle size with prevailing particles less than 2 μm. The BET-surface area was found to be 15 m^2g^{-1} [Zoz et al. 2004].

In order to compare the strength of both cements, samples were tested using a universal testing machine at a compressive rate of 900 Ns^{-1}. Tests were applied after 1, 3, 7 and 28 days (Fig. 6.51).

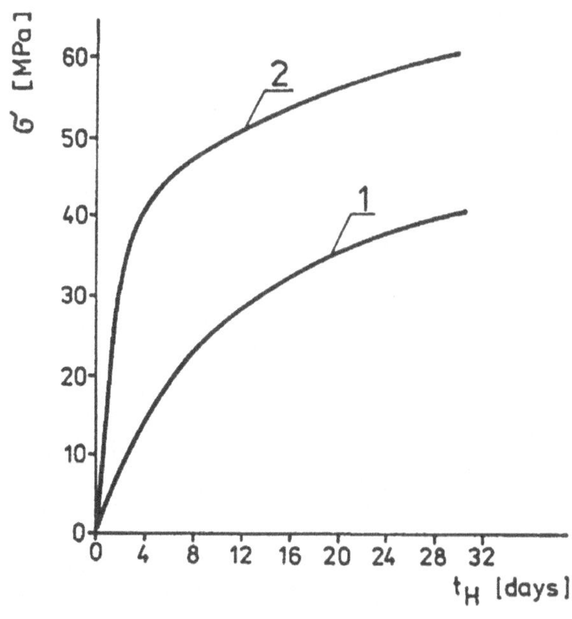

Fig. 6.49 Compressive strength, σ of concretes made from non-activated (1) and mechanically activated (2) Portland cement 400, as a function of solidification time, t_H [Meyer 1977]

Fig. 6.50 Optical micrographs of the as received (*left*) and of the processed cement (*right*) [Zoz et al. 2004]

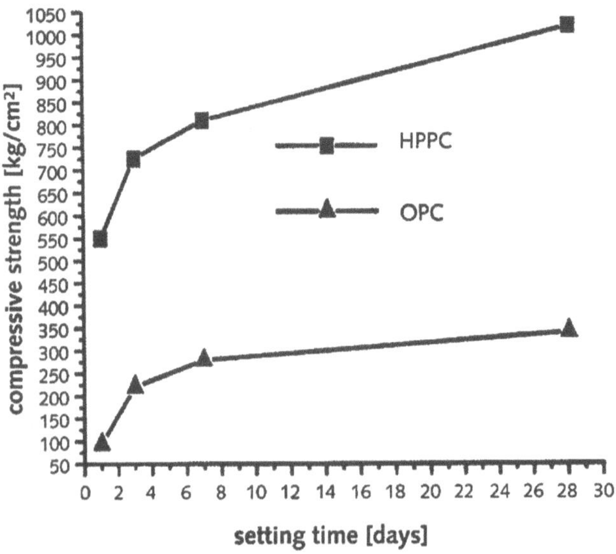

Fig. 6.51 Compressive strength of concretes made from non-activated (OPC) and mechanically activated (HPPC) Portland cement as a function of setting (solidification) time [Zoz et al. 2004]

The compressive strength results for the non-activated material from around 9.4 MPa after the first day to 32 MPa after 28 days curring. The mechanically activated material achieves a compressive strength of 55 MPa after the first day and of 102 MPa after 28 days. The obtained results can answer positively the key question of high-energy milling availability and scaling for the large amount of material produced in cement industry [Zoz et al. 2004].

6.7.3 Solidification of Silicate Concretes with Mechanically Activated Precursors

Solidification of concrete prepared from cement proceeds through several (at least three) steps:

- the hydrolysis of an anhydrous phase (constituents of cement clinker)
- nucleation and growth of nuclei of hydrosilicate (CHS phase) and
- diffusion of water through the film of hydrated products.

In cases where the first step determines the rate of setting, the hydraulic properties of alite (tricalcium silicate) and belite (dicalcium silicate) are of great importance [Boldyrev 1986; Tkáčová 1989].

6.7.4 Modification of Concrete Properties by Application of Waste Materials After Milling

In certain cases cement can be diluted with substances that have latent hydraulic properties, such as various industrial waste materials. They are present is ash formed by coal combustion, in the wastes of casting and abrasive production, etc. The quality of binders prepared from these materials depends upon the correct selection of the mixture composition (it should correspond to the composition of the cement) and on proper mixing providing the components to give hydrated compounds when coming in contact with water [Pavlenko et al. 2000]. It was shown that mechanical activation of mixtures of different industrial wastes with composition close to that cement allows a substantial increase in their binding properties [Sulimenko et al. 1994]. Moreover, the other properties of cement can be improved by mechanical activation such as behaviour, bending strength, etc.

However, keeping in mind that addition of waste materials can degrade the cement quality, particularly its mechanical characteristics, the amount of additive is usually restricted to only several percentages [Avvakumov et al. 2001].

There have been some interesting recent studies concerning mechanical activation of ashes from coal-fired plants and their use in clinker production of cement industry and concrete production [Johansson et al. 1999; Paya et al. 2000; Mezencevová and Števulová 2002; Števulová et al. 2005; Števulová and Sverák 2005].

Mechanical activation of cement with addition of fly ash (20% wt.) for only 3 min improved mechanical properties of obtained product. After 28-day solidification the compressive strength was higher for 57.9% in comparison with inactivated mixture composed of Portland cement and fly ash. Mechanical activation in this case significantly improved the quality of Portland cement with addition of fly ash. This result indicates that controlled mechanical activation would enable substitution of cement with other (waste) materials, while retaining the quality of obtained product to the desired level [Sekulic et al. 1999; Kumar et al. 2004, 2007].

Mechanical activation of fluidised coal fly ash with addition of various surfactants and an influence on its pozzolanic activity was studied [Števulová et al. 2005; Števulová and Sverák 2005]. The highest compressive strength values of hardened fly ash based concrete (25 wt.%) after 28- and 90-days curing time were obtained for the cement/fly ash composite containing the finest fly ash prepared by mechanical activation of coal fly ash for 1.5 h with 0.2% triethanolamine in a vibratory mill. This result is in accordance with data from literature [Berdov et al. 1983; Shi and Day 2003] according to the particle size, specific surface area and activity of surface of fly ash particles play a significant role at the hardening of cement/fly ash pastes. A number of fly ash based geopolymer products (60–90% fly ash) with unique properties have been developed, for example high strength (up to 120 MPa) geopolymer cements, self-glazed geopolymer tiles, and geopolymer pavement tiles based on synergistic utilization of waste from different sources. Mechanical activation can be used to add flexibility in processing and tailoring the properties of geopolymer products [Kumar et al. 2007].

Recently, it the role of water addition during milling for binder preparation has been estimated [Avvakumov et al. 2000]. An initial mixture for binder preparation included high-calcium ash (the product of coal combustion at heat plant), worked-out mixture of casting industry containing silica as a main component, and high-alumina waste of the abrasive industry, i.e. all cement-forming compounds, as well as the products of their interaction. This mixture was mechanically activated in presence of small amounts of water (~2–5 wt.%) [Avvakumov et al. 2001]. This method provides a decrease in the amount of non-bound CaO that causes worsening of long-time strength characteristics of product in water. The properties of the binder thus prepared were similar to those of standard. The results are shown in Table 6.25.

Table 6.25 Changes in CaO content, compression strength and water proof resistance for mechanically activated industrial waste Avvakumov et al. 2001

Mechanical activation (min)	CaO content (%)	Compression strength (MPa)	Water proof resistance
–	12.6	27.8	Low
1	6.33	35.6	Low
3	4.82	41.4	Low
6	3.92	52.5	High
10	2.80	56.8	High
15	2.02	43.5	High

6.7.5 Manufacture of Exterior Wall Panels with Intervention of Mechanical Activation

In previous Soviet Union, hydrothermally solidificated lime-sand materials termed Silicalcite has been developed [Hint 1962]. As an example a Silicalcite-wall section is shown in Fig. 6.52.

Fig. 6.52 Production of a Silicalcite exterior wall panel in building industry [Heinicke 1972]

The initial mixture consisted of 88% SiO_2 and 12% CaO, was mechanically activated in a high-speed disintegrator. Material with strength 3.103 MPa and density $1.9.10^3$ kg/m^3 was prepared by solidification in an autoclave. Due to its high strength, the material was aimed for the manufacture of large construction units in the building industry [Hint 1962].

In summary, the results discussed in Sect. 6.7 show that the most different building materials may be varied by through mechanical activation. Pecularities of a mechanical approach open new possibilities to incorporate in cement pre-determined amounts of waste materials like slags and ashes and contribute to the development of waste-free technologies. Utilization of waste is not only beneficial from the point of view of resource conservation, energy saving and CO_2 emisions decrease, but also solves the problems associated with disposal of wastes [Števulová and Sverák 2005].

6.8 Agriculture

Phosphates that occur in volcanic rocks as apatites or in sedimentary rocks as phosporites are used in agriculture for production of phosphate fertilizers. Natural phosphates have first to be converted to the water-soluble form by their decomposition in acids. Superphosphate is a classical phosphorus fertilizer, produced by decomposition of phosphates in sulphuric acid. A higher grade of superphosphate with 2–3 times higher content of water-soluble phosphatic component can be prepared by decomposition of phosphates in a mixture of sulphuric and phosphoric acids [Tkáčová 1989]. Traditionally, the following reactions are used:

$$Ca_3F(PO_4)_3 + 5H_2SO_4 + 2.5H_2O \rightarrow 5(CaSO_4.0.5H_2O) + 3HPO_4 + HF \quad (6.23)$$

$$Ca_3F(PO_4)_3 + 7H_3PO_4 + 5H_2O \rightarrow 5[Ca(H_2PO_4).H_2O) + HF \quad (6.24)$$

Acid soils may fertilized by natural phosphates in the as-milled state. In view of their low immediate effect, they are used as long-term fertilizers. In an attempt to enhance the efficiency extensive research was carried out [see references in monographs Heinicke 1984; Tkáčová 1989]. Several apatites have been studied as Kola-apatite which is relatively undisturbed fluoroapatite, or Morocco apatite which is a carbonate-fluoroapatite having a relatively strongly disturbed structure and an increased reactivity in comparison to Kola apatite.

Under the influence of mechanical activation, the reactivity of apatites is strongly increased. The reactivity is usually characterized by their solubility in citric acid. Figure 6.53 shows a strong increase of solubility for several natural phosphate after mechanical activation.

The best solubilities have been obtained for soft-earth sedimentary African phosphorites. However, under appropriate milling and dissolution conditions, complete solubilities of magmatic Kola apatite may be achieved [Paudert et al. 1979].

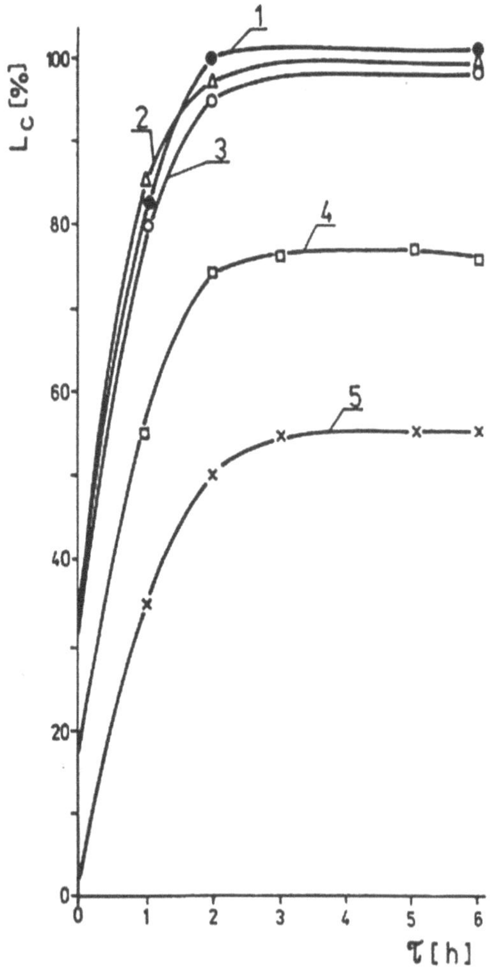

Fig. 6.53 Solubility of phosphates in citric acid, L_C vs. time of vibratory milling, τ. 1 – Anaphos, 2 – Tunisian phosphorite, 3 – Egyptian phosphorite, 4 – Morocco phosphorite, 5 – Kola apatite [Heinicke et al. 1977]

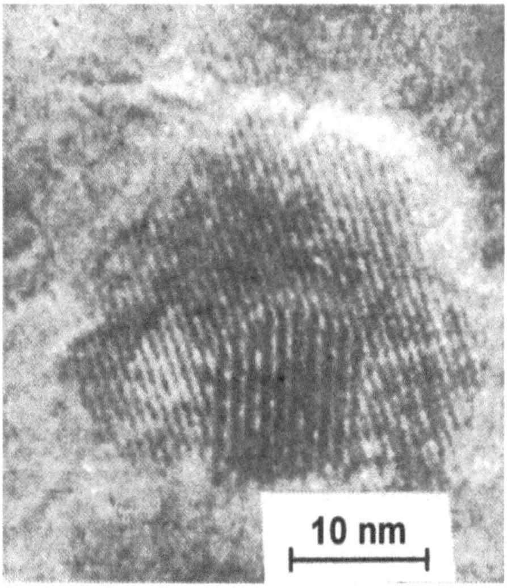

Fig. 6.54 HRTEM of apatite mechanically activated by wet milling [Chaikina 1996]

It was determined that the surface area increase during milling only contributed to the increase in solubility at the beginning of the treatment. Generally, the defects (lattice strain and reduction of crystallite size) and nanostructures (Fig. 6.54) generated in mechanically activated samples are the fundamental reason for increases in reactivity.

The increased degree of apatite disordering has to be attributed to the formation of X-ray amorphous regions. Activation in a planetary mill results in the formation of highly active X-ray amorphous regions that are completely soluble in ammonium citrate solution [Paudert et al. 1979]. The decisive effect of amorphization was confirmed by experiments in which defects in activated solid were gradually annealed at various temperatures. Figure 6.55 shows that the solubility decreases with the increasing extent of recrystallization at increasing temperature of annealing.

Based on the results of investigation of the structure reactivity correlation, commercial phosphate fertilizers have been developed. In Fig. 6.56 practical application of mechanically activated Morocco phosphorite is shown. The positive influence of high-energy milling on plant growth is unambigous.

The low solubility can be increased by co-milling with abrasive mineral. Apatite milled after adding a small amount of SiO_2 changed, so that more than 85% of its phosphate content became soluble in the citric acid solution. This occured because the apatite structure of the phosphate rock was mechanochemically decomposed during milling by reacting with SiO_2 and volatilizing fluor as SiF_6, which are both contained in the rock [Arai et al. 1976; Arai 1976]. Since then, this process has been applied to the magmatic Kola apatite which is hardly decomposed by H_2SO_4 [Kolosov et al. 1979].

The different approach has been applied in order to modify the properties of phosphates fertilizers [Jusupov et al. 2000; Jusupov and Shumskaya 2002]. Zeolites have been used in a common milling with phosphates. In Fig. 6.57 the effect of mechanical activation of clinoptilolite-superphosphate mixtures on solubility of

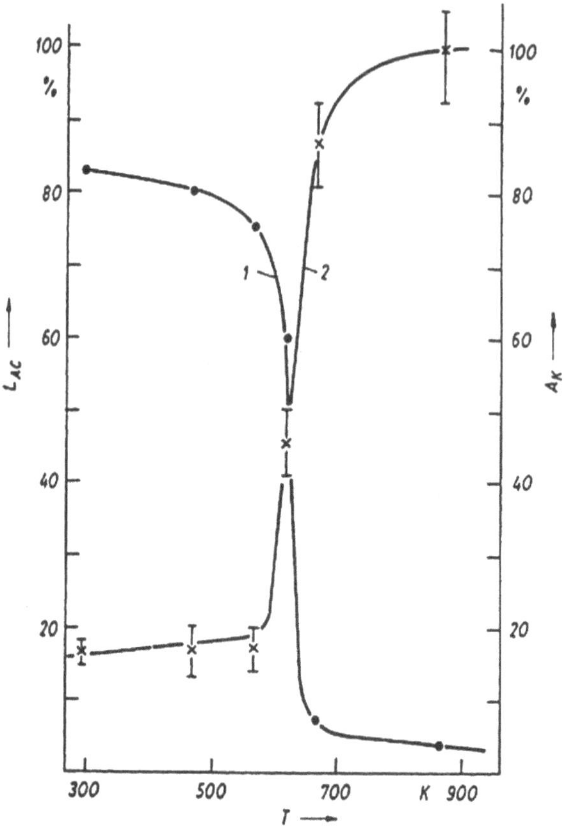

Fig. 6.55 Solubility, L_{AC} (1) vs. crystalline portion, A_K (2) at annealing by temperature T for Marocco phosphorite [Paudert et al. 1979]

phosphate in citric acid and water is illustrated. In particular, the water solubility decreases from 40 to 5%.

The Al-bearing active centers of zeolite interact with phosphate, preserving its ability to practically complete solubility in the 2% solution of citric acid, which may serve as a test of agrochemical activity. About 50% of phosporus from fertilizers is known to be transformed into forms inaccessible for plants. The obtained zeolite-superphosphate may be considered as a new form of phosphate fertilizer capable of correlating mobile phosphorus supply for soil with agrochemical demands.

Based on the laboratory studies, commercial phosphate fertilizers have been developed from Kolaapatite under the trade name Tribophos. A plant physiology study showed that the mechanically activated Tribophos has the same fertilizing effect as superphosphate in acid soils. The necessary degree of its activation increases with decreasing pH of the soil. The maximum milling energy is required for neutral and weakly alkaline soils [Heinicke et al. 1977; Paudert et al. 1979].

Optimal parameters of mechanical activation have been determined for the phosphate ores of different mineral composition [Yaneva et al. 2005]. Mills of continuous

Fig. 6.56 Influence of mechanical activation of Morocco phosphorite (deposit Youssoufia) on a plantgrowth. From left to right: without phosphorite, non-activated phosphorite (−125 μm), phosphorite milled at milling energy 50 and 150 kWh/t, respectively [Gock 1987]

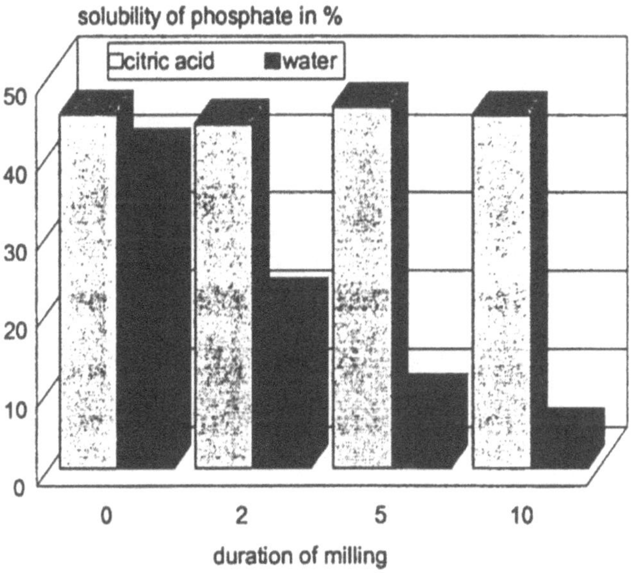

Fig. 6.57 Solubility of phosphate vs. milling time (in min) [Jusupov et al. 2000]

operation with low energy consumption have been designed, their productive capacity being, at the beginning, 3 t/h. Long-term agrochemical tests (for 15 years) of mechanically activated phosphate ores of different composition have been carried out at wide range of soils and with various crops. These tests demonstrated high fertilizing effect [Chaikina 1986; Boldyrev et al. 1996]. The technology was installed at the Burenkhan deposit of phosphorite ores in Mongolia. It was proven that the

Table 6.26 Comparison of traditional and mechanochemical methods for production of phosphate fertilizers [Suryanarayana 2004]

Process characteristics	Traditional method	Mechanochemical method
Pittrman's probe in % citric acid S	80–85	55–60
Ammonium citrate	60	40–50
Increase of harvest related to superphosphate	1.00	0.95
Duration of production process	20 d (periodical)	20–24 h (continuous)
Consumption of energy per ton	450 kWh/t	500 kWh/t
Consumption of reagents per ton	0.7 t H_2SO_4 (60%)	–
Duration of agrochemical action of the fertilizer	1–2 years	5–7 years
Release of fluorine into atmosphere	Yes	No
Increase of acidity of soils	Yes	No
Possibility to use in small deposits (<1 million tons) of phosphorus rocks	No	Yes

mechanochemicaly pretreated phosporite fertilizers have a prolonged action and are comparable with fertilizer based on superphosphate [Amgalan et al. 1998].

The technological advantages of mechanochemical methods over traditional methods are summarized in Table 6.26.

Different machines consume different amounts of power for production of the fertilizers, and the capacity of the mills is also different. It was reported that the vibrocentrifuge mill was the best suited for this purpose and it could also be used continuously. As mentioned above, the first factory to produce the phosphate fertilizer was built in Mongolia. Since then more than 1.5 million tons of the fertilizer has been produced.

6.9 Pharmacy

6.9.1 Solubility of Drugs

Dissolution testing of drug compounds is of fundamental importance in the design and development of pharmaceutical products [Bisrat and Nyström 1988]. Dissolution rate is normally described by Noyes-Whitney equation

$$\frac{\mathrm{d}W}{\mathrm{d}t} = \frac{D}{h_D} S_A \left(C_s - C_t\right) \tag{6.25}$$

where the dissolution rate ($\mathrm{d}W/\mathrm{d}t$) at time is a function of three types of material and experimental parameters, namely specific surface area S_A, concentration gradient (C_s–C_t) and diffusional transport D/h_D (D – diffusion coefficient, h_D – the distance over which diffusion is the dominant transport mechanism).

The rate of absorption of slightly soluble drugs from the gastrointestinal tract and other sites is often limited by the rate of dissolution of the drug substance. For dissolution of poorly soluble drugs some general remarks can be given

- the drug should be as fine as possible
- the drug should be deagglomerated form already in the dosage form, and
- the carrier should be highly soluble in water and thereby rapidly delivering individual drug particles to the dissolution liquid [Nyström 1998].

At present about 40% of the drugs being in the development pipelines are poorly soluble, even up to 60% of compounds coming directly from synthesis are poorly soluble [Merisko-Liversidge 2002]. Poor solubility is in most cases associated with poor bioavailability. Bioavailability is how well a drug will reach an effective therapeutical level in the body, and may be influenced by various factors. A drug may be safe and effective, but never reach the therapeutic level in the body if the bioavailability is poor. Bioavailability essentially depends on three factors: solubility, permeability and dose [Lipinski et al. 1997], and the question of the minimum acceptable solubility can only be answered if the other two factors are known. According to biopharmaceutics classification system, a drug substance is considered highly soluble when the highest strength dosage is soluble in 250 mL of aqueous media over the range pH=1.0–7.5 [Yu et al. 2002].

There are two basic approaches to overcome the bioavailability problems of the drugs

- increase of equilibrium solubility (e.g. by complex formation), and
- increase of dissolution velocity

The first approach was a limited success as clearly demonstrated by the low number of products on the market based on such technologies. A much more straight forward way is increasing the dissolution velocity by increasing the surface area of the drug powder, i.e. micronisation leading to mean particle sizes approximately 3–5 μm. However, many of the new compounds show such a low solubility that micronisation does not lead to a sufficient increase in bioavailability after oral administration. Therefore the next step taken was nanonization. The drug powder is transferred to drug nanocrystals, typically sizes are around 200–600 nm [Bushrab and Müller 2003].

Drug nanocrystals can be produced by various methods principally the top down and bottom up techniques (Fig. 6.58).

The bottom up technique is the classical precipitation approach, the drug is dissolved in a solvent which is subsequently added to a nonosolvent to precipitate the crystals. A priori this technique is difficult to handle, the crystal growth needs to be stopped to avoid formation of microcrystals. In addition this technology cannot be applied to the increasing number of drugs being poorly soluble in all media [Bushrab and Müller 2003]. On the other hand the top down technique (milling) is the method of choice for industrial production in pharmacy. In addition to size reduction, milling may alter the crystalline structure and cause chemical changes in some drugs [Parrott 1974, 1990].

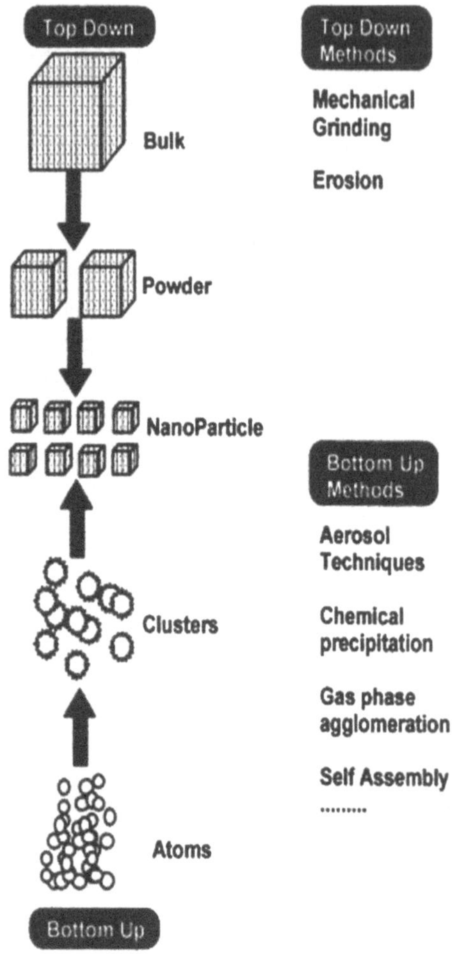

Fig. 6.58 Schematic representation of the top down and bottom up techniques for nanomaterials preparation [Dutta and Hoffmann 2003]

Micronisation and nanonisation by milling can increase the equilibrium solubility level of drugs. Such a thermodynamically metastable elevation is probably due to the creation of a thin surface layer of disordered material around a more crystalline core [Nyström 1998]. In general, drugs in solid dosage forms are used in their crystalline state. The amorphous state, however, has also become of interest in preparing solid dosage forms because it improve the dissolution behaviour of drug.

Fine milling have been used to micronize the pharmaceutical powder of orsodeoxycholic acid (UDCA) which is bad soluble and shows a pronounced efficacy for gallstone dissolution [Choi et al. 2004]. The particle size was reduced in the order of vibration mill < planetary mill < jet mill. The UDCA crystals were changed into amorphous state with increasing milling time. The in vivo tests performed with mice showed more significantly hepato-protective effect compared with intact UDCA.

Fine milling has been also used for obtaining respirable size fraction of drugs (typically 2–10 μm) [Patton and Platz 1992]. Drugs administered by inhalation have become increasingly popular, and dry powders inhalers are now commonly in use [Hilfiker 2006]. The influence of crystal habit on disorder produced by milling was investigated using β-succimid acid, in plate like and needle like morphologies [Chikhalia et al. 2006]. A layer decrease of crystallinity were seen in the plate-like morphology compared to the needle-like morphology.

Sodium chloride NaCl is used to reduce the particle size of drugs. This ingredient is frequently applied as a tonicity adjusting agent. A typical formulation may contain hydrophobic drug, surfactant, NaCl, and other ingredients. Steroid drug dexamethasone milling in the presence of NaCl was tested. Inclusion of the chloride in the milling slurry containing dexamethasone and surfactant polysorbate 20 significantly decreases volume mean diameter of milled dexamethasone. A reduction in particle size could potentially translate into a more robust manufacturing process, as well as improved bioavailability of the suspended solid [Apte et al. 2003].

6.9.2 Polymorphism and Amorphization

Polymorphism, which is very common in connection with drug substances, is the ability to crystallize as more than one distinct crystal species. Since different inter- and intramolecular interactions, such as van der Waals interactions and hydrogen bonds will be present in different crystal structures, different polymorphs will have different free energies and therefore different physical properties such as solubility, chemical stability, melting point, density, etc. [Hilfiker 2006].

Polymorphism is very common in connection with drug substances as small organic molecules. Literature values concerning the prevalence of true polymorphs range up to 51% [Storey et al. 2004]. In the absence of solvents and humidity, the thermodynamically stable polymorph is the only one that is guaranteed not to convert into another polymorphic form. This is why this form is most often chosen for the drug product. The disadvantage of the thermodynamically stable form is that is always the least soluble polymorph and therefore has the lowest bioavailability [Hilfiker 2006]. The difference in the solubility of various polymorphs are typically lower than a factor of 2 [Pudipeddi and Serajuddin 2005], but sometimes as much as 5-fold difference can be observed [Chemburkar et al. 2000].

If the solubility of the most stable polymorph is critically low and no salt is feasible liquid-like formulations (emulsions, microemulsions, liposomal formulations) of soft gelatine capsules filled with solutions of the drug in a non-aqueous solvent may be used. Alternatively, a metastable solid form, a solvate or a co-crystal might be selected for development. Sometimes it is better to use amorphous form rather than a metastable polymorph [Hilfiker 2006]. Different polymorphs will also have different mechanical properties, such as hardness, powder flow properties, compressibility and bonding strength. A well-known example is paracetamol, where the thermodynamically stable form (monoclinic form I) cannot be compressed into stable states while the metastable form II (orthorombic) can as it shows more favourable properties with respect to plastic deformation [Nichols and Frampton 1998]. Generally, the thermodynamically most stable polymorph is also the most stable chemically [Singhal and Curatolo 2004]. This has been attributed to the fact that its density is typically higher, but it could also be explained by lower free energy.

The influence of mechanical activation on polymorphism of sulphathiazole has been studied [Shakhtshneider 1997]. According to the literature data sulphathiazole exists at least in three crystalline modifications and an amorphous form [Lagas and Lerk 1981]. In the initial stage of mechanical treatment of polymorph III, a broadening of X-ray diffraction peaks and a decrease in their intensity was observed. It was suggested that part of the substance had converted into an amorphous state and later into a high temperature form I. The quantitative estimates showed that the amorphous part of the substance might achieve up to 50 wt.% and after prolonged treatment, recrystallization of the drug was observed. Modification I is the most suitable form for pharmaceutical use owing to its higher dissolution rate but strict conditions have to be applied to preserve this form. It appears that the following scheme (Fig. 6.59) was realized.

The problem of polymorphism and amorphization in drugs has been thoroughly reviewed from the point of view of mechanical activation [Shakhtshneider and Boldyrev 1999]. Activation can occur throughout the bulk of the particle (as in the mechanical treatment of ionic crystals) or only in the surface layer (as in the shear deformation of quartz) [Steinike and Linke 1982]. During milling of griseofulvin drug, the thickness of amorphous surface layer did not exceed 5.8 wt.% of the substance [Elamin et al. 1994]. Whether an amorphous state is formed during mechanical treatment or not depends to a large extent on the conditions under which the mechanical treatment is performed. As a result of mechanical activation of cephalexin [Otsuka and Kaneniwa 1988] and cephalothin sodium [Otsuka et al. 1994], the originally crystalline solids became amorphous. However, the effect of the subsequent prolonged mechanical treatment on the already formed amorphous cephalothin sodium was opposite: partial crystallization was observed.

In some cases an increase in milling time causes not only amorphization but also chemical transformation. For example, the mechanical activation of ampicillin trihydrate [Takahashi et al. 1984], lactose monohydrate, and citric acid monohydrate

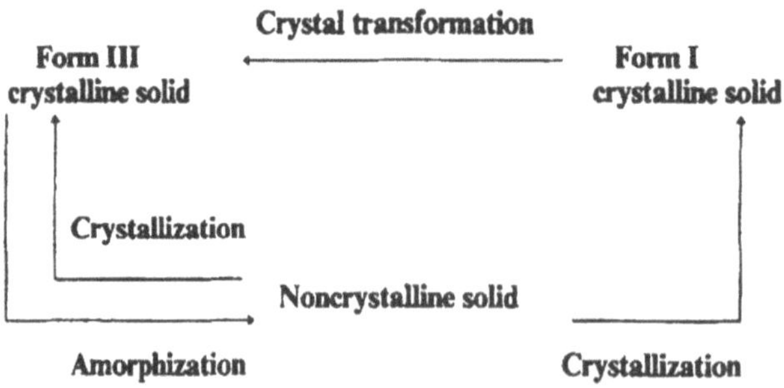

Fig. 6.59 The pathway of crystalline form of sulphathiazole during mechanical activation [Shakhtshneider 1997]

[Hüttenrauch and Fricke 1981] results in partial dehydratation accompanied by disordering of the structure.

The mechanical activation of molecular crystals can serve as a method to produce amorphous, metastable states. This affects the solubilization of solids and is an important factor if molecular solids are used as drugs [Shakhtshneider and Boldyrev 1999].

Mechanical activation can lead to changes not only in the crystal structure, but also in molecular structure of drugs. In this case a special type of polymorphism is observed. Results on the influence of mechanical treatment on physico-chemical properties of piroxicam are a typical example [Shakhtshneider 1997]. Piroxicam is an efficient nonsteroidal antinflammatory agent which exist in several polymorphs and in a monohydrate form [Vrečer et al. 1991]. Piroxicam is practically insoluble in water, and as a result its bioavailability is low. After mechanical activation not only the structure of piroxicam but also its reactivity is changed. The solubility of mechanically activated piroxicam was 4-times higher than that of the non-activated sample (Fig. 6.60).

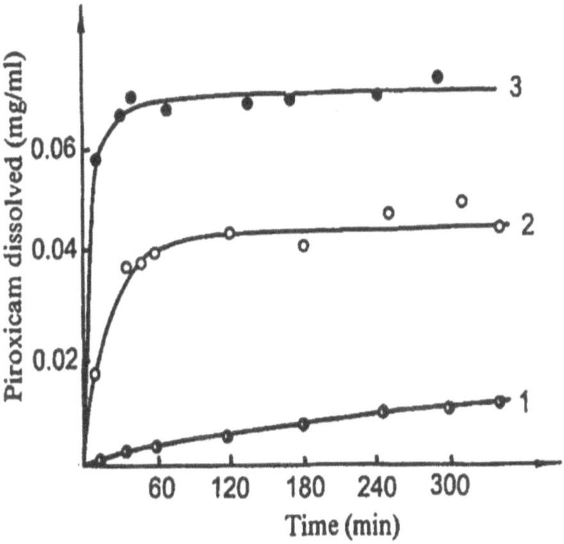

Fig. 6.60 Dissolution curves of piroxicam, 1 – non activated, 2 – mechanically activated alone, 3 – mechanically activated in the presence of surfactant polyvinylpyrrolidone [Shakhtshneider 1997]

However, in some cases the changes induced by mechanical activation are not always favourable and less solubility is observed.

6.9.3 Biological Activity of Drugs

It has been already established that a method of mechanical activation can be applied to modify drugs so as to increase their therapeutic efficiency.

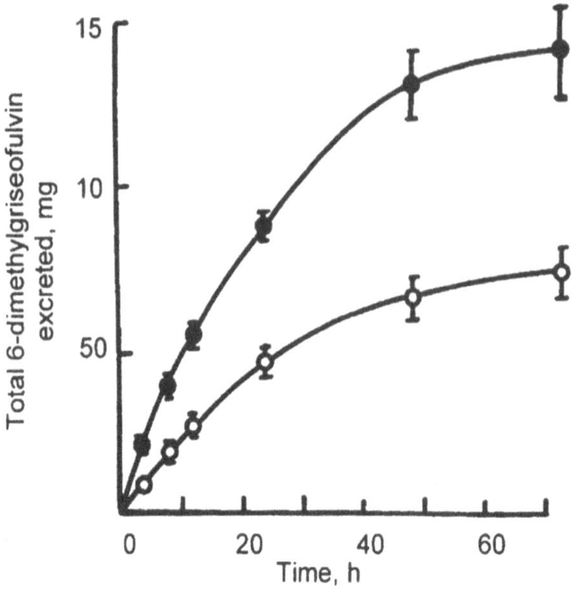

Fig. 6.61 Urinary excretion of 6-dimethylgriseofulvin after oral administration: (•) milled griseofulving with polymer, (○) micronized griseofulvin [Yamamoto et al. 1974]

For example, mechanical activation of griseofulvin, a drug widely used to cure fungus diseases, causes along with solubilization an increase in its biological availability in comparison with non-activated sample by 75% due to an increase in the rate of absorption into blood plasma [Yamamoto et al. 1974]. At the same time, the rate of excretion of the products of griseofulvin metabolism (principally 6-dimethylgriseofulvin) from organism increases (Fig. 6.61).

As a result of milling sulfamonomethoxin, a pronounced effect on the chemotherapeutic activity has been observed. Experiments, using streptococcal septicemia in white mice, demonstrated the advantages of milled drug (Table 6.27).

The solubility of mechanically activated sulphamonomethoxin was 1.53 times higher than that of non milled sample. When mice were given the mechanically activated sample, the number of mice who recovered 10 days after administration was almost two times higher than for the reference group that received non milled sample. The average life duration increased by 13% [Shakhtshneider and Boldyrev 1999].

Table 6.27 Chemotheraupetic efficiency of sulfamonomethoxin, commercially available and milled in a jet mill, for the model of streptococcal septicemia in mice [Savitskaya et al. 1990]

Preparation	Animals in group	Revived animals (10th day) (%)	Overall lifetime (%)
Non milled SMM	20	30	69
Milled SMM	20	55	82
Reference (without curing)	20	0	17

Mechanical activation of phenytoin with cellulose or chitin also leads to an increase in the rate of excretion of metabolism products from organism and to an increase in the rate of phenytoin transfer into the blood plasma nearly by a factor of three [Sawaynagi et al. 1983].

6.9.4 Mechanochemical Synthesis of Drugs

The advantages of mechanochemical approach for synthesis of drugs can be illustrated for sodium benzoate, sodium salicylate and acetylsalicylic acid production.

Sodium benzoate is one of the most important products in the pharmaceutical industry. It is usually obtained by the neutralization of benzoic acid by soda in an aqueous solution. The technological process consists of includes six stages. To obtain 500 kg of the benzoate, 3000 l of water is required and the process takes 60 h. Another example is the synthesis of sodium salicylate. Six technological stages are also required and the entire process takes 70 h. To obtain 500 kg of sodium salicylate, a mixture of water (500 L) and ethyl alcohol (100 L) is required [Boldyrev 1996; Shakhtshneider and Boldyrev 1999, Boldyrev 2004].

The application of mechanochemical methods has enabled the neutralization of various organic acids (benzoic, salicylic, citric, sebacic, indolacetic, ascorbic) in one stage without any solvents [Dushkin et al. 1974; Dushkin 2004] using NaOH, KOH, Na_2CO_3, K_2CO_3 or $CaCO_3$ as the neutralizing agents.

The traditional and mechanochemical processes are compared in Table 6.28. An important advantage of mechanochemical technology is that enormous amounts of water and its subsequent discharge as effluent are no longer needed. Also the technological rate of the process is higher.

Acetylsalicylic acid (ASA) is non-steroidal drug often used as an analgesic, antipyretic and as an anti-inflammatory. The most popular and used brand name is Aspirin. It has also antiplatelet effect and is used in long-term, low doses to prevent

Table 6.28 Synthesis of sodium benzoate and sodium salicylate by traditional and mechanochemical technology [Boldyrev 1996]

Traditional technology	Mechanochemical technology
1. Dissolution of the acid in water	1. Mixing of the acid with neutralization agent
2. Neutralization	2. Mechanochemical treatment
3. Filtration	3. Packing
4. Drying	
5. Milling	
6. Packing	
The process is carried out in a batch mode. The duration is 50–70 h. The yield: 450–500 kg/cycle.	The process is carried out in a continous mode at a rate up to 100 kg/h.

heart attach and cancer. Its solubility in water is 1 mg/mL at 20°C. The low solubility of ASA in water decreases its pharmacological efficiency and causes undesirable side effects when the substance is used in drugs. The salts of ASA posses increased solubility. However, with all advantages of ASA salts, these are expensive drugs manufactured on a small scale. During the production, ASA undergoes partial decomposition into salicylic and acetic acid as a result of hydrolysis. The existing requirements to the purity of the product make the production process more complex and the product more expensive [Dushkin 2004].

Alternative forms are so called effervescent pharmaceuticals. The salt formation effect is achieved during the dissolution of tablets or granules containing the ASA substance and a relatively large amounts of neutralizing fillers-sodium bicarbonate or carbonate. In addition to the indicated components, these compositions should always include a solid-water-soluble organic acid (citric, malic, or ascorbic) to accelerate the destruction of a tablet or granule due to the evolution of CO_2 during the interaction with carbonates and bicarbonates. However, the presence of an acid of this kind requires that the composition should incorporate neutralizing agents in substantial excess over the amount necessary to neutralize pure ASA. A typical disadvantage of the preparations is low mass fraction of ASA: 10–16% for the above samples, the total mass of the tablet being 3–3.5 g [Dushkin 2004].

There is an innovative approach to apply mechanochemistry for obtaining soluble materials based on ASA for their subsequent use as drugs. The effervescent tablets "Aspirin+C" (Bayer, Germany) were chosen as a reference drug of soluble aspirin with identical characteristics. Several pharmacokinetic characteristics have been followed, e.g. the maximal concentration in blood, bioavailability and rate of absorption by application of laboratory animals (rabbits). Pharmacological tests showed that the developed drugs and the reference drug are similar in pharmacological activity. However, the advantage of the mechanochemical approach is for example the decreasing of the total mass of the tablet (by a factor of 5), number of auxiliary substances, gas evolution during dissolution and the overall cost of the preparation. The disadvantages include the lower rate of dissolution in cold water (10–25°C). The preparation has successfully passed preclinical and clinical pharmacological tests and was registered for medical application. The obtained fast-dissolving substances can be used in compositions with other pharmaceutical or biologically active substances [Dushkin 2004].

For mechanochemical technology, special vibrocentrifugal mills with productivity 10 g/h–1 t/h have been designed. In the most intense mode, centrifugal acceleration acting on the milling bodies reached 40 g. In order to obtain especially pure products (for example, for pharmaceutical purposes), the inner surface of the drum and the surface of milling bodies can be coated with an inert material. The schematic diagram and a pilot plant unit is given in Fig. 6.62.

The design allows smooth variations of the feeder productivity and mill rate. The technological regulations for production of fast-dissolved substance (brand names Aspinat and later Askopirin) have been developed and coordinated with the Ministry of Health in Russia [Dushkin 2004].

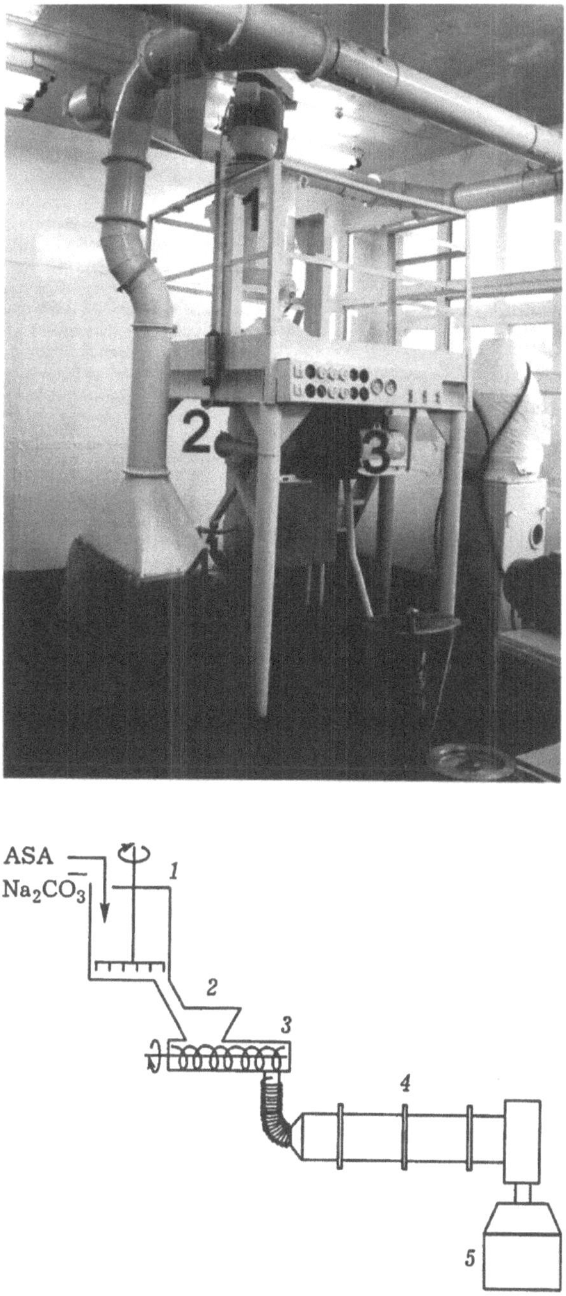

Fig. 6.62 Schematic diagram (*down*) and a mechanochemical pilot plant unit (*above*) for the production of fast-dissolving substances of ASA, 1 – turbo mixer, 2,3 – accumulating tank with feeding screen, 4 – continuous vibrocentrifugal mill, 5 – product collector [Dushkin 2005]

6.10 Waste Treatment

6.10.1 Mineral Waste

6.10.1.1 Asbestos

Based on its properties, asbestos has been applied in many industrial branches. The mechanical strength, chemical and thermal stability, good isolation properties and fibreforming play the main role in its application in building, textile and maschine industry. The world production reached 4.8 bilion tons/years in 1980s of the last century. However, because of its toxic properties (see later) the production of asbestos was later on strongly reduced. In European Union, some of asbestos forms (blue and brown) asbestos has even been forbidden.

Asbestos is in principle a group of serpentine and amphibole minerals (Table 6.30).

The most important industrial minerals are chrysotile and crocydolite. The danger of some asbestos forms lies in the form of fine particles. From this point of view the fiber forms are the most dangerous. A length (L) longer then 5 μm, the width (D) lower then 3 μm, and the ratio L/D>3 of asbestos particles lead to asbestosis, lung cancer and other lethal diseases connected with breast and abdomen malignancies. In this connection crocydolite and, amosite, as representatives of fiber asbestos are most dangerous. SEM of chrysotile (A) and crocydolite (B) are given in Fig. 6.63. Combination of filaments with various lengths can be clearly seen. In case of crocydolite, the fibers amount 70% of the total mass.

There are several approaches in industry how to minimize the dangerous character of asbestos fibers. Chemical and thermal treatment has been applied with some success [Adie 1993; Jovanovic 1994].

Asbestos, mainly crocydolite, is a very refractory mineral. Its structure can be destroyed by an action of hydrogenfluoride acid (HF) which selective formation of SiF_4 and as a consequence, fiber structure of asbestos is destroyed. $Ca(OH)_2$, CaF_2, metal oxides and hydroxides as non-soluble compounds are formed after neutralization step involving addition of $Ca(OH)_2$. The process is selective but economy is not favourable and application of HF is environmentally non-friendly [Swidersky 1994; Vogt et al. 1996].

Table 6.30 Asbestos minerals [Tröger 1969]

Group	Mineral	Formula
Serpentine asbestos	Chrysotile (white asbestos)	$Mg_3[Si_2O_5](OH)_4$
	Authophyllite asbestos	$(Fe,Mg)_7Si_8O_{22}(OH)_2$
	Aktinolite	$Ca_2(Fe,Mg)_7Si_8O_{22}(OH)_2$
Amphibole asbestos	Amosite (brown asbestos)	$Fe_7Si_8O_{22}(OH)_2$
	Tremolite	$Ca_2Mg_5Si_8O_{22}(OH,F)_2$
	Crocydolite (blue asbestos)	$(Na,K)_2(Fe^{2+},Mg)_3(Fe^{3+},Al)_2Si_8O_{22}(OH)_2$

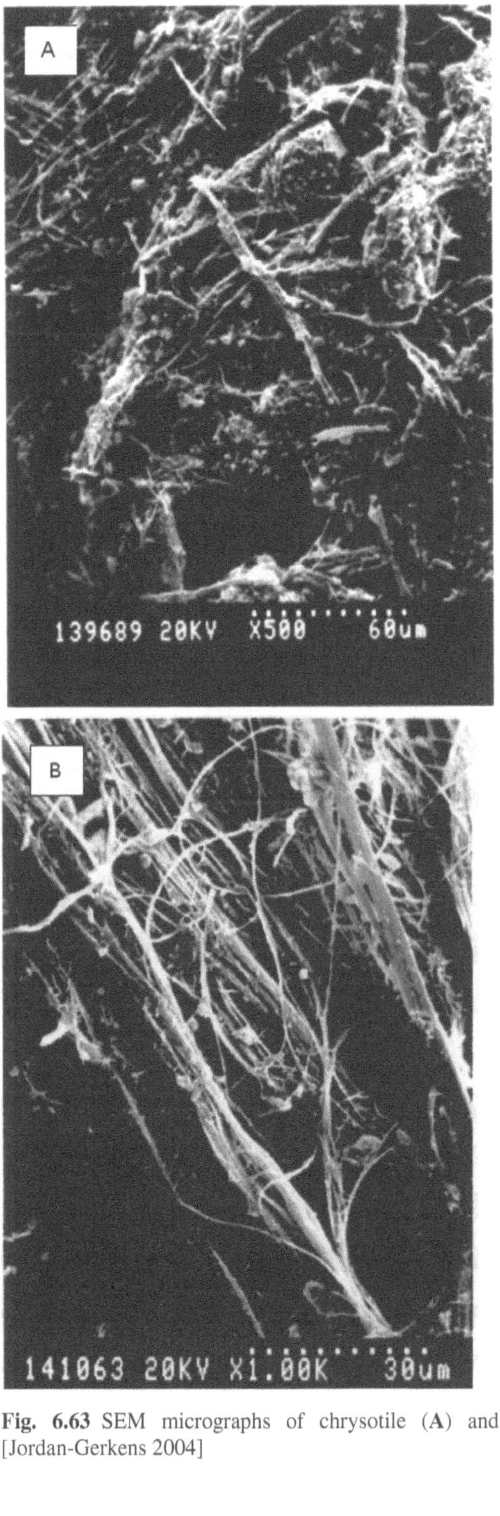

Fig. 6.63 SEM micrographs of chrysotile (**A**) and crocydolite (**B**), magnification-1000x [Jordan-Gerkens 2004]

Thermal treatment of asbestos minerals (Table 6.30) is based on their transformations to less dangerous forms. In a temperature interval 400–1050°C, hydroxide groups are loosing and hydrates of amorphous compounds are formed [Ströbel 1993]. Forsterite Mg_2SiO_4 and quartz SiO_2 together with water has been observed bei thermal treatment of chrysotile at 1100°C. The solid products are not harmful and can be utilized in the building industry. As a product, well accepted transformation of fibres into glassy granules can be obtained. However, there is very great demand for electricity in this process as well as metalforming waste, salts and products of gas cleaning have to be further treated.

The new technical concept has been developed and tested in a pilot plant unit in Germany [Gock et al. 1994]. Laboratory testing has shown the possibility of mechanical destruction of asbestos fibres by milling (Fig. 6.64).

Several processing variants have been tested, e.g. mechanical activation in a separate step or mechanical activation as pretreatment step in combination with thermal treatment which was applied after milling. Various processing parameters were tested, e.g. milling and roasting time and temperature. In both variants industrial eccentric vibratory mill ESM (Siebtechnik, Germany) has been applied. Asbestos cement and asbestos wool were used as typical representatives of harmful asbestos waste.

An example of combined mechano-thermal treatment is given in Fig. 6.65. Waste is mechanically activated in a vibratory mill and then thermally treated by temperature slower than theoretically needed for asbestos transformation. Throughoutput is

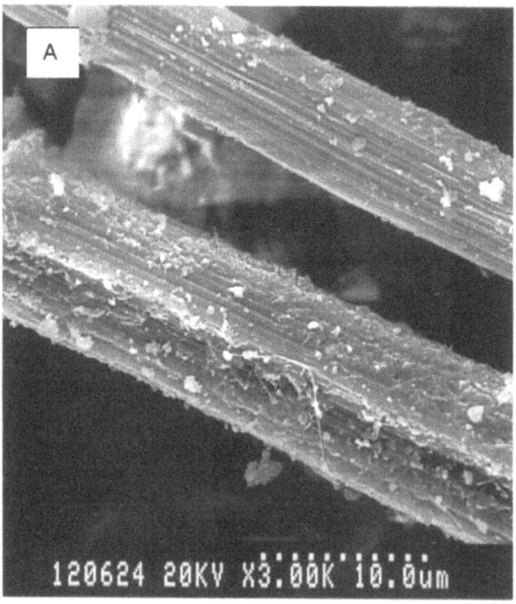

Fig. 6.64 SEM micrographs of asbestos: magnification 3000x. **A** – before milling, **B** – after milling [Vogt et al. 1996]

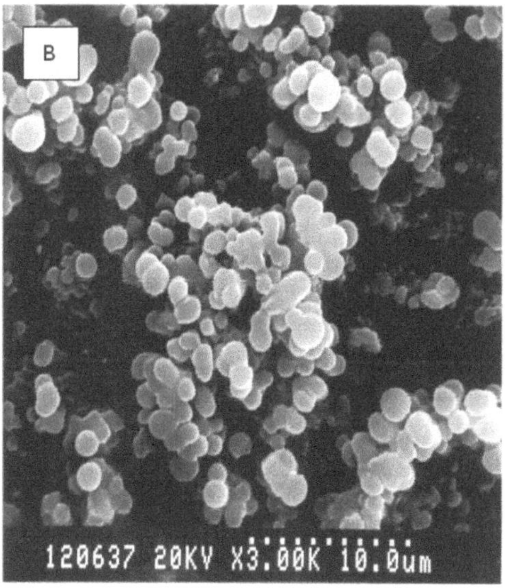

Fig. 6.64 (continued)

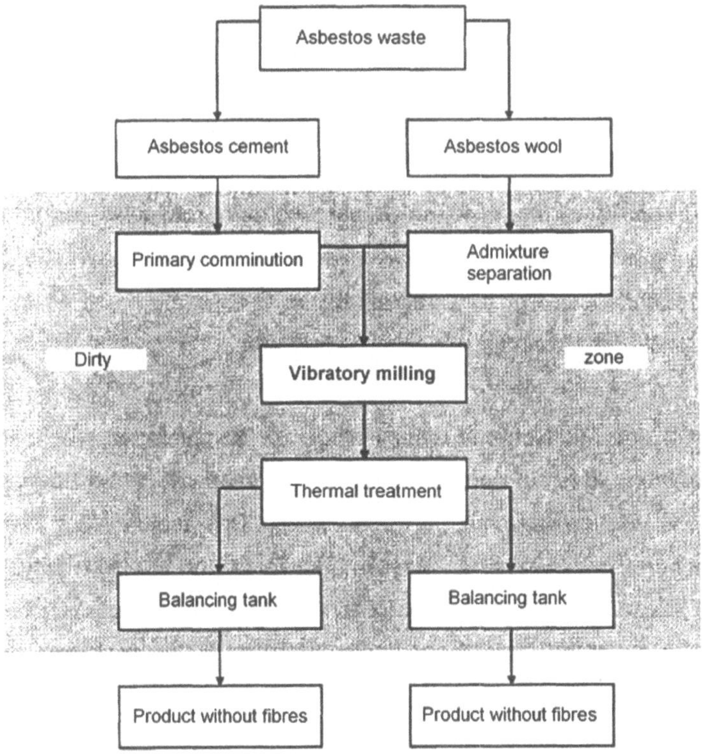

Fig. 6.65 Combined mechano-thermal treatment of asbestos waste [Jordan-Gerkens 2004]

Fig. 6.66 Dirty zone in the combined mechano-thermal treatment of asbestos waste [Vogt et al. 1996]

1.55 t/h and two ESM 765–2 ks mills are needed. Photograph of the dirty zone is in Fig. 6.66 where a material tank on left side and vibratory mill on right side can be seen.

Comparison of the process economy is given in Table 6.31. Data for the thermal treatment were obtained from AsbestEx company for comparison [AsbestEx 1994]. The best cost data were calculated for mechano-thermal treatment where the total cost for asbestos cement waste is only 75% of the cost calculated for AsbestEx company.

After processing, the asbestos waste exhibited remarkable hydraulic properties. The compressive strength of the test specimens is in the range of conventional cements. Possible utilization of the milled asbestos fibres could be in the form of an additive for cement, as material to be used in pressure grouting by the construction industry, or in the rehabilitation of soils and buildings.

Table 6.31 Comparison of total costs for several asbestos waste treatments [Jordan-Gerkens 2004]

Parameter	Mechanical activation	Mechano-thermal treatment		Thermal treatment
Milling time (min)	45	10	5	–
Roasting time (min)	–	15	10	60
Roasting temperature (°C)	–	900	700	1000
Total cost (euro/t)	239	172	146	194

6.10.1.2 Metallurgical Calcine

Oxidative roasting of the major gold-bearing refractory sulfides pyrite FeS_2 and arsenopyrite FeAsS results in the volatalisation of sulphur and arsenic oxides. The liberated gold particles remain prevailingly with the calcine and the subsequent cyanide leaching of gold is improved. Ideally hematite (γ-Fe_2O_3) should be the principal product of roasting [Haque 1987]. However, the generation of sulfur, arsenic and other metals with volatile oxides gives rise to serious environmental concerns. Mechanical activation can substantial improve the treatment of metallurgical calcines.

A calcine leach residue with the content of 28 gt^{-1} Au and having P_{90} (the size at which 90% passes) of 82 microns was milled in the presence of cyanide and lime [Liddell and Dunne 1988]. The P_{90} after milling was 4 microns, and 59% of the gold was dissolved during milling procedure. A subsequent 24 h leaching of the milled calcine resulted in the gold dissolution increasing to 60%, indicating that essentially all of the leaching has occurred during milling.

Arsenopyrite and pyrite residues from the cyanide leaching of calcines were treated mechanochemically [Corrans and Angove 1991]. These residues can often contain appreciable quantities of gold encapsulated within the hematite matrix and can only be liberated by milling the calcine to sizes below about 10 microns. The results of pyritic and arsenopyritic calcines milling are shown in Fig. 6.67.

In the case of the arsenopyrite calcine the head grade can approach 100 gt^{-1} Au, so that mechanochemical treatment to reduce the value of the cyanidation residue can be economic. In the case of the pyrite calcine, the curve is flatter whilst the head grade is also lower (50–70 gt^{-1} Au), so that this procedure is less attractive.

Tetrahedrite calcines after thermal treatment is economically interesting waste because of content of Cu, Sb, Ag and Au metals.

The investigations were carried out on a tetrahedrite calcine from Rudňany (Slovakia). This material is a rich source of copper and silver, but it is not suitable for pyrometallurgical processing because of the high content of antimony. The chemical composition of as received calcine is given in Table 6.32.

The results of mechanochemical leaching of this tetrahedrite calcine are shown in Fig. 6.68. The process proceeds in two regimes: mechanochemical (I) and chemical (II). The mill temperature reached 84°C and during subsequent chemical leaching rose to 96°C [Baláž et al. 1995].

The liquid/solid ratio in the described experiments was 4.8, 3.3 and 2.5 and resulted in power input of 157, 108 and 82 kWh per tone of concentrate, respectively. These examined liquid/solid ratios were acceptable for plant operation. The almost complete extraction of Sb in 15–20 min and of As in 20–40 min can be achieved (see also Table 6.32). The process is highly selective for Sb and As with only 0.03–0.12% of the copper dissolving. The chemical analysis performed after mechanochemical treatment (Table 6.32) shows that the described process enables us to obtain a Cu-Ag-Au concentrate of high quality. The decrease in antimony content to below 1% show that this material may be processed by pyrometallurgy.

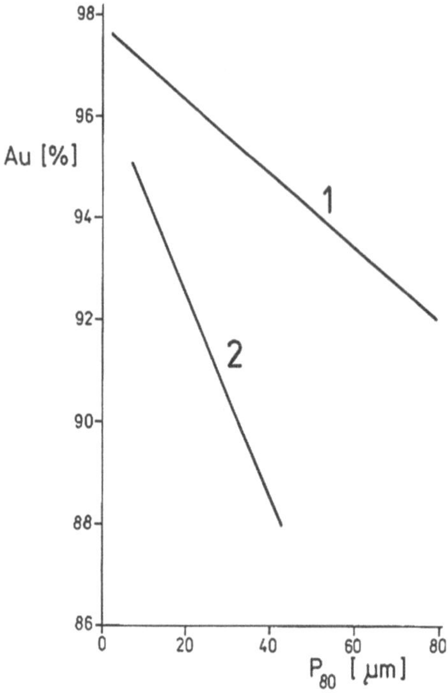

Fig. 6.67 The gold recovery in cyanide leaching from mechanically activated calcines. 1 – pyrite calcine, 2 – arsenopyrite calcine [Corrans and Angove 1991]

A study was undertaken to examine the effect of attrition milling on the leaching properties of titanium slag waste composed of iron magnesium titanate and a silicate phase. It was shown that attrition milling was a very effective method in accelerating the rate of transformation of titanium slag into synthetic rutil or synthetic anatase [Toromanoff and Habashi 1986].

Table 6.32 Chemical composition of tetrahedrite calcine [Baláž et al. 1995]

Element	Me (%)	
	As received	After mechanochemical treatment
Cu	17.60	16.27
Sb	6.03	0.19
S	19.80	21.82
Fe	24.50	25.63
Na	0.44	2.85
Bi	0.28	0.13
As	0.93	0.14
Hg	0.66	< 0.01
Ag	1040 gt^{-1}	1213 gt^{-1}
Au	10.4 gt^{-1}	11.5 gt^{-1}

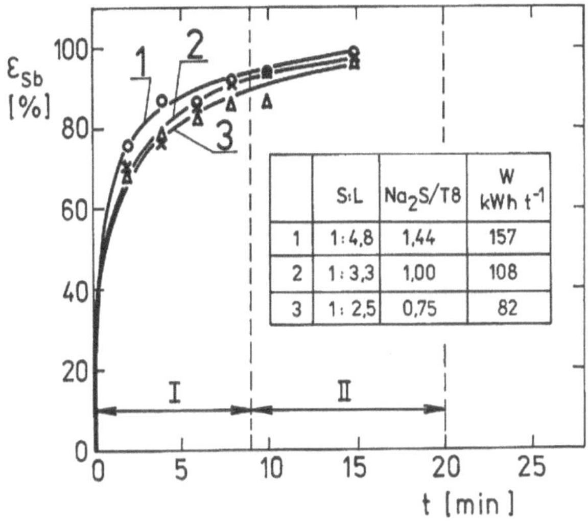

Fig. 6.68 Recovery of antimony into leach, ε_{Sb} vs. time of leaching, t. I – mechanochemical leaching, II – chemical leaching. Leaching conditions: 300 gl^{-1} Na_2S + 53 gl^{-1} NaOH, rate of pulp feeding 95–105 $l\,h^{-1}$ [Baláž et al. 1995]

6.10.1.3 Red Mud

Red mud is an industrial solid waste from Bayer process for alumina production [Havlík 2005]. Red mud constitutes a highly complex system due to the large number of mineral phases. There is a tendency to evaluate the possibility for recycling the red mud as a slimmer in the production of bricks, laterites and tile kilns, an also as material for production of clay-based ceramics [Sglavo et al. 2000]. It has been estimated that mechanical activation can selectively influence the solid state properties of the red mud waste [Budroni et al. 2000]. In the presence of suitable additional phases, the activated red mud may become reactive to an industrially interesting level. Some high-temperature reactions of the red mud occur earlier and at lower temperature, as well as the magnetic properties being drastically changed. Hematite γ-Fe_2O_3 as the component can be effectively removed through magnetic separation.

6.10.2 Materials Waste

Industrial planetary mills working in continuous mode has been applied for treatment of inorganic waste materials. The mill MP-4 works with installed power 200 kW at productivity 3 t/h for obtained size fraction-10 microns (Fig. 6.69).

The new planetary mill of this type has been installed with an energy consumption rate 50 kW/ton. The new generation of the mill would increase the production capacity to $>$ 50 t/h [Akkerman 2006].

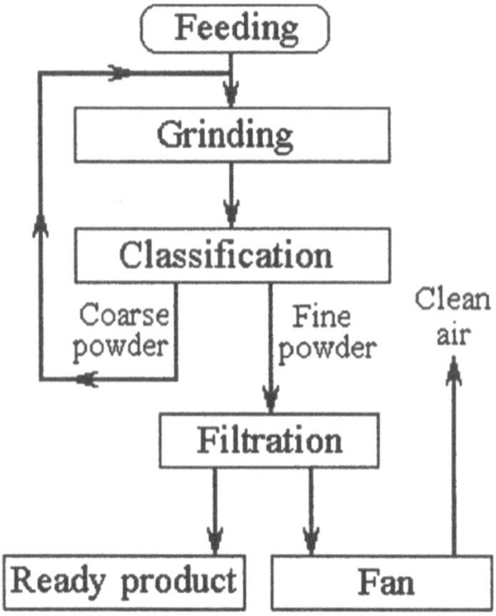

Fig. 6.69 The flowsheet of application of planetary milling in a continuous mode (*above*) and the industrial planetary mill MP-4 (*down*) [http://www.ttd.spb.ru]

The mechanochemical technology of obtaining powders of tungsten carbide was developed for processing of materials based on caked carbides. Cutting tools on a base of caked carbides are produced using complicated, multi-step methods with creation of a large amount of waste with a tungsten content. To this, waste cutting tools after the end of their life, are added. The main component is tungsten carbide (WC), which usually forms 92% of the weight of the cutting material. The binder is metallic cobalt, whose content determines the mechanical properties of the cutting tool.

Mechanical processing of waste from cutting is problematic. Conventional milling requires 10–100 h, and the iron wear can be tens of percents. Use of wet methods create a large amount of liquid waste with a content of heavy metal salts, which worsen the economics of the process and has a negative influence upon the environment.

Perspective method of processing waste cutting tools is the application of mechanochemical technology whilst using high-intensive mills. Planetary mills, working in a batch regime, were successfully tested as well as vibratory working in a continuous mode. The latter reached a production of 50 kg of milled product per hour. The advantage of mechanochemical processing of waste is, in this case, that the product is a powder with a wide range of particle size distribution. These particles are further classified in sieves as various applications require various grades of granularity. For example, the particles with a size of 200–1500 micrometers are suitable for induction melting of hard surfaces, the particles with a size smaller than 20 micrometers are suitable for electro-erosive materials, and particles smaller than 1 micrometer are used for dispersion hardening of metals and alloys [Lomovskij and Boldyrev 2006].

Mechanochemical methods allow to obtain the rare metals present in used catalysts which are used in the automobile industry [Mruck and Ameling 1988]. After their solubilization, metals of the platinum group are obtained from their solutions using mechanochemically synthesized sorbents on a CaO base [Klescev 1983; Domarev 1985].

It has been shown that mechanochemical methods can also be applied for obtaining rare earth metals from waste such as used television and computer monitors, displays, light bulbs, etc. Compounds of europium, lanthanum, cerium and terbium pay an active role in this luminescence-using equipment. Even short term mechanical activation provided, in the further stage of extraction, transfer of 70–80% of yttrium and europium into a solution using an application of HCl at laboratory temperature [Filio et al. 1994; Kano and Saito 1998].

6.10.3 Organic Waste

Disposal of organic chemicals like polyvinyl chloride (PVC) plastics, polychlorinated biphenyls (PCBs), chlorinated pesticides, polychlorinated dibenzodioxins (PCDD) and – furans (PCDF) has become a serious problem of present technology. Traditional oxidative degradation, such as incineration in waste combustors, represents a significant environmental hazard. Alternative remedy technologies have not been as effective as incineration and have met only limited success. Research therefore is seeking for development of new disposal technologies [Monagheddu et al. 1999].

From organics, PVC has excellent properties such as corrosion and chemical resistance, flexible hardness, high cost performance in comparison to other plastics, that lead to production of commodities and industrial applications in the construction, agriculture and chemical fields. However, after the usage of such PVC products, much will end up as waste. Combustion of PVC waste produces hydrochloric acid, chlorine gas and dioxins which causes severe environmental problems. Since more than half of weight in PVC contains chlorine (56.7 wt.%), it causes a serious problem when it is burned out, even if as fuel. It is therefore required that safe and novel method for disposing the waste PVC to prevent air and soil pollution would be developed [Tongamp et al. 2006; Tongamp et al. 2007].

Several mechanochemical approaches has been tested to treat organic waste [Rowlands et al. 1994; Hall et al. 1996]. In the first paper published in 1994, the importance of a low-temperature process involving milling of toxic material has been stressed. Saito et al. have studied a process for dechlorination of PVC and other chlorinated compounds [Zhang et al. 1999; Mio et al. 2002; Inoue et al. 2004; Saeki et al. 2004; Kano et al. 2006; Zhang et al. 2007; Tongamp et al. 2006, 2007].

Mechanochemical decomposition of these substances has attracted much attention because of the easy operation and evident merit for the disposal of vast amounts of samples contaminated by chlorinated polymers. Inorganic compounds such as CaO, $CaCO_3$, MgO, ZnO, $Ca(OH)_2$, KOH, NaOH, $LiCoO_2$ and other have been investigated. From the additive investigated, complete decomposition of PVC was achieved with CaO within two hours of milling. Under similar condition, more than 95% of chlorine in PVC was removed by $CaCO_3$ and $Ca(OH)_2$ in 4 and 12 h, respectively [Tongamp et al. 2006].

The mechanochemically induced reaction when PVC is milled with $CaCO_3$ can be represented by the following reactions [Tongamp et al. 2007].

$$[-CH_2-CHCl-]+CaCO_3 \rightarrow [-CH=CH-]+CO_2+H_2O+CaCl_2 \quad (6.26)$$

$$[-CH_2-CHCl-]+CaCO_3 \rightarrow [-CH_2-CHOH-]+CO_2+CaCl_2 \quad (6.27)$$

Organic compounds in the milled products after washing could be recovered as energy source and $CaCl_2$ could be recovered for further application [Tongamp et al. 2006]. Recently small company has been established in Japan to apply this mechanochemical technology. The company is applying mechanochemical reactor (Fig. 6.70) for treatment of kilograms of PVC-CaO powder mixture [Zhang et al. 2000].

The results of detoxication tests on dioxine-contaminated soil with help of mechanochemistry are given in Fig. 6.71. The results indicate that it is possible to reduce dioxine toxicity to zero (i.e., not detectable) [Zhang et al. 2001].

Besides PVC, the mechanochemical approach has been applied to treatment of other organic waste such as hexabrombenzene, trichlorbenzene, dioxine, chlorobiphenyl, polytetrafluoroethylene [Saito 2004].

The combination of milling with autopropagating high-temperature reactions for hexachlorobenzene has been documented. No residual traces of the organohalide compound were found in the reacted powders or in the gaseous phase sampled from the headspace of the milling reactor. The solid products of the reaction were restricted to CaHCl and $CaCl_2$ salts and graphite. Similar results were obtained from a

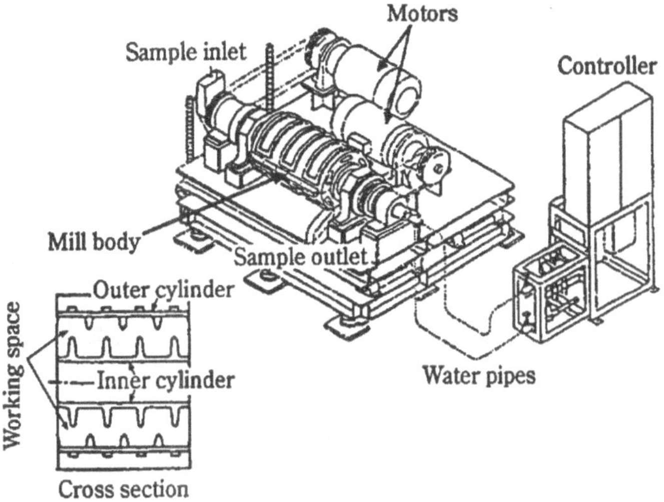

Fig. 6.70 The mechanochemical reactor for PVC-CaO treatment [Zhang et al. 2000]

waste with content of traces of polychlorodibenzodioxins and polychlorodibenzofurans. A transformation yield greater than 99.6% was obtained with a decrease of the toxic equivalent concentration from 2492 to 0.788 μg/kg [Monagheddu et al. 1999].

There is a permanent interest in an application field for such activities. For example DMCR technology (dehalogenation by mechanochemical reaction) has been elaborated [Birke et al. 2004]. By applying mechanochemistry to contaminated materials, polyhalogenated pollutants like HCH, DDT, PCB, dioxine and PCP are

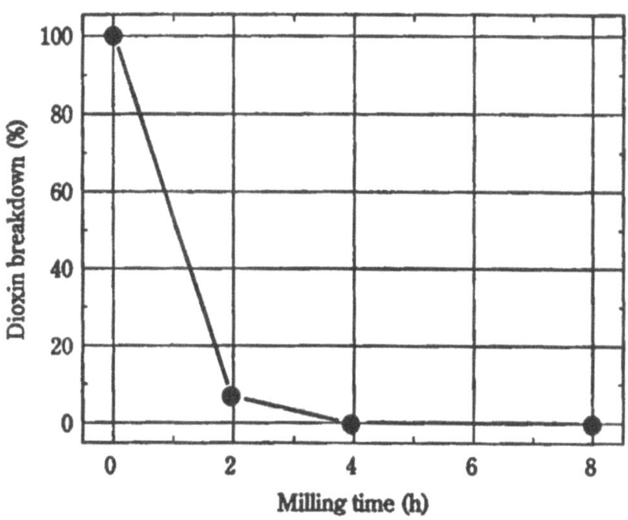

Fig. 6.71 Dioxin breakdown percentage according to milling time [Zhang et al. 2001]

reductively dehalogenated directly inside the contaminated matrix. A ball mill is used, a highly effective mixing device and a mechanochemical reactor in one common operation step. The efficiency fot treatment of polychlorinated biphenyls in natural soils, filter dusts and transformer oils in an industrial mill has been successfully proven [http://www.tribochem.com].

References

Ackerman, J.B., Anderson, C.G., Nordwick, S.M. and Krys, L.E. (1993) Hydrometallurgy at the Sunshine Mine metallurgical complex. In: B. Hiskey and M. Wadsworth (Eds.) *Proceedings of the AIME Meeting "Hydrometallurgy – Fundamentals, Technology and Innovations"*, Littleton, Colorado, pp. 477–498.

Adam, K., Natarajan, K.A., Riemer, S.C. and Iwasaki, I. (1986a) Electrochemical aspects of grinding media-mineral interaction in sulphide ore grinding. Corrosion (Houston) 42, 440–446.

Adam, K., Natarajan, K.A. and Iwasaki, I. (1986b) Grinding media wear and its effect on the flotation of sulfide minerals. International Journal of Mineral Processing 12, 39–54.

Adams, M. and Johnson, G.D. (2001) Advances in nickel sulfide process treatment. In: *Proceedings of the International Conference ALTA 2001 Nickel/Cobalt-7*, Melbourne, Australia.

Adie, D. (1993) Verlaglasung, die saubere Lösung, *UTECH-Proc. of Faserkongress,* Berlin, pp. 165–197.

Akkerman, E. (2006) Out of this world. Industrial Minerals, June, pp. 61–62.

Allen, J.L. (1998) Sunshine mine responding to mineral changes in ore feed. Engineering and Mining Journal 199, pp. 46–50.

Amer, A.M. (1995) Investigation of the direct hydrometallurgical processing of mechanically activated complex sulphide, Akarem area, Egypt. Hydrometallurgy 38, pp. 225–234.

Amgalan, J., Chaikina, M.V., Dulamsuren, M. and Bilegbaatar, A. (1998) Mechanical activation of phosphorites as an ecologically pure technology to prepare phosphorous-containing fertilizers. Chemistry for Sustainable Development 6, pp. 221–226.

Anderson, C.G., Harrison, K.D. and Krys, L.E. (1993) Process integration of sodium nitrite oxidation and fine grinding in refractory precious metal concentrate pressure leaching. In: R.K. Mishra (Ed.) *Proc. XVIIth IPMI Conference "Precious Metals 1993"*, Rhode Island.

Anderson, C.G. and Nordwick, S.M. (1996) Pretreatment using alkaline sulfide leaching and nitrogen species catalyzed pressure oxidation on a refractory gold concentrate. In: G.W. Warren (Ed.) *Proc. EPD Congress*, Anaheim, California, pp. 323–341.

Angove, J. (1993) The Activox™ process for refractory gold ores. In: *Proc. Int. Conf. Randol Gold Forum*, Beaver Creek, Australia, pp. 1–12.

Ani-Metprotech, stirred vertical (UFM) mills-installation list 1998.

Apte, S.P., Cagle, M., Bryant, R., Weidenbach, A. and Kulshreshtha, A.K. (2003) Use of sodium chloride to facilite reduction of particle size of dexamethasone during ball milling. Drug Development and Industrial Pharmacy 29, pp. 367–373.

Arai, Y. (1976) *Chemistry of Powder Production.* Chapman and Hall, London.

Arai, Y., Yasue, T. and Honada, T. (1976) Journal of Chemical Society Japan, pp. 591–603.

AsbestEx System GmbH, Dortmund (1994): Asbestensorgung-Wertstoffgewinnung statt Deponierung, Entsorgungs Technik 10, pp. 12–13.

Auerbach, H., Dorr, H., Haefner, E. and Nebe, R. (1987) Zum Einfluss der Trockenmahlung eines sulfidischen Erzen in einer Planetenmühle auf die Flotation. Erzmetall 40, pp. 614–617.

Avvakumov, E.G., Pavlenko, S.I. and Kosova, N.V. (2000) Compositional binder from mechanochemical activated industrial wastes. Chemistry for Sustainable Development 8, pp. 636–440.

Avvakumov, E.G., Senna, M. and Kosova, N. (2001) *Soft Mechanochemical Synthesis, A Basis for New Chemical Technologies.* Kluwer Academic Publishers, Boston.

Bailey, A.E.W., Kimberley, I.E. and Ward, S.G. (1954) Chemical constitution of coal. I. Preparation and properties of humic acids from coal. Fuel 33, pp. 209–221.

Baláž, P., Tkáčová, K., Mišura, B., Paholič, G. and Briančin, J. (1989) Use of mechanochemical processing principles for intensifying the leaching of complex sulphide ores. In: D.S. Fleet (Ed.) *Proc. Int. Symp. "Extractive Metallurgy'89"*, Institute of Mining and Metallurgy, London, pp. 751–762.

Baláž, P., Kammel, R., Kušnierová, M. and Achimovičová, M. (1994a) Mechanochemical treatment as a new non-pollution method of metals recovery. In: *Proc. Int. Conf. "Hydrometallurgy '94"*, Cambridge 1994, Chapman and Hall, London 1994, pp. 209–218.

Baláž, P., Kammel, R. and Achimovičová, M. (1994b) Mechanochemical treatment of tetrahedrite as a new ecologically non-polution way of metals recovery. Metall 48, pp. 217–220.

Baláž, P., Sekula, F., Jusko, F., Kočí, M., Dugas, V. and Lauko, L. (1994c) Process for tetrahedrite ore treatment. Slovak Patent No. 081–94.

Baláž, P., Sekula, F., Jakabský, Š. and Kammel, R. (1995) Application of attrition grinding in alkaline leaching of tetrahedrite. Minerals Engineering 8, pp. 1299–1308.

Baláž, P. and Sekula, F. (1996) Mechanochemical preparation of Cu-Ag-Au concentrate from tetrahedrite raw materials. In: M.A. Sánchez, P. Vergara and S.H. Castro (Eds.) *Proc. IIIrd Int. Symp. "Clean Technology for the Mining Industry"*, Concepción, pp. 469–475.

Baláž, P. Kammel, R., Havlík, T., Achimovičová, M. and Števulová, N. (1997a) Hydrometallurgical treatment of calcined tetrahedrite concentrate. Metall 51, pp. 386–389.

Baláž, P., Kammel, R., Sekula, F. and Jakabský, Š. (1997b) Mechanochemical leaching: the possibility to influence the rate of metals extraction from refractory ores. In: H. Hoberg, and H. von Blottnitz (Eds.) *Proc. XXth Int. Miner. Proc. Congr.*, Vol. 4, Aachen, pp. 149–159.

Baláž, P., Števulová, N., Kammel, R. and Malmström, R. (1998a) Extraction of Ni, Cu and Co from mechanically activated pentlandite concentrate. Metall 11, pp. 620–623.

Baláž, P., Frommel, E.A., Kern, D.G., LaCount, R.B., Martello, D.V. and Turčániová, L'. (1998b) Chemical cleaning of Slovak brown coal by GACL (grinding and aqueous caustic leaching). In: *Proc. XVth Ann. Int. Pittsburgh Coal Conference*, Pittsburgh, USA, CD.

Baláž, P. (2000) *Extractive Metallurgy of Activated Minerals*. Elsevier, Amsterdam.

Baláž, P., LaCount, R.B., Kern, D.G. and Turčániová, L'. (2001) Chemical treatment of coal by grinding and aqueous caustic leaching. Fuel 80, pp. 665–671.

Baláž, P., Achimovičová, M., Ficeriová, J., Godočíková, E., Villachica, C. and Sanchez, M. (2003) Mechanochemistry of copper minerals: new challenges in extractive metallurgy. In: P.A. Riveros, D. Dixon, D.B. Dreisinger, and J. Menacho (Eds.) *Proc. Int. Conf. "Copper 2003/Cobre2003"*, Canadian Institute of Mining, Metallurgy and Petroleum, Santiago, Chile, pp. 353–365.

Baláž, P. and Achimovičová, M. (2006) Mechano-chemical leaching in hydrometallurgy of complex sulphides. Hydrometallurgy 84, pp. 60–68.

Berdov, G.J., Balachnin, M.W. and Protalinskij, A.N. (1983) Izvestija Sibirskovo Otdelenija AN ZSSR ser. chim. nauk 45–51 (in Russian).

Biangardi, S. and Pietsch, H. (1975) The treatment of sulphidic copper ores by the LM (Lurgi-Mitterberg) process. In: G.A. Davies, and J.B. Scuffhan (Eds.) *Symp. Hydrometallurgy*, Manchester, The Institution of Chemical Engineers Symposium Series No. 42, pp. 5.1–5.7.

Biangardi, S. and Pietsch, H. (1976) Verarbeitung sulfidischer Kupfererze nach dem Lurgi-Mitterberg (LM)-Verfahren. Erzmetall 29, pp. 73–80.

Birke, V., Mattik, J. and Runne, D. (2004) Mechanochemical reductive dehalogenation of hazardous polyhalogenated contaminants. Journal of Materials Science 39, pp. 5111–5116.

Bisrat, M. and Nyström, Ch. (1988) Physicochemical aspects of drug release. VIII The relation between particle size and surface specific dissolution rate in agitated suspensions. International Journal of Pharmaceutics 47, pp. 223–231.

Boldyrev, V.V. (1986) Mechanical activation and its application in technology. Journal of Chimica Physics 83, 821–829.

Boldyrev, V.V. (1996) Mechanochemistry and mechanical activation of solids. Materials Science Forum 225–227, 511–520.

Boldyrev, V.V., Lyakhov, N.Z., Chaikina, M.V. and Denisov, M.G. (1996) Mechanochemical technology of phosphorus fertilizers production. Chemistry for Sustainable Development 4, 97–99.

Boldyrev, V.V. (2004) Mechanochemical modification and synthesis of drugs. Journal of Materials Science 39, 5117–5120.

Boldyrev, V.V. (2006) Mechanochemistry and mechanical activation of solids. Russian Chemical Reviews 75, 177–189.

Bond, G.C. (1962) *Catalysis by Metals*. Academic Press, London.

Boudart, M. and Burwell, R.L. Jr. (1974) Mechanism in heterogeneous catalysis. In: E.S. Lewis (Ed.) *Techniques in Chemistry*, John Wiley and Sons, New York, pp. 693–740.

Braga, D., Grepioni, F. and Orpen, A.G., (1999) *Crystal Engineering: From Molecules and Crystals to Materials*. Kluwer Academic Publishers, Dordrecht.

Braga, D. and Grepioni, F. (2004) Reactions between or within molecular crystals. Angewandte Chemie International Edition 43, 4002–4011.

Braga, D. and Grepioni, F. (2005) Making crystals from crystals: a green route to crystal engineering and polymorphism. Chemical Communications, 29, 3635–3645.

Braga, D., Curzi, M., Johansson, A., Polito, M., Rubini, K. and Grepioni, F. (2006) Simple and mechanochemical preparation of porous crystalline material based on a 1D coordination network for uptake of small molecules. Angewandte Chemie 118, 148–152.

Budroni, G, Cocco, G., Jiang, J.Z., Carturan, G. and Enzo, S. (2000) X-ray powder diffraction and Mössbauer study of red mud residue from alumina production. Materials Science Forum 343–346, 695–700.

Bushrab, F.N. and Müller, R.H. (2003) Nanocrystals of poorly soluble drugs for oral administration. New Drugs 5, 20–23.

Cases, I.M., Doerler, N. and Francois, M. (1985) Influence of different types de broyage fin sur lex mineraux: I cas du quartz. In: *Proceedings of the XVth International Minerals Processing Congress*, Vol.1, Cannes, pp. 169–178.

Cevasco, G. and Thea, S. (1999) Mechanism of alkaline hydrolysis of some HO-π-COOAr acyl derivates. The Journal of Organic Chemistry 64, 5422–5426.

Chaikina, M.V. (1986) Natural phosphates: structure-chemical classification and waste free mechanochemical technology. Chemistry for Sustainable Development 4, 71–95.

Chaikina, M.V. (1996) Physico-chemical fundamentals of mechanical activation of complex phosphate-containing systems and their technological application, DSc. Thesis, Institute of Solid State Chemistry Siberian Branch of Russian Academy of Sciences, Novosibirsk (in Russian).

Chemburkar, S.R., Bauer, J., Dening, K., Spiwek, H., Patel, K., Morris, J., Henry, R., Spanton, S., Dziki, W., Porter, W., Quick, J., Bauer, P., Donaubauer, J., Narayanan, B.A., Soldani, M., Rilley, D. and McFarland, K. (2000) Organic Process Research and Development 4, 413–417.

Chikhalia, V., Forbes, R.T., Storey, R.A. and Ticehurst, M. (2006) The effect of crystal morphology and mill type on milling induced crystal disorder. European Journal of Pharmaceutical Sciences 27, 19–26.

Chiotti, P. and Markuszewski, R. (1985) Reaction of pyrite with fused sodium hydroxide. Industrial and Engineering Chemistry Process Design and Development 24, 1137–1140.

Čanturija, V.A. and Vigdergauz, V.E. (1993) *Electrochemistry of Sulphides. Theory and Practice of Flotation*. Nauka, Moscow (in Russian).

Choi, W.S., Kim, H.I., Kwak, S.S., Chung, H.Y., Hang Y., Chung Hae Y., Yamamoto K., Oguchi T., Tozuka Y., Yonemochi, E. and Terada, K. (2004) Amorphous ultrafine particle preparation for improvement of bioavailability of insoluble drugs: grinding characteristics of fine grinding mills. International Journal of Minerals Processing 74S, S165–S172.

Chrenkova, T.M. (1993) *Mechanochemical Activation of Coal*. Nedra, Moscow (in Russian).

Chriswell, C.D., Markuszewski, R. and Jewell, D.W. (1994) Improved caustic coal desulfurization by a float-sink/leaching process employing 50% aqueous NaOH solutions. Fuel Processing and Technology 37, 19–32.

Corrans, I.J. and Angove, J.E., (1991) Ultrafine milling for the recovery of refractory gold. Minerals Engineering 4, 763–776.

Corrans, I.J., Johnson, G.D. and Angove, J.E. (1993) The recovery of nickel and gold from sulfide concentrates. In: *Proceedings of the XVIIIth International Mineral Processing Congress*, Sydney, pp. 1227–1231.

Corrans, I.J. and Angove, J.S. (1993) Activation of mineral species, Australian Patent 663 523, April 29.

Corrans, I.J., Angove, J.E. and Johnson, G.D. (1995) The treatment of refractory copper-gold ores using Activox[R] processing. In: *Proceedings of the International Conference "Randol Gold Forum"*, Perth, pp. 221–224.

Crawford, A. (1951) The preparation of ultra-clean coal in Germany. Transactions of the Institute of Mining Engineers 111, 204–219.

Čanturija, V.A. and Lunin, V.D. (1983) *Electrochemical Methods of Flotation Processes Identification*. Nauka, Moscow (in Russian).

Čanturija, V.A. and Vigdergauz, V.E. (1994) *Electrochemistry of Sulfides*. Nauka, Moscow (in Russian).

Dayton, S. (1982) Equity silver on line with leach plant. Engineering Mining Journal, January, 78–83.

Domarev, N.F. (1985) Synthesis of adsorbent for platinum metals recovery, Russian patent 3867675.

Dincer, H., Önal, G. and Gock, E. (2000) The influence of different types of grinding stress on the production of chromium chemicals. In: G. Özbayoglu, C. Hostin, M.Ü. Ataloy, C. Hicyilmaz and A.I. Arol (Eds.) In: *Proceedings of the International Mineral Processing Conference "Mineral Processing on the Verge of the 21st Century"*, Istanbul, Balkema, Rotterdam 2000, pp. 47–53.

Dushkin, A.V., Rykova, Z. Yu, Shakhtshneider, T.P. and Boldyrev, V.V. (1974) To the possibility of solid state mechanochemical low molecular organic synthesis. International Journal of Mechanochemistry and Mechanical Alloying 1, 48–55.

Dushkin, A.V. (2004) Potential of mechanochemical technology in organic synthesis and synthesis of new materials. Chemistry for Sustainable Development 12, 251–273.

Dushkin, A.V. (2005) Possibilities of mechanochemical technology in organic synthesis and enhancement of materials solubility, DSc. Thesis, Novosibirsk.

Dutrizac, J.E. and Morrison, R.M. (1984) The leaching of some arsenide and antimonide minerals in ferric chloride media. In: R.C. Baustista (Ed.) *Proceedings of the International Conference. "Hydrometallurgical Process Fundametals"*. Plenum Press, New York, pp. 77–112.

Dutta, J. and Hoffmann, H. (2003) http://www.mxsg3.epfl.ch/ltp/Cours/Nanomat/pdf

Elamin, A.A., Ahlneck, C., Alderborn, G. and Nyström, C. (1994) Increased metastable solubility of milled griseofulvin, depending on the formation of a disordered surface structure. International Journal of Pharmaceutics 111, 159–170.

Elisejev, N.I., Dolženkova, A.N. and Epelman, M.L. (1980) Investigation of oxygen behaviour by mineral processing of copper-zinc ores. Obogaščenije rud 1, 11–14 (in Russian).

Enderle, U., Woodall, P., Duffy, M. and Johnson, N.W. (1997) Stirred mill technology for regrinding Mc Arthur River and Mount Isa zinc/lead ores. In: H. Hoberg and H. von Blattnitz (Eds.) *Proceedings of the XXth International Mineral Processing Congress*, Aachen, Vol. 2, pp. 71–78.

Evans, H.E. and Johnson, G.D. (1999) Activox technology for the treatment of sulfide concentrates. In: *Copper Hydromet Roundtable '99*, Randol International Phoenix, Arizona.

Fábry, O. (1992) Development of mineral processing at Železnorudné bane Rudvnany. Rudy- Uhlí 4, 120–126 (in Slovak).

Filio, J.M, Kasai, E., Umetsu, Y., Saito, F. and Chung, H.S. (1994) Grinding of EP Dust and its effect on solubility of metal compounds in water. Journal of Chemical Engineering of Japan 27, 492–497.

Franke, N.W., Kiebler, M.W., Ruof, C.H., Savich, T.R. and Howard, H.C. (1952) Water soluble polycarbolic acids by oxidation of the coal. Industrial and Engineering Chemistry 44, 2784–2792.

Friedman, L.D. and Kinney, C.R. (1950) Humic acids from coal. Controlled air-oxidation of coals and carbons at 150°C to 400°C. Industrial and Engineering Chemistry 42, 2525–2529.
Gasparrini, C. (1995) *Gold and Other Precious Metals, from Ore to Market.* The Space Eagle Publishing, Tucson.
Gärtner, C.E., Kammel, R., Kheiri, M.J., Pawlek, F. and Simon, M. (1991) Untersuchungen über das Laugungsverhalten feinstgemahlener Zinkblende-und Kupferkies-Flotationskonzentrate. Metall 45, 886–890.
Gerlach, J. and Gock, E. (1973) Verfahren zur Aufbereitung sulfidischer, arseninidischer order antimonischer Materialen, Austrian Patent No. A7963/71-3.
Gerlach, J.K., Gock, E. and Ghosh, S.K. (1973) Activation and leaching of chalcopyrite concentrates by diluted H_2SO_4 solutions. In: D.J.I. Evans and R.S. Schoemaker (Eds.) *Proceedings of theIInd International Symposium on Hydrometallurgy*, Chicago, pp. 87–94.
Gerlach, J. (1982) Untersuchungen zur direkten Laugung komplexer sulfidischer Erze. Metall, 36, 518–523.
Glembockij, V.A. (1972) *Physical Chemistry of Flotation.* Nedra, Moscow (in Russian).
Gock, E. (1977) Beinflussung des Löseverhaltens sulfidischer Rohstoffe durch Festkörperreaktionen bei der Schwingmahlung, Habilitationschrift, Technische Universität Berlin.
Gock, E. (1978) Beeinflussung der Löseverhaltens von Kupferkies durch Festkörperreactionen bei der Schwingmahlung. Erzmetall 31, 282–288.
Gock, E. and Asiam, E. (1986) Process for extraction of precious metals. Patent Zimbabwe 191/86(1986); South Africa Patent 86/7/138 (1986); Australian Patent 595 236 (1986).
Gock, E. (1987) Mechanische Aktivierung von Phosporitkonzentraten im Hinblick auf den direkten Einsatz als Düngemittel. Research report, Technical University Berlin, 35p.
Gock, E. and Asiam, E. (1988) Gewinnung von Edelmetallen aus Kohlenstoffhaltigen Arsenopyritkonzentraten nach mechanischer Aktivierung. In: K. Tkáčová (Ed.) *Proceedings of the Vth International Symposium on"Theoretical and Technological Aspects of Comminution and Mechanical Activation of Minerals"*, Košice, pp. 60–65.
Gock, E., Florescu, R. and Betgovarez, W. (1994) Verfahren zur Aufbereitung von faserstoffhaltigen Materialen, German Patent No. 4332031.
Gould, V.G. (1955) Sunshine's tetrahedrite ores yield electrolytic antimony. Engineering Mining Journal 156, 91–94.
Greenberg, S.A. (1961) Reaction between silica and calcium hydroxide solutions. I. Kinetics in the temperature range 30–85°C. Journal of Physical Chemistry 65, 12–16.
Habashi, F. (1978) *Chalcopyrite Its Chemistry and Metallurgy.* McGraw Hill, New York.
Habashi, F. (1993) *A Textbook of Hydrometallurgy.* Metallurgie Extractive Quebec, Sainte Foy, Quebec.
Hall, A.K., Harrowfield, J.M., Hart, R.J. and McCormick, P. (1996) Mechanochemical reaction of DDT with calcium oxide. Environmental Science and Technology 30, 3401–3407.
Haque, K.E. (1987) Gold leaching from refractory ores-literature survey. Mineral Processing and Extractive Metallurgy Review 2, 235–253.
Havlík, T., Škrobian, M. and Dudáš, D. (1991) Study of acid oxidative leaching of tetrahedrite. Hutnické listy 11/12, 76–80 (in Slovak).
Havlík, T., Škrobian, M. and Baláž, P. (1994) Study of acid oxidizing leaching of tetrahedrite. Erzmetall 47, 112–119.
Havlík, T., Škrobian, M. and Kammel, R. (1998) The leaching behaviour of tetrahedrite concentrate in nitrate solution. Metall 52, 210–213.
Havlík, T. and Škrobian, M. (2000) Hydrometallurgy. In: F.H. Chung and D.K. Smith (Eds.) *Industrial Applications of X-ray Detection*, Marcel Dekker, Inc., New York, pp. 303–315.
Havlík, T. (2005) *Hydrometallurgy*. Emilena, Košice (in Slovak).
Heinicke, G., Harenz, H. and Sigrist, K. (1967) Zeitschrift für Anorganische und Allgemeine Chemie 352, 168.

Heinicke, G., Bock, N. and Harenz, H. (1970) Zum Mechanismus der tribomechanisch aktivierten Metallcarbonylbildung unter Einfluss schmefelhaltiger Substanz. Zeitschrift für Anorganische und Allgemeine Chemie 372, 162–170.

Heinicke, G. (1972) Wissenschaft und Fortschritt 22, 11.

Heinicke, G., Bock, N. and Eckstein, H. (1975) Metalformen aus der Gasphase. Wissenschaft und Fortschritt 25, 10–15.

Heinicke, G., Paudert, H., Harenz, H., Steinicke, U. and Pöthig, R. (1977) Žurnal prikladnoj chimiji 50, 969–977.

Heinicke, G. (1984) *Tribochemistry*. Academie-Verlag, Berlin.

Hilfiker, R. (2006) Relevance of solid-state properties for pharmaceutical products. In: R. Hilfiker (Ed.) *Polymorphism: in the Pharmaceutical Industry*, Wiley-VCh, Weinheim, pp. 1–19.

Hint, I.A. (1962) *Fundamentals of Silicalcite Production*. Strojizdat, Moscow (in Russian).

Hiskey, J.B. (1986) Technical developments in hydrometallurgy. Journal of Metals July, 41–46.

Hiskey, J.B. and Atluri, V.P. (1988) Dissolution chemistry of gold and silver in different lixiviants. Mineral Processing and Extractive Metallurgy Reviews 4, 95–134.

Holmes, W.C. (1944) How electrolytic antimony is made at Sunshine plant. Engineering Mining Journal 45, 5–58.

Hourn, M., Turner, D.W. and Holzberger, I.R. (1999) Atmospheric mineral leaching process, US Patent 5,993,635.

Hourn, M. and Halbe, D. (1999) The NENATECH process: Results on Frieda River copper gold concentrates. In: *Proceedings of the International Conference of Randol Copper Hydromet Roundtable '99*, Golden, Colorado, Randol International, pp. 97–102.

http://www.albionprocess.com

http://www.isamill.com

http://www.nishineng.com

http://www.tribochem.com

http://www.ttd.spb.ru

http://www.wmt.com.au

http://www.zoz.com

http://www.zoz.de

Hüttenrauch, N. and Fricke, S. (1981) Influence of lattice defects on the course and rate of drying of particulate solids. International Journal of Pharmaceutical Technology and Product Manufacture 2, 35–37.

Inoue, T., Miyazaki, M., Kamitani, M., Kano, J. and Saito, F. (2004) Mechanochemical dechlorination of polyvinylchloride by co-grinding with various metal oxides. Advanced Powder Technology 15, 215–225.

Isupova, L.A., Sadykov, V.A., Tikhov, S.F., Kimhai, O.N., Kovalenko, O.N., Kustova, G.N., Ovsyannikova, I.A., Dovbii, Z.A., Kryukova, G.N., Rozovskii, A. Ya, Tretyakov, V.F. and Lunin, V.V. (1996) Monolith perovskite catalysts for environmentally benign fuels combustion and toxic waste incineration. Catalysis Today 27, 249–256.

Ivanov, E., Patton, V. and Grigorieva, T. (1996). In: C. Suryanarayana (Ed.) *Processing and Properties of Nanocrystalline Materials*, Warrendale, PA, TMS, pp. 189–197.

Johansson, K., Larsson, C., Antzutkin, O.N., Forsling, W., Rao, H.K. and Romin, V. (1999) Kinetics of the hydratation reaction in the cement paste with mechanochemically modified cement ^{29}Si-magic-angle-spinning NMR study. Cement and Concrete Research 29, 1575–1581.

Johnson, G., Becker, G. and Dredge, K. (1995) New technology for the treatment of base metals. In: *Proc. The Aus/MM Conference*, Newcastle, Australia, pp. 115–119

Johnson, N.W., Gao, M., Young, M.F. and Cronin, B. (1998) Application of IsaMill (a horizontal stirred mill) to the lead-zinc concentrator (Mount Isa Mines Ltd.) and the cycle. In: *Proceedings of the AusIMM Annual Conference; The Mining Cycle,* Mount Isa, Australia, pp. 291–297.

Johnson, G.D., Evans, H.A. and Turner, J.H. (2000) Autoclave availability: Activox^{R} the advantages of operating at reduced temperatures and pressures. In: *Proceedings of the International Conference ALTA 2000 Nickel/Cobalt-6*, Melbourne, Australia.

Jordan-Gerkens, A. (2004) Entsorgung von Asbestabfällen durch mechanische Faserzerstörung, PhD.Thesis, Technical University Clausthal.

Jovanovic, I. (1994) Stand der Behandlung und Entsorgung asbesthaltiger Reststoffle. Kernforscherzentrum Karlsruhe GmbH, Karlsruhe.

Juhász, Z. and Opoczky, L. (1982) *Mechanical Activation of Silicates by Fine Grinding*. Akademia Kiado, Budapest.

Juhász, Z. and Opoczky, L. (1990) *Mechanical Activation of Minerals by Grinding: Pulverizing and Morphology of Particles*. Ellis Horwood, London.

Jusupov, T.S., Shcherbakova, M.Y. and Koroleva, S.M. (1977) The influence of milling intensity on physico-chemical properties of quartz. Obogaščenije rud 5, 77–79 (in Russian).

Jusupov, T.S., Koroleva, S.M., Shcherbakova, M.Y. and Pantjukova, L.P. (1987) Influence of mechanical activation on flotability of minerals. Fiziko-techničeskoje problemy rozrabotki poleznych isskopajemych 5, 94–97 (in Russian).

Jusupov, T.S. and Heegn, H.P. (1997) Influence of mechanical activation of minerals on the physico-chemical properties of minerals surfaces and flotability. In: H. Hoberg and H. von Blottnitz (Eds.), *Proceedings of the XXth International Mineral Proceesing Congress*, Vol. 3, Aachen, pp. 141–150.

Jusupov, T.S., Anoshin, G.N., Kirillova, E.A. and Čanturija, V.A. (1998) Improvement of the technological processes on the basis of activation ore grinding. In: S. Atak, G. Önal and M.S. Celik (Eds.) *Proceedings of the VIIth International Mineral Processing Symposium. "Innovations in Mineral and Coal Processing"*, Istanbul, Balkema, Rotterdam 1998, pp. 9–12.

Jusupov, T.S., Heegn, H.P. and Shumskaya, L.G. (2000) Beneficiation and mechanical-chemical alterations of natural zeolites. In: P. Massacci (Ed.) *Proceedings of the XXIst International Mineral Processing Congress*, Vol.A, Rome 2000, Elsevier, Amsterdam 2000, pp. A144–A151.

Jusupov, T.S. and Shumskaya, L.G. (2002) Control of zeolites cationic exchange with phosphates by soft mechanochemical treatment. Fiziko-techničeskije problemy razrabotki poleznych iskopajenych 2, 83–88 (in Russian).

Kammel, R., Pawlek, F., Simon, M. and Li Ximing (1987a) Oxidizing leaching of sphalerite under atmospheric pressure. Metall 41, 158–161.

Kammel, R., Pawlek, F., Simon, M. and Li Ximing (1987b) Leaching of attritor ground sphalerite under atmospheric pressure. In: *Proceedings of the AIME Annual Meeting*, Denver, pp. 1–22.

Kano, J. and Saito, F. (1998) Correlation of ball impact energy with yield of soluble vanadium compound produced in EP dust by dry mechanochemical treatment. Journal of Chemical Engineering of Japan 31, 1014–1015.

Kano, J., Zhang, Q., Saito, F., Baron, M. and Nzihou, A. (2006) Synthesis of hydroxyapatite with the mechanochemical treatment products of PVC and CaO. Process Safety and Environmental Protection 84, 309–312.

Kaufmann, D.E., Namyslo, J.C., Gock, E. and Florescu, R. (2006) Verfahren zur Herstellung organischer Spurenelementverbindungen, Offenleg. DE 10 2004 039 486 A1.

Kaupp, G., Schmeyers, J., Naimi-Jamal, M.R., Zoz, H. and Ren, H. (2002a) Reactive milling with the Simoloyer^R: environmentally benign quantitative reactions without solvent and wastes. Chemical Engineering Science 57, 763–765.

Kaupp, G., Naimi-Jamal, M.R., Ren, H. and Zoz, H. (2002b) Environmentally protecting reactive milling. Chemie Technik 31, 206–208.

Kaupp, G., Naimi-Jamal, M.R., Ren, H. and Zoz, H. (2003a) Reactive dry-milling for environmental protection-encouraging industrial applications for high kinetic processing. Process Worldwide 4, 24–27.

Kaupp, G., Naimi-Jamal, M.R. and Schmeyers, J. (2003b) Solvent-free Knoevenagel condensations and Michael additions in the solid state and in the melt with quantitative yield. Tetrahedron 59, 3753–3760.

Kaupp, G. (2005) *Organic Solid-State Reactions with 100% Yield.* Springer Berlin/Heidelberg, Vol. 254 in Book Series-Topics in Current Chemistry, pp. 95–183.

Kaupp, G. (2006) Waste free large-scale synthesis without auxiliaries for sustainable production omitting purifying workup. Crystal Engineering Communications 8, 794–804.

Kähler, J., Friedrich, J. and Gock, E. (1996) Processing of molybdenite concentrate containing rhenium without exhaust gas generation. Erzmetall 49, 415–425.

Kharitonov, A.S., Fenelonov, V.B., Voskresenskaya, T.P., Rudina, N.A., Molchanov, V.V., Plyasova, L.M. and Panov, G.I. (1995) Mechanism of FeZSM-5 milling and its effect on catalytic performance in benzene to phenol oxidation. Zeolites 15, 253–258.

Klescev, N.F. (1983) Adsorbent for platinum metals recovery and its recovery, Russian Patent 35844 37.

Klimpel, R.R. (1997) *Introduction to the Principles of Size Reduction of Particles by Mechanical Means*. Engineering Research Center at the University of Florida, Gainesville, pp. 1–41.

Kloiber, O., Johnson, G., Palmer, Ch., Swarts, A. and Brits, J. (2005) Tati hydrometallurgical demonstration plant project. In: *Proceedings of the International Conference ALTA 2005 Nickel/Cobalt*, Perth, Australia, pp. 1–21.

Koch, E. and Schrader, R. (1970) Giessereitechnik 16, 406–412.

Kolosov, A.S., Boldyrev, V.V., Chaikina, M.V., Tarancova, M.I., Gordejeva, G.I., Paudert, R., Harenz, H., Heinicke, G., Dünkel, L. and Pöthig, R. (1979) Reactivity of mechanically activated natural zeolites. Izvestija SO AN SSSR, seria chimičeskich nauk 6, 148–155 (in Russian).

Kostova, N.G., Spojakina, A.A., Dutková, E. and Baláž, P. (2007) Mechanochemical approach for preparation of Mo-containing β-zeolite. Journal of Physics and Chemistry of Solids. 68, 1169–1172.

Kostovic, M. and Pavlica, J. (1995) Effect of grinding media on galena floatability. Erzmetall 48, 538–541.

Kulebakin, V.G., Terechova, O.G., Molčanov, V.I. and Žižajev, A.M. (1999) *Activation of Raw Materials*. Nauka, Novosibirsk (in Russian).

Kumar, R., Bandopadhyay, A., Rajinikanth, V., Alex, T.C. and Kumar, S. (2004) Improved processing of blended slag cement through mechanical activation. Journal of Materials Science 39, 3449–3452.

Kumar, R., Kumar, S. and Mehrotra, S.P. (2007) Towards sustainable solutions for fly ash through mechanical activation. Resources, Conservation and Recycling 52, 157–179.

Kumar, R., Alex, T.C., Jha, M.K., Khan, Z.H., Makaptra, S.P. and Mishra, C.R. (2004) Mechanochemistry and the Bayer process of alumina production. In: A.T. Tabereaux (Ed.) *Proceedings of the International Conference "Light Metals 2004"* The minerals, Metals & Materials Society, Warendale 2004, pp. 31–34.

Kumar, R., Alex, T.C., Khan, Z.H., Mahapatra, S.P. and Mehrotra, S.P. (2005) Mechanical activation of bauxite potential and prospects in the Bayer process. In: H. Kvande (Ed.) *Proceedings of the International Conference "Light Metals 2005"*, The Minerals, Metals & Materials Society, Warendale 2005, pp. 77–79.

Kušnierová, M., Šepelák, V. and Briančin, J. (1993) Effects of biodegradation and mechanical activation on gold recovery by thiourea leaching. Journal of Metals December, 54–56.

La Brooy, S.R., Linge, H.G. and Walker, G.S. (1994) Review of gold extraction from ores. Minerals Engineering 7, 1213–1241.

Lagas, M. and Lerk, C.F. (1981) The polymorphism of sulphathiazole. International Journal of Pharmacy 8, 11–24.

Lagerblad, B., Jennings, H.M. and Chen, J.J. (2004) Modification of cement paste with silica fame-a NMR study. In: *Proceedings of the 1st International Symposium on Nanotechnology in Construction*, Royal Society of Chemistry, Paislery, Scotland.

Lagerblad, B. and Vogt, C. (2004) *Ultrafine Particles to Save Cement and Improve Concrete Properties*. Swedish Cement and Concrete Research Institute, Stockholm, 40 p.

Liddell, K.S. (1997) Design of ultra-fine grinding mills and their potential application in copper hydrometallurgy. In: *Proceedings of the International Conference of ALTA 1997 Copper Hydrometallurgical Forum*, Brisbane, Australia 1997, pp. 1–13.

Lipinski, C.A., Lombardo, F., Domini, B.W. and Freney, P.J. (1997) Experimental and computational approaches to estimate solubility and permeability in drug discovery and development settings. Advanced Drug Delivery Reviews 23, 3–25.

Lomovskij, O.I. (1994) Mechanochemical methods for solving ecological problems. Chemistry for Sustainable Development 2, 473–482 (in Russian).
Lomovskij, O.I. and Boldyrev, V.V. (2006) *Mechanochemistry for Solving Environmental Problems*. GPNTB SO RAN, Novosibirsk (in Russian).
Lorentzen, L. and van Deventer, J.S.L. (1992) Electrochemical interactions between gold and its associated minerals during cyanidation. Hydrometallurgy 30, 177–194.
Madey, T.E., Yates, Jr. J.T., Sandstrom, D.R. and Voorhoeve, R.J.H. (1976) Catalysis by Solid Surfaces. In: N.B. Hannay (Ed.) *Treatise on Solid State Chemistry*, Vol. 6B, Surfaces II, Plenum Press, New York, pp. 1–124.
Marsden, J. and House, I. (1992) *The Chemistry of Gold Extraction*. Ellis Horwood, New York.
Masciantonio, P.X. (1965) The effect of molten caustic on pyritic sulphur in bituminous coal. Fuel 44, 269–275.
Melnikov, S.M. (1977) *Antimony*. Metallurgia, Moscow (in Russian).
Merisko-Liversidge, E. (2002) Nanocrystals: Resolving pharmaceutical formulation issues associated with poorly water-soluble compounds. In: *Particles 2002*, Orlando, Florida, Paper No. 45.
Meyer, K. (1977) Physico-chemical Crystallography. VEB Deutscher Verlag für Grundstoffindustrie, Leipzig.
Mezencevová, A. and Števulová, N. (2002) Characterization of fluidised ashes and possibilities of modification of their properties by mechanical activation. Chemical Papers 96, 515–516.
Minejev, G.G., Syrtlanova, T.S. and Skobejev, I.K. (1976) Leaching of gold from refractory sulfidic ores by alkaline solutions of aminoacids. Obogaščenije rud 4, 171–176 (in Russian).
Minejev, G.G. (1985) Concentration and recovery of precious metals from refractory gold-arsenic concentrates by jet milling. Obogaščenije rud 12, 27–31 (in Russian).
Mio, H., Saeki, S., Kano, J. and Saito, F. (2002) Estimation of mechanochemical dechlorination of polyvinylchloride. Environmental Science and Technology 36, 1344–1348.
Molchanov, V.V. and Buyanov, R.A. (2000) Mechanochemistry of catalysts. Russian Chemical Reviews 69, 435–450.
Molchanov, V.V. and Buyanov, R.A. (2001) Scientific grounds for the application of mechanochemistry to catalyst preparation. Kinetics and Catalysis 42, 366–374.
Molchanov, V.I. and Jusupov, T.S. (1981) *Physical and Chemical Properties of Fine Ground Minerals*. Nedra, Moscow (in Russian).
Monagheddu, M., Mulas, G., Doppiu, S., Cocco, G. and Raccanelli, S. (1999) Reduction of polychlorinated dibenzodioxins and dibenzofurans in contaminated muds by mechanically induced combustion reactions. Environmental Science and Technology 33, 2485–2488.
Mruck, H.J. and Ameling, R. (1988) Comminution of ceramic catalysts such as automobile catalytic convertors for metal recovery, German Patent 380 38 04 DE.
Mullov, V. and Lodejščikov, V.V. (1979) Experience with planetary mills for processing of sulfide minerals. Izvestija SO AN SSSR 3, 71–73 (in Russian).
Münster, H.P. (1997) *Taschenbuch des Metallhandels*, Hüthig BmbH, Heidelberg.
Natarajan, K.A. and Iwasaki, I. (1984) Electrochemical aspects of grinding media-interactions in magnetite ore grinding. International Journal of Mineral Processing 13, 53–71.
Netzsch (1994a) Feinmahltechnik GmbH – Programme.
Netzsch (1994b) Feinmahltechnik GmbH – Technical Information FT 002.
Nichols, G. and Frampton, C.S. (1998) Physicochemical characterization of the orthorhombic polymorphs of paracetamol crystallized from solution. Journal of Pharmaceutical Sciences 87, 684–693.
Nordwick, S.M. and Anderson, C.G. (1993) Advances in antimony electrowinning at the Sunshine mine. In: B. Hiskey and M. Wadsworth (Eds.) *Proceedings of the AIME Meeting "Hydrometallurgy – Fundamentals, Technology and Innovations"*, Littleton, Colorado, pp. 1107–1128.
Nowak, A.M., Meyers, R.A., Wan, E.I. and Hucko, R.E. (1994) Development of the molten-caustic leaching (MCL) process. In: *Proceedings of the XIIth International Coal Preparation Congress, Cracow, Poland*, Vol. 1, pp. 417–424.
Nyström, Ch. (1998) Dissolution properties of poorly soluble drugs: theoretical background and possibilities to improve the dissolution behaviour. In: R.H. Müller, S. Benita and

H.L. Böhm (Eds.) *Emulsions and Nanosuspensions for the Formulation of Poorly Soluble Drugs*, Medpharm, Scientific Publishers, Stuttgart, pp. 143–148.
Ocepek, D. and Milosavljkevic, R. (1977) About the influence of liberation degree to surface properties and flotation efficiency of some Yugoslavian copper ores. Erzmetall 30, 113–118.
Otsuka, M. and Kaneniwa, W. (1988) A kinetic study of the crystallization process of nocrystalline indomethacin under isothermal conditions. Chemical and Pharmaceutical Bulletin 36, 4026–4032.
Otsuka, M., Otsuka, K. and Kaneniwa, W. (1994) Relation between polymorphic transformation pathway during grinding and the physicochemical properties of bulk powders for pharmaceutical preparations. Drug Development and Industrial Pharmacy 20, 1649–1660.
Palmer, C.M. and Johnson, G.D. (2005) The Activox process: Growing significance in the nickel industry. Journal of Metals, July, 40–47.
Parrott, E.L. (1974) Milling of pharmaceutical solids. Journal of Pharmaceutical Sciences 63, 813–829.
Parrott, E.L. (1990) Comminution. In: J. Swarbrick and J.C. Boylan (Eds.) *Encyclopedia of Pharmaceutical Technology*, Vol. 3, Marcel Dekker, New York, pp. 101–121.
Parton, R.F., Jacobs, J.M., Huybrechts, D.R. and Jacobs, P.A. (1989) Zeolites as catalysts, sorbents and detergent builders: applications and innovations. In: H. Karge and J. Weitkamp, J. (Eds.) *Studies in Surface Science and Catalysis*, Vol. 46, Elsevier, Amsterdam, pp. 163.
Patton, J.S. and Platz, R.M. (1992) Penetration enhancement for polypeptides through epithelia. D. Routes of delivery-case studies. 2. Pulmonary delivery of peptides and proteines for systemic action. Advances in Drug Delivery Reviews 8, 179–196.
Pattrick, R.A.D. and Hall, A.J. (1983) Silver substitution into zinc, cadmium and iron tetrahedrites. Mineralogical Magazine 47, 441–451.
Paudert, R. (1965) Chemie und Technik 17, 449.
Paudert, R., Harenz, H., Heinicke, G., Dünkel, L., Pöthig, R., Kolosov, A.S., Boldyrev, V.V., Chaikina, M.V., Taranzova, M.I. and Gordeeva, G.I. (1979) Zur Erhöhung der Reaktivität von apatischen Rohphosphaten durch mechanische Aktivierung. Zeitschrift für Anorganische and Allgemeine Chemie 452, 141–150.
Paul, I.C. and Curtin, D.Y. (1975) Reaction of organic crystals with gases. Science 187, 19–26.
Pauli, I.A. Avvakumov, E.G., Isupova, L.A., Poluboyarov, V.A. and Sadykov, V.A. (1992) Sibirskij Chimičeskij Žurnal 3, 133 (in Russian).
Pavlenko, S., Malyshkin and Bazhenov, Yu. (2000) *Cement-Free Small Grain Concrete from Secondary Mineral Resources*. Siberian Branch of Russia Academy of Sciences, Novosibirsk (in Russian).
Pawlek, F. (1976a) The influence of grain size and mineralogical composition on the leachability of copper concentrates. In: (J.C. Yannopoulos, J.C. Agarwall (Eds.) *Proc. Int. Symp. Extractive Metallurgy of Copper*, Vol. II, Metallurgical Society AIME, Las Vegas, pp. 690–705.
Pawlek, F. (1976b) Verfahren zur hydrometallurgischen Gewinnung von Kupfer aus sulfidischen Kupferkonzentraten, German Patent No. 2629691.
Pawlek, F. (1977) Verfahren zur hydrometallurgischen Aufarbeitung von Kupferkies, Kupfer-Nickel-Konzentraten und Rückstanden, Zwischenprodukten und Flugstäuben, Austrian Patent No. 341231.
Pawlek, F. (1978) Verfahren zur Beschleunigung von Laugungsprozessen bei der hydrometallurgischen Verarbeitung sulfidischer Erzkonzentrate, Austrian Patent No. 358290.
Pawlek, F. (1983) *Metallhüttenkunde*. Walterde Gruyter, Berlin.
Pawlek, F., Kheiri, M.J. and Kammel, R. (1989) Optimierung der Feinstmahlung mit einer Rührwerkskugelmühle gezeigt am Beispiel eines Zinkkonzentrates. Metall 43, 838–842.
Pawlek, F., Kheiri, M.J. and Kammel, R. (1992) Mechano-chemischen Verfahren zur Gewinnung von Tonerdehydrat aus Bauxit. Metall 46, 575–580.
Paya, J., Monzo, J., Borrachero, M.V., Peris-Mora, E. and Amahjour, F. (2000) Mechanical treatment of fly ashes. Part IV: Strength development of ground fly ashes-cement mortars cured at different temperatures. Cement and Concrete Research 30, 543–551.

Peace, J.D., Young, M.F. and Curry, D.C. (2005) Fine grinding as enabling technology-the IsaMill. In: *Proc. VIth Annual Conf. "Crushing and Grinding 2005"*, Perth, pp. 1–20.

Peace, J.D., Curry, D.C., Barns, K.E., Young, M.F. and Rule, C. (2006) Transforming flowsheet design with inert grinding-the IsaMill. In: *Proceedings of the Canadian Mineral Processors XXXVIIIth Annual Operator Conference*, Ottawa, pp. 1–19.

Peacey, J., Guo, X.J. and Robles, E. (2003) Copper hydrometallurgy-current status, preliminary economics, future directions and positioning versus smelting. In: *Proceedings of the Vth International Conference Copper 2003/Cobre 2003*, Santiago, Chile, pp. 1–18.

Peng, Y., Grano, S., Ralston, J. and Fornasiero, D. (2002) Towards prediction of oxidation during grinding. I. Galena flotation. Minerals Engineering 15, 493–498.

Plaksin, V.A. and Šafejev, R.S. (1962) The influence of galena particle size on its flotability. Doklady AN SSSR 142, 131–134 (in Russian).

Pudipeddi, M. and Serajuddin, A.T.M. (2005) Trends in solubility of polymorphs. Journal of Pharmaceutical Science 94, 929–939.

Püpke, I. (1971) Construction and operating method of an attrition mill in powder metallurgy. Powder Metallurgy International 3, 94–96.

Rastogi, R.P., Bassi, P.S. and Chadha, S.L. (1963) Journal of Physical Chemistry 67, 2569.

Rastogi, R.P. and Singh, N.B. (1966) Journal of Physical Chemistry 70, 3315.

Rastogi, R.P. and Singh, N.B. (1968) Journal of Physical Chemistry 72, 4446.

Ray, M. and Formanek, V. (1960) Some factors affecting selectivity in the differential flotation of lead-zinc ores, particularly in the presence of oxidized lead minerals. In: *Proceedings of the International Mineral Processing Congress*, The Institution of Mining and Metallurgy, London, pp. 343–352.

Reggel, L., Raymond, R., Wender, I. and Blaustein, B.D. (1992) Preparation of ash-free coal by mild chemical treatment. American Chemical Society Division for Fuel Chemistry, Preprint 17, 44–48.

Rowlands, S.A., Hall, A.K., McCormick, P.G., Street, R., Hart, R.J., Ebell, G.F. and Donecker, P. (1994) Destruction of toxic materials. Nature 367, 223.

Saeki, S., Lee, J., Zhang, Q. and Saito, F. (2004) Co-grinding $LiCoO_2$ with PVC and water leaching of metal chlorides formed in ground products. International Journal of Mineral Processing 74S, S373-S378.

Saito, F. (2004) Research works for 30-years: report, Institute of Multidisciplinary Research for Advanced Materials. Tohoku University, Japan.

Savitskaya, A.V., Khantimer, E.I., Artemova, L.G. and Ezersky, M.L. (1990) Pharmatsiya 39, 26–30.

Sawaynagi, Y., Nambu, N. and Nagai, T. (1983) Dissolution properties and bioavailability of phenytoin from mixtures with chitin or chitosan. Chemical and Pharmaceutical Bulletin 31, 2064–2068.

Schlegel, E. (1976) Dissertation B, Institute for Mineral Processing, Freiberg.

Schrader, R. and Tetzner, G. (1961) Die mechanische Aktivierung von natürlichen Magnetit gegenüber Schwefelwassestoff. Zeitschrift für Anorganische und Allgemeine Chemie 309, 55–64.

Schrader, R. (1966) Über die Darstellung Aktiver Zustände an festen Stoffen durch mechanische Energie. Freiberger Forschungshefte A 392 (1966) 81–97.

Schrader, R., Tetzner, G. and Grund, H. (1966) Zeitschrift für Anorganische und Allgemeine Chemie 342, 204.

Schrader, R. and Hoffmann, B. (1973) Änderung der Reaktionsfähigkeit von Festkörpern durch vorhergehende mechanische Bearbeitung. In: V. Boldyrev and K. Meyer (Eds.) *Festkörperchemie*, VEB Deutscher Verlag für Grundstoffindustrie, Leipzig, pp. 522–543.

Sekula, F., Baláž, P., Jusko, F., Molnár, F. and Jakabský, Š. (1998) Hydrometallurgical technology of tetrahedrite concentrate processing from the Mária mine locality in Rožvnava. In: T. Sasvári and M. Jančura (Eds.) *Rožvnavské rudné pole*, Acta Montanistica Slovaca-Monography 3, pp. 149–156 (in Slovak).

Sekulic, Ž., Popov, S. and Milosevic, S. (1998) Comminution and mechanical activation of portland cement in different mill types. Ceramics 42, 25–28.

Sekulic, Ž, Popov, S., Duričic, M. and Rosic, A. (1999) Mechanical activation of cement with addition of fly ash. Materials Letters 39, 115–121.

Sglavo, V.M., Maurina, S., Conci, A., Salviati, A., Carturan, G. and Cocco, G. (2000) Bauxite red mud in the ceramic industry. Part 2. Production of clay-based ceramics. Journal of European Ceramic Society 20, 245–252.

Shakhtshneider, T.P. (1997) Phase Transformations and stabilization of metastable states of molecular crystals under mechanical activation. Solid State Ionics 101–103, 851–856.

Shakhtshneider, T.P. and Boldyrev, V.V. (1999) Mechanochemical synthesis and mechanical activation of drugs. In: E. Boldyreva and V.V. Boldyrev (Eds.) *Reactivity of Molecular Solids*, John Wiley & Sons, New York, pp. 271–311.

Shan, N.D., Chriswell, C.D. and Markuszewski, R. (1989) Separation by countercurrent washing of coal-caustic mixtures during chemical coal cleaning. Separation Science and Technology 24, 79–95.

Shi, C. and Day, R.L. (2003) Comparison of different methods for enhancing reactivity of pozzolans. Cement and Concrete Research 31, 813–813.

Singhal, D. and Curatolo, W. (2004) Drug polymorphism and dosage form design: a practical perspective. Advances in Drug Delivery Reviews 56, 335–347.

Steinike, U. and Linke, E. (1982) Die Rolle von Anregungszuständen und Strukturdefekten bei mechanisch aktivierten Festkörperreaktionen. Zeitschrift für Chemie 22, 397–401.

Storey, R., Docherty, R., Higginson, P., Dallman, C., Gilmore, C., Barr, G. and Dong, W. (2004) Automation of solid form screening procedures in the pharmaceutical industry-how to avoid the bottlenecks. Crystallographical Reviews 10, 45–56.

Ströbel, G. (1993) Neue Verfahren zur Behandlung asbesthaltiger Abfälle-Thermische Behandlung, *UTECH-Proc. of Faserkongress*, Berlin, pp. 199–214.

Sulimenko, L.M., Schalunenko, N.J. and Urkhanova, L.A. (1994) Mechanochemical activation of binder composition. Cement 2, 38–40 (in Russian).

Suryanarayana, C. (1999) *Non-equilibrium Processing of Materials*. Pergamon Press, Oxford.

Suryanarayana, C. (2001) Mechanical alloying and mixing. Progress in Materials Science 46, 1–184.

Suryanarayana, C., Ivanov, E. and Boldyrev, V.V. (2001) The science and technology of mechanical alloying. Materials Science and Engineering A 304–306, 151–158.

Suryanarayana, C. (2004) *Mechanical Alloying and Milling*. Marcel Dekker, New York.

Svedala Grinding-Technical Information: SAM mills-refference list 1996.

Swidersky, H.W. (1994) Verfahren zur Chemischen Zersetzung von Asbest. Entsorgungspraxis 4, 52–54.

Syrtlanova, L.S., Minejev, G.G., Smagunov, V.N. and Perepelica, L.S. (1979). Intensification of arsenopyrite-pyrite concentrate alkaline leaching by dry milling. Izvestija SO AN SSSR 3, 50–55 (in Russian).

Szanto, E.V. and Schultes, W. (1969) Beitrag zur mechanischen Aktivierung von Bauxiten durch Schwingmahlung. Erzmetall 22, 465–518.

Števulová, N., Bálintová, M., Briančin, J. and Szeghyová, Z. (2005) Mechanochemical formation of dicalcium silicate. Chemical Papers 99, 420–422.

Števulová, N. and Sverák, T. (2005) Application of mechanochemical methods in environmental engineering. Chemical Papers 99, 423–424.

Takahashi, Y., Nakashima, K., Nakagawa, H. and Sugimoto, I. (1984) Effect of grinding and drying on the solid state stability of ampicillin trihydrate. Chemical and Pharmaceutical Bulletin 32, 4963–4970.

Tanaka, K. (2003) *Solvent-free Organic Synthesis*. Wiley-VCH, Weinheim.

Tkáčová, K. (1989) *Mechanical Activation of Minerals*. Elsevier, Amsterdam.

Tkáčová, K., Baláž, P., Mišura, B., Vigdergauz, V.E. and čanturija, V.A. (1993) Selective leaching of zinc from mechanically activated complex Cu-Pb-Zn concentrate. Hydrometallurgy 33, 291–300.

Toda, F. (Ed.) (2002) *Organic Solid-State Reactions*. Kluwer, Amsterdam.
Tongamp, W., Zhang, Q. and Saito, F. (2006) Mechanochemical decomposition of PVC by using La_2O_3 as additive. Journal of Hazardous Materials B137, 1226–1230.
Tongamp, W., Kano, J., Zhang, Q. and Saito, F. (2007) Simultaneous treatment of PVC and oyster-shell wastes by mechanochemical means. Waste Management 28, 484–488.
Toromanoff, I. and Habashi, F. (1986) Activation of titanium slag by attrition grinding. Minerals and Metallurgical Processing August, 191–197.
Tröger, W.E. (1969) *Optischer Bestimmung der gesteinsbildenden Minerale. Teil 2.* Schweizerbart'sche Verlagsbuchhandlung, Stuttgart.
Turčániová, L'. and Baláž, P. (2000) Reactivity of coal activated by mechanochemical treatment. Journal of Materials Synthesis and Processing 8, 365–367.
Turčániová, L'., Škvarla, J. and Baláž, P. (2000) A contribution to the mechanism of formation of humic acid in coal. Journal of Materials Synthesis and Processing 8, 359–363.
Turčániová, L'., Bežovská, M., Baláž, P., Vidlář, J., Jewel, D.M. and Fečko, P. (2001) Contribution of mechanical activation of coal to the development of theory and practical aspects of advanced coal technologies. In: *Proceedings of the XVIIIth Annual International Pittsburgh Coal Conference*, Newcastle, Australia, CD.
Turčániová, L'., Kádarová, J., Imrich, P., Liptaj, T., Vidlář, J., Vašek, J., Foldyna, F., Sitek, J., and Baláž, P. (2004) Reactivity of mechanically activated coals for special utilization. Journal of Materials Science 39, 5467–5470.
Türke, W. and Fischer, P. (1978) Hydrometallurgical treatment of complex copper sulfide concentrates with special reference to the Lurgi-Mitterberg process. In: M.J. Jones (Ed.) *Proceedings of the. International Symposium "Complex Metallurgy'78"*, Bad Harzburg 1978, pp. 101–112.
Vogt, V., Florescu, R. and Gock, E. (1996) Verwertung von Asbest nach mechanischer Faserzerstörung. Erzmetall 49, 443–454.
Vrečer, F., Srčič, S. and Šmid-Korbar, J. (1991) Investigation of piroxicam polymorphism. International Journal of Pharmacy 68, 35–41.
Verbich, F. (1998) Geochemistry and mineralization of Nováky coal seam. Acta Montanistica Slovaca 3, 378–383 (in Slovak).
Wall, N.C. and Pattison, D. (1997) Minerals International, March, 16–19.
Weisser, O. and Landa, S. (1972) *Sulphide Catalysts, Their Properties and Applications.* Academia, Prague.
Welham, N. (1997) The effect of extended milling on minerals. Canadian Institute of Metallurgy Bulletin 90, 64–68.
Welham, N. and Llewellyn, D.J. (1998) Mechanical enhancement of the dissolution of ilmenite. Minerals Engineering 11, 827–841.
Wills, B.A. (1984) The separation by flotation of copper-lead-zinc sulphides. Mining Magazine 150, 36–41.
Wyslouzil, D.M. and Salter, R.S. (1990) Silver leaching fundamentals. In: T.S. Makey, R.D. Prengaman (Eds.) *Proceedings of the International Symposium "Lead-Zinc'90"*, TMS Meeting, pp. 87–107.
Yamamoto, K., Nakano, M., Arita, T. and Nakai, Y. (1974) Dissolution rate and bioavailability of griseofulvin from a ground mixture with microcrystalline cellulose. Journal of Pharmacokinetics and Pharmacodynamics 2, 487–493.
Yaneva, V., Petkova, V. and Dombalov, D. (2005) Structural transformations accompaying mechanochemical activation of natural phosphorite from Syria (2005). Chemistry for Sustainable Development 13, 347–354.
Yu, L.X., Amidon, G.L., Polli, J.E., Zhao, H., Mehta, M., Conner, D.P., Shah, V.P., Lesko, L.J., Chen, M.L., Lee, V.H.L. and Hussain, A.S. (2002) Biopharmaceutics classification system: the scientific basis for biowaiver extensions. Pharmaceutical Research 19, 921–925.
Zhang, Q., Saito, F., Shimme, K. and Masuda, S. (1999) Dechlorination of PVC by a mechanochemical treatment under atmospheric condition. Journal of Powder Technology Japan 36, 468–473.

Zhang, Q., Kano, J. and Saito, F. (2001) Mechanochemical technology: application to materials synthesis and to the separation and processing of recyclable materials from wastes. KONA 19, 7–15.

Zhang, Q., Saito, F., Shimme, K. and Masuda, S. (2000) Dechlorination of PVC by mechanochemical treatment. KONA 18, 207–211.

Zhang, Q., Saeki, S., Tanaka, Y., Kano, J. and Saito, F. (2007) A soft-solution process for recovering rare metals from metal/alloy wastes by grinding and washing with water. Journal of Hazardous Materials A 139, 438–442.

Zoz, H. (1995a) Attritor Technologie-Stand der Technik. Metall 49, 196–198.

Zoz, H. (1995b) Attritor technology-latest developments. Materials Science Forum 179–181, 419–424.

Zoz, H., Ernst, D., Mizutani, T. and Okouchi, H. (1997) Semi-continuously mechanical alloying in a production scale using cycle operation-Part I. Metall 51, 568–572.

Zoz, H., Reichardt, R. and Ernst, D. (1998) Semi-continuously mechanical alloying in a production scale using cycle operation-Part II. Metall 51, 521–527.

Zoz, H., Jaramillo, V., Tian, Z., Trindade, B., Ren, H., Chimal, V.O. and Diaz de la Torre, S. (2004) High performance cements and advanced ordinary Portland cement manufacturing by HEM-rwfinement and activation. Cement Lime Ggypsum 57, 60–70.

Author Index

Arnold, M., 78
Arrhenius, 68–69, 181, 182, 183
Averbach, B.L., 52, 60, 158
Avrami, M., 24, 25, 26, 28

Bacon, F., 2
Baláž, P., 116
Bayer, 318–319
Becher, 212
Bernhardt, C., 109
Binning, 31, 146, 148
Boldyrev, V.V., 13
Bond, F.C., 105–106
Bowden, F., 6
Bragg, 140, 153, 157, 161, 208–209
Brus, L.E., 38
Butyagin, P.Yu, 14

Cauchy, 155
Clark, G.L., 7
Cohen, 31
Cottrell, A.M., 161
Curie, 75
Curl, 31

Dachille, F., 7

Einstein, 138–139
Ekimov, 31
Erofeev, B.V., 25, 26, 28

Faraday, M., 3, 20, 31
Fert, A., 75
Feynman, R., 31, 76
Fink, M., 6
Flavickij, F.M., 5
Freundlich, H., 289–290
Furuya, E.G., 290

Gauss, 156
Gerber, 148
Gibbs, 16
Gock, E., 109
Griffith, A.A., 128
Gründberg, P., 75

Hall, W.H., 52, 59, 70–71, 155–156, 158–159, 241
Hedvall, J.A., 6, 186
Heegn, H.P., 109
Heinicke, G., 10
Helmholtz, 16
Hooke, 104
Hüttig, G.F., 17, 122

Iijima, 31

Jimbo, G., 122
Juhász, A.Z., 15, 128

Kelvin, 78
Kick, F., 105–106
Kirpičev, V.L., 105–106
Knoll, 143
Kroto, 31

Lea, 3
Lindemann, F.A., 73
Lyakhov, N.Z., 15, 28
Lycurgus, 31

Mie, G., 31
Mössbauer, R.L., 24, 162, 200

Noyes, 370

Ohlberg, S.M., 160
Onuschenko, 31
Opoczky, L., 128
Ostwald, W., 1, 4

Parker, L.H., 4
Petch, N.J., 70–71
Peters, K., 8

Quate, 148

Rayleigh, 137
Rebinder, P.A., 128
Rittinger, P.R., 105–106, 123
Rockenberger, J., 162
Rohrer, 31, 146
Rose, H.E., 108
Rowan, R., 7
Roy, H., 7
Rumpf, H., 124
Ruska, 134

Saito, F., 390
Scherrer, P., 52, 158–159
Schrader, R., 18
Shi, F.G., 73
Siegbahn, K., 139
Smalley, R., 31
Smékal, A., 14
Strickler, D.W., 160
Szegvari, 114

Tabor, D., 6
Tamman, G., 6, 186
Tanaka, T., 122
Theophrastus of Eresus, 2
Thiessen, P.A., 9

Urakajev, F.Ch., 205

Van der Waals, 124–125, 148, 373
Von Laue, 153

Wanetig, P., 6
Warren, B.E., 52, 60, 158
Wenzel, C.F., 3
Whitney, 370
Williamson, G.K., 52, 59, 155, 158, 241

Subject Index

AAC/UBC process, 322
Activox process, 321–322, 324–330
Adhesion, 124
Adsorption, 126–129
Aerosol processing, 39
AFM (atomic force microscopy), 53, 133, 148–150
Agglomeration, 26, 60, 81, 122–125, 241–242
Aggregation, 62, 80–81, 103, 122–125, 236, 270
Agriculture, 365–370
Albion process, 330–332
Alloying metal, 186–187
Aluminothermic process, 186
Amorphization, 15, 121, 373–375
Ampicillin, 374–375
Analogy model, 14
Anatase, 80
Anilite, 242–243
Ankerite, 266
Antimone oxide, 203–204
Antimonite (stibnite), 217, 227–228, 240
Apatite, 365–367
Arsenic, 290
Arsenic sulphide, 230–231
Arsenopyrite, 137, 266–268, 311–314, 335, 357, 385–386
Asbestos, 380–384
Aspirin, 377–378
Atomic force microscopy (AFM), 53, 133, 148–150
Attritor, 108, 114–116, 285, 303, 318–321
Auripigment, 231–232, 357

Bactech/Mintek process, 322
Bacterium, 82
Ball-to-powder ratio, 118
Bastnaesite, 248
Battery, 90
Bauxite, 318
Bearing, 355
BET, 152
Bioavailability, 371
Biocop process, 322
Bismuth, 273–276
Blue shift, 38–39, 62
Böhmit, 318
Bohr magneton, 168
Bohr radius, 37
Boltzmann distribution, 168
Bornite, 261–262, 332
Bottom up technique, 40, 371
Building industry, 358–365

Cadmium sulphide, 38–39, 55–62, 73, 90, 162
Calcite, 6, 9, 18, 266, 288–290
Calcium oxide, 183–184
Cancer, 83–89
Carbon dioxide, 134
Carbon nanotube, 33
Cassiterite, 300
Catalyst, 90, 389
Celestite, 285–287
Cellulose, 6
Cement, 358–365
Cephalexin, 374
Cerussite, 9
Chalcantite, 246–248
Chalcocite, 232, 236, 242–244, 260–263, 322, 332
Chalcopyrite, 116, 144, 185, 218, 259–260, 264, 278, 291, 303–308, 322, 330, 332
Chemical capping, 49
Chemical shift, 139, 141, 164–165

Chemical vapor deposition, 39, 41
Chemisorption, 49
Chemotherapy, 86
Chromite, 314–317
Chromium, 75
Chrysocolla, 245
Cinnabar, 2, 180–181, 218
Cluster, 32, 35, 80–81, 161
Coal, 355–358
Coating, 90
Cobalt, 75, 326–328
Cobalt aluminide, 219–220
Cobaltite, 332
Cobalt sulphide, 219–220
Coercivity, 73–74
Colloidal mill, 1
Colloidal particles, 32
Colour center, 169
Comminution, 3, 103
Composite, 90
Compression, 107
Concrete, 358–365
Contamination, 121
Copper, 48, 71, 75, 321–323
Copper oxide, 194–197, 203
Copper sulphide, 184–185, 219, 231–236
Crack formation, 104
Crack propagation, 104–105
Crystal engineering, 341–345
Crystal growth, 24
Crystallinity, 160
Crystallite, 34
  size, 59, 79, 157
Cutting tool, 90
Cyanide, 276–283, 323–324, 327

Destructive adsorbent, 90
Diaspor, 318
Diffusion, 70, 187
Diffusivity, 70
Digenite, 242–243
Dislocation, 11, 71–72, 105
Display, 90
Dissolution, 68
Djurleite, 232, 242–244
DNA, 82, 84
Drug, 82, 370–379
Dynatec process, 322

Elastic constant, 70
Elastic deformation, 105–106
Electric discharge assisted mechanical milling, 117
Electric resistivity, 73
Electrodeposition, 49
Electron energy-loss spectroscopy (EELS), 151
Electron magnetic resonance (EMR), 167
Electron paramagnetic resonance (EPR), 167–171
Electrostatic stabilization, 127
Enargite, 268–273, 332
Endothermic reaction, 4
Energy dispersive X-ray spectroscopy (EDS), 151
Exchange reaction, 188
Excitation process, 10
Exciton, 37
Extended X-ray absorption fine structure (EXAFS), 162
Extractive metallurgy, 302–341

Feldspar, 300
Ferrite, 207–209
Ferromagnet, 74
Fertilizer, 365–370
Fine milling, 103
Flotation, 299–301
Flux, 186
Fly ash, 363–364
FTIR, 136
Fuel cell, 90

GACL (grinding alkaline chemical leaching) process, 356–358
Galena, 181–182, 217–220, 278, 291, 298–299
Geerite, 259
Geocoat process, 322
Giant magnetoresistance, 75–76
Gibbs energy, 16–18
Gold, 18, 31, 73, 82, 87, 276–280, 308–314, 327–329
Grain
  boundary, 69–70
  size, 46, 70–71
GRAVIMELT process, 357–358
Greenockite, 57–58
Griseofulvin, 374, 376

Halide, 3–4
Hardness, 70–71
Hawleyite, 57–58
Hematite, 24, 26, 158–159, 160, 197–198, 201, 318
Hercynite $FeAl_2O_4$, 198, 219
Heterogeneous catalysis, 345–352
Hierarchic model, 10, 11

High-energy milling, 44–45, 56–57, 103, 107, 111, 116, 121
High-resolution transmission electron microscopy (HRTEM), 53–54, 133, 152
Hollandite, 81
Hot spot theory, 6, 8–9
Humic acid, 355–358
Hydargillit, 318
Hydrogen peroxide, 305–307
Hydrothermal route, 49
Hydroxide, 318

Ilmenite, 212–216, 259
Impact, 12, 107
Impulse model, 13–14
Inert gas condensation, 49
Infrared spectroscopy, 133–138
Irigetmet process, 323–324
Iron, 72, 75
Iron carbide, 24
Iron oxide, 197–202
Iron sulphide, 218–219, 234–236
Isomer shift, 164–165

Jamesonite, 268–271
Jarosite, 167
Jet mill, 311

Kinetic model, 13
Kinetics, 19–20, 23

Laser ablation, 41
Lattice strain, 45–47, 59, 157, 161
Lead oxide, 205
Lead sulphide, 111, 138, 142–143, 148, 154, 223–227, 274–275
Litharge, 7, 23–24
Lurgi-Mitterberg process, 322–323, 324

Maghemite, 200
Magic number, 36
Magma-plasma model, 9–11
Magnesite, 9
Magnetic domain, 74
Magnetic material, 90
Magnetic splitting, 164–165
Magnetite, 200, 299
Magnetization, 221–222, 225–229
Magnetocaloric effect, 75
Magnetoresistance, 75–76
Malachite, 245
Manganese oxide, 81, 202
Massicot, 7, 23–24
Materials engineering, 352–355
Materials waste, 387–389
Mechanical activation, 1, 14–16, 18, 28–29, 109, 137, 257, 280, 283–284, 298
Mechanical alloying, 39, 352–355
Mechanical energy, 1, 10
Mechanical stress, 4, 7
Mechanochemical equilibrium, 121–122
Mechanochemical leaching, 280, 283–287
Mechanochemical processing, 39, 48, 92, 259
Mechanochemical reactor, 5
Mechanochemical reduction, 2, 15
Mechanochemical synthesis, 27, 49
Mechanometallurgy, 218
Melting point, 73
Melting temperature, 72
Melt process, 285, 333–341
Mercurous chloride, 4
Mercury sulphide, 140
Metallothermic reaction, 186–187
Metallurgical calcine, 385–387
Metathesis reaction, 188
Micelle, 43–44, 126–128
Milling
  atmosphere, 119
  medium, 118
  speed, 119
  time, 119
Mineral processing, 297–301
Mixer mill, 113
Molybdenite, 220–221, 239, 308
Mössbauer spectroscopy, 162–167, 221, 223, 226–227, 234–235

Nanocapsule, 86–87
Nanocluster, 32, 35
Nanocomposite, 26, 188–190
Nanocrystal, 32, 371
Nanocrystalline semiconductor, 43–44
Nanofiber, 32
Nanogeoscience, 76–81
Nanomaterial, 68–69
Nanoparticle, 26, 32, 36, 63, 72, 82–89, 90–91, 161
Nanorod, 32
Nanoscience, 3, 29–76, 92
Nanoshell, 82
Nanosize effect, 35
Nanosphere, 86–87
Nanostructure, 32–34
Nanotechnology, 30, 90–91
Nanotube, 31
Nanovector, 89
Nanowire, 32
Nenatech process, 321–322, 330–332

Nickel, 75, 292, 325–328, 346, 352
Nickel sulphide, 65–66
Nucleation, 24

Organic solid-state reaction, 341–345
Organic waste, 389–392

Paint, 90
Particle size, 64–65, 103
Pentlandite, 264–265, 291, 308, 332
Perovskite, 348–350
Pharmacy, 370–379
Phenytoin, 377
Phonon theory, 11
Phosphate, 365–370
Phosphorite, 366
Physical vapor deposition, 39, 40–41
Pin mill, 108
Piroxicam, 375
Placer Dome process, 322
Planetary mill, 108, 111–113
Plastic deformation, 45, 71, 106, 125
Polymer, 6
Polymorphism, 373–375
Polyvinyl chloride, 389–390
Precipitation, 44
Proustite, 280
Pyrargyrite, 280
Pyrite, 137, 214–215, 221–222, 226, 264, 278, 290–291, 311–313, 330, 332, 385–386
Pyrrhotite, 226–227, 230, 259, 291, 299

Quadrupole splitting, 164–165
Quantum
  confinement, 37
  dot, 32, 36–37, 38, 49
  size effect, 35, 37
Quartz, 177, 222, 230, 259, 266, 278, 300–301, 359–360, 365, 374, 382

Rate of leaching, 258
Realgar, 231, 357
Red mud, 387
Redox reaction, 187–188, 207
Relaxation
  curve, 15
  process, 15–16
  time, 10, 15
Reverse micelle, 43–44
Rolling mill, 108
Room temperature solder, 354–355
Rutile, 212

Saturation magnetization, 75
Scaling-up the process, 236–239
Scanning electron microscopy (SEM), 143–146, 236
Scanning tunneling microscopy (STM), 146–148
Scheelite, 210, 259
Self-organization, 80
Self-sustaining reaction, 3, 240
Semiconductor, 37, 48–50
  nanocrystal, 49
Sensor, 91
Shear, 107
Silicate, 167
Silicon, 242–244
Silver, 280–283, 308–314, 354
Sodium carbonate, 5
Sodium chloride, 103–104, 373
Sodium thiosulphate, 274
Sol-gel process, 39, 43
Solid state reaction, 4–5
Solubility, 78
Solution growth technique, 49
Sorption, 288–292
Specific heat, 72
Spex mill, 113
Sphalerite, 50, 182–185, 219–220, 278, 291, 298
Spherical model, 11
Spray processing, 42, 49
Sputtering, 41, 49
Stealth^{TM} particles, 88–89
Stellite, 355
Steric stabilization, 126
Stibnite, 179–180
Stirred ball mill, 114–116
Stored energy, 47
Strain energy, 104
Strength, 70
Strontianite, 285, 287
Strontium ferrate, 116
Structural relaxation, 15
Successive ion layer deposition, 49
Sulphamonomethoxin, 376
Sulphathiazole, 374
Sulphur, 114, 180–181, 183
Sunshine process, 332–333, 335, 339–341
Surface area, 21, 36, 62, 67–68, 103
Surfactant, 121, 126–129

Temperature effect, 119–120
Tennantite, 334–335
Tetrahedrite, 268–271, 278, 280, 282–286, 334–341, 385–386

Textile fibre, 91
Theory of energy balance, 14
Theory of short-live active cluters, 11–13
Thermal activation, 257
Thermal booster, 186
Thermal evaporation, 41
Thermodynamic equilibrium, 17
Thermodynamics, 16–19
Thermoelectric power generator, 354
Thiosulphate, 277–283
Tin oxide, 203
Tin sulphide, 66–69
Tip of a propagating crack, 8–9
Titanate, 212–216
Titanium carbide, 41
Titanium oxides, 205–206
Top down technique, 40, 371
Transmission electron microscopy (TEM), 133, 150–152, 159, 225, 230
Triboplasma, 11
Troilite, 221–223, 233
Tumor, 85–86, 88–89
Tungsten carbide, 209–211, 388–389

Ultrafine milling, 103
UV-VIS, 61–62, 64–65

Vacuum evaporation, 49
Vibration mill, 107–111, 116, 303–306, 315

Water, 134
Wet chemical synthesis, 39
Wolframate, 209–211
Wurtzite, 50, 219–220
Wüstite, 262

X-ray absorption near edge structure (XANES), 161
X-ray absorption spectroscopy (XAS), 161
X-ray amorphous phase, 121
X-ray diffraction (XRD), 51–52, 63, 67, 152–162
X-ray photoelectron spectroscopy (XPS), 138–143

Zeolite, 348–352, 368
Zinc, 288–290
Zinc cadmium sulphide, 63–65
Zinc sulphide, 50–55, 90, 136, 170–171, 219–220
ZOZ, 117

SPRINGER NATURE

GPSR Compliance

The European Union's (EU) General Product Safety Regulation (GPSR) is a set of rules that requires consumer products to be safe and our obligations to ensure this.

If you have any concerns about our products, you can contact us on ProductSafety@springernature.com

In case Publisher is established outside the EU, the EU authorized representative is:

Springer Nature Customer Service Center GmbH
Europaplatz 3
69115 Heidelberg, Germany

FSC
www.fsc.org
MIX
Papier | Fördert
gute Waldnutzung
FSC® C083411

Zeitfracht Medien GmbH
Ferdinand-Jühlke-Straße 7
99095 Erfurt, Deutschland
produktsicherheit@kolibri360.de